CONTEMPORARY
MATHEMATICS

Volume 85

Banach Space Theory

Proceedings of a Research Workshop
held July 5–25, 1987
with support from
the National Science Foundation

Bor-Luh Lin, Editor

AMERICAN MATHEMATICAL SOCIETY
Providence · Rhode Island

The Research Workshop on Banach Space Theory was held at the University of Iowa, Iowa City, on July 5–25, 1987 with support from the National Science Foundation, Grant DMS-8604481.

1980 *Mathematics Subject Classification* (1985 *Revision*). Primary 46-06, 46B10, 46B20, 46B22, 46B25.

Library of Congress Cataloging-in-Publication Data

Research Workshop on Banach Space Theory (1987: University of Iowa)
Banach space theory: proceedings of a research workshop held July 5–25, 1987, with support from the National Science Foundation/Bor-Luh Lin, editor.
p. cm. –(Contemporary mathematics, ISSN 0271-4132; v. 85)
"The Research Workshop on Banach Space Theory was held at the University of Iowa, Iowa City"–T.p. verso.
Includes bibliographies.
ISBN 0-8218-5092-x (alk. paper)
1. Banach spaces–Congresses. I. Lin, Bor-Luh. II. American Mathematical Society. III. Title. VI. Series: Contemporary mathematics (American Mathematical Society); v. 85.
QA322.2.R47 1987 88-38106
515.7′32–dc 19 CIP

Printed in the United States of America.
This volume was printed directly from author-prepared copy.
The paper used in this book is acid-free and falls within the guidelines established to ensure permanence and durability. ♾

CONTENTS

PREFACE

These Proceedings are from a Research Workshop on Banach Space Theory held at the University of Iowa in Iowa City, Iowa from July 5th to July 25th, 1987.

The intent of the Workshop was to provide participants with a close, working atmosphere, one in which the exchange of ideas could be promoted at an informal level. This included two hours daily seminars on current problems. In addition, participants were provided with adjacent offices to facilitate collaboration and consultation. At the conclusion of the Workshop, these Proceedings were compiled. This volume includes several papers initiated during the Workshop, which are presented here in their final form. The volume also includes some papers whose contributors would like to participate in the Workshop but were unable to come. No papers in this volume will be published elsewhere.

We would like to express our thanks to Professor A. Pelczynski for attending the Workshop at our invitation as an Ida Beam Distinguished Visiting Professor. We are deeply indebted to Professor J. Lindenstrauss for his tremendous help in organizing the Workshop and to Professors W.B. Johnson, G. Pisier and H.P. Rosenthal for their valuable suggestions and help. We would also like to thank those colleagues and participants, in many cases coming a great distance, whose encouragement and contributions led to the success of the project.

The Workshop was supported by a grant from the National Science Foundation, the Department of Mathematics, and the Ida Beam Foundation at the University of Iowa. We thank them for their generous help in making the project possible. Finally, we would like to thank Professor Irwin Kra for including the Proceedings in the series of Contemporary Mathematics.

Bor–Luh Lin

Workshop Participants*

Dale E. Alspach, Oklahoma State University, Stillwater, Oklahoma

S. Argyros, University of Crete, Iraklion, Crete, Greece

Peter G. Casazza, University of Missouri, Columbia, Missouri

William J. Davis, Ohio State University, Columbus, Ohio

Guanggui Ding, Nankai University, Tianjin, China

Catherine Finet, Université de l'Etat à Mons, Mons, Belgium

Ji Gao, University of Pittsburgh, Pittsburgh, Pennsylvania

Nassif Ghoussoub, University of British Columbia, Vancouver, British Columbia, Canada

Sylvie Guerre, Université Paris VI, Paris, France

Richard Haydon, Brasenose College, Oxford, England

William B. Johnson, Texas A & M University, College Station, Texas

Nigel J. Kalton, University of Missouri, Columbia, Missouri

M. Amine Khamsi, Université Paris VI, Paris, France

Pei–Kee Lin, Memphis State University, Memphis, Tennessee

Joram Lindenstrauss, The Hebrew University, Jerusalem, Israel

Gun–Marie Lovblem, University of Stockholm, Stockholm, Sweden

Aleksander Pelczynski, Polish Academy of Sciences, Warszawa, Poland

Gilles Pisier, Université Paris VI, Paris, France

Edward Odell, University of Texas, Austin, Texas

Yves Raynaud, Université Paris VI, Paris, France

Haskell P. Rosenthal, University of Texas, Austin, Texas

*Other than those from the University of Iowa

Gideon Schechtman, Weismann Institute of Sciences, Rehovot, Israel

Abderrazak Sersouri, Université Paris VI, Paris, France

Stanimir L. Troyanski, University of Sofia, Sofia, Bulgaria

J. Barry Turett, Oakland University, Rochester, Michigan

Xintai Yu, East China Normal University, Shanghai, China

Junfeng Zhao, Wuhan University, Wuhan, China

TITLES OF THE SEMINARS GIVEN BY THE PARTICIPANTS

Seminars were held daily at 2:30–3:30 p.m. and 3:45–4:45 p.m. during the Workshop. The following is the list of titles in the seminars arranged in chronological order.

SOME CHARACTERIZATIONS OF SIMPLICES WITH RADON–NIKODYM PROPERTY
H.P. Rosenthal

WEAK HILBERT SPACES
P.G. Casazza

APPROXIMATING ZONOIDS BY ZONOTOPES I
J. Lindenstrauss

ALMOST ISOMETRIC METHODS IN SOME ISOMORPHIC EMBEDDING PROBLEMS
Yves Raynaud

SUMS OF INDEPENDENT RANDOM VARIABLES IN REARRANGEMENT INVARIANT FUNCTION SPACES
W.B. Johnson

ON THE GEOMETRY OF SPACES WHERE ANALYTIC MARTINGALES CONVERGE
N. Ghoussoub

APPROXIMATING ZONOIDS BY ZONOTYPES II
J. Lindenstrauss

ON MATHEMATICAL OLYMPIADS
A. Pelczynski

COMPLEXITY OF WEAKLY NULL SEQUENCES
D. Alspach

NON RNP SETS WITH PCP
S. Agyros

OPERATORS INDUCED BY H^1 MULTIPLIERS AND C^*–ALGEBRAS I
A. Pelczynski

OPERATORS INDUCED BY H^1 MULTIPLIERS AND C^*–ALGEBRAS II
A. Pelczynski

WEAK HILBERT SPACES I
G. Pisier

WEAK HILBERT SPACES II
G. Pisier

SHELAH'S SPACE WITH FEW OPERATORS
R.G. Haydon

EMBEDDING FINITE METRIC SPACES IN NORMED SPACES
G. Schechtman

CONTRACTIVE PROJECTIONS ON l_p
W.J. Davis

ON BAIRE 1/4–FUNCTIONS
E. Odell

ON FRECHET DIFFERENTIABILITY IN ORLICZ SPACES
S. Troyanski

ISOMETRICS ON $L_{p,1}$
J.B. Turett

ON THE CLOSEDNESS OF THE SUMS OF CLOSED OPERATORS
S. Guerre

THE APPROXIMATION PROBLEM OF ALMOST ISOMETRIC OPERATORS BY ISOMETRIC OPERATORS
Guanggui Ding

ON QUASI–REFLEXIVE BANACH SPACES
C. Finet

ON BANACH SPACES WHICH ARE ISOMETRIC TO THEIR BIDUALS
A. Sersouri

AN EXAMPLE OF AN IRREGULAR RANDOM MEASURE
S. Guerre

ON THE WEAK* FIXED POINT PROPERTY
A. Khamsi

ON KUR SPACES
Xintai Yu

THE ORDERING STRUCTURES ON BANACH SPACES
Junfeng Zhao

Contemporary Mathematics
Volume **85**, 1989

ISOMETRIES OF JAMES SPACE

Steven F. Bellenot

ABSTRACT

For one of the norms commonly used for James' quasi–reflexive space J, the only onto isometries of J are plus or minus the identity. Some of the results of Azimi and Hagler are generalized to the case of spaces with shrinking neighborly bases.

We show that for certain "natural norms" on some Banach spaces X, that X has only trivial isometries. (Trivial isometries means the only (onto) isometries are ± identity). In particular, one of the commonly used norms on James space (but not one in which is isometric to its bidual) has these properties (Theorem 1). The class of "natural norms" is much larger than the quasi–reflexive spaces. Indeed we show these conditions are satisfied for "natural norms" for some space with non–separable duals (Proposition 2 and Example 5). Along the way we generalize some of the results of Azimi and Hagler [3] to spaces with shrinking neighborly bases (Proposition 4).

The existence of Banach spaces with only trivial isometries has been known since 1971 (Davis [9]). Indeed, each separable Banach space and even a non–separable Hilbert

---

AMS(MOS) Subject Classification (1980): 46B20

Keywords and phrases: James' space, isometries, neighborly basis, extreme points.

space can be re–normed to have only trivial isometries [6]. Pelczynski constructed a K, so that C(K) has only trivial isometries (this is also in [9]). However, James space is the first space with "natural norm" shown to have trivial isometries.

James' quasi–reflexive space is unique in that J is used to represent any Banach space *isomorphic* J. In [13], Semenov and Skorik show there is a "commonly used norm" on J in which J has 4 onto isometries and a simple (but not natural) renorming in which J has trivial isometries. Sersouri [14] has shown that J in the $|||\cdot|||$ (defined below) has only trivial isometries. Sersouri's result depends on the facts that J is quasi–reflexive and that J is isometric to its bidual in this norm. Also note that J is isomorphic to J ⊕ Hilbert space and hence J can be re–normed to have lots of isometries. For other results on J see [10], [12, pp. 25, 103, 132], [8], [11], [1], [2] and [4].

A version of this paper was presented at the conference at Kent State University in the summer of 1985. The author would like to acknowledge A. Andrew, J.E. Jamison, J.N. Hagler and R.C. James for suggestions, help and questions.

§0. Preliminaries

Our notation is mostly standard and follows that of [12] where undefined terms may be found. An exception is we will say a vector x in a Banach space X is *extreme* if x is an extreme point of the ball $\{y \in X: ||y|| \leq ||x||\}$ or equivalently either $x = 0$ or $x/||x||$ is an extreme point of the closed unit ball of X. The case $x = 0$ is unimportant.

The space J is the set of real null sequences with bounded squared variation. We will often identify the sequence (a_n) with the (sometimes formal) sum $\Sigma a_n e_n$. We consider two norms on J:

$$2||\sum a_n e_n||^2 = \sup\left\{ \sum_{i=1}^{k} |a_{p(i)} - a_{p(i+1)}|^2 \right\} \text{ and}$$

$$2|||\sum a_n e_n|||^2 = \sup\left\{\sum_{i=1}^{k} |a_{p(i)} - a_{p(i+1)}|^2 + |a_{p(k+1)} - a_{p(1)}|^2\right\},$$

where both sups are over all integer choices of $\{p(i)\}_{i=1}^{k+1}$ with $0 \leq p(1) < p(2) < \ldots < p(k+1)$. By convention, $a_0 = 0$ always. In either norm, $J^{**} = \left[J \cup \left\{\Sigma\, e_n\right\}\right]$ and if $\Sigma\, a_n e_n \in J^{**}$, $\rho\left[\Sigma\, a_n e_n\right] = \lim_N \rho\left[\sum_1^N a_n e_b\right]$ ($\rho = ||\cdot||$ or $|||\cdot|||$). Furthermore, in $|||\cdot|||$, J and J^{**} are isometric. These results can be found in [12, p. 25].

The important property of these two norms is that the basis $\{e_n\}$ is *neighborly*, that is $||\Sigma\, a_n e_n|| \geq ||\sum_{n\neq j} a_n e_n + be_j||$ where b is either a_{j-1} or a_{j+1} and $j \geq 1$. Basic properties of neighborly norms are given in [4] and [11]. For us, the main tool is the facts if $\Sigma\, a_i e_i \in J^{**}$ and for some integers $n < m-1$ we have $a_n > a_{n+1} = \cdots = a_{m-1} > a_m$ or $a_n < a_{n+1} = \cdots = a_{m-1} < a_m$ then $\Sigma\, a_i e_i$ is not extreme. Indeed, $y = \sum_1^n a_i e_i + a_n\left[\sum_{n+1}^{m-1} e_i\right] + \sum_m^\infty a_i e_i$ and $z = \sum_1^n a_i e_i + a_m\left[\sum_{n+1}^{m-1} e_i\right] + \sum_m^\infty a_i e_i$ both have norms no bigger than $||\Sigma\, a_i e_i||$ and $\Sigma\, a_i e_i = tz + (1-t)y$ where $0 < t < 1$.

Let $s_n = \sum_{i=n}^\infty e_n \in J^{**}$. It is easy to check that for each n, s_n is extreme in both the norms $||\cdot||$ and $|||\cdot|||$. In the norm $||\cdot||$, the following are also extreme (in J^{**}): $s_n - ts_m$ for $n < m$ and $t \geq 1$ and $s_n - s_m + ts_q$ for $n < m < q$ and $t \geq 0$. However, in $|||\cdot|||$ the vector $s_2 - 2s_3$ is not extreme.

We note there are neighborly norms on J in which even s_n, $n \geq 2$ are not extreme. (See [5, p. 105]).

§1. The main result

Theorem 1. *In the norm* $\|\cdot\|$, *the only onto isometries of* J *are* $\pm$ *identity.*

Before proving the theorem, we state a more general proposition. The proofs of these two results will proceed in parallel. Although there is a gap in the proof of Proposition 2 which will be filled in the next section.

Proposition 2. Suppose

(i) $\{e_n\}$ is a normalized shrinking neighborly basis for X;

(ii) $\{e_{2n}\}_n$ is not equivalent to the usual basis for c_0;

(iii) If for some $p < q$, $a_p \neq 0 = a_q$ and $\lim_j a_j = 1$, then $\|\Sigma a_j e_j\| > 1$;

(iv) Let $s_n = \sum_{i=n}^{\infty} e_i \in X^{**}$ and suppose the following vectors are extreme in X^{**}:

(a) s_n for $n \geq 1$,

(b) $s_n - ts_m$ for $n < m$ and $t \geq 1$, and

(c) $s_p - s_q + ts_r$ for $p < q < r$ and $t \geq 0$;

then the only onto isometries of X are $\pm$ identity.

Proof. (of Theorem 1 and Proposition 2). We start with J. Let $T: J \longrightarrow J$ be an onto isometry. It follows that $T^{**}: J^{**} \longrightarrow J^{**}$ and the induced map $\tilde{T}: J^{**}/J \longrightarrow J^{**}/J$ are both onto isometries. But since J^{**}/J is the reals, either $\tilde{T} = \text{identity}$ or $-\tilde{T} = \text{identity}$. Replacing T by $-T$ if necessary, we may assume $\tilde{T} = \text{identity}$. Restated in terms of T^{**}, if $T^{**}\left[\Sigma a_i e_i\right] = \Sigma b_i e_i$ then $\lim_i a_i = \lim_i b_i$. Although this seems to make critical use of J being quasi–reflexive, Proposition 4 in the next section, says this is true for X as well.

The two proofs are the same from this point on. But we will continue to prove it for J be just using the properties listed in Proposition 2.

We start by making some notation. Already we have defined $s_n = \sum_{j=n}^{\infty} e_j$.

(1) $T^{**}s_n = \Sigma\, a_j^n e_j$ note $0 \le a_j^n \le 1$, $\lim_j a_j^n = 1$, and $\Sigma\, a_j^n e_j$ are extreme.

(2) $B_n = \{m: a_j^m = 0$ for $j < n$ but $a_n^m \neq 0\}$,

(3) $\pi: N \longrightarrow N$ is defined by $\pi(n) = m$ exactly when $n \in B_m$.

We establish the results by proving the following sequence of claims. We prove (A) last, since it is by far the longest.

Claims

(A): If $j,k \in B_n$, and $j < k$ then $a_n^j \le a_n^k$. Hence the sets B_n are finite and π is unbounded.

(B): If $i < j < k$, $p < q$, $p < r$, $p \in B_i$, $q \in B_j$ and $r \in B_k$, then $q < r$. Hence π is eventually non–decreasing.

(C): If $i < j$, $p < q$, $p < r$, $p \in B_i$ and $q,r \in B_j$, then $q = r$. Hence π is eventually one to one.

(D): The number of elements in $\bigcup_{i<k} B_i$ is at least $k-1$. Hence eventually $\pi(n) \le n$ and thus for some K, $\pi(K+i) = \pi(K) + i$, for $i \ge 0$.

(E): If $n \ge K$ and $m < n$, then $a_{\pi(n)}^m = 1$. Thus for $n \ge K$, $T^{**}s_n = s_{\pi(n)}$. Therefore for large n, $\pi(n) = n$.

(F): For all n, $T^{**}s_n = s_n$ and therefore $T = $ identity.

Proof. (B): Suppose $r < q$, now $s_p - ts_q$ is extreme for $t \ge 1$ thus we may assume for some $m < j-1$, $a_m^p < a_{m+1}^p = \cdots = a_{j-1}^p \ge a_j^p$, because for large t,

$a^p_{j-1} > a^p_j - ta^q_j$. On the other hand, $s_p - s_r + ts_q$ is extreme for $t \geq 0$ and for large t we have $a^p_{j-1} < a^p_j + ta^q_j$, a contradiction.

Thus if $M = \max\{j: \pi(j) \leq \pi(1)\}$, then $i > j > M$ implies $\pi(i) \geq \pi(j)$. since $\pi(1) < \pi(i)$, $\pi(j)$. This follows from the first part of (B).

(C): Suppose $q < r$, again $s_p - ts_q$ is extreme for $t \geq 1$. Thus we may assume for some $m < j-1$, $a^p_m < a^p_{m+1} = \dots = a^p_j \geq a^p_j$, because for large t, $a^p_{j-1} > a^p_j - ta^q_j$. On the other hand, $s_p - s_q + ts_r$ is extreme for $t \geq 0$ and for large t we have $a^p_{j-1} < a^p_j - a^q_j + ta^r_j$, a contradiction.

Again if $M = \max\{j: \pi(j) \leq \pi(1)\}$, then $i > j > M$ implies $\pi(i) > \pi(j)$ now. Thus eventually B_k is always a singleton.

(D): This is a counting result. The set of vectors $V_k = \left\{\Sigma\, b_j e_i \in J^{**}: b_j = 0 \text{ for } j < k\right\}$ has codimension $k-1$ in J^{**}. If the number of elements in $\bigcup_{i<k} B_i$ is less than $k-1$, then $T^{**}(J)/V_k$ would have too small a dimension.

Thus if $M = \max\{j: (j) < \pi(1)\}$ then $i > M$ implies $\pi(i) \leq i$. Since i is the number of elements in $\bigcup_{j<\pi(i)} B_j \geq \pi(i) + 1 - 1$. Now the function $f(i) = i - \pi(i)$ is non–increasing for $i > M$, since $\pi(i+1) \geq \pi(i) + 1$. Since $f(i) \geq 0$, eventually $f(i)$ is constant. If $i,j \geq K$ implies $f(i) = f(j)$ then π maps $i \geq K$ onto $i \geq \pi(K)$.

(E): If K,m and n are given and suppose $a^m_{\pi(n)} < 1$. Since $\lim a^m_j = 1$ and s_m is extreme, there are $q > n$, $p < \pi(q) - 1$ so that $a^m_p > a^m_{p+1} = \dots = a^m_{\pi(q)-1} < a^m_{\pi(q)}$. However $s_m - ts_q$ is extreme for $t \geq 1$ and for large t, $a^m_{\pi(q)-1} > a^m_{\pi(q)} - ta^q_{\pi(q)}$. This is a contradiction.

Now if $K \leq m < n$, we have $a^m_{\pi(n)} = 1$. Since $0 < a^m_{\pi(m)} \leq 1$, and s_m is extreme, we must also have $a^m_{\pi(m)} = 1$. In other words $T^{**}s_m = s_{\pi(m)}$.

To see $\pi(i) = i$ for large i, apply the results obtained so far to T^{-1}. Thus we obtain a π' with the properties for $i \geq K'$, both $\pi'(i) \leq i$ and $T^{-1**}s_i = s_{\pi'(i)}$. Thus $i = \pi(\pi'(i)) \leq \pi(i)$, since $\pi'(i) \leq i$ and π is increasing.

(F): Let m be defined by $T^{**}s_m \neq s_m$ but for $n > m$ $T^{**}s_n = s_n$. Observe by (E) if $k \leq m < j$, then $a^k_j = 1$, and it follows that $\pi(k) < \pi(m+1)$ since T maps $[e_j]_{j>m}$ onto $[e_j]_{j>m}$. If $\pi(m) = m$, then again by (E) $T^{**}s_m = s_m$, so we may assume $\pi(m) < m$, and $m > 1$.

Now for $1 \leq k \leq m$, we must have $a^k_m = 1$. Otherwise, since s_k is extreme, there is an $n < m$, so that $a^k_n > a^k_{n+1} = \dots = a^k_m < a^k_{m+1} = 1$. But this can't be since $s_k - 2s_{m+1}$ is extreme and $a^k_m > a^k_{m+1} - 2a^{m+1}_{m+1} = -1$.

Next for $1 \leq j < k \leq m$, we must have $a^j_{m-1} \geq a^k_{m-1}$. Otherwise since $s_j - s_k + s_{m+1}$ is extreme, we would have $a^j_{m-1} - a^k_{m-1} < 0 = a^j_m - a^k_m < a^j_{m+1} - a^k_{m+1} + a^{m+1}_{m+1} = 1$, a contradiction.

Let b_j be defined by $T^{**}(\Sigma b_j e_j) = s_m$. Using $e_j = s_j - s_{j+1}$, we have for $1 \leq n < m$, $\sum_1^m b_j(a^j_n - a^{j+1}_n) = 0$ and $\sum_1^m b_j(a^j_m - a^{j+1}_m) = b_m = 1$, since these are the coefficients of s_m. But $a^j_{m-1} - a^{j+1}_{m-1} \geq 0$ for $j < m$, $b_m(a^m_{m-1} - a^{m+1}_{m-1}) = a^m_{m-1} \geq 0$ and $0 \leq b_j \leq 1$ by the same reasoning as in (1). Therefore $a^m_{m-1} = 0$.

For J, this is enough for a contradiction. Since $\pi(m) < m$ and thus $a^m_{\pi(m)} > 0$, $\pi(m) < m-1$ and $\|\sum_{j=m}^{\infty} a^m_j e_j\| = 1$ and so $\|\Sigma a^j_m e_j\| > 1$. But this contradicts the fact that T^{**} is an isometry. For X this is just condition (iii).

(A): First we show that for each i, $\lim_n a_i^n = 0$. Let (f_i) be the coefficient functionals to (e_i). Now $a_i^n = (T^{**}s_n)(f_i) = s_n(T^*f_i) = \sum_n^\infty b_j^i$, where $T^*f_i = \sum b_j^i f_j$. Since (e_i) is shrinking, (f_i) is boundedly complete and so $|a_i^n| \le \|\sum_n^\infty b_j^i f_j\| \longrightarrow 0$ as $n \longrightarrow \infty$. Thus the first statement in (A) implies the others.

Before proving the first statement in (A) we need a lemma. This lemma is stated in a manner so that its proof is clear.

Lemma 3.

(I) Suppose $0 < a_1, a_2, b_1, b_2$,; $a_1 \ge a_2$ and $b_1 \ge b_2$.

If $a_1 = \lambda b_1$, with $\lambda > 1$ but $a_2 \ne \lambda b_2$ then for small enough $\epsilon > 0$ either

$a_2 > \lambda b_2$ and $0 < a_1 - (\lambda-\epsilon)b_1 < a_2 - (\lambda-\epsilon)b_2$, or

$a_2 < \lambda b_2$ and $0 < a_1 - (\lambda+\epsilon)b_1 > a_2 - (\lambda+\epsilon)b_2$.

(II) Suppose $0 < a_1, a_2, a_3, b_1, b_2, b_3$; $a_1 > a_2 \le a_3$ and $b_1 > b_2 \le b_3$.

If $a_1 = \lambda b_1$, $a_2 = \lambda b_2$ with $\lambda > 1$ but $a_3 \ne \lambda b_3$ then for small enough $\epsilon > 0$ either

$a_3 > \lambda b_3$ and $a_1 - (\lambda+\epsilon)b_1 < a_2 - (\lambda+\epsilon)b_2 < a_3 - (\lambda+\epsilon)b_3$, or

$a_3 < \lambda b_3$ and $a_1 - (\lambda-\epsilon)b_1 > a_2 - (\lambda-\epsilon)b_2 > a_3 - (\lambda-\epsilon)b_3$.

(III) Suppose $0 < a_1, a_2, a_3, b_1, b_2, b_3$; $a_1 < a_2 \ge a_3$ and $b_1 < b_2 \ge b_3$.

If $a_1 = \lambda b_1$, $a_2 = \lambda b_2$ with $\lambda > 1$ but $a_3 \ne \lambda b_3$ then for small enough $\epsilon > 0$ either

$a_3 > \lambda b_3$ and $a_1 - (\lambda-\epsilon)b_1 < a_2 - (\lambda-\epsilon)b_2 < a_3 - (\lambda-\epsilon)b_3$, or

$a_3 < \lambda b_3$ and $a_1 - (\lambda+\epsilon)b_1 > a_2 - (\lambda+\epsilon)b_2 > a_3 - (\lambda+\epsilon)b_3$.

We are now ready to prove if $j, k \in B_n$ and $j < k$ then $a_n^j \le a_n^k$. Suppose otherwise that $a_n^j > a_n^k$ and let $\lambda > 1$ so that $a_n^j = \lambda a_n^k$. We will show by an inductive procedure that $a_m^j = \lambda a_m^k$ for all m. To start, we already have $a_m^j = \lambda a_m^k$ for $m \leq n$. Go to step 1.

Step 1: Since s_j, s_k are extreme points, $a_n^j \geq a_{n+1}^j$ and $a_n^k \geq a_{n+1}^k$. Since $s_j - (\lambda \pm \epsilon) s_k$ is an extreme point and since Lemma 3(I) applies, we have $a_{n+1}^j = \lambda a_{n+1}^k$. If $a_n^j > a_{n+1}^j$, then go to step 2. Otherwise $a_n^j = a_{n+1}^j$, and $a_{n+1}^j \geq a_{n+2}^j$. Thus we can reapply Lemma 3(I) with $a_1 = a_n^j = a_{n+1}^j$, $a_2 = a_{n+2}^j$, $b_1 = a_n^k = a_{n+1}^k$, $b_2 = a_{n+2}^k$. Thus $a_{n+2}^j = \lambda a_{n+2}^k$. Again if $a_{n+1}^j > a_{n+2}^j$, then go to step 2. Otherwise, we repeat step 1 until $a_m^j > a_{m+1}^j$ and then go to step 2 or we run out of integers and then go to step 4.

Step 2: When we enter this step we have $a_p^j > a_{p+1}^j > 0$ and $a_i^j = \lambda a_i^k$ for $i \leq p+1$. Since s_j, s_k are extreme points $a_{p+1}^j \leq a_{p+2}^j$ and $a_{p+1}^k \leq a_{p+2}^k$. Since $s_j - (\lambda \pm \epsilon) s_k$ is an extreme point and since Lemma 3(II) applies, we have $a_{p+2}^j = \lambda a_{p+2}^k$. If $a_{p+1}^j < a_{p+2}^j$, then go to step 3. If $a_{p+1}^j = a_{p+2}^j$, then reapply Lemma 3(II) until $a_{p+1}^j = a_{p+2}^j = \ldots = a_p^j < a_{q+1}^j$ (eventually this must be true by (1)) and then go to step 3.

Step 3: When we enter this step we have $0 < a_p^j < a_{p+1}^j$ and $a_i^j = \lambda a_i^k$ for $i \leq p+1$. Since s_j, s_k are extreme points $a_{p+1}^j \geq a_{p+2}^j$ and $a_{p+1}^k \geq a_{p+2}^k$. Since $s_j - (\lambda \pm \epsilon) s_k$ is an extreme point and since Lemma 3(III) applies, we have $a_{p+2}^j = \lambda a_{p+2}^k$. If $a_{p+1}^j > a_{p+2}^j$ then go back to step 2. If $a_{p+1}^j = a_{p+2}^j$, then we reapply Lemma 3(III) until $a_{p+1}^j = a_{p+2}^j = \ldots = a_q^j > a_{q+1}^j$ and return to step 2 or until we run out of integers and go on to step 4.

Step 4: Thus $a_i^j = \lambda a_i^k$ for all i. But this is impossible since $\lambda > 1$ and both $\Sigma\, a_i^j e_i = \lambda \Sigma a_i^k e_i$ and $\Sigma\, a_i^k e_i$ have norm one. This completes (A) and hence both Theorem 1 and Proposition 2 are done. □

§2. A non–quasi–reflexive example.

The goal of this section is to show the Proposition 2 applies to some separable space X with X^{**} non–separable. First we fill in the gap in the proof of Proposition 2 via the proposition below. The proofs of (A) and (B) of Proposition 4 are very similar to the proofs of similar statements (for a special case) in [3].

Proposition 4. Let $\{e_n\}$ be a shrinking neighborly basis for X with coefficient functionals $\{f_n\}$. If $\{e_{2n}\}_n$ is not equivalent to the usual basis of c_0, then

(A) The sequence (f_n) is a weak $(\sigma(X^*,X^{**}))$ Cauchy sequence. Let L be its limit in X^{**}.

(B) The weak*–sequential closure $(\sigma(X^{***},X^{**}))$ of X^* in X^{***} is $[\{f_n\} \cup \{L\}]$.

(C) If $T\colon X \longrightarrow X$ is an onto isometry, then $T^{***}(L) = \pm L$. Hence if $\Sigma\, a_n e_n$, $\Sigma\, b_n e_n \in X^{**}$ and $T^{**}(\Sigma\, a_n e_n) = \Sigma\, b_n e_n$, then $\lim a_n = \pm \lim b_n$.

Proof. (A): Suppose (f_n) is not weak* Cauchy, then we have $\delta > 0$, $F \in X^{**}$ with $||F|| = 1$ and sequences $(n(i))$ and $(m(i))$ with $n(i) < m(i) < n(i+1)$, so that $F(f_{n(i)} - f_{m(i)}) > \delta$. Thus $||\sum_1^N (f_{n(i)} - f_{m(i)})|| > N\delta$ and since (f_n) is invariant under spreading [4, p. 557], $||\sum_1^N (f_{2i-1} - f_{2i})|| > N\delta$. Now $\{f_{2i-1} - f_{2i}\}_i$ is a subsymmetric basic sequence [7] and hence it is equivalent to the usual basis of ℓ_1 [12, p. 120].

Now the projection $P: X \longrightarrow X$ given by

$Px = \sum_i ((f_{2i-1}(x) - f_{2i}(x))/2)(e_{2i-1} - e_{2i})$ has $||P|| \le 2$ [4, p. 557]. Hence $(f_{2i-1} - f_{2i})$ is equivalent to the coefficient functionals for $(e_{2i-1} - e_{2i})$. Thus $(e_{2i-1} - e_{2i})$ and the equivalent basic sequence (e_{2i}) [4, p. 557] are equivalent to the usual basis of c_0. We note that the converse is also true. That is if (e_{2i}) is equivalent to the usual basis of c_0, then $F = \sum_i e_{2i} \in X^{**}$ shows (f_n) isn't weak Cauchy.

(B): Suppose (u_n) is a weak Cauchy sequence in X^* and that its limit in X^{***} is u_∞. Since $X^{***} = X^* \oplus X^\perp$ we have $u_\infty = u + v_\infty$ with $u \in X^*$ and $v_\infty \in X^\perp$. Hence $v_n = u_n - u$ is also weak Cauchy in X^* and the limit of (v_n) is v_∞. Furthermore, for each i, $\lim v_n(e_i) = 0$. If $\liminf ||v_n|| = 0$, then $v_\infty = 0$ and we are done so we may assume that $\{||v_n||\}$ is bounded and bounded below by some $\delta > 0$.

By standard perturbations on the gliding hump we may assume there are sequences $(n(i))$, $(m(i))$ with $n(i) \le m(i) < n(i+1) - 1$ and scalars (a_i) so that $v_i = \sum_{j=n(i)}^{m(i)} a_j f_j$. Now since $\sum e_n \in X^{**}$, the number $\alpha = \lim_i \sum_{j=n(i)}^{m(i)} a_j$ exists. We may assume $a_{m(i)+1}$ is defined so that $\sum_{j=n(i)}^{m(i)+1} a_j = 0$, and hence $\lim a_{m(i)+1} = -\alpha$. Let $w_i = \sum_{j=n(i)}^{m(i)+1} a_j f_j$, and note that in the $\sigma(X^{***},X^{**})$–topology, w_i converges to $v_\infty - \alpha L$. Thus to complete the proof it suffices to show that (w_i) is weakly null.

Suppose we knew (w_i) was an unconditional basis. Let $W = [w_i]$ and note (w_i) is a $\sigma(W,W^*)$ Cauchy sequence. From [12, p. 22], (w_i) is $\sigma(W,W^*)$–null and hence (w_i) is $\sigma(X^*,X^{**})$–null. So it suffices to show (w_i) is an orthogonal sequence in X^*.

The property of (f_n) being invariant under spreading now comes into play. The notion can be a mess, so we will illustrate the proof by showing $||w_1 + w_2 + w_3|| \ge ||w_1 + w_3||$. Let N be real large, then for $0 \le k \le M$

$$\|w_1 + \sum_{j=n(2)}^{m(2)+1} a_j f_{j+k} + \sum_{j=n(3)}^{m(3)+1} a_j f_{j+N}\| = \|w_1 + w_2 + w_3\|.$$

Now the average of these $M + 1$ vectors will have norm $\leq \|w_1 + w_2 + w_3\|$, but this average will zero out most of the middle term. The non–zero part of the middle term is $(M+1)^{-1}\left[\sum_{j=n(2)}^{m(2)} \left[\sum_{i=n(2)}^{j} a_i\right] e_j + \sum_{j=n(2)+1}^{m(2)+1} \left[\sum_{i=j}^{m(2)+1} a_i\right] e_{j+M}\right]$, whose norm converges to zero as M tends to ∞. Hence $\|w_1 + w_3\| \leq \|w_1 + w_2 + w_2\|$. (This is the same proof as in [7] used to shown $(f_{2i-1} - f_{2i})$ is orthogonal.)

(C): If $T^*: X^* \to X^*$ is an onto isometry, then T^{***} must map the weak*–sequential closure of X^* in X^{***} onto itself. Thus $T^{***}(L) = \alpha L + b_n f_n$, where $\Sigma\, b_n f_n \in X^*$. Also the induced map on $[\{f_n\} \cup \{L\}]/X^*$ is an isometry onto itself and hence $\alpha = \pm 1$. Replacing T with $-T$ if necessary, we may assume $T^{***}(L) = L + \Sigma\, b_n f_n$.

On the other hand, since $L \in X^\perp$ and $T(X) = X$ we have $T^{***}(L)(x) = L(T^{**}x) = L(Tx) = 0$. Hence $T^{***}(L) \in X^\perp$ and thus $\Sigma\, b_n f_n = 0$. Therefore $T^{***}(L) = L$. The second statement follows from the first. □

Example 5. There is a separable X with X^{**} non–separable which satisfies the hypotheses of Proposition 2.

Construction: X is constructed from certain Orlicz sequence space in the same way J is constructed for ℓ_2. The Orlicz function M (defined below) does not satisfy the Δ_2–condition at zero. Let

$$M(t) = \begin{cases} 0 & t = 0 \\ 10e^{-1/t} & 0 < t \leq 1/2 \\ M'(1/2)(t-1/2)+M(1/2) & t > 1/2 \end{cases}$$

and note that since $M(1/2) > 1$, the range $t > 1/2$ is never used for computing norms.

The space Y with unit vectors $\{y_i\}$ is the Orlicz sequence space h_M [10, p. 137ff]. That is

$$||\Sigma\, a_i y_i||_y = \inf\{\lambda > 0\colon \Sigma\, M(|a_i|/\lambda) \leq 1\}$$

and Y is $[y_i]$ in this norm. It is known [12, p. 138] that Y contains c_0 as a subspace. Note that the unit vectors y_i do not have norm one. Indeed $||\sum_1^n y_i|| = \ln 10n$.

Since $\{y_i\}$ is a symmetric basis for Y, we can define a neighborly norm on $[e_n] = X$ by

$$K||\sum a_n e_n|| = \sup\{||\sum_1^k |a_{p(i)} - a_{p(i+1)}| y_{i+1}||_Y\}$$

where the sup is over integer sequences $\{p(i)\}_1^{k+1}$ with $0 \leq p(1) < p(2) < \dots < p(k+1)$ (again $a_0 = 0$). The constant K is chosen to be $||y_1 + y_2|| = \ln 20$ so that $||e_i|| = 1$. It follows that $X = [e_i]$ is separable, and since (e_{2n}) is equivalent to (y_n) [4, p. 556], we have $c_0 \subset X$ and hence X^{**} is non–separable.

We start showing that the hypothesis of Proposition 2 is satisfied. Note that we have already shown that (ii) is true, since (e_{2n}) is equivalent to (y_n) and $\lim ||\sum_1^n y_i|| = \infty$. Also the only part of (i) not immediate from the construction is to show that (e_n) is shrinking. There does not seem to be a general result that implies that (e_n) is shrinking. Although it is known that if (e_n) is not shrinking then X contains ℓ_1 [4, p. 555] and that Y does not contain ℓ_1 [12, p. 143].

Suppose (e_n) is not shrinking. By standard methods we can obtain sequences $(n(k))$, $(m(k))$ so that $n(k) \leq m(k) < n(k+1) - 1$ and a sequence (a_i) so that $a_i = 0$ if $m(k) < i < n(k+1)$ and $\ln 20 \|\sum_{i=n(k)}^{m(k)} a_i e_i\|_X = 1$ and $\ln 20 \|\sum_{k=1}^{2n} \sum_{i=n(k)}^{m(k)} a_i e_i\|_X > N$. Let $E = \{p(i)\colon 1 \leq i \leq t+1\}$ so that

$$\|\sum_{i=1}^{t} |a_{p(i)} - a_{p(i+1)}| y_i\|_y > N.$$

For each k with $1 \leq k \leq 2N-1$, either there is or is not an i with $1 \leq i \leq t$ so that

$$(*) \quad n(k) \leq p(i) \leq m(k) \text{ and } n(k+j) \leq p(i+1) \leq m(k+j), \text{ for some } j \geq 1.$$

If $(*)$ is true then we define $c_k = |a_{p(i)} - a_{p(i+1)}|$ and put $m(k)+1$ into the set G. That is $G = \{m(k)+1\colon (*) \text{ is true for } k\}$. If $(*)$ is false for k let $c_k = 0$. Let $\{q(i)\}_{i=1}^{s+1}$ be the list of elements in $E \cup G$, and let $b_i = |a_{q(i)} - a_{q(i+1)}|$. We have

$$\|\sum_1^{s} b_i y_i + \sum_1^{2N-1} c_i y_{i+s}\|_y > N.$$

The advantage to this last form is that the $\{b_i\}_1^s$ can be divided into 2N subsets each of which estimate and hence are $\leq \|\sum_{n(k)}^{m(k)} a_i e_i\| \ln 20$.

We have $\sum_1^s M(b_i) \leq 2N$ and $\sum_1^s M(b_i/N) + \sum_1^{2N-1} M(c_i/N) > 1$. Note that each $|a_i| \leq 1/\ln 20$ hence each $|b_i|$, $|c_i| \leq 1$. Thus $\sum_1^{2N-1} M(c_i/N) \leq 2N\, M(1/N) = 2Ne^{-1/N}$.

Since each $M(b_i)$ is part of an estimate for some $\ln 20 \,||\sum_{n(k)}^{m(k)} a_i e_i||$ we have $M(b_i) \leq 1$ and at most $2N$ of the i's have $M(b_i) > 1/2$. Let $F = \{i\colon M(b_i) > 1/2\}$. If $i \notin F$, then $M(b_i/N) = M(b_i)^N \leq 2^{-N+1} M(b_i)$. If $i \in F$, then $M(b_i/N) \leq M(1/N) = e^{-1/N}$. Therefore

$$\begin{aligned}\sum_1^s M(b_i/N) &= \sum_{i\in F} M(b_i/N) + \sum_{i\notin F} M(b_i/N)\\ &\leq 2Ne^{-1/N} + 2^{-N+1}\sum_{i\notin F} M(b_i)\\ &\leq 2Ne^{-1/N} + 2N2^{-N+1}.\end{aligned}$$

Combining these results yields the contradiction $1 < 4Ne^{-1/N} + 2N\,2^{-N+1}$. Therefore (e_n) is shrinking.

The last conditions to check depend strongly on the norm. Before checking them the following lemma is needed. This lemma is perhaps well known for a larger class of Orlicz functions and the proof is included for the readers aid and completeness.

Lemma 6. Properties of the norm in Y.

(A) If $0 \leq a_1,\dots,a_n$, $a_i \leq b_i$ and for some j with $1 \leq j \leq n$, $a_j < b_j$, then $||\sum_1^n a_i y_i|| < ||\sum_1^n b_i y_i||$.

(B) If $\|\sum_1^n a_iy_i\| = \|\sum_1^n (a_i + b_i\epsilon)y_i\|$ for all small enough $\epsilon \geq 0$ then $b_1 = b_2 = \ldots = b_n = 0$.

Proof. (A): The function $f(\lambda) = \sum_1^n M(a_i/\lambda)$ is continuous and strictly decreasing and strictly less than $g(\lambda) = \sum_1^n M(b_i/\lambda)$.

(B): We may assume all the norms are equal to one and hence $|a_i + b_i\epsilon| < 1/2$ for all i and small ϵ. We may assume $a_i \geq 0$ (but not the b_i's). Hence we have for small $\epsilon \geq 0$

$$\sum_1^n M(a_i + b_i\epsilon) = 1.$$

(We may assume $a_i + b_i\epsilon \neq 0$ for all i and $\epsilon > 0$, and $b_i > 0$ if $a_i = 0$.). Take $d^2/d\epsilon^2$ of both sides obtaining

$$\sum_1^n M(a_i + b_i\epsilon) \left[\frac{1-2(a_i+b_i\epsilon)}{(a_i+b_i\epsilon)^4}\right] b_i^2 = 0.$$

But since $(a_i + b_i\epsilon) < 1/2$ all the factors but b_i^2 are > 0. Therefore $b_i = 0$. □

Now we check condition (iii) of Proposition 2. We are given $p < q$, $a_p \neq 0 = a_q$ and $\lim_j a_j = 1$. Let $p(1) = 0$, $p(2) = p$, $p(3) = q$ and $q < p(4) < p(5)$. We have

$$\|\sum a_je_j\| \geq \|\sum_1^{p(4)} a_je_j\| \geq \| |a_p|y_1 + |a_p|y_2 + |a_{p(4)}|y_3 + |a_{p(4)}|y_4 \|/\ln 20.$$

Letting $p(4) \longrightarrow \infty$ and using Lemma 6(A) we have

$$\|\sum a_j e_j\| \geq \| \, |a_p|(y_1 + y_2) + y_3 + y_4\|/\ln 20 > \|y_3 + y_4\|/\ln 20 = 1.$$

Thus condition (iii) is satisfied.

The proofs that all the collections in (iv) are extreme are similar so we will only show (c) for $t > 0$ and $p > 1$. If $x = s_p - s_q + ts_r$ is not extreme for some $1 < p < q < r$ and $t > 0$, then there are vectors $\xi = \Sigma\, \xi_n e_n$ and $\eta = \Sigma\, \eta_n e_n$ with $\xi \neq \eta$, $\|\xi\| = \|\eta\| = \|x\|$ and $x = (\xi+\eta)/2$. Pick $p(1) = 0$, $p(2) < p$, $p \leq p(3) < q$, $q \leq p(4) < r$, $r \leq p(5) < p(6)$ so that $x_{p(i)} \neq \xi_{p(i)}$ for at least one $i \leq 5$, where $x = \Sigma\, x_i e_i$. Since $\{p(i)\}_{i=1}^{6}$ norms $\sum_{1}^{p(5)} x_i e_i$, it must also norm $\sum_{1}^{p(5)} \xi_i e_i$ and $\sum_{1}^{p(5)} \eta_i e_i$. Let $a_i = x_{p(i)} - x_{p(i+1)}$ for $i \leq 4$ and $a_5 = x_{p(5)}$. Write $a_i + b_i = \xi_{p(i)} - \xi_{p(i+1)}$ for $i \leq 4$ and $a_5 + b_5 = \xi_{p(5)}$. We have $\|\sum_{1}^{5} a_i y_i\| = \|\sum_{1}^{5} (a_i + b_i \epsilon) y_i\|$ for $0 \leq \epsilon \leq 1$. By Lemma 6(B) we have $b_i = 0$ for all i, which implies $x_{p(i)} = \xi_{p(i)}$ for all i, a contradiction. □

REFERENCES

1. Andrew, A. *Spreading basic sequences and subspaces of James' quasi–reflexive space*, Math Scand. 48 (1981), 276–282.

2. Andrew, A. *James' quasi–reflexive space is not isometric to any subspace of its dual*, Israel J. Math. 38 (1981), 109–118.

3. Azimi, P. and Hagler, J.N. *Examples of hereditarily* ℓ_1*–Banach spaces failing the Schur property*, Pac. J. Math. 122 (1986), 287–297.

4. Bellenot, S.F. *Transfinite duals of quasi–reflexive Banach spaces*, Trans. Amer. Math. Soc. 273 (1982), 551–577.

5. Bellenot, S.F. *The J–sum of Banach spaces*, J. Funct. Anal. 48 (1982), 95–106.

6. Bellenot, S.F. *Banach spaces with trivial isometries*, Israel J. Math. 56 (1986), 89–96.

7. Brunel, A. and Sucheston, L. *Equal signs additive sequences in Banach spaces.* J. Funct. Anal. 21 (1976), 286–304.

8. Casazza, P.G. *James' quasi–reflexive space is primary*, Israel J. Math. 26 (1977), 294–305.

9. Davis, W.J. *Separable Banach spaces with only trivial isometries*, Rev. Roum. Math. Pures et Appl. 16 (1971), 1051–1054.

10. James, R.C. *A non–reflexive Banach space isometric with its second conjugate space*, Proc. Nat. Acad. Sci., USA 37 (1951), 174–177.

11. James, R.C. *Banach spaces quasi–reflexive of order one*, Studia Math. 60 (1977), 157–177.

12. Linderstrauss, J. and Tzafriri, L., Classical Banach Spaces I Sequence Spaces, Springer Verlag, New York/Berlin, 1977.

13. Semenov, P.V. and Skorik, A.I. *Isometries of James spaces*, Math Notes 38 (1985), 804–808. (Translated from Russian – Mat. Zametki 38 (1985), 537–544).

14. Sersouri, A. *On James' type spaces*, Trans. Amer. Math. Soc. (to appear).

Author's Address: Department of Mathematics
University of Texas at Austin
Austin, TX 78712

Currently: Department of Mathematics
The Florida State University
Tallahassee, Florida 32306–3027

Contemporary Mathematics
Volume **85**, 1989

Quasi-reflexive and tree spaces constructed in the spirit of R.C. James

STEVEN F. BELLENOT, RICHARD HAYDON and EDWARD ODELL*

Abstract. Given a normalized basis (e_i), we construct two spaces $J(e_i)$ and $JT(e_j)$. When $[(e_i)]$ is reflexive the space $J(e_i)$ is quasi-reflexive of order one and $JT(e_i)$ is a dual space not containing ℓ_1 with a non-separable dual. If (f_i) is the unconditional basis naturally obtained from (e_i), then $[(e_i)]$ reflexive implies $[(f_i)]$ is also reflexive. Similar results are obtained for right dominant bases, a weak-subsymmetric-basis-like property, introduced to help describe the structure of $J(e_i)$.

AMS Classification Numbers: 46B10, 46B15

Key Words and Phrases: basis, quasi-reflexive Banach space, right dominant basis, Tsirelson space.

0. Introduction

In this paper we consider generalizations of the famous quasi-reflexive Banach space J of R.C. James [**J1**] and generalizations of the James tree space [**J2**]. The space J is constructed by a certain recipe applied to the unit vector basis of ℓ_2 and J is known [**HW**] to give a positive solution (for $X = \ell_2$) to the following problem.

Problem. Let X be an infinite dimensional reflexive Banach space. Does there exist a quasi-reflexive Banach space Q such that every infinite dimensional subspace of Q contains a further (infinite dimensional) subspace which embeds into X? (Note that $J \oplus X$ trivially provides a quasi-reflexive space which contains X.)

We define a Banach space $J(e_i)$ which generalizes the construction of J to an arbitrary normalized basis (e_i). Of course a number of other authors have considered such generalizations (see *e.g.*, [**B1**], [**CL**], [**CLL**]) but those generalizations imposed more restrictive conditions (such as symmetry) on (e_i). Our object in studying

*Research partially supported by NSF Grant DMS-8601752.

$J(e_i)$ was twofold. First we wished to address the problem above, and secondly we wanted to investigate the question as to whether $J(e_i)$ is quasi-reflexive of order one provided that $[(e_i)]$ is reflexive. This second question was raised by A. Pełczyński following a talk by one of the authors at the Iowa Workshop on Banach Spaces (June 1987). (This talk concerned one particular example of $J(e_i)$ [**HOR**].) Theorem 4.1 gives an affirmative answer to the second question. The problem was originally raised by H. Rosenthal. We show (Corollary 5.2) that $J(e_i)$ gives an affirmative solution for Tsirelson's space [**FJ**] (see also [**CS**]) but we do not know if it solves the problem in general. One obvious obstruction is the fact that $J(e_i)$ has some unconditionality built into it. For example it is easy to show that every subspace of $J(e_i)$ contains an unconditional basic sequence (Proposition 2.1).

We now describe the organization of this paper and mention some other results contained herein. Section 1 contains the definition of $J(e_i)$. We also introduce the notion of a right dominant basis and show how to obtain a right dominant basis (g_i) from (e_i). We show (Proposition 1.1) that if (f_i) is the unconditionalization of (e_i), then $J(e_i)$, $J(f_i)$ and $J(g_i)$ are naturally isomorphic. A number of our results have analogues for the spaces $F = [(f_i)]$ and $G = [(g_i)]$. For example (Theorem 4.1) shows that F and G are reflexive if $E = [(e_i)]$ is reflexive. Our proof of Theorem 4.1 involves a combinatorial lemma (Lemma 3.1) which has a simple proof using Banach space theory. Theorem 2.2 shows that $J(e_i)$ is either quasi-reflexive (of order one) or it contains an isomorph of either c_0 or ℓ_1. Section 5 contains a result related to the problem. Namely every subspace of $J(e_i)$ contains an unconditional basic sequence which is dominated by another unconditional basic sequence (d_i) that is obtained from (e_i) (Proposition 5.1). Section 6 contains a tree version $JT(e_i)$ of the construction which generalizes JT, the James' tree space [**J2**]. We show (Theorem 6.1) that much of the structure of JT carries over to $JT(e_i)$. For example if E is reflexive, $JT(e_i)$ does not contain a isomorph of ℓ_1. A number of examples are included throughout the text.

We use standard Banach space notation as may be found in the book [**LT**]. Following their notation, by a subsymmetric basis we mean an *unconditional* basis which is equivalent to all of its subsequences. We shall say a basis is *spreading* if it's equivalent to all of its subsequences.

1. Construction of $J(e_i)$ and other normings

Let (e_i) be a normalized basis for a Banach space E. We shall define several norms on c_{00}, the linear space of all finitely supported sequences of reals.

The space $J(e_i)$

For $(a_i) \in c_{00}$, define

$$\Big\|\sum a_i u_i\Big\| = \sup\Big\{ \Big\|\sum_{i=1}^{k}\Big(\sum_{j=p(i)}^{q(i)} a_j\Big) e_{p(i)}\Big\| : \\ k \in \mathbb{N} \text{ and } 1 \le p(1) \le q(1) < p(2) \le q(2) < \cdots < p(k) \le q(k) \Big\}. \tag{1.1}$$

We let $J(e_i)$ be the completion of c_{00} under this norm. Clearly (u_i) is a normalized monotone basis for $J(e_i)$. Of course if (e_i) is the unit vector basis of ℓ_2, then $J(e_i)$ is just James' quasi-reflexive space J [**J1**].

If $x = \sum_{i=1}^{k}(\sum_{j=p(i)}^{q(i)} a_j)e_{p(i)}$ is as in (1.1), we shall say *x is a representative of* $\sum a_i u_i$ in E.

The "unconditionalized" space $F = F(e_i)$

There is a standard way to obtain an unconditional basis (f_i) from (e_i) (see [**LT**, pp.19–20]). For $(a_i) \in c_{00}$ define

$$\Big\|\sum a_i f_i\Big\| = \sup\Big\{ \Big\|\sum_{i\in F} a_i e_i\Big\| : F \subseteq \mathbb{N} \Big\}. \tag{1.2}$$

$F = F(e_i)$ is the completion of c_{00} under the norm in (1.2). (f_i) is a normalized suppression-1-unconditional basis for F. (f_i) is equivalent to (e_i) if and only if (e_i) is unconditional. We say $x = \sum_{i\in F} a_i e_i$ is a *representative* of $\sum a_i f_i$ in E.

The "right dominate" space $G = G(e_i)$

For $(a_i) \in c_{00}$ define

$$\left\|\sum a_i g_i\right\| = \sup\Big\{ \left\|\sum a_{n(i)} e_{m(i)}\right\| : \quad 1 \le m(1) \le n(1) < m(2) \le n(2) < \cdots \Big\}. \tag{1.3}$$

$G = G(e_i)$ is the completion of c_{00} under (1.3). (g_i) is a normalized suppression-1-unconditional basis for G. The basis (g_i) also has a one-sided subsymmetric property which we now define. As usual, $x = \sum a_{n(i)} e_{m(i)}$ is called a *representative* of $\sum a_i g_i$ in E.

Definition. An unconditional basis (x_i) is called *right dominant* if there exists a constant $C < \infty$ so that whenever $1 \le m(1) \le n(1) < \cdots < m(i) \le n(i) < \cdots$, we have

$$\left\|\sum a_{n(i)} x_{m(i)}\right\| \le C \left\|\sum a_{n(i)} x_{n(i)}\right\| . \tag{1.4}$$

Thus $(x_{n(i)})$ C-dominates $(x_{m(i)})$ whenever $(x_{n(i)})$ is, in the restrained sense above, to the right of $(x_{m(i)})$. The smallest such C is called the *right dominant constant of* (x_i).

Remarks 1. The basis (e_i) is right dominant if and only if it is equivalent to the basis (g_i) given by the norm in (1.3) (see Proposition 1.1).

2. The condition (1.4) alone does not imply that (x_i) is unconditional. Indeed it is easy to check that the basis (u_i) of any $J(e_i)$ satisfies (1.4).

3. It is known that a normalized spreading weakly null basis is unconditional. The standard proof of this fact (see *e.g.*, [**O2**]) yields that a normalized weakly null basis (x_i), which satisfies (1.4), has an unconditional subsequence.

Our first proposition yields that (g_i) is right dominant and that the three spaces $J(e_i)$, $J(f_i)$ and $J(g_i)$ all have equivalent bases.

Proposition 1.1.

1. *$G(e_i)$ and $G(f_i)$ are (naturally) isometric,*
2. *$J(e_i)$ and $J(f_i)$ are (naturally) isometric,*
3. *The norms on $J(e_i)$ and $J(g_i)$ are 4-equivalent,*
4. *(g_i) is a right dominant basis with right dominant constant not exceeding 2.*

Proof. 1. and 2. are immediate.

3. Let $\|\cdot\|$ be the norm in $J(e_i)$ and let $|||\cdot|||$ denote the norm in $J(g_i)$. Since (g_i) 1-dominates (e_i) we have $\|\cdot\| \le |||\cdot|||$. It remains to prove that if $(a_i) \in c_{00}$, then $|||\sum a_i u_i||| \le 4\|\sum a_i u_i\|$. Choose $1 \le p(1) \le q(1) < \cdots < p(k) \le q(k)$ such that

$$\Big|\Big|\Big|\sum a_i u_i\Big|\Big|\Big| = \Big\|\sum_{i=1}^{k}\Big(\sum_{j=p(i)}^{q(i)} a_j\Big) g_{p(i)}\Big\| .$$

Define a sequence (b_n) of reals by $b_{p(i)} = \sum_{j=p(i)}^{q(i)} a_j$ and $b_n = 0$ if $n \notin \{p(1), \ldots, p(k)\}$. Thus

$$\Big|\Big|\Big|\sum a_i u_i\Big|\Big|\Big| = \Big\|\sum_{i=1}^{k} b_i g_i\Big\| = \Big\|\sum_{i=1}^{\ell} b_{n(i)} e_{m(i)}\Big\|$$

for some $1 \le m(1) \le n(1) < \cdots < m(\ell) \le n(\ell)$. We may assume that $n(i) \in \{p(1), \ldots, p(k)\}$ for $i \le \ell$ (otherwise $b_{n(i)} = 0$). Hence we may assume (for perhaps a smaller k and by relabeling $p(i)$ and $q(i)$) that

$$\Big|\Big|\Big|\sum a_i u_i\Big|\Big|\Big| = \Big\|\sum_{i=1}^{k}\Big(\sum_{j=p(i)}^{q(i)} a_j\Big) e_{m(i)}\Big\|$$

where $1 \le p(1) \le q(1) < \cdots < p(k) \le q(k)$ and $1 \le m(1) \le p(1) < \cdots < m(k) \le q(k)$. Note that $m(2i-1) \le p(2i-1) \le q(2i-1) < p(2i) < m(2i+1)$ and $m(2i) \le p(2i) \le q(2i) < p(2i+1) < m(2i+2)$. Thus by the triangle inequality in E,

$$\Big\|\sum_{\substack{i=1\\ i\,\mathrm{odd}}}^{k}\Big(\sum_{j=p(i)}^{q(i)} a_i\Big) e_{m(i)}\Big\| \le \Big\|\sum_{\substack{i=1\\ i\,\mathrm{odd}}}^{k}\Big(\sum_{j=m(i)}^{q(i)} a_j\Big) e_{m(i)}\Big\| + \Big\|\sum_{\substack{i=1\\ i\,\mathrm{odd}\\ p(i)\neq m(i)}}^{k}\Big(\sum_{j=m(i)}^{p(i)-1} a_j\Big) e_{m(i)}\Big\|$$

$\le 2\Big\|\sum a_i u_i\Big\|$ since both terms in the line above are representatives of $\sum a_i u_i$.

Similarly

$$\Big\| \sum_{\substack{i=1\\ i\,\text{even}}}^{k} \Big(\sum_{j=p(i)}^{q(i)} a_j \Big) e_{m(i)} \Big\| \le 2 \Big\| \sum a_i u_i \Big\|$$

and 3. follows by the triangle inequality.

4. Let $(a_i) \in c_{00}$ and $1 \le m(1) \le n(1) < \cdots < m(i) \le n(i) < \cdots$. We need to show that $\| \sum a_i g_{m(i)} \| \le 2 \| \sum a_i g_{n(i)} \|$. Since (g_i) is suppression-1-unconditional, we may assume (by perhaps re-labeling) that $\| \sum_i a_i g_{m(i)} \| = \| \sum_i a_i e_{r(i)} \|$ where $1 \le r(1) \le m(1) < \cdots < r(i) \le m(i) < \cdots$. By breaking $\sum_i$ into $\sum_{i\,\text{even}}$ and $\sum_{i\,\text{odd}}$ we may suppose $\| \sum_i a_i g_{m(i)} \| \le 2 \| \sum_i a_{2i} e_{r(2i)} \|$. But $r(2i) \le m(2i) \le n(2i) < m(2i+1) < r(2i+2) \le n(2i+2)$ and so $\sum_i a_{2i} e_{r(2i)}$ is a representation of $\sum_i a_{2i} g_{n(2i)}$ and thus

$$\Big\| \sum_i a_i g_{m(i)} \Big\| \le 2 \Big\| \sum a_{2i} g_{n(2i)} \Big\| \le 2 \Big\| \sum a_i g_{n(i)} \Big\| . \qquad \blacksquare$$

We conclude this section with a few simple examples.

Example 1. Let (e_i) be the natural basis of $E = \ell_2 \oplus \ell_4$ with (e_{2i-1}) equivalent to the unit vector of ℓ_2 and (e_{2i}) equivalent to the unit vector basis of ℓ_4. Then the right dominant basis (g_i) obtained from (e_i) is equivalent to the unit vector basis of ℓ_2. Thus by Proposition 1.1(3), $J(e_i)$ is isomorphic to J. The ℓ_4-part of $\ell_2 \oplus \ell_4$ is overpowered by the ℓ_2-part. In particular E does not embed into $J(e_i)$, since ℓ_2 is a subspace of each infinite dimensional subspace of J [**HW**].

2. J has another basis (e_i) given by

$$\Big\| \sum a_i e_i \Big\| = \sup \Big\{ \Big(\sum_{i=1}^{k} |a_{p(i)} - a_{p(i+1)}|^2 \Big)^{1/2} : 1 \le p(1) < p(2) < \cdots < p(k+1) \Big\} .$$

In this case, the basis (f_i) of $F(e_i)$ is equivalent to the unit vector basis of ℓ_2 and thus, again by Proposition 1.1, $J(e_i)$ is isomorphic to J.

3. Let (e_i) be the unit vector basis of Tsirelson's space T [**FJ**] (or [**CS**]) or one of its superspaces as considered in [**B2**]. This is an example of a right dominant basis which is not subsymmetric.

The Leftist View

It is possible to replace $e_{p(i)}$ by $e_{q(i)}$ in (1.1), obtaining a space $K(e_i)$. In this case one would then consider *left dominant* bases defined as in (1.4) except that we would now require $1 \leq n(1) \leq m(1) < n(2) \leq m(2) < \cdots$. The "left dominant" space $H = H(e_i)$ would be defined by

(1.3′)

$$\left\|\sum a_i h_i\right\| = \sup\left\{\left\|\sum a_{n(i)} e_{m(i)}\right\| : 1 \leq n(1) \leq m(1) < n(2) \leq m(2) < \cdots\right\}.$$

One would then obtain a leftist version of Proposition 1.1.

We do not have any example of a basis which is left dominant but not right dominant (or vice versa). Nor do we have an example where $K(e_i)$ and $J(e_i)$ differ.

Our final two examples show that a right dominant basis (left dominant basis) cannot in general be renormed so that its domination constant is 1.

Example 4. The unit vector basis (e_i) of Tsirelson's space T (Example 3) is right dominant with constant 1 and also left dominant. However T cannot be renormed so that its left domination constant is 1. Indeed if (g_i) is right dominant (respectively, left dominant) with constant C, then for all $k \in \mathbb{N}$, $\|\sum_{i=1}^k a_i g_i\| \leq C^k \|\sum_{i=1}^k a_i g_{n(i)}\|$ for any subsequence $(g_{n(i)})$ (respectively, $C^k \|\sum_{i=1}^k a_i g_i\| \leq \|\sum_{i=1}^k a_i g_{n(i)}\|$). It follows that (g_i) is subsymmetric if it has an equivalent norm in which (g_i) is right dominant with constant 1 and (perhaps) a different equivalent norm in which its left dominant with constant 1.

5. For $n \in \mathbb{N}$, let $d_n = \sqrt{n} - \sqrt{n-1}$ and define

$$\left\|\sum a_i g_i\right\| = \sup\left\{\sum |a_{n(i)}| d_{m(i)} : 1 \leq m(1) \leq n(1) < \cdots < m(i) \leq n(i) < \cdots\right\}.$$

Using the proof of Proposition 1.1(4) it easy to show that (g_i) is right dominant with constant 2. Furthermore, since (d_n) is decreasing, (g_i) is left dominant with constant 1. However (g_i) cannot be renormed to have right dominance constant 1. This follows from our observations in Example 4, provided we note (g_i) is not subsymmetric. Indeed, $\|\sum_1^n g_i\| = \sqrt{n}$, but $\|\sum_{i=k}^{k+n} g_i\| = 1 + \sqrt{k+n} - \sqrt{k} \to 1$ as $k \to \infty$.

Remarks. It is easy to show that

1) A right dominant basis (g_i) is left dominant if and only if every subsequence $(g_{m(i)})$ is equivalent to its shift $(g_{m(i+1)})$. Note also that $(g_{m(i)})$ is right dominant whenever (g_i) is right dominant. Thus if (g_i) is equivalent to (g_{i+1}) for each right dominant basis (g_i) then right dominant bases will also be left dominant.
2) (g_n) is right dominant (respectively, left dominant) if and only if (g_n^*) (the biorthogonal functions of (g_n)) is left dominant (respectively, right dominant).

2. Adding to the left and unconditional sequences

For each $(n(i))$, a sequence of integers, with $1 \le n(1) < n(2) < \cdots$ define an "adding to the left" operator L on $J(e_i)$ by

$$L\Big(\sum a_i u_i\Big) = \sum_i \Big(\sum_{j=n(i)}^{n(i+1)-1} a_j\Big) u_{n(i)} \; . \tag{2.1}$$

Proposition 2.1.

1. *There exists an absolute contant C such that for all $(n(i))$, $||L|| \le C$.*
2. *If $b_i = \sum_{j=m(i)}^{m(i+1)-1} a_j u_j$ is a normalized block basis of (u_j) with*

$$\sum_{j=m(i)}^{m(i+1)-1} a_j = 0 \quad \textit{for all} \quad i \; ,$$

then (b_i) is unconditional.

More generally,

3. *If $b_i = \sum_{j=m(i)}^{m(i+1)-1} a_j u_j$ is a normalized block basis of (u_j) with*

$$\sum_{i=1}^{\infty} \Big| \sum_{j=m(i)}^{m(i+1)-1} a_j \Big| < \infty \; ,$$

then (b_i) is unconditional.

Proof. 3. follows immediately from 2. by the standard perturbation lemma.

1. By (3) and (4) of Proposition 1.1, we may assume that (e_i) is right dominant with right dominance constant C. Let L be given by (2.1), let $(a_i) \in C_{00}$ and choose integers $1 \leq p(1) \leq q(1) < \cdots < p(k) \leq q(k)$ so that

$$\left\|L\Big(\sum a_i u_i\Big)\right\| = \left\|\sum_{i=1}^{k}\Big(\sum_{p(i)\leq n(s)\leq q(i)}\Big(\sum_{j=n(s)}^{n(s+1)-1} a_j\Big)\Big)e_{p(i)}\right\| .$$

By throwing out any zero terms and relabeling, we can increase $p(i)$ to $p'(i)$ and $q(i)$ to $q'(i)$ so that

$$\sum_{p(i)\leq n(s)\leq q(i)} \sum_{j=n(s)}^{n(s+1)-1} a_j = \sum_{j=p'(i)}^{q'(i)} a_j$$

and $1 \leq p'(1) \leq q'(1) < \cdots < p'(k) \leq q'(k)$. Thus by right dominance,

$$\begin{aligned}\left\|L\Big(\sum a_i u_i\Big)\right\| &= \left\|\sum_{i=1}^{k}\Big(\sum_{j=p'(i)}^{q'(i)} a_j e_{p(i)}\Big)\right\| \\ &\leq C\left\|\sum_{i=1}^{k}\Big(\sum_{j=p'(i)}^{q'(i)} a_j\Big)e_{p'(i)}\right\| \leq C\left\|\sum a_i u_i\right\| .\end{aligned}$$

2. It suffices by 1. to construct for each $F \subseteq \mathbb{N}$ an increasing sequence of integers $(n(i))$ so that the corresponding adding to the left operator L satisfies $L(\sum c_i b_i) = \sum_{i\in F} c_i b_i$. To do this we just let $n(i)$ be the increasing listing of the set $\{(m(i))\}_{i=1}^{\infty} \cup \bigcup_{i\in F}\{j : m(i) < j < m(i+1)\}$. Note that if $i \in F$, then $L(b_i) = b_i$, but if $i \notin F$, then $L(b_i) = (\sum_{j=m(i)}^{m(i+1)-1} a_j)u_{m(i)} = 0$ by hypothesis. ∎

The basis (u_i) is never a shrinking basis for $J(e_i)$. Indeed the sum functional S given by $S(\sum a_i u_i) = \sum a_i$ is a normalized element in $J(e_i)^*$. Furthermore, unless (u_i) is equivalent to the unit vector basis of ℓ_1, (u_i) is not unconditional. However our next result is a close analogue of a well known result for unconditional bases [**LT**, pp.21–23]. We let (u_i^*) denote the biorthogonal functionals to (u_i) in $J(e_i)^*$.

Theorem 2.2.

1. *Either* (u_i) *is boundedly complete or* c_0 *embeds into* $J(e_i)$.
2. *Either* $J(e_i)^* = [\{S\} \cup \{u_i^*\}_{i=1}^\infty]$ *or* ℓ_1 *embeds into* $J(e_i)$.
3. *Therefore* $J(e_i)$ *is either quasi-reflexive (of order one) or* $J(e_i)$ *contains a subspace isomorphic to either* c_0 *or* ℓ_1.

Proof. 1. Suppose there exists a sequence $(a_i) \subseteq \mathbb{R}$ with $\sum a_i u_i$ divergent, yet $\sup_N \|\sum_1^N a_i u_i\| < \infty$. Since $S \in J(e_i)^*$, there exists $(m(i))$ so that $b_i = \sum_{j=m(i)}^{m(i+1)-1} a_j u_j$ is a semi-normalized block basis with $\sum_i |\sum_{j=m(i)}^{m(i+1)-1} a_j| < \infty$. Thus by Proposition 2.1, (b_i) is unconditional. Since $\sup_N \|\sum_1^N b_i\| < \infty$, (b_i) is equivalent to the unit vector basis of c_0.

2. Suppose $x^* \in J(e_i)^* \setminus [\{S\} \cup \{u_i^*\}]$. Let $x_i = x^*(u_i)$ and let x_∞ be any cluster point of (x_i) in $\mathbb{R}$. By replacing x^* by $x^* - x_\infty S$ we may assume $x_\infty = 0$. Furthermore, by replacing x^* by $x^* - \sum x_{n(i)} u_{n(i)}^*$ where $\sum_i |x_{n(i)}| < \infty$, we may assume $x_{n(i)} = 0$ for some subsequence $(n(i))$ of $\mathbb{N}$. Since $x^* \notin [(u_i^*)]$ we may also suppose that $(\sum_{j=n(i)+1}^{n(i+1)-1} x_j u_j^*)_{j=1}^\infty$ is a semi-normalized block basis of (u_i^*). Thus we can find scalars (a_j) so that

$$b_i \equiv \sum_{j=n(i)+1}^{n(i+1)-1} a_j u_j - \left(\sum_{j=n(i)+1}^{n(i+1)-1} a_j\right) u_{n(i)}$$

is a semi-normalized block basis of (u_i) and satisfies

$$x^*(b_i) = \sum_{j=n(i)+1}^{n(i+1)-1} x_j a_j \geq \delta > 0$$

for all i and some $\delta > 0$. Since (b_i) is unconditional (by Proposition 2.1) it is therefore equivalent to the unit vector basis of ℓ_1.

3. If c_0 does not embed in $J(e_i)$, then by 1, $[(u_i^*)]^* \cong [(u_i)]$. If ℓ_1 does not embed in $J(e_i)$, then by 2, $[(u_i^*)]^{**} \cong [(u_i)]^* = [\{S\} \cup \{u_i^*\}]$. Thus $[(u_i^*)]$ and its dual $[(u_i)] = J(e_i)$ are both quasi-reflexive of order 1. ∎

Remark. From Proposition 2.1(2) it follows that if (b_i^*) is a skipped block basis of (u_i^*), say $b_i^* = \sum_{j=n(i)+1}^{n(i+1)} x_j u_j^*$ for some subsequence $(n(i))$ of $\mathbb{N}$, then (b_i^*) is unconditional. In particular if (x_j) and $(n(i))$ are as in 2., then (b_i^*) is equivalent to the unit vector basis of c_0 and thus $[(b_i)]$ is complemented in $J(e_i)$ via $P(x) \equiv \sum b_i^*(x)b_i$. Thus, just as in the case of an unconditional basis, if $J(e_i)$ contains ℓ_1 it contains a complemented copy of ℓ_1.

Our last proposition of Section 2 gives some information, in certain cases, on the subspace of $J(e_i)$ spanned by $(u_{2i-1} - u_{2i})$.

Proposition 2.3. *If (e_i) is right dominant then the unconditional basic sequence $(u_{2i-1} - u_{2i})$ in $J(e_i)$ is equivalent to (e_{2i}). Furthermore if the shift $T(\sum a_i e_i) = \sum a_i e_{i+1}$ is bounded on E, then $[(u_{2i-1} - u_{2i})]$ is complemented in $J(e_i)$.*

Proof. We may assume that (e_i) is 1-unconditional. Setting $p(i) = q(i) = 2i$ in (1.1) yields $||\sum a_i e_{2i}|| \le ||\sum a_i(u_{2i-1} - u_{2i})||$. For the reverse inequality, let $(a_i) \in c_{00}$ and choose $1 \le p(1) \le q(1) < \cdots < p(k) \le q(k)$ so that

$$\Big\|\sum_i a_i(u_{2i-1} - u_{2i})\Big\| = \Big\|\sum_{i=1}^k \Big(\sum_{j=p(i)}^{q(i)} b_j\Big) e_{p(i)}\Big\|,$$

where $b_{2j-1} = a_j$ and $b_{2j} = -a_j$.

Now for $i \le k$,

$$\sum_{j=p(i)}^{q(i)} b_j = \begin{cases} 0 & \text{if } p(i) \text{ is odd and } q(i) \text{ is even} \\ b_{q(i)} & \text{if } p(i) \text{ is odd and } q(i) \text{ is odd} \\ b_{p(i)} & \text{if } p(i) \text{ is even and } q(i) \text{ is even} \\ b_{p(i)} + b_{q(i)} & \text{if } p(i) \text{ is even and } q(i) \text{ is odd.} \end{cases}$$

Thus

$$\Big\|\sum a_i(u_{2i-1} - u_{2i})\Big\| \le \Big\|\sum_{\substack{i=1 \\ p(i)\,\text{even}}}^k b_{p(i)} e_{p(i)}\Big\| + \Big\|\sum_{\substack{i=1 \\ q(i)\,\text{odd}}}^k b_{q(i)} e_{p(i)}\Big\|.$$

Since (e_i) is 1-unconditional,

$$\Big\|\sum_{\substack{i=1 \\ p(i)\,\text{even}}}^k b_{p(i)} e_{p(i)}\Big\| \le \Big\|\sum a_i e_{2i}\Big\|.$$

Using twice that (e_i) is C-right dominant and the unconditionality again, we have

$$\Big\| \sum_{\substack{i=1 \\ q(i)\,\mathrm{odd}}}^{k} b_{q(i)} e_{p(i)} \Big\| \le C \Big\| \sum_{\substack{i=1 \\ q(i)\,\mathrm{odd}}}^{k} b_{q(i)} e_{q(i)} \Big\|$$

$$\le C \Big\| \sum a_i e_{2i-1} \Big\| \le C^2 \Big\| \sum a_i e_{2i} \Big\| .$$

Thus

$$\Big\| \sum a_i (u_{2i-1} - u_{2i}) \Big\| \le (C^2+1) \Big\| \sum a_i e_{2i} \Big\| .$$

To prove the furthermore statement, let $R(\sum a_i u_i) = \sum (a_{2i-1} + a_{2i}) u_{2i}$ and $L(\sum a_i u_i) = \sum (a_{2i-1} + a_{2i}) u_{2i-1}$. By Proposition 2.1(1), L is bounded. We shall see shortly that R is bounded. Thus $I-(R+L)/2$ is a projection onto $[(u_{2i-1}-u_{2i})]$.

Define $\widetilde{T}(\sum a_i u_i) = \sum a_i u_{i+1}$. Since T is bounded on E, $\widetilde{T}$ is bounded on $J(e_i)$. Indeed, if x is a representative of $\sum a_i u_{i+1}$, then $x = Ty$ for some representative y of $\sum a_i u_i$, and so $\|\widetilde{T}\| \le \|T\|$. Thus $R = \widetilde{T} L$ is also bounded on $J(e_i)$. ■

Remarks 1. Under the hypothesis of the proposition, (e_{2i-1}) is equivalent to (e_{2i}) and so (e_i) is eqivalent to (e_{i+1}). Indeed, (e_{2i}) dominates (e_{2i-1}) by right dominance, and (e_{2i-1}) dominates (e_{2i}) since T is bounded. The equivalence of (e_i) and (e_{i+1}) follows by unconditionality.

2. In the case where (e_i) is subsymmetric many of these results are known. Indeed "dual versions" are contained in [**B1**].

3. If (e_i) is the unit vector basis for certain Tsirelson superspaces [**B2**], then (e_i) is not equivalent to (e_{2i}), yet (e_i) is right dominant and equivalent to (e_{i+1}).

4. Clearly there is a leftist viewpoint for this section as well.

3. A combinatorial lemma

Our next goal is to prove that $J(e_i)$ is quasi-reflexive whenever E is reflexive. Toward that end we require a combinatorial lemma. Recall that a collection $\mathcal{F} \subseteq 2^{\mathbb{N}}$ is said to be *adequate* if whenever $G \subseteq F \in \mathcal{F}$, then $G \in \mathcal{F}$.

Lemma 3.1. *Let $\mathcal{F}$ be an adequate (defined below) collection of finite subsets of $\mathbb{N}$. Let $\delta > 0$ and suppose that whenever $(a_i)_{i=1}^k \subseteq \mathbb{R}$ with $\sum_{i=1}^k a_i = 1$ and $a_i \geq 0$ $(1 \leq i \leq k)$, then there exists $F \in \mathcal{F}$ with $\sum_{i \in F} a_i \geq \delta$. Then there exists an infinite subsequence $M \subseteq \mathbb{N}$ such that $F \in \mathcal{F}$ for all finite $F \subseteq M$.*

Remark. The hypothesis on $\mathcal{F}$ implies that for all finite $H \subseteq \mathbb{N}$, there exists $F \subseteq H$, $F \in \mathcal{F}$ with $|F| \geq \delta |H|$ ($|F|$ denotes the cardinality of F). But this is strictly weaker than the hypothesis. To see this one need only consider the Schreier sets $\mathcal{F} = \{F \subseteq \mathbb{N} : F \text{ is finite and } \min F \leq |F|\}$.

Proof. Define a norm on c_{00} by $||\sum a_i h_i|| = \sup\{|\sum_{i \in F} a_i| : F \in \mathcal{F}\}$. The hypothesis yields $||\sum a_i h_i|| \geq \delta/2 \sum |a_i|$ and thus (h_i) is equivalent to the unit vector basis of ℓ_1.

Furthermore each $F \in \mathcal{F}$ may be naturally identified with an element in $Ba[(h_i)]^*$ via $F(\sum a_i h_i) = \sum_{i \in F} a_i$. $\mathcal{F}$ is thus a 1-norming set and so $[(h_i)]$ is isometric to a subspace of $C(\overline{\mathcal{F}})$, where $\overline{\mathcal{F}}$ denotes the w^*-closure of $\mathcal{F}$ in $Ba[(h_i)]^*$. Since $[(h_i)]$ is isomorphic to ℓ_1 and ℓ_1 does not embed into $C(K)$ when K is countable, $\overline{\mathcal{F}}$ must be uncountable. Of course $\overline{\mathcal{F}}$ may be identified with the closure of $\mathcal{F}$ in $2^{\mathbb{N}}$ given the product (or pointwise) topology. Thus there exists an infinite $M \subseteq \mathbb{N}$ with $M \in \overline{\mathcal{F}}$. Since $\mathcal{F}$ is adequate, every finite subset of M is in $\mathcal{F}$. ∎

4. Bases and duality

Let (e_i) be a normalized basis for E and let (f_i), (g_i) and (u_i) be the derived bases of Section 1 for F, G and $J(e_i)$, respectively.

Theorem 4.1.

1. *If (e_i) is boundedly complete, then the bases (f_i), (g_i) and (u_i) are all boundedly complete.*
2. *If (e_i) is shrinking, then ℓ_1 does not embed into any of the spaces F, G or $J(e_i)$. In particular the bases (f_i) and (g_i) are both shrinking.*
3. *If E is reflexive, then F and G are reflexive and $J(e_i)$ is quasi-reflexive (of order one).*

Proof. 1. Let (x_i) be any of the three bases, (f_i), (g_i) or (u_i).

Claim. *If $y = \sum_{i=p}^q a_i x_i$, there exists a representative z of y in $[e_i]_{i=p}^q$, with $||y|| \leq 2||z||$.*

The claim yields that (x_i) is boundedly complete. Indeed let $y_i \in [x_j]_{j=p(i)}^{p(i+1)-1}$ with $p(1) < p(2) < \cdots$ and $||y_i|| > \varepsilon > 0$ for all i. Choose a representative $z_i \in [e_j]_{j=p(i)}^{p(i+1)-1}$ of y_i by the claim. Then for all k, $\sum_{i=1}^k z_i$ is a representative of $\sum_1^k y_i$ and so $||\sum_1^k y_i|| \geq ||\sum_1^k z_i|| \to \infty$ as $k \to \infty$ since (e_i) is boundedly complete.

It remains to prove the claim. If $(x_i) = (f_i)$ this is obvious. If $y = \sum_{i=p}^q a_i u_i$, choose $1 \leq r(1) \leq s(1) < \cdots < r(k) \leq s(k)$ so that $||y|| = ||\sum_{i=1}^k (\sum_{j=r(i)}^{s(i)} a_j) e_{r(i)}||$. We may assume $s(k) \leq q$ and $p \leq s(1)$. If $p \leq r(1)$ we are done. Otherwise

$$||y|| \leq \left\|\left(\sum_{j=p}^{s(1)} a_j\right) e_{r(1)}\right\| + \left\|\sum_{i=2}^{k}\left(\sum_{j=r(i)}^{s(i)} a_j\right) e_{r(i)}\right\| .$$

If the latter vector has norm at least as large as $2^{-1}||y||$ we set it equal to z. Otherwise let $z = (\sum_{j=p}^{s(1)} a_j) e_p$.

The case $(x_i) = (g_i)$ is similar since if $\sum a_{n(i)} e_{m(i)}$ is a representative of y, then only $e_{m(1)}$ can lie outside of $\{e_i\}_{i=p}^q$.

2. Again, let (x_i) be one of the bases (f_i), (g_i) or (u_i) and suppose that ℓ_1 embeds in $[(x_i)]$. Then there is a normalized block basis $y_i = \sum_{j=m(i-1)+1}^{m(i)} b_j x_j$

which is equivalent to the unit vector basis of ℓ_1. Thus for some $\delta > 0$, $||\sum a_i y_i|| \geq 4\delta \sum |a_j|$. If $(x_i) = (g_i)$, then by replacing (y_i) by a normalized block basis of long averages we may assume $\sup |b_j| < \delta$. If $(x_i) = (u_i)$, then $\sum_{j=m(i-1)+1}^{m(i)} b_j$ is bounded. Thus by replacing (y_i) by a normalized block basis of long averages of differences $y_{k(i)} - y_{\ell(i)}$ we may suppose that $|\sum_{j=p}^{q} b_j| < \delta$ for all $p \leq q$, and $\sum_{j=m(i-1)+1}^{m(i)} b_j = 0$ for all i.

Let $a_i > 0$ with $\sum_1^k a_i = 1$.

Claim. *There exist representatives* $z_i \in [e_j]_{j=m(i-1)+1}^{m(i)}$ *for* y_i *with* $||\sum_1^k a_i z_i|| > 2\delta$.

This follows from our above restrictions on the b_i's by letting z be a representative of $\sum_1^k a_i y_i$ with $||z|| > 4\delta$ and throwing away (with a loss of at most 2δ) those terms in z which result from crossing over the boundary between the supports of two (or more) y_i's.

Next we apply the combinatorial Lemma 3.1. Let $\mathcal{F} = \{F \subseteq \mathbb{N} : F$ is finite and there exists $f \in \text{Ball } E^*$ with $||f|_{[e_j]_{j=m(i-1)+1}^{m(i)}}|| > \delta$ for $i \in F\}$. We shall show that $\mathcal{F}$ satisfies the hypothesis of Lemma 3.1. Let $\sum_{i=1}^n a_i = 1$ with $a_i \geq 0$ and let $z_i \in [e_j]_{j=m(i-1)+1}^{m(i)}$ be a representative of y_i with $||\sum_{i=1}^n a_i z_i|| > 2\delta$. Let $f \in \text{Ball } E^*$ with $f(\sum_{i=1}^n a_i z_i) > 2\delta$ and let $F = \{i \leq n : |f(z_i)| > \delta\}$. Then

$$2\delta < \sum_{i \in F} a_i f(z_i) + \sum_{\substack{i \leq n \\ i \notin F}} a_i f(z_i)$$

$$\leq \sum_{i \in F} a_i + \delta \sum_{\substack{i \leq n \\ i \notin F}} a_i \leq \sum_{i \in F} a_i + \delta ,$$

whence $\sum_{i \in F} a_i > \delta$.

By Lemma 3.1 there exists an infinite set $M \subseteq \mathbb{N}$ so that $F \in \mathcal{F}$ for all finite $F \subseteq M$. Thus for all n, there exists $f_n \in \text{Ball } E^*$ with

$$||f_n|_{[e_j]_{j=m(i-1)+1}^{m(i)}}|| > \delta$$

for $i \in M$ with $i \leq n$. By w^* compactness there exists $f \in \text{Ball } E^*$ with

$$||f|_{[e_j]_{j=m(i-1)+1}^{m(i)}}|| > \delta$$

for all $i \in M$. But this is impossible since (e_i) is shrinking.

3. If E is reflexive, then (f_i) and (g_i) are shrinking and boundedly complete by 1. and 2. Therefore F and G are reflexive. Also $J(e_i)$ is quasi-reflexive by Theorem 2.2. ∎

Remark. The converse of 1. is false. Indeed if (e_i) is the square variation basis of J (see example 2 in Section 1), then (e_i) is not boundedly complete yet (f_i), (g_i) and (u_i) are boundedly complete. If $(f_i) = (e_i)$ is the unit vector basis of $\ell_2 \oplus c_0$ (see example 1 of Section 1), then (f_i) is not boundedly complete, but both (g_i) and (u_i) are boundedly complete.

C. Schumacher has shown that (f_i) shrinking need not imply (e_i) is shrinking. In fact (e_i) can be chosen as a basic sequence in c_0 and in her example (f_i) is right dominant. We do not know if (g_i) shrinking implies that (f_i) is shrinking.

5. Subspaces of $J(e_i)$

There is another way to obtain an unconditional basis from (e_i). let $(m(i))_{i=0}^{\infty}$ be an increasing sequence of integers with $m(0) = 0$. For $(a_i) \in c_{00}$, let

$$\left\| \sum_{i=1}^{\infty} a_i d_i \right\| = \sup \left\{ \left\| \sum_{i=1}^{\infty} a_i y_i \right\| : y_i \in \text{Ball}\, [e_j]_{j=m(i-1)+1}^{m(i)} \right\} .$$

(d_i) is clearly a 1-unconditional basis, which we shall call the *basis determined by* $(m(i))$. Our next proposition yields that every infinite dimensional subspace of $J(e_i)$ contains a normalized unconditional basic sequence which is dominated by some (d_i). A corollary of this is a positive solution to the problem mentioned in Section 0 for $E =$ Tsirelson's space.

Proposition 5.1. *For each infinite dimensional subspace Z of $J(e_i)$, there is a normalized unconditional basic sequence $(z_i) \in Z$ and integers $0 = m(0) < m(1) < \cdots$ so that (d_i) dominates (z_i), where (d_i) is the basis determined by $(m(i))$.*

Proof. By Proposition 2.1, we may assume Z contains (w_i), a normalized block basis of (u_i) satisfying condition 2. of the proposition. In particular, (w_i) is unconditional. The proposition is trivial if (w_i) is equivalent to the unit vector basis of c_0, and thus we may suppose $\sup_k \|\sum_{i=1}^k w_i\| = \infty$.

By considering elements of the form $z = \sum_p^q w_i / \|\sum_p^q w_i\|$ where $q - p$ is large, we obtain a block basis $z_i = \sum_{j=m(i-1)+1}^{m(i)} b_j u_j$ of (w_i) satisfying $|\sum_{j=p}^q b_j| < 2^{-i-1}$ for $m(i-1)+1 \le p \le q \le m(i)$. Let (d_i) be the basis determined by $(m(i))$.

We shall show (d_i) dominates (z_i). Let $(a_i) \in c_{00}$ with $\|\sum a_i d_i\| \le 1$. This implies that $|a_i| \le 1$ for all i. Now $\sum a_i z_i = \sum c_j b_j u_j$ where $c_j \equiv a_i$ if $m(i-1)+1 \le j \le m(i)$. Choose $1 \le p(1) \le q(1) < p(2) \le q(2) < \cdots < p(k) \le q(k)$ so that $\|\sum c_j b_j u_j\| = \|\sum_{i=1}^k (\sum_{j=p(i)}^{q(i)} c_j b_j) e_{p(i)}\|$. Let $F = \{i : p(i) \le m(j) < q(i)$ for some $j\}$. Thus

$$\begin{aligned}
\Big\|\sum a_i z_i\Big\| &\le \Big\|\sum_{\substack{i=1\\ i\notin F}}^{k} \Big(\sum_{j=p(i)}^{q(i)} c_j b_j\Big) e_{p(i)}\Big\| + \Big\|\sum_{\substack{i=1\\ i\in F}}^{k} \Big(\sum_{j=p(i)}^{q(i)} c_j b_j\Big) e_{p(i)}\Big\| \\
&\le \Big\|\sum a_i d_i\Big\| + \sum_{\substack{i=1\\ i\in F}}^{k} \Big|\sum_{j=p(i)}^{q(i)} c_j b_j\Big| \\
&\le 1 + \sum_{i=1}^{k} |a_i| \Big|\sum_{j=p(i)}^{q(i)} b_j\Big| \\
&\le 1 + \sum_{i=1}^{k} 2^{-i} \le 2\,.
\end{aligned}$$

Thus (z_i) is dominated by (d_i). ∎

Corollary 5.2. *Let (e_i) be the unit vector basis of Tsirelson's space T (or its dual T^*). Then every infinite dimensional subspace of $J(e_i)$ contains an infinite dimensional subspace which embeds into T (respectively, T^*).*

Proof. Let (d_i) and (z_i) be given by the proposition. There exists an absolute constant C such that (x_i) is C-equivalent to $(e_{m(i)})$ if x_i is any norm one vector in $[e_j]_{j=m(i-1)+1}^{m(i)}$ **[CS]**. Thus (d_i) is equivalent to $(e_{m(i)})$ which implies that $(e_{m(i)})$ dominates (z_i). Also (z_i) dominates $(e_{m(i)})$ since each z_i has a representative in $[e_j]_{j=m(i-1)+1}^{m(i)}$ of norm at least 2^{-1}. ∎

Remark 5.3. Our proof of Theorem 4.1(2) showed that if (e_i) is shrinking and (d_i) is the corresponding unconditional basis determined by $(m(i))$, then (d_i) is shrinking.

6. Tree Spaces

Let (e_i) be a normalized basis for E. In this section we define the tree space $JT(e_i)$ and show that $JT(e_i)$ often enjoys much of the structure of JT, the original tree space of James **[J2]** (see also **[LS]**). Some of the arguments are modifications of ones which appeared in **[O1]** where $JT(e_i)$ was defined for $E = T$, the Tsirelson space.

$\mathcal{D}$ shall denote the dyadic tree. Precisely, let $\mathcal{D} = \{\phi\} \cup \bigcup_{n\in\mathbb{N}}\{0,1\}^n$ with the usual partial order. Thus for $\alpha, \beta \in \mathcal{D}$, $\alpha \le \beta$ exactly when β extends α. We shall also require a linear order on $\mathcal{D}$ given by a 1–1 function "o" of $\mathcal{D}$ onto $\mathbb{N}$. This function is inductively defined by $o(\phi) = 1$ and $o(\alpha) = 2o(\alpha^-) + \varepsilon_{n+1}$ for $\alpha = (\varepsilon_1, \ldots, \varepsilon_n, \varepsilon_{n+1})$. Here α^- denotes the immediate predecessor of α, $\alpha^- = (\varepsilon_1, \ldots, \varepsilon_n)$. When convenient we shall use "o" to write $(u_\alpha)_{\alpha\in\mathcal{D}} = (u_{o(\alpha)})_{\alpha\in\mathcal{D}} = (u_i)_{i\in\mathbb{N}}$.

The *length* or *depth* of $\alpha = (\varepsilon_1, \ldots, \varepsilon_n)$ is $|\alpha| \equiv n$ ($|\phi| = 0$). Equivalently, for $\alpha \ne \phi$, $|\alpha| = [\![\log_2 o(\alpha)]\!]$, where $[\![\cdot]\!]$ is the greatest integer function. A *segment* $S \subseteq \mathcal{D}$ is an order interval $[\alpha, \beta] = \{\gamma \in \mathcal{D} : \alpha \le \gamma \le \beta\}$. A branch is a maximal linearly ordered subset of $\mathcal{D}$. We use Γ to denote the set of all branches of $\mathcal{D}$.

If $S = [\alpha, \beta]$ is a segment we define $o(S) = o(\alpha)$ and for $x : \mathcal{D} \to \mathbb{R}$ we define $S^*(x) = \sum_{\gamma\in S} x(\gamma)$.

Definition of $JT(e_i)$

$JT(e_i)$ is the completion of the linear space of all finitely supported functions $x : \mathcal{D} \to \mathbb{R}$ under the norm

$$\|x\| = \sup\Big\{ \Big\| \sum_{i=1}^{k} S_i^*(x) e_{o(S_i)} \Big\| : k \in \mathbb{N} \text{ and } \tag{6.1}$$
$$(S_i)_{i=1}^k \text{ are disjoint segments in } \mathcal{D} \Big\}.$$

Thus if (e_i) is the unit vector basis of ℓ_2, then $JT(e_i) = JT$. As usual, $\sum_{i=1}^k S_i^*(x) e_{o(S_i)}$ is called a *representative* of x. For $\alpha \in \mathcal{D}$ we let $u_\alpha(\beta) = \delta_{\alpha,\beta}$. Thus $(u_\alpha)_{\alpha\in\mathcal{D}}$ is a normalized monotone basis (in the $o(\alpha)$ ordering) for $JT(e_i)$. If $\gamma \in \Gamma$, then $[(u_\alpha)_{\alpha\in\gamma}]$ is just the space $J(e_{o(\alpha)})_{\alpha\in\gamma}$ considered in the earlier sections. Let $(u_\alpha^*)_{\alpha\in\mathcal{D}}$ be the coefficient functionals of (u_α).

Note that by (6.1), S^* is a norm one element of $JT(e_i)^*$ when S is a segment. Similarly, the *branch functionals* γ^*, given by $\gamma^*(x) = \sum_{\alpha\in\gamma} x(\alpha)$, are also norm one elements of the dual for $\gamma \in \Gamma$. If γ_1 and γ_2 are distinct elements of Γ, then $\|\gamma_1^* - \gamma_2^*\| \geq 1$ and thus $\Gamma^* \equiv \{\gamma^* : \gamma \in \Gamma\}$ is a nonseparable subset of $JT(e_i)^*$. However, in the w^*-topology Γ^* is naturally homeomorphic to the Cantor set $\Delta = \{0,1\}^{\mathbb{N}}$. This induces a norm one map, hat : $JT(e_i) \to C(\Delta)$ given by $\text{hat}(y) = \hat{y}$ where $\hat{y}(\gamma) = \gamma^*(y)$.

Each $\alpha \in \mathcal{D}$ determines a clopen set $K_\alpha \subseteq \Delta$ given by $K_\alpha = \{\beta \in \Delta : \beta$ extends $\alpha\}$. Each $\alpha \in \mathcal{D}$ also determines a probability measure μ_α on Δ satisfying

$$\int f\, d\mu_\alpha = m(K_\alpha)^{-1} \int_{K_\alpha} f\, dm$$

where m is Lebesgue measure on Δ.

Theorem 6.1. *Let (e_i) be a normalized basis for E.*

a) *If (e_i) is boundedly complete, then $(u_\alpha)_{\alpha\in\mathcal{D}}$ (in the $o(\alpha)$ ordering) is a boundedly complete basis for $JT(e_i)$.*

b) *If E is reflexive, then ℓ_1 does not embed into $JT(e_i)$ and $JT(e_i)^* = [(u_\alpha^*)_{\alpha\in\mathcal{D}} \cup \Gamma^*]$.*

Proof. We may assume that (e_i) is suppression-1 unconditional. Indeed, just as in Proposition 1.1, $JT(e_i)$ and $JT(f_i)$ are naturally isometric where (f_i) is given by (1.2). Furthermore, Theorem 4.1 yields that (f_i) is boundedly complete (respectively, $[(f_i)]$ is reflexive) if (e_i) is boundedly complete (respectively, E is reflexive).

(a) Suppose (e_i) is boundedly complete and (y_n) is a semi-normalized block basis of (u_α). We want to show that $||\sum_{n=1}^N y_n|| \to \infty$ as $N \to \infty$. We first note that we may assume that for all $\alpha \in \mathcal{D}$.

$$\lim_{n\to\infty} \sup\{\ |S^*(y_n)| : S = [\alpha, \beta] \text{ for some } \beta \geq \alpha\ \} = 0\ . \tag{6.2}$$

Otherwise there exist $\delta > 0$, $(\beta_i) \subseteq \mathcal{D}$ and a subsequence $(y_{n(i)})$ of (y_n) so that $|S_i^*(y_{n(i)})| > \delta$ for $S_i = [\alpha, \beta_i]$. Since (y_n) is a block basis of (u_α), there exists $(\alpha_i) \subseteq \mathcal{D}$ so that the segments $T_i \equiv [\alpha_i, \beta_i]$ are disjoint, for all i, $T_i^*(y_{n(i)} = S_i^*(y_{n(i)})$ and for $n \neq n(i)$, $T_i^*(y_n) = 0$. Thus by (6.1),

$$\Big\|\sum_{n=1}^N y_n\Big\| \geq \Big\|\sum_{\{i:n(i)\geq N\}} T_i^*(y_{n(i)})e_{o(\alpha_i)}\Big\| \to \infty$$

as $N \to \infty$ since (e_i) is boundedly complete.

Let $(m(n))$ be defined so that $y_n \in [u_i]_{i=m(n-1)+1}^{m(n)}$ (in the $o(\alpha)$-order) and let $||y_n|| > \delta > 0$ for all n. Fix n and let $p > n$. Let D be a finite collection of disjoint segments such that $||\sum_{S\in D} S^*(y_p)e_{o(S)}|| > \delta$. We can and do assume that if $S = [\alpha, \beta] \in D$, then $o(\beta) \leq m(p)$. $D = A \cup B \cup C$ where

$$A = \{\ S \in D : o(S) \leq m(n)\ \}\ ,$$

$$B = \{\ S \in D : m(n) < o(S) \leq m(p-1)\ \} \quad \text{and}$$

$$C = \{\ S \in D : m(p-1) < o(S) \leq m(p)\ \}\ .$$

By (6.2) if p is sufficiently large, then for $S \in A$, $|S^*(y_p)| < \delta/2m(n)$. Thus $||\sum_{S\in B\cup C} S^*(y_p)e_{o(S)}|| > \delta/2$.

Now either $||\sum_{S\in C} S^*(y_p)e_{o(S)}|| > \delta/4$ or $||\sum_{S\in B} S^*(\sum_{j=n+1}^{p-1} y_j)e_{o(S)}|| > \delta/8$ or $||\sum_{S\in B} S^*(\sum_{j=n+1}^{p} y_j)e_{o(S)}|| > \delta/8$. In any case we can find a set of disjoint segments D' so that each $S \in D'$ is contained within the support of $\sum_{j=n+1}^p y_j$ and

$$\Big\|\sum_{S\in D'} S^*\Big(\sum_{j=n+1}^{p} y_j\Big)e_{o(S)}\Big\| > \delta/8\ .$$

Thus by blocking (y_n) we may assume that each y_n has a representative z_n in $[e_j]_{n=m(n-1)+1}^{m(n)}$ of norm exceeding $\delta/8$. Since (e_i) is boundedly complete,

$$\Big\|\sum_1^N y_n\Big\| \geq \Big\|\sum_1^N z_n\Big\| \to \infty \quad \text{as} \quad N \to \infty\,.$$

(b) We first prove the following

Lemma 6.2. *If (e_i) is shrinking, (x_i) is a normalized block basis of (u_α) and $(\hat{x}_i)$ converges weakly to zero in $C(\Delta)$, then (x_i) is weakly null.*

Proof. If (x_i) is not weakly null, then by passing to a subsequence we may assume $F(x_i) > \delta > 0$ for all i and some norm one element $F \in JT(e_i)^*$. Replacing (x_i) by a convex block basis of (x_i) of convex combinations we may also assume that $||\hat{x}_i||_{C(\Delta)} \to 0$. This just means that $\sup\{|\gamma^*(x_i)| : \gamma \in \Gamma\} \to 0$. However we require a stronger property, $\sup\{|S^*(x_i)| : S \text{ is a segment }\} \to 0$. We obtain this by once more replacing (x_i) by a block basis of long averages.

Now (x_i) is semi-normalized $(1 \geq ||\gamma_i|| > \delta)$ and thus by normalizing and passing again to a subsequence we can obtain: (x_i) is a normalized non weakly null sequence with $x_i \in [u_n]_{n=m(i-1)+1}^{m(n)}$ and $\sup\{|S^*(x_{i+1})| : S \text{ is a segment }\} \leq \varepsilon_{i+1}$, where (ε_i) satisfies $\sum_{i=1}^\infty m(i)(\sum_{j=i+1}^\infty \varepsilon_j) < 1/2$. Let (d_i) be the normalized unconditional basis determined by $(m(i))$ (see Section 5). Thus

$$\Big\|\sum a_i d_i\Big\| = \sup\Big\{\Big\|\sum a_i z_i\Big\| : z_i \in \text{Ball}\,[e_n]_{n=m(i-1)+1}^{m(i)}\Big\}\,.$$

We claim that (d_i) dominates (x_i). Since (d_i) is shrinking (see Remark 5.3), (x_i) is then weakly null, a contradiction.

To see that (d_i) dominates (x_i), let $||\sum_1^k a_i x_i|| = 1$ and let us prove that $||\sum_1^k a_i d_i|| > 1/2$. Choose D, a finite set of disjoint segments, such that $||\sum_{S\in D} S^*(\sum_{i=1}^k a_i x_i) e_{o(S)}|| = 1$. For $i \leq k$ let D_i be those segments in D which originate in the support of x_i. Thus $D_i = \{S \in D : m(i-1) < o(S) \leq m(i)\}$. For $S \in D_i$, since $|a_j| \leq 1$,

$$\Big|S^*\Big(\sum_{j=i+1}^k a_j x_j\Big)\Big| \leq \sum_{j=i+1}^k |S^*(x_j)| \leq \sum_{j=i+1}^\infty \varepsilon_j\,.$$

Thus since $^{\#}D_i \leq m(i)$,

$$\sum_{i=1}^{k} \sum_{S \in D_i} \left| S^*\left(\sum_{j=i+1}^{k} a_j x_j \right) \right| \leq \sum_{i=1}^{\infty} m(i) \sum_{j=i+1}^{\infty} \varepsilon_j < \frac{1}{2} .$$

It follows that if for $S \in D_i$ we let $\overline{S}$ be the initial segment of S whose support is contained entirely within the support of x_i, then

$$\left\| \sum_{i=1}^{k} \sum_{S \in D_i} \overline{S}\left(\sum_{j=1}^{k} a_j x_j \right) e_{o(\overline{S})} \right\| > \frac{1}{2} . \quad \blacksquare$$

Next we observe that if (e_i) is shrinking and ℓ_1 does not embed into $JT(e_i)$, then $JT(e_i)^* = [(u_\alpha^*)_{\alpha \in \mathcal{D}} \cup \Gamma^*]$. Indeed if this were not the case, then there exists a norm one element $x^{**} \in JT(e_i)^{**}$ such that $x^{**}|_{[(u_\alpha^*) \cup \Gamma^*]} = 0$. By [**OR**] (or [**LT**, p.101]) there is a normalized sequence $(x_n) \subseteq JT(e_i)$ which converges weak* (in $JT(e_i)^{**}$) to x^{**}. Since $x^{**}|_{[(u_\alpha^*)]} = 0$, we may assume (by passing to a subsequence and perturbing) that (x_n) is a block basis of (u_α). Since $x^{**}|_{\Gamma^*} = 0$, $(\hat{x}_n)$ is weakly null in $C(\Delta)$. By Lemma 6.2, (x_n) is weakly null and thus $x^{**} = 0$, a contradiction.

It remains to prove that ℓ_1 does not embed into $JT(e_i)$. Suppose, to the contrary, that (x_i) is a normalized block basis of (u_α) which is equivalent to the unit vector basis of ℓ_1. If $(\hat{x}_i)$ had a weak Cauchy subsequence then, by passing to a subsequence and replacing x_i by $x_{2i} - x_{2i-1}$, we could assume $(\hat{x}_i)$ was weakly null. But then Lemma 6.2 yields (x_i) is weakly null which is a contradiction.

Thus by Rosenthal's theorem [**LT**, p.99] we may assume that $(\hat{x}_i)$ is equivalent to the unit vector basis of ℓ_1, and moreover that there exist $r \in \mathbb{R}$ and $\delta > 0$ so that if $A_i = \{\gamma^* \in \Delta : \hat{x}_i(\gamma^*) \geq r + \delta\}$ and $B_i = \{\gamma^* \in \Delta : \hat{x}_i(\gamma^*) \leq r\}$, then (A_i, B_i) is Boolean independent. This means that $\bigcap_{i=1}^{N} C_i \neq \emptyset$ for any of the 2^N choices of $C_i = A_i$ or B_i. We may assume $r + \delta > 0$ (or else replace (x_i) by $(-x_i)$).

Assume $x_i \in [u_n]_{n=m(i-1)+1}^{m(i)}$ where $1 = m(0) < m(1) < \cdots$. Let $B_0 = \Delta$. Now since $\hat{x}_i$ is continuous, A_i and B_i are both closed and thus by compactness we can find $\gamma_i \in \Gamma$ and $\alpha_i, \beta_i \in \gamma_i$ so that for all i,

(1) $$\gamma_i \in \bigcap_{j=0}^{i-1} B_j \cap \bigcap_{j=1}^{\infty} A_j \quad \text{and}$$

(2) $$\gamma_i \cap \{ \alpha : m(i-1) < o(\alpha) \leq m(i) \} = [\alpha_i, \beta_i] .$$

Since $\hat{x}_i\big|_{K_{\beta_i}} = \hat{x}_i(\gamma_i) \geq r + \delta$ and $\hat{x}_i(\gamma_j) \leq r$ for $j > i$, the infinite "tail segments" $T_i = \{\alpha \in \gamma_i : \alpha \geq \alpha_i\}$ are pairwise disjoint and $T_i^*(x_j) \geq r + \delta$ for $j \geq i$. Fix k and for $i \leq k$, let $S_i^k = \{\alpha \in T_i : o(\alpha) \leq m(k)\}$. Thus $S_i^{k*}(x_k) = T_i^*(x_k) \geq r + \delta$ for $i \leq k$ and so $||x_k|| \geq ||\sum_{i=1}^k S_i^{k*}(x_k) e_{o(\alpha_i)}|| \to \infty$ as $k \to \infty$ since $(e_{o(\alpha_i)})$ is boundedly complete. But $||x_k|| = 1$, a contradiction. ∎

Remark. Our proof of (b) actually yields a stronger result: If (e_i) is a shrinking unconditional basis and no (incomparable) subsequence of (e_i) is equivalent to the unit vector basis of c_0, then ℓ_1 does not embed into $JT(e_i)$ and $JT(e_i)^* = [(u_\alpha^*) \cup \Gamma^*]$. This is best possible as the next example shows.

Example 6.3. Let (e_i) be the unit vector basis of c_0. Then ℓ_1 embeds isometrically into $JT(e_i)$. Indeed $\left(\sum_{i=2^n}^{2^{n+1}-1} (-1)^i u_i\right)_{n=1}^\infty$ is 1-equivalent to the unit vector basis of ℓ_1.

Moreover $JT(e_i)$ is, in this case, isomorphic to $C(\Delta)$. We shall show this by proving that $JT(e_i)$ is isomorphic to $C(\Delta) \oplus c_0$. Note that for $\gamma \in \Gamma$, $J(e_i)_{i\in\gamma}$ is isomorphic to c_0 since $(u_i)_{i\in\gamma}$ is equivalent to the summing basis of c [**LT**, p.20].

Let $x_i = u_i - (u_{2i} + u_{2i+1})$ for $i \in \mathbb{N}$. Clearly $\hat{x}_i \equiv 0$ and so $[(x_i)] \subseteq$ Ker hat. If S_j is the initial segment, $[1, j]$ (in the $\mathcal{D}$-order), then $S_j^*(\sum_{i=1}^\infty a_i x_i) = a_j$. Since any segment is the difference of two such S_j's we have $||\sum a_i x_i|| \leq 2 \sup |a_i|$, and it follows that (x_i) is equivalent to the unit vector basis of c_0.

Next we shall show that $[(x_i)] =$ Ker hat. Let $y = \sum b_i u_i \in JT(e_i)$. It is easy to check that $y - \sum_{j=1}^n S_j^*(y) x_i = w_n + z_n$ where

$$z_n = \sum_{i=2n+2}^{\infty} b_i u_i \quad \text{and} \quad w_n = \sum_{i=n+1}^{2n+1} S_i^*(y) u_i \ .$$

Since hat $(\sum_{j=1}^n S_j^*(y) x_j) = 0$, $\hat{y} = \hat{w}_n + \hat{z}_n$ and so $||w_n|| = ||\hat{w}_n|| \leq ||\hat{y}|| + ||\hat{z}_n||$. Now let $||\hat{z}_n|| \leq ||z_n|| \to 0$ and so if $y \in$ Ker hat, then $||w_n|| \to 0$ as well. Hence $y = \sum_{j=1}^\infty S_j^*(y) x_j \in [(x_i)]$.

Finally we show that hat has a right inverse unhat: $C(\Delta) \to JT(e_i)$. Once done, it follows that $P =$ unhat $\circ$ hat is a projection $JT(e_i)$. Moreover unhat is an into isomorphism and thus Ker P = Ker hat $= [(x_i)]$ and range P = range unhat is isomorphic to $C(\Delta)$. Unhat is defined as follows. For $f \in C(\Delta)$ let

$$\text{unhat}(f) = \sum_{\alpha \in \mathcal{D}} f_\alpha u_\alpha$$

where $f_\phi = \int f\, dm$ and $f_\alpha = \int f\, d\mu_\alpha - \int f\, d\mu_{\alpha^-}$ for $|\alpha| \geq 1$. Note that if $S = [\alpha, \beta]$, then $|S^*(\text{unhat}(f))| = |\int f\, d\mu_\beta - \int f\, d\mu_{\alpha^-}|$. This shows that the series $\sum f_\alpha u_\alpha$ converges and moreover $||\text{unhat}\ (f)|| \leq 2||f||_{C(\Delta)}$. Furthermore,

$$\text{hat}\Big(\sum_{|\alpha| \leq n} f_\alpha u_\alpha\Big) = \sum_{|\alpha| = n} \Big(\int f\, d\mu_\alpha\Big) 1_{K_\alpha} \to f$$

in $C(\Delta)$, and so hat o unhat$(f) = f$.

Remark. In view of the problem asked in the introduction, one can raise the following

Question. If E is a separable (infinite dimensional) reflexive space does there exist a separable space X with non separable dual such that every infinite dimensional subspace of X contains an infinite dimensional subspace which embeds into E?

The space $JT(e_i)$ yields an affirmative solution if (e_i) is the unit vector basis of T (this solves a problem in [**O1**]). Indeed this follows from the following result which can be obtained from the proof of Theorem 6.1.

Proposition 6.4. *If E is reflexive and* hat: $JT(e_i) \to C(\Delta)$ *is strictly singular, then every infinite dimensional subspace of $JT(e_i)$ contains (up to a perturbation) a normalized block basis $x_i \in [u_n]_{n=m(i-1)+1}^{m(i)}$ which is dominated by the (d_i) sequence determined by $(m(i))$.*

The fact that hat is strictly singular in the case where (e_i) is the unit vector basis of T follows from the fact that the only spreading models in $JT(e_i)$ are ℓ_1 [**O1**] and Pajour's localization of Rosenthal's ℓ_1 theorem [**P**] by localizing our proof that $JT(e_i)$ does not contain ℓ_1. We do not know, in general, if hat : $JT(e_i) \to C(\Delta)$ is strictly singular whenever E is reflexive.

References

[B1] S.F. Bellenot, *Transfinite duals of quasi-reflexive Banach spaces*, Trans. Amer. Math. Soc. **273** (1982), 551–577.

[B2] S.F. Bellenot, *Tsirelson superspaces and ℓ_p*, J. Funct. Anal. **69** (1986), 207–228.

[CS] P.G. Casazza and T. Shura, *Tsirelson space*, Lecture Notes in Math., Springer-Verlag, Berlin/New York, in press.

[CL] P.G. Casazza and R.H. Lohman, *A general construction of spaces of the type of R.C. James*, Canad. J. Math. **27** (1975), 1263–1270.

[CLL] P.G. Casazza, B.L. Lin and R.H. Lohman, *On nonreflexive Banach spaces which contain no c_0 or ℓ_p*, Canad. J. Math. **33** (1980), 1382–1389.

[FJ] T. Figiel and W.B. Johnson, *A uniformly convex Banach space which contains no ℓ_p*, Composito Math. **29** (1974), 179–190.

[HOR] R. Haydon, E. Odell and H. Rosenthal, in preparation.

[HW] R. Herman and R. Whitley, *An example concerning reflexivity*, Studia Math. **28** (1967), 289–294.

[J1] R.C. James, *Bases and reflexivity of Banach spaces*, Ann. of Math. **52** (1950), 518–527.

[J2] R.C. James, *A separable somewhat reflexive Banach space with non-separable dual*, Bull. Amer. Math. Soc. **18** (1974), 738–743.

[LS] J. Lindenstrauss and C. Stegall, *Examples of separables spaces which do not contain ℓ_1 and whose duals are non-separable*, Studia Math. **54** (1975), 81–105.

[LT] J. Lindenstrauss and L. Tzafriri, "Classical Banach Spaces I, Sequence Spaces", Springer-Verlag, Berlin, 1977.

[O1] E. Odell, *A non-separable Banach space not containing a subsymmetric basic sequence*, Israel J. Math. **52** (1985), 97–109.

[O2] E. Odell, *Applications of Ramsey theorems to Banach space theory*, in "Notes in Banach Spaces" (H.E. Lacey, ed.), University of Texas Press, Austin, Texas (1980), 379–404.

[OR] E. Odell and H. Rosenthal, *A double dual characterization of separable Banach spaces containing ℓ_1*, Israel J. Math. **20** (1975), 375–384.

[P] A. Pajour, "Sous-espaces ℓ_1^n des Espaces de Banach", Travaux En Cours, Hermann, Paris, 1985.

Steven F. Bellenot
Florida State University
Tallahassee, Florida 32306

Richard Haydon
Brasenose College
Oxford, England

Edward Odell
The University of Texas
Austin, Texas 78712

Contemporary Mathematics
Volume **85**, 1989

PROJECTIONS ONTO BLOCK BASES

Peter G. Casazza

Abstract: We consider the problem: If P is a projection of a Banach space X with a unconditional basis onto one of its block bases, then does $(I-P)(X)$ have a unconditional basis? We discover the surprising result that, in most cases, not only does $(I-P)(X)$ have a unconditional basis, but $(I-P)(X)$ is isomorphic to the span of a subsequence of the original unconditional basis. We conjecture, that this result holds in general.

1. Introduction. If $\{x_n\}$ is a basis of a Banach space X, a **block basis** of $\{x_n\}$ is a sequence

$$y_n = \sum_{i=P_n+1}^{p_{n+1}} a_i x_i, \text{ where } p_o = 0 < p_1 < p_2 < \ldots.$$

M. Zippin [7] has shown that every block basis of a basis for X can be extended to a basis for X. That is, given $\{x_n\}$ and $\{y_n\}$ as above, there is a basis $\{z_m\}$ for X and $m_1 < m_2 < \ldots$ so that $z_{m_n} = y_n$, for all n. A major open problem in Banach space theory is whether a complemented subspace of a space with a unconditional basis has a (unconditional) basis?

1980 Mathematics Subjection classification: Primary 46B15
Key words and phrases: Projection unconditional basis, isomorphism, equivalent bases.
The research in this paper was supported by NSF DMS 8500938.

The answer to this question is more than likely no, in light of a recent paper of S. J. Szarek [6] which shows that there exist Banach spaces with bases which have complemented subspaces without bases. There is a special case of this problem which goes back to Zippin [7] which is also unanswered. That is, if $\{y_n\}$ is a P-complemented block basis of a unconditional basis $\{x_n\}$ for a Banach space X, then does $(I-P)(X)$ have a unconditional basis? We will show that in many cases the answer to this problem is "yes" in a very strong sense. Namely, $(I-P)(X)$ will be isomorphic to $[\{x_{n_i}\}]$ for some $n_1 < n_2 < \ldots$. This leads us to the following conjecture:

Conjecture: If $\{x_n\}$ is a unconditional basis for a Banach space X and if $\{y_n\}$ is a P-complemented block basis of $\{x_n\}$, then there are natural numbers $n_1 < n_2 < \ldots$ so that $(I-P)(X) \approx [\{x_{n_i}\}]$. In particular, $(I-P)(X)$ has a unconditional basis.

Several examples which show that our results are best possible are given in the sequel. We use the standard definitions and notation as may be found in [3]. A sequence $\{x_n\}$ in a Banach space X is a **basis** for X if each $x \in X$ has a unique representation of the form

$$x = \sum_{n=1}^{\infty} a_n x_n,$$

for scalars $\{a_n\}$. If $\{x_n\}$ is a basis for X, the **dual functionals of** $\{x_n\}$, denoted x_n^*, are the uniquely defined elements $x_n^* \in X^*$ given by: $x_n^*(x) = a_n$. The **natural projections of the basis** are the operators

$$P_n(x) = \sum_{i=1}^{n} a_i x_i.$$

It is known [3] that

$$K = \sup_{n \geq 1} \| P_n \| < \infty$$

and we call K the basis constant of X. Moreover, a sequence $\{x_n\}$ in a Banach space X is a basis for $[\{x_n\}]$ if and only if it has a finite basis constant. If $\{E_n\}$ is a sequence of finite dimensional subspaces of X, we say $\{E_n\}$ forms a **finite dimensional decomposition** for X, and write

$$X = \sum_{n=1}^{\infty} \oplus E_n,$$

if for every $x \in X$ there are unique elements $x_n \in E_n$ so that

$$x = \sum_{n=1}^{\infty} x_n.$$

As before, if

$$Q_n(x) = \sum_{i=1}^{n} x_i$$

then

$$K = \sup_{n \geq 1} \| Q_n \| < \infty$$

and the Q_n's are called the **natural projections of the finite dimensional decomposition** and K is called the decomposition constant. It is easily seen that if

$$X = \sum_{n=1}^{\infty} \oplus E_n$$

with decomposition constant K and each E_n has a basis $\{x_i^n\}_{i=1}^{m_n}$ with basis constant $\leq M$, then $\{x_i^n\}_{n=1,\, i=1}^{\infty\ m_n}$ is a basis for X with basis constant $\leq KM$. We say that $\{x_n\}$ is a **unconditional basis** for X if the series

$$X = \sum_{n=1}^{\infty} a_n x_n$$

converges unconditionally for every $x \in X$. In this case

$$K = \sup_{\|x\|=1} \sup_{\epsilon_n = \pm 1} \| \sum_{n=1}^{\infty} \epsilon_n a_n x_n \| < \infty$$

and is called the **unconditional basis constant for** $\{x_n\}$ and we say $\{x_n\}$ is a K-unconditional basis for X. We can similarly have the unconditional F.D.D. constant. If $\{x_n\}$ and $\{y_n\}$ are bases for Banach spaces X and Y respectively, we say $\{x_n\}$ is **equivalent** to $\{y_n\}$, and write $\{x_n\} \approx \{y_n\}$, if the linear operator $Tx_n = y_n$ is an isomorphism of X onto Y. Finally, if there is a projection P of a Banach space X onto its subspace Y, we call Y a P-complemented subspace of X. If $\{x_n\}$ is a basis for X and $\{y_n\}$ is a block basis, and P a projection of X onto $Y = [\{y_n\}]$ then we call $\{y_n\}$ a P-complemented block basis of $\{x_n\}$.

2. Block bases of conditional bases. We start with a result which has not appeared in print before but was known to people working in the area. We will sketch the proof for completeness.

Proposition 1: If X is a Banach space with a basis $\{x_n\}$ and

$$y_n = \sum_{i=p_n+1}^{p_n+1} a_i x_i$$

is a block basis of $\{x_n\}$, then $X/[\{y_n\}]$ has a basis.

Proof: Let $E_n = [\{x_i\}_{i=p_n+1}^{p_{n+1}}]$ for every n and let $Q : X \to X/[\{y_n\}]$ be the quotient map. Then $X/[\{y_n\}] \approx \sum \oplus QE_n$, a finite dimensional decomposition for QX. Also, the QE_n are uniformly isomorphic to a $(\dim E_n - 1)$-dimensional subspace of E_n. Since each E_n has a basis with the same basis constant as $\{x_n\}$, it follows (see [3,7]) that each QE_n has a basis with basis constant $\leq 2K$. Now, by the discussion in section 1,

$$\sum_{n=1}^{\infty} \oplus QE_n$$

has a basis. ∎

If P is a projection of a Banach space X onto a subspace M, then it is an elementary result that $(I - P)(X) \approx X/M$. This yields immediately:

Corollary 2: If X is a Banach space with a basis $\{x_n\}$ and if $\{y_n\}$ is a P-complemented block basis of $\{x_n\}$, then $(I - P)(X)$ has a basis.

In the next section, we will consider when $(I-P)(X)$ is isomorphic to a subsequence of the basis $\{x_n\}$. Next, we show that this may fail for conditional bases. To show this, we prove a lemma which may be of independent interest.

Lemma 3: If X is a Banach space with a basis and ℓ_1 embeds complementably into X, then X has a basis $\{z_n\}$ so that $\{z_{2n}\} \approx \{z_{2n-1}\} \approx \{e_n\}_{\ell_1}$.

Proof: Since ℓ_1 embeds complementably into X, we have $X \approx X \oplus \ell_1$. Now let $\{x_n\}$ be any normalized basis for X and define:

$$z_{2n} = (x_n, e_n),$$

$$z_{2n-1} = (x_n, -e_n).$$

Then it is easily checked that this $\{z_n\}$ works. ∎

A similar argument shows that if $\{x_n\}$ is a basis for X, then there is a basis $\{z_n\}$ for $X \oplus c_0$ so that $\{z_{2n}\} \approx \{z_{2n-1}\} \approx \{x_n\}$.

Example 4: There is a Banach space X with a conditional basis $\{x_n\}$ which has a block basis $\{y_n\}$ so that $[\{y_n\}]$ is P-complemented in X but $(I-P)(X)$ is not isomorphic to $[\{x_{n_i}\}]$ for any $n_1 < n_2 < \ldots$.

PROOF: Let $\{z_n\}$ be the basis for $X = c_o \oplus \ell_1$ given in lemma 3. Then $y_n = z_{2n} - z_{2n-1}$ is a block basis of $\{z_n\}$, $[\{y_n\}] = \ell_1$ and so is P-complemented and $(I-P)(X) \approx c_o$. Since every subsequence of $\{z_n\}$ has a further subsequence equivalent to $\{e_n\}_{\ell_1}$, it follows that $(I-P)(X) \not\approx [\{z_{n_i}\}]$ for any $n_1 < n_2 < \ldots$. ∎

3. Block bases of unconditional bases.

In this section we will make use of the following trivial fact:

Fact A: If M is a closed subspace of a Banach space X and P_1, P_2 are bounded linear projections of X onto M, then $(I-P_1)(X) \approx (I-P_2)(X) \approx X/M$.

We will also need a generalization of a result of Cassazza, Kalton, and Tzafriri [1]. To formulate these results more compactly, we introduce some notation. If $\{x_n\}$ is a normalized K-unconditional basis for X and

$$y_n = \sum_{i=p_n+1}^{p_{n+1}} b_i x_i$$

is a normalized P-complemented block basis of $\{x_n\}$,then there is a bounded block basis

$$y_n = \sum_{i=p_n+1}^{p_{n+1}} b_i x_i^*$$

of $\{x_i^*\}$ so that [1]:

$$Px = \sum_{n=1}^{\infty} y_n^*(x) y_n,$$

for all $x \in X$. We call the y_n^* the **projection functionals** of $\{y_n\}$. In particular,

$$1 = y_n^*(y_n) = \sum_{i=p_n+1}^{p_{n+1}} a_i b_i.$$

We will say that $\{y_n\}$ is a **non-trivial P-complemented block basis of** $\{x_n\}$ if

$$\sup_i |a_i b_i| < 1.$$

The motivation for this definition is the following: If we can choose $p_n + 1 \leq i_n \leq p_{n+1}$ so that

$$\lim_{n \to \infty} a_{i_n} b_{i_n} = 1,$$

then it follows easily that $\{x_{i_n}\} \approx \{y_n\}$ and the projection is getting its "support"from x_{i_n}. That is, the "block" $\{y_n\}$ is basically just a subsequence of $\{x_n\}$ and the projection P is basically just the natural projection onto this subsequence.

Our next lemma is quite elementary and so we will only sketch the proof. It will allow us to assume, in the sequel, that a_i, $b_i \geq 0$ for all i and block bases $\{y_n\}$, $\{y_n^*\}$.

Lemma 5: Let $\{x_n\}$ be a K-unconditional basis for X and let

$$y_n = \sum_{i=p_n+1}^{p_{n+1}} a_i x_i$$

be a block basis of $\{x_n\}$. For any choice of signs $\varepsilon_i = \pm 1$, $\varepsilon = \{\varepsilon_i\}$, define

$$y_n^\varepsilon = \sum_{i=p_n+1}^{p_{n+1}} \varepsilon_i a_i x_i,$$

for all n. Let $Y = [\{y_n\}]$ and $Y^\varepsilon = [\{y_n^\varepsilon\}]$.

Then:

(i) $\{y_n\} \approx \{y_n^\varepsilon\}$,

(ii) $X/Y \approx X/Y^\varepsilon$,

(iii) If $\{y_n\}$ is P-complemented in X then $\{y_n^\varepsilon\}$ is P_o-complemented in X for some $\| Po \| \leq K \| P \|$ and $(I - P)(X) \approx (I - Po)(X)$.

Proof: Define $\phi : X \to X$ by:

$$\phi \sum_{n=1}^{\infty} \alpha_n x_n = \sum_{n=1}^{\infty} \varepsilon_n \alpha_n x_n.$$

Then ϕ is an isomorphism of X onto X mapping y_n to y_n^ε. The conclusions of the lemma are now immediate. ∎

Our next Theorem is a generalization of Theorem 1.8 from [1].

Theorem 6: Let $\{x_n\}$ be a normalized K-unconditional basis for X and let $\{y_n\}$ be a normalized P-complemented block basis of $\{x_n\}$. Let

$$y_n = \sum_{i=p_n+1}^{p_{n+1}} a_i x_i \text{ and } y_n^* = \sum_{i=p_n+1}^{p_{n+1}} b_i x_i^*$$

be chosen so that $a_i, b_i \geq 0$ for all i and

$$Px = \sum_{n=1}^{\infty} y_n^*(x) y_n$$

for all $x \in X$.

(1) If there is an $\varepsilon > 0$ and natural numbers $p_n + 1 \leq i_n \leq p_{n+1}$ so that $\varepsilon \leq a_{i_n} b_{i_n}$, then $\{y_n\} \approx \{x_{i_n}\}$ and $(I - P)(X) \approx [\{x_k\}_{k \neq i_n}]$.

(2) If $\{y_n\}$ is non-trivial then

(i) $[\{y_n\}]$ is isomorphic to a complemented subspace of $(I - P)(X)$,

(ii) There is a partition $\{A_1, A_2\}$ of $\mathbf{N}$ so that if $X_i = [\{x_n\}_{n \in A_i}]$ for i=1,2, then $X \approx X_1 \oplus X_2$ and X_i embeds complementably into $(I - P)(X)$.

Proof:

(1) By our assumptions, $\frac{\varepsilon}{K} \leq a_{i_n}$ for all n, so for all choices of scalars $\{\alpha_n\}$,

$$\frac{\varepsilon}{K^2} \| \sum_{n=1}^{\infty} \alpha_n x_{i_n} \| \leq \| \sum_{n=1}^{\infty} \alpha_n y_n \| .$$

Also, $Px_{i_n} = b_{i_n} y_n$ and $\frac{\varepsilon}{K} \leq b_{i_n}$ implies,

$$\| \sum_{n=1}^{\infty} \alpha_n y_n \| \leq \frac{K}{\varepsilon} \| \sum_{n=1}^{\infty} \alpha_n P x_{i_n} \|$$

$$\leq \frac{K}{\varepsilon} \| P \| \| \sum_{n=1}^{\infty} \alpha_n x_{i_n} \| .$$

So $\{y_n\} \approx \{x_{i_n}\}$. If we define $P_o : X \to [\{y_n\}]$ by:

$$P_o\Big(\sum_{n=1}^{\infty} \alpha_n x_n\Big) = \sum_{n=1}^{\infty} \Big(\frac{\alpha_{i_n}}{a_{i_n}}\Big) y_n,$$

then P_o is a bounded linear projection of X onto $[\{y_n\}]$ and clearly $(I - P_o)(X) = [\{x_k\}_{k \neq i_n}]$. By fact A, $(I - P)(X) \approx (I - P_o)(X)$ and this concludes the proof of (1).

(2) Assume

$$\sup_i a_i b_i \leq 1 - \varepsilon.$$

Then we may choose partitions $\{\sigma_n^i\}_{i=1}^2$ of $\{p_n + 1, p_n + 2, \ldots, p_{n+1}\}$ for each n, so that

$$\varepsilon \leq \sum_{j \in \sigma_n^i} a_j b_j \leq 1, \text{ for } i = 1,\ 2, \text{ and all } n.$$

Let

$$y_n^i = \sum_{j\in\sigma_n^i} a_j x_j,$$

and

$$y_n^{i*} = \sum_{j\in\sigma_n^i} b_j x_j.$$

Since

$$y_n^*(y_n^i) = \sum_{j\in\sigma_n^i} a_j b_j \geq \varepsilon,$$

by the proof of part (1), $\{y_n^1\}_{n=1}^\infty \approx \{y_n^2\}_{n=1}^\infty \approx \{y_n\}_{n=1}^\infty$. Letting

$$c_n^i = \sum_{j\in\sigma_n^i} a_j b_j$$

we may define projections $P_i : X \to [\{y_n\}]$, $i = 1, 2$, by:

$$P_i(x) = \sum_{n=1}^{\infty} \left(\frac{y_n^{i*}(x)}{c_n^i}\right) y_n.$$

Since $C_n^i \leq \varepsilon^{-1}$, we have that P_i is a bounded linear projection of X onto $[\{y_n\}]$. Since $P_1(y_n^2) = 0$, for all n, it follows that $[\{y_n\}] \approx [\{y_n^2\}] \subset (I - P_1)(X) \approx (I - P)X$, by fact A. Also, for i=1,2,

$$Q_i(x) = \sum_{n=1}^{\infty} \left(\frac{y_n^{i*}(x)}{c_n^2}\right) y_n^i$$

is a bounded linear projection of X onto $[\{y_n^i\}]$. Hence, $[\{y_n\}]$ embeds complementably into $(I - P)(X)$. If we let

$$A_i = \bigcup_{n=1}^{\infty} \sigma_n^i \text{ and } X_i = [\{x_j\}_{j\in A_i}]$$

for i=1,2, we see that $P_1(X_2) = 0 = P_2(X_1)$. Since $(I - P_1)(X) \approx (I - P_2)(X) \approx (I - P)X$, we see that X_i embeds into

$$(I - P)(X)$$

for i=1,2. The natural projections

$$R_i\Big(\sum_{n=1}^{\infty} \alpha_n x_n\Big) = \sum_{n=1}^{\infty} \sum_{j\in\sigma_n^i} \alpha_j x_j$$

of the basis $\{x_n\}$ show that the X_i's are isomorphic to complemented subspaces of $(I-P)(X)$. ∎

Theorem 6 gives the interesting fact that, at least in pieces, the whole space X embeds complementably into the complement of a non-trivial complemented block basis of X. The proof also shows the following unusual result.

Corollary 7: If

$$y_n = \sum_{i=p_n+1}^{p_{n+1}} a_i x_i$$

is a P-complemented block basis of a unconditional basis $\{x_n\}$ for X, with projection functionals

$$y_n^* = \sum_{i=p_n+1}^{p_{n+1}} b_i x_i^*,$$

and if

$$\sup_i |a_i b_i| \le \frac{1}{4},$$

then $[\{y_n\}] \oplus [\{y_n\}]$ embeds complementably into $(I-P)(X)$.

This seems unusual because at first, it isn't clear why $[\{y_n\}]^2$ should even be embeddable into X. We now consider several interesting consequences of Theorem 6. To simplify the argument, we first prove a lemma.

Lemma 8: Let $X \approx X_1 \oplus X_2$ and let P_i, i=1,2, be projections of X onto subspaces $Y_i \subset X_i$. If $Y_1 \approx Y_2$, then $(I-P_1)(X) \approx (I-P_2)(X)$.

Proof: By fact A we may assume $P_1(X_2) = 0 = P_2(X_1)$. Let $L : Y_1 \to Y_2$ be an isomorphism. Now, $(I-P_1)(X) \approx \Big(I|_{X_1} - P_1\Big)(X) \oplus X_2 \approx \Big(I|_{X_1} - P_1\Big)(X) + Y_2 \oplus \Big(I|_{X_2} -$

$P_2\big)(X) \approx \big(I|_{X_1} - P_1\big)(X) \oplus Y_1 \oplus \big(I|_{X_2} - P_2\big)(X) \approx X_1 \oplus \big(I|_{X_2} - P_2\big)(X) \approx (I - P_2)(X).$

∎

It follows that if Y_i, i=1,2, are P_i-complemented subspaces of X and $Y_1 \approx Y_2$ then $X \oplus (I - P_1)(X) \approx X \oplus (I - P_2)(X)$. We now consider some consequences of Theorem 6.

Corollary 9: If $\{x_n\}$ is a normalized unconditional basis of a Banach space X, $\{y_n^i\}$ are P_i-complemented block basis of $\{x_n\}$, i=1,2, $\{y_n^1\} \approx \{y_n^2\}$ and support $y_n^1 = \text{support}\, y_n^2$, then $(I - P_1)(X) \approx (I - P_2)(X)$.

Proof: We will do this in three steps.

$$\text{Let } y_n^i = \sum_{j=p_n+1}^{p_{n+1}} a_j^i x_j$$

and

$$y_n^{i*} = \sum_{j=p_n+1}^{p_{n+1}} b_j^i x_j^*,$$

for i=1,2 and all n. We may assume $a_j^i,\ b_j^i \geq 0$ and

$$\sum_{j=p_n+1}^{p_{n+1}} a_j^i b_j^i = 1.$$

Step I: If

$$\sup_j a_j^1 b_j^1 \leq 1 - \varepsilon,$$

then the corollary holds. In this case, we may partition $\{p_n + 1, p_n + 2, \ldots, p_{n+1}\}$ into σ_n^i, $i = 1, 2$, so that

$$\sum_{j \in \sigma_n^i} a_j^1 b_j^1 \geq \varepsilon.$$

Since

$$\sum_{j \in \sigma_n^1 \cup \sigma_n^2} a_j^2 b_j^2 = 1,$$

We may assume without loss of generality that

$$\sum_{j\in\sigma_n^2} a_j^2 b_j^2 \geq \frac{1}{2}.$$

If we let for i=1,2,

$$z_n^i = \sum_{j\in\sigma_n^1} a_j^i x_j$$

and

$$w_n = \sum_{j\in\sigma_n^2} a_j^2 x_j,$$

then by the proof of Theorem 6, $\{z_n^i\} \approx \{y_n^1\}$ and $\{w_n\} \approx \{y_n^2\}$. Now let

$$X_i = [\{x_j\}_{j\in\bigcup_{n=1}^{\infty}\sigma_n^i}],$$

$Y_1 = [\{z_n\}] \subset X_1$ and $Y_2 = [\{w_n\}] \subset X_2$. Let R_i, Q be the bounded linear projections on X given by:

$$R_i\Big(\sum_{n=1}^{\infty}\alpha_n x_n\Big) = \sum_{n=1}^{\infty}\frac{\left(\sum_{j\in\sigma_n^i}\alpha_j a_j^i\right)}{\left(\sum_{j\in\sigma_n^i} a_j^1 b_j^1\right)} z_n^i,$$

and

$$Q\Big(\sum_{n=1}^{\infty}\alpha_n x_n\Big) = \sum_{n=1}^{\infty}\frac{\left(\sum_{j\in\sigma_n^2}\alpha_j a_j^2\right)}{\left(\sum_{j\in\sigma_n^2} a_j^2 b_j^2\right)} w_n.$$

Now, by lemma 8, $(I - R_1)(X) \approx (I - Q)(X)$. So it suffices to show that $(I - P_1)(X) \approx (I - R_1)(X)$ and $(I - P_2)(X) \approx (I - Q)(X)$. Since these are symmetrical results, we will prove the first. By considering the projection $R = R_1 + R_2$ on X we see that $X \approx (I - R)(X) \oplus [\{z_n^1, z_n^2\}]$. Since $y_n^1 \in [z_n^1, z_n^2]$ for every n, by Theorem 6 and fact A, we have that $(I - P_1)(X) \approx (I - R)(X) \oplus [\{z_n^1\}] \approx (I - R_1)(X)$. This completes the proof of step I.

Step II: For each n there are natural numbers $p_n + 1 \leq i_n, m_n \leq p_{n+1}$ so that $\frac{1}{2} \leq a_{i_n}^1 b_{i_n}^1$ and $\frac{1}{2} \leq a_{m_n}^2 b_{m_n}^2$. By Theorem 6, $\{x_{i_n}\} \approx \{y_n^1\} \approx \{y_n^2\} \approx \{x_{m_n}\}$ and $(I -$

$P_1)(X) \approx [\{x_j\}_{j \neq i_n}]$, $(I-P_2)(X) \approx [\{x_j\}_{j \neq m_n}]$. It is trivial now to check that $[\{x_j\}_{j \neq i_n}] \approx [\{x_j\}_{j \neq m_n}]$.

Step III: The general case. Partition $\mathbf{N}$ into $\{A_1, A_2, A_3\}$ by:

$$A_1 = \{n | a_j^1 b_j^1 \leq \frac{1}{2}, \text{ for all } p_n + 1 \leq j \leq p_{n+1}\},$$
$$A_2 = \{n \notin A_1 | a_j^2 b_j^2 \leq \frac{1}{2}, \text{ for all } p_n + 1 \leq j \leq P_{n+1}\},$$
$$A_3 = \mathbf{N} \backslash (A_1 \cup A_2).$$

Let

$$X_i = \Big[\bigcup_{n \in A_i} \{x_j | p_n + 1 \leq j \leq p_{n+1}\} \Big]$$

for i=1,2,3. By step I,

$$\Big(I|_{X_1} - P_1\Big)(X_1) \approx \Big(I|_{X_1} - P_2\Big)(X_1),$$

and

$$\Big(I|_{X_2} - P_1\Big)(X_2) \approx \Big(I|_{X_2} - P_2\Big)(X_2).$$

By step II,

$$\Big(I|_{X_3} - P_1\Big)(X_3) \approx \Big(I|_{X_3} - P_2\Big)(X_3).$$

It follows that $(I - P_1)(X) \approx (I - P_2)(X)$. ∎

Now we can give some strong evidence to support our conjecture.

Theorem 10: Let $\{x_n\}$ be a unconditional basis for a Banach space X and

$$y_n = \sum_{i=p_n+1}^{p_{n+1}} a_i x_i$$

a P-complemented block basis of $\{x_n\}$.

(1) If $\{y_n\} \approx \{x_{i_n}\}$ for some $p_n + 1 \leq i_n \leq p_{n+1}$, then $(I - P)(X) \approx [\{x_j\}_{j \neq i_n}]$

(2) If for every $n_1 < n_2 < \ldots$, $[\{y_{n_i}\}] \approx [\{y_{n_i}\}]^2$, then $(I - P)(X) \approx [\{x_{m_i}\}]$ for some $m_1 < m_2 < \ldots$.

Proof:

(1) If we define $P_o\Big(\sum_{n=1}^{\infty} \alpha_n x_n\Big) = \sum_{n=1}^{\infty} \alpha_{i_n} x_{i_n}$, to be the natural projection of X onto $[\{x_{i_n}\}]$, then $(I - P_o)(X) = [\{x_j\}_{j \neq i_n}]$. Since $\{y_n\} \approx \{x_{i_n}\}$, by corollary 9, $(I - P)(X) \approx (I - Po)(X)$.

(2) Let $A_1 = \{n|_{|} a_j b_j| \leq \frac{1}{2}$, for all $p_n + 1 \leq j \leq p_{n+1}\}$. Let $X_1 = [\{x_j | p_n + 1 \leq j \leq p_{n+1}, n \in A_1\}]$, $X_2 = [\{x_{|}j\, x_j \notin X_1\}]$ and let $Y = [\{y_n\}_{n \in A_1}]$. By Theorem 6 and our assumption that $Y \approx Y \oplus Y$ we have some space Z so that,

$$\begin{aligned}\Big(I|_{X_1} - P\Big)(X_1) &\approx Y \oplus Z \\ &\approx Y \oplus Y \oplus Z \\ &\approx Y \oplus \Big(I|_{X_1} - P\Big)(X_1) \\ &\approx X_1.\end{aligned}$$

By part (1), $\Big(I|_{X_2} - P\Big)(X_2)$ is isomorphic to the span of a subsequence of $\{x_j | j \in X_2\}$. Hence $(I - P)(X) \approx \Big(I|_{X_1} - P\Big)(X_1) \oplus \Big(I|_{X_2} - P\Big)(X_2) \approx [\{x_{m_i}\}]$ for some $m_1 < m_2 < \ldots$ ∎

Every unconditional basis for $c_o, \ell_p.and\ell_p \oplus \ell_q$ satisfy (2) of Theorem 10. Symmetric and subsymmetric bases also do. Another interesting case is the Haar system $\{h_n\}$ for L_p, $1 \leq p < \infty$. Since $Y = [\{h_{n_i}\}] \approx L_p\ or \ell_p$ for all $n_1 < n_2 < \ldots$, it follows that $Y \approx Y^2$. Theorem 10 yields that the Haar system is "block complementably reproducible" in any unconditional basis.

The proof of Theorem 10 above and a technique of P. G. Casazza and B. L. Lin [2] yields:

Corollary 11: Let X be a Banach space with a unconditional basis $\{x_n\}$ and let P be a projection on X. Then either PX or (I-P)X contains a P_o-complemented subspace Z so that $P_o X = Z \approx [\{x_{n_i}\}]$ and $(I - P_o)X \approx [\{x_{m_i}\}]$ for some $n_1 < n_2 < \ldots$ and $m_1 < m_2 < \ldots$.

Another immediate consequence of the proof of Theorem 6 is:

Corollary 12: If $\{(x_n, o),\ (o, y_n)\}$ is a unconditional basis of $X \oplus Y$, and $\{z_n\}$ is a P-complemented block basis, then $(I-P)(X \oplus Y) \approx G \oplus H$ where G(resp. H) is isomorphic to a complemented block basis of $\{x_n\}$ (resp. $\{y_n\}$.)

Finally, we give some examples to show that these results are best possible.

Example 13: Corollary 9 Fails if the blocks $\{y_n^1\}$ and $\{y_n^2\}$ do not have the same supports. In $\ell_p \oplus \ell_q$, (where e_n is the unit vector basis) The complemented block bases $\{(e_n, o),\ (o, e_{2n})\}$ and $\{(e_{2n}, o), (o, e_n)\}$ are equivalent (after a permutation) but their complements are ℓ_q and ℓ_p respectively.

Example 14: Corollary 9 fails if one of the blocks is not complemented. Let X_p be the subspace of $\ell_p \oplus \ell_2$ constructed by H. P. Rosenthal [4]. Then $X_p = [\{e_n^p \oplus w_n e_n^2\}]$ for appropriate weights $\{w_n\}$. It is known [4] that X_p is not isomorphic to a complemented subspace of $\ell_p \oplus \ell_2$, while its basis $y_n = e_n^p \oplus w_n e_n^2$ is clearly a block basis of the unit vector basis of $\ell_p \oplus \ell_2$. Since $[(\ell_p \oplus \ell_2)/X_p]^* \approx X_p^\perp = [\{w_n e_n^p \oplus (-e_n^2)\}]$, it follows that $\ell_p \oplus \ell_2/X_p$ is not isomorphic to the span of a subsequence of the unit vector basis of $\ell_p \oplus \ell_2$. If we go one step further and let $X = X_p \oplus (\ell_p \oplus \ell_2)$, then $z_n = (y_n, o)$ is a P-complemented block basis of the natural basis of X, and $(I-P)(X) = \ell_p \oplus \ell_2$. But, $\{(o, y_n)\}$ is an uncomplemented block basis in X which is equivalent to $\{z_n\}$ and supp $z_n =$ supp (o, y_n). But if $Y = [\{(o, y_n)\}]$ then X_p embeds complementably into X/Y and so $X/Y \not\approx (I-P)(X)$.

BIBLIOGRAPHY

1. P. G. Casazza, N. J. Kalton, and L. Tzafriri, *Decompositions of Banach Lattices into Direct Sums*, Trans. A.M.S. (**304**), 1987, pp. 771–800.

2. P. G. Casazza and B. L. Lin, *Projections on Banach spaces with symmetric bases*, Studia Math (**52**) 1974, pp. 189–193.

3. J. Lindenstrauss and L. Tzafriri, *Classical Banach spaces I*, Sequence spaces, Springer-Verlag, (1979).

4. H. P. Rosenthal, *On the subspaces of* $L_p(p > 2)$ *spanned by sequences of independent random variables*, Israel J. Math. (**8**) 1970, pp. 273–303.

5. I. Singer, *Bases in Banach spaces I*, Springer-Verlag, (1970).

6. S. J. Szarek, *A Banach space without a basis which has the bounded approximation property*, Acta. Math. (**159**) 1987, pp. 81–98.

7. M. Zippin, *A remark on Bases and reflexivity in Banach spaces*, Israel J. Math. (**6**) 1968, pp. 74–79.

Department of Mathematics
The University of Missouri
Columbia, Missouri 65211

Contemporary Mathematics
Volume **85**, 1989

The Schroeder-Bernstein Property for Banach Spaces*

Peter G. Casazza

ABSTRACT: We review the known results and open problems related to the Schroder-Bernstein problem for Banach spaces. We also show that "large complements" in a Banach space are unique and derive several consequences of this. We end by examining some likely candidates for counter examples to the general problem.

1. INTRODUCTION: The Schroeder-Bernstein Theorem in set theory asserts: given two sets X and Y for which there are one-to-one mappings of X into Y and Y into X, then there is a one-to one mapping of X onto Y. The natural reformulation of this property into Banach spaces would be If two Banach spaces X and Y are Isomorphic to subspaces of each other, must they be isomorphic?

However, this formulation of the Theorem is well known to fail for Banach spaces. An easy example is obtained by letting $X = C[0,1]$ and $Y = C[0,1] \oplus \ell_1$. Since $C[0,1]$ is a universal separable Banach space (i.e. every separable Banach space is isomorphic to a subspace of $C[0,1]$) it follows that X and Y are isomorphic to subspaces of one another.

AMS subject classification: Primary 46B20
Keywords and phrases: Schroeder-Bernstein property, complemented subspaces, primary.
*Supported in part by NSF DMS 8410975

To see that X and Y are not isomorphic (and in fact, Y is not isomorphic to any complemented subspace of X) we use a result of Pelczynski [32] which states that every complemented subspace of $C[0,1]$ contains a (necessarily complemented [35]) subspace isomorphic to c_0. So if Y is complemented in X, then ℓ_1 is complemented in X and hence c_0 is complemented in X and hence c_0 is complemented in ℓ_1 (which is difficult to accomplish since C_0 does not even embed in ℓ_1). For later reference, let us note that both X and Y are isomorphic to their squares (i.e. $X \approx X \oplus X$ and $Y \approx Y \oplus Y$), X embeds complementably in Y and Y embeds in X, but never complementably.

The current reformulation of the Schroeder-Bernstein property into Banach space Theory is:

Definition 1.1: A Banach space X is said to satisfy **The Schroeder-Bernstein Property** (SBP) if for any Banach space Y, if X and Y are isomorphic to **complemented** subspaces of one another, then X and Y are isomorphic.

The important question is:

Problem 1.2: Does every Banach space have SBP?

At this time, there are no known Banach spaces which fail SBP although most people seem to feel that such spaces exist. We will discuss some possible examples in sections (5) and (6). In sections (3) and (4) we examine spaces which are known to have SBP and spaces for which the property is unknown. SBP is known to hold in such elementary spaces as ℓ_p (Pelczynski [31]) and such complex spaces as James' famous quasi-reflexive space (Casazza [10]), while it is unknown in such elementary spaces as the Lorentz sequence spaces d(a,p).

Some very elementary questions about SBP are unanswered.

Problem 1.3: If X has SBP, does X* have SBP?

Problem 1.4: If X* has SBP, does X have SBP?

Problem 1.5: If X and Y have SBP, does $X \oplus Y$ have SBP?

We show later that the answer to problem 1.5 is "yes" if X and Y are totally incomparable. This is a result of Edelstein and Wojtaszczyk [18] and Wojtaszczyk [37]. Other natural problems would imply a negative answer to SBP which may help to indicate why they are also so difficult. For example, it is unknown if there is a Banach Space X which is isomorphic to its cube but not its square. i.e. $X \approx X^3 = X \oplus X \oplus X$ but $X \not\approx X^2 = X \oplus X$. If such an X exists, it fails SBP since X and X^2 must embed complementably in one another without being isomorphic.

SBP is a very basic and natural property for a space which arises most of the time when one is trying to show that two Banach spaces are isomorphic. At this time, there is only one large class of spaces known to have it (Banach spaces which are ℓ_p-decomposable). So far, this has been a very intractable problem in almost any general setting. It has been solved in some particular cases with the use of force and some special properties of the space. Any solution to the problem for a large class of spaces would be worthwhile.

2. DECOMPOSITION METHODS: A Class of Spaces with SBP.

To check SBP, we must show that two Banach spaces which embed complementably in one another are isomorphic. Pelczynski once observed that having a little information about each space (namely, they are isomorphic to their squares) is all that is needed.

PELCZYNSKI'S DECOMPOSITION METHOD I: If X and Y embed complementably in one another and $X \approx X^2$, $Y \approx Y^2$, then X and Y are isomorphic.

Since we will want to generalize the technique later, let us now check why this decomposition method works. If $X \approx X^2$ and X embeds comlementably in Y, then X has a complement, say Z, in Y.

Hence

$$Y \approx X \oplus Z \approx X \oplus X \oplus Z \approx X \oplus Y.$$

So if Y is also isomorphic to its square and embeds complementably in X we have $X \approx X \oplus Y \approx Y$.

At the time Pelczynski made this observation, there were few known Banach spaces which were not isomorphic to their squares. James' quasi-reflexive space J (see[10]), $C(\omega)$ and variations of them were basically it. There did not even exist an example of a Banach space with a unconditional basis which was not isomorphic to its square. This void was later filled by T. Figiel [19] who showed that the space

$$X = (\sum_{n=1}^{\infty} \oplus \ell_{p_n}^{r_n})_{\ell_2}$$

has the property that X^{m+1} does not embed into X^m, for any $m = 1, 2, .., $ for appropriately chosen p_n decreasing to 2 and r_n increasing to infinity.

This decomposition method has the drawback that it requires knowing something about both spaces X and Y. But to show that X has SBP, we need to show that X is isomorphic to **every** Y which contains a complemented subspace isomorphic to X and which embeds complementably into X. Since there are no general conditions for showing a space is isomorphic to its square, this method works very well for showing two concrete spaces (which are known to be isomorphic to their squares) are isomorphic, but doesn't help with SBP.

Pelczynski [31] then gave another decomposition method which requires no information about Y but a much stronger assumption on X.

Pelczynski's Decomposition Method II:

If $X \approx (\sum \oplus X)_{\ell_p}$ or $(\sum \oplus X)_{c_o}$, then X has SBP.

To prove this , let Y be any Banach space so that X and Y embed complementably in one another. Since $X \approx X^2$, we know that $Y \approx X \oplus Y$. On the other hand, since Y is complemented in X, there is a space Z so that $X \approx Y \oplus Z$. Now,

$$X \approx (\sum \oplus X)_{\ell_p} \approx (\sum \oplus (Y \oplus Z))_{\ell_p} \approx (\sum \oplus Y)_{\ell_p} \oplus (\sum \oplus Z)_{\ell_p}$$

$$\approx Y \oplus (\sum \oplus Y)_{\ell_p} \oplus (\sum \oplus Z)_{\ell_p} \approx Y \oplus X.$$

This proof works equally well if we merely assume $(\sum \oplus X)_{\ell_p}$ embeds complementably in X instead of being isomorphic to X. But this is not a more general result since by the same argument, if $(\sum \oplus X)_{\ell_p}$ embeds complementably into X then it is already isomorphic to X. We now have:

Corollary II.1: c_0 and ℓ_p, $1 \leq p < \infty$ have SBP.

Pelczynski [31] actually showed more than this. That is, he showed that c_0 and ℓ_p, $1 \leq p < \infty$ are prime Banach spaces. **A prime** space is a Banach space which is isomorphic to all of its complemented infinite dimensional subspaces. It is clear then that every prime Banach space has SBP. Unfortunately, only one other prime Banach space has ever been found. That is the space ℓ_∞, shown by Lindenstrauss [24] to be prime. It is an open question whether any other prime spaces exist. The two best possibilities for new prime spaces are the minimal Orlicz spaces [28], [29], [30], and the Kalton-Peck space [23] (see section (6) for a discussion of this space).

By Pelezynski's method II, if Y is any Banach space, then $X = (\sum \oplus Y)_{\ell_p}$ has SBP. This is the only large class of spaces ever found to have SBP.

Let us consider again the proof of Pelczynski's decomposition method I. If $Y \approx Y^2$ and Y embeds complementably in X then $X \approx X \oplus Y$. The converse of this clearly fails since for any $Y \not\approx Y^2$, if $X \approx (\sum \oplus Y)_{\ell_p}$, then $X \approx X \oplus Y$. However, if X also embeds complementably in Y, then the converse does hold.

Observation 2.1: Let X and Y embed complementably in one another. Then $X \approx X \oplus Y$ if and only if $Y \approx Y^2$. Hence, $X \approx X \oplus Y \approx Y$ if and only if $X \approx X^2$.

Proof: We assume $X \approx X \oplus Y$. Since X embeds complementably in Y, there is a Banach space Z so that $Y \approx X \oplus Z$. It follows that:

$$Y^2 \approx Y \oplus Y \approx (X \oplus Z) \oplus Y \approx (X \oplus Y) \oplus Z \approx X \oplus Z \approx Y. \blacksquare$$

I think of our next observation as the "uniqueness of large complements" in a Banach space. Among other things, it will give us a reformulation of SBP. It is well known that complements are not unique in a Banach space. For example,

$$\begin{aligned} C[0,1] &\approx C[0,1] \oplus C[0,1] \\ &\approx C[0,1] \oplus c_o, \end{aligned}$$

and,

$$\begin{aligned} d(a,p) &\approx d(a,p) \oplus d(a,p) \\ &\approx d(a,p) \oplus \ell_p, \end{aligned}$$

while c_0 is not isomorphic to $C[0,1]$ and ℓ_p is not isomorphic to d(a,p).

However, our next observation shows that complements are unique if they are big enough.

Observation 2.2: Let

$$X \approx H \oplus Y \approx H \oplus Z.$$

(1) Then $X \oplus Y^n \approx X \oplus Z^n$, for all $n = 1, 2, \ldots$,

(2) If X embeds complementably into both Y and Z, then $Y \approx Z$.

Proof:

(1) $X \oplus Y \approx H \oplus Z \oplus Y \approx H \oplus Y \oplus Z \approx X \oplus Z$. Now, $X \oplus Y^2 \approx X \oplus Z \oplus Y \approx X \oplus Z^2$, and continue.

(2) By our assumption, there are Banach spaces K and W so that, $Y \approx X \oplus W$ and $Z \approx X \oplus K$. Now,

$$Y \approx X \oplus W \approx H \oplus Z \oplus W \approx H \oplus X \oplus K \oplus W,$$

and

$$Z \approx X \oplus K \approx H \oplus Y \oplus K \approx H \oplus X \oplus W \oplus K.$$

Therefore, $Y \approx Z$. ∎

This observation has several interesting consequences.

Corollary 2.3: If X^2 embeds complementably in X then there is a Banach space Y so that $Y \approx Y^2$ and $X \approx X \oplus Y$.

Proof: Since X^2 embeds complementably in X, there is a Banach space Z so that $X \approx X \oplus X \oplus Z$. Let $Y = X \oplus Z$ and note that

$$Y \approx X \oplus Z \approx (X \oplus X \oplus Z) \oplus Z \approx X^2 \oplus Z^2 \approx Y^2.$$

Hence,

$$X \approx X \oplus Y. \blacksquare$$

Note that there do exist Banach spaces X and Y so that X embeds complementably in Y but X^2 does not even embed into Y. To see this, we note that it is easy to write Figiel's space F (which does not contain a copy of its square [19]) as $F \approx Y \oplus Y$, for some space Y. If F embeds into Y then F^2 embeds into $Y \oplus Y \approx F$, which is impossible.

Corollary 2.4: If $X^n \approx X^{n+1}$, for some n, then $X \approx X^2$ if and only if X^2 embeds complementably in X.

Proof:

$$\begin{aligned} X^n &\approx X^{n+1} \\ &\approx X^n \oplus X. \end{aligned}$$

By observation 2.1, $X \approx X^2$ if X^n embeds complementably into X. ∎

We can easily generalize 2.4. If $X^n \approx X^{n+k}$ for some n and k, then $X^n \approx X^{n+k} \approx X^n \oplus X^k \approx X^{n+k} \oplus X^k \approx X^{n+2k} \approx \ldots \approx X^{n+mk}$ for all $m = 1, 2 \ldots$. But there is no way of knowing in general if $X^n \approx X^{n-k} \approx X^{n-2k} \approx \ldots$. However, if X contains a complemented copy of X^2, then we may iterate in both directions.

Corollary 2.5: If X^2 embeds complementably into X and $X^n \approx X^{n+k}$ for some n, k, then $X^m \approx X^{m+k}$ for all $m = 1, 2, \ldots$

Proof: Since X^2 embeds complementably into X, there is a Banach space Y so that $X^m \approx X^n \oplus Y$. Then

$$X^m \approx X^n \oplus Y \approx X^{n+k} \oplus Y \approx (X^n \oplus Y) \oplus X^k \approx X^m \oplus X^k \approx X^{m+k}. \blacksquare$$

Finally, let us mention that observation 2.1 can be generalized also.

Corollary 2.6: If X and Y^2 embed complementably in each other, then $X \approx X \oplus Y$ if and only if $Y^2 \approx Y^3$.

Proof: For the "if" part, choose a Banach space Z so that $X \approx Z \oplus Y^2$. Then
$X \approx Z \oplus Y^2 \approx Z \oplus Y^3 \approx (Z \oplus Y^2) \oplus Y \approx X \oplus Y$.
For the "only if " part, we note that if $X \approx X \oplus Y$, then $X \approx X \oplus Y \approx X \oplus Y^2 \approx X \oplus Y^3$. Now $Y^3 \approx Y^2 \oplus Y$ embeds complementably into $X \oplus Y \approx X$. Since X embeds complementably into Y^2, so does Y^3 and so the result follows by observation 2.2. ∎

We can also prove the decomposition methods of Pelczynski by using observation 2.2. For method I, put $Y = Y, Z = X$ and $H = X \oplus Y$ into observation 2.2. For method II, put $Y = Y, Z = X$ and $H = (\sum \oplus (X \oplus Y))_{\ell_p}$ into observation 2.2.

Finally, we mention that observation 2.2 gives us a reformulation of SBP for Banach spaces containing their squares complementably.

Observation 2.9: If X^2 embeds complementably into X, then X has SBP if and only if for any Banach space Y for which X and Y embed complementably into one another, there is a Banach space Z so that $Z \oplus X \approx Z \oplus Y$ and Z embeds complementably into X.

Proof: For one direction let $Z = 0$ and the other is immediate by observation 2.2. ∎

3. SPACES WHICH HAVE SBP:

We saw earlier that c_0 and $\ell_p, 1 \leq p < \infty$, were shown to have SBP by Pelczynski [31] and ℓ_∞ has SBP is due to Lindenstrauss [24]. Also Pelczynski has shown [31] that for any Banach space $X, (\sum \oplus X)_{\ell_p}, 1 \leq p < \infty$, and $(\sum \oplus X)_{c_0}$ have SBP. In particular, $(\sum \oplus \ell_p^n)_{\ell_q}, (\sum \oplus \ell_p^n)_{c_0}, (\sum \oplus \ell_\infty^n)_{\ell_q}, 1 \leq p, q < \infty$, all have SBP. It is also known that $C[0,1]$ has SBP (Lindenstrauss and Pelczynski, [25]) and that $C(\alpha)$ has SBP as long as α is not regular (Alspach and Benjamini [1]). Pelczynski's decomposition method II shows that L_p has SBP for $1 < p < \infty$. Casazza [10] showed that James' quasi-reflexive space J has SBP and Andrews [4] generalized this to James Tree. J. T. Arazy [5] has shown that C_p has SBP for $1 < p < \infty$. Pelczynski [33] (see also [34] for a simpler construction) showed that there is a universal basis for all normalized unconditional bases and the space U this universal basis spans has SBP. Capon [8], [9] has shown that L_p (X) has SBP.

Each of the spaces discussed above actually has another related property to SBP. That is, each of these spaces is primary. A Banach space X is **primary** if whenever $X \approx Y \oplus Z$, either $X \approx Y$ or $X \approx Z$. Sufficient conditions for a space to be primary (and it is unknown if they are necessary) are:

(1) Whenever $X \approx Y \oplus Z$, then X embeds complementably in either Y or Z.

(2) X has SBP.

Pelczynski [33] showed that his universal space U has SBP while Casazza and Lin [13] showed that it is primary. Pelczynski's decomposition method II shows that L_p has SBP while Alspach, Enflo and Odell [2] showed that the L_p-spaces are primary $(1 < p < \infty)$.

Bourgain [6] has shown that H^∞ is primary. Wojtaszczyk [38] has shown that the disc algebra A is isomorphic to $(\sum \oplus A)_{C_p}$, that H^1 is isomorphic to $(\sum \oplus H^1)_{\ell_1}$, and that H^∞ is isomorphic to $(\sum \oplus H^\infty)_{\ell_\infty}$. It follows by Pelczynski's decomposition method II that A and H^1 have SBP. We do not know if Pelczynski's method works for ℓ_∞- sums but an examination of [24] and [12] where ℓ_∞ - decompositions are used might shed some light on this. There are spaces with SBP which are not primary (e.g. $c_0 \oplus \ell_1$) which we will discuss later). It is unknown if there is any space X failing SBP but if such a space were primary then it would follow that there is a Banach space Y so that X and Y embed complementably in one another, X is primary but not isomorphic to its square while Y is isomorphic to its square but not primary.

A result of Edelstein and Wojtaszczyk [18] (see Wojtaszczyk [37] for a stronger result) shows that if X and Y are totally incomparable (i.e. they have no isomorphic infinite dimensional subspaces), and have SBP, then $X \oplus Y$ has SBP. In particular, $\ell_p \oplus \ell_q$ has SBP for all $1 \leq p, q \leq \infty$, as does $c_0 \oplus \ell_p, 1 \leq p \leq \infty$ and $c_0 \oplus J$, etc. None of these spaces can be primary since they are direct sums of totally incomparable spaces.

4. SPACES FOR WHICH SBP IS UNKNOWN:

SBP is unknown for the Orlitz spaces (see [26], [28], [29], [30], and the Lorentz spaces d(a,p) (see [3], [14]). In fact, it is not known if a single one of these spaces has SBP (except the degenerate cases where they reduce to ℓ_p). It is not known if **any** of the Tsirelson-Type spaces has SBP (see [7], [11], [15], [16, [20], [21], [36]. That is, SBP is not known for Tsirelson's space T, convexified Tsirelson's space $T_c = T^{(2)}$, symmetric Tsirelson's space T_s, etc. An important space for which SBP is unknown is the Kalton-Peck space Z_2 (see [22], [23]). This space discussed in section (6) is believed to fail SBP but no one has been able to prove it yet. The space $C(\alpha)$, for α regular, is unknown to have or fail SBP. The recent sophisticated examples on the Radon-Nikodgn property do not seem to be understood well enough yet to even think about SBP for them.

5. BANACH SPACES WITH SYMMETRIC BASES AND SBP.

A normalized basis $\{x_n\}_{n=1}^{\infty}$ of a Banach space X is **symmetric** if it is equivalent to all of its permutations. i.e. For every permutation π of N, the operator $L : X \to X$ defined by $Lx_n = x_{\pi(n)}$ is an isomorphism. Casazza and Lin [13] proved an important theorem about projections on Banach spaces with symmetric bases.

Theorem 5.1: If X has a symmetric basis and $X \approx Y \oplus Z$, then X embeds complementably into either Y or Z.

Recall from section 3 that a space must be primary if it has SBP and the property stated in Theorem 5.1. So it follows that,

Theorem 5.2: A Banach space with a symmetric basis is primary if and only if it has SBP.

Using observation 2.2 and its consequences we can state.

Proposition 5.3: Let X be a Banach space with a symmetric basis.

(1) If $X \approx Z \oplus X$, then whenever $X \approx Z \oplus Y$, either $X \approx Z$ or $X \approx Y$,

(2) If Y embeds complementably into X but X does not embed complementably in Y, then Y has a unique complement in X. i.e. If $X \approx Y \oplus Z \approx Y \oplus H$, then $Z \approx H$.

(3) If Z is isomorphic to the span of a constant coefficient block basic sequence in X, then whenever $X \approx Z \oplus Y$, either $X \approx Z$ or $X \approx Y$.

It follows that if X has a symmetric basis and fails SBP then:

(α) There are Banach spaces Y and Z so that $X \approx Y \oplus Z$ while $Y \not\approx Y^2$ and $Z \not\approx Z^2$.

(β) There is a Banach space Y which embeds complementably in X but $X \not\approx X \oplus Y$.

(γ) There is a Banach space Y which embeds complementably in X but Y is not isomorphic to PX or (I - P)(X) for any projection onto a constant coefficient block basic sequence in X (i.e. for any averaging projection P on X).

(θ) There is a Banach space Y which embeds complementably into X but whenever $Z = (\sum \oplus Y)$ embeds into X, it is not complemented.

Perhaps (θ) needs some explanation. Since X has a symmetric basis $\{x_n\}_{n=1}^{\infty}$, if $\{J_n\}_{n=1}^{\infty}$ is a partition of the natural numbers into infinite disjoint sets, say $J_n = \{(n,i); i = 1,2\ldots\}$, then $L(x_{(n,i)}) = x_i$ is an isomorphism. So if Y is complemented in X, Y sits naturally as a complemented subspace of span $\{x_{(n,i)};\ i = 1,2\ldots\}$, for all $n = 1,2\ldots$. So the formal sum $Z = (\sum +Y)$ is a subspace of

$$X = (\sum_{n=1}^{\infty} + \text{ span } \{X_{(n,i)}; i = 1,2,\ldots\}).$$

Then (θ) says that this formal Z cannot be complemented in X.

Until recently, there was no known example of a space with a symmetric basis that had all of these necessary conditions for failing SBP. However, Carothers, Casazza and

Flinn (unpublished example) have exhibited an example of a special Lorentz sequence space d(a,2) with the following properties:

(a) Figiel's space F [19] (for which F has an unconditional basis and $F \not\approx F^2$) is isomorphic to a complemented subspace of d(a,2),

(b) The formal sum $Z = (\sum +F)$ does not embed complementably into d(a,2). In particular, F is not isomorphic to P(d(a,2)) or (I - P)(d(a,2)) for any projection P of d(a,2) onto a constant coefficient block basic sequence.

To show that this space fails SBP then, we need to show that the complement of F in d(a,2) is not isomorphic to its square or, equivalently, it is not isomorphic to d(a,2). This seems to be very difficult and it is possible that all Lorentz sequence spaces are primary and hence have SBP anyway.

6. THE KALTON-PECK SPACE: A possible counter-example to SBP.

At this time, the most likely candidate for a counter-example to SBP is the wonderful example of Kalton and Peck of the reflexive space Z_2 [23]. The Kalton-Peck space is the space of all sequences $\{(a_n, b_n)\}_{n=1}^{\infty}$ of couples of real numbers such that

$$\| \{(a_n, b_n)\}_{n-1}^{\infty} \| = (\sum_{n=1}^{\infty} b_n^2)^1/2 + \Big(\sum_{n=1}^{\infty} a_n - b_n \log(\frac{|{}^b n|}{(\sum |b_n|)^{2\frac{1}{2}})^2)^{\frac{1}{2}}})\Big)$$

is finite. The above expression is not a norm, but satisfies a weak triangle inequality and is equivalent to a norm. Among the very many interesting properties of this space are the following (discovered by Johnson, Lindenstrauss, and Schechtman [22]):

(1) Every infinite dimensional complemented subspace of Z_2 contains a complemented subspace isomorphic to Z_2.

(2) No infinite dimensional complemented subspace of Z_2 has an unconditional basis.

It follows from observation 2.2 that all complements are unique (infinite dimensional, of course) in Z_2. i.e. $Z_2 \approx H \oplus X \approx H \oplus Y$ imples $X \approx Y$. Also, by (1), if Z_2has SBP then it is a new prime Banach space (see section 1). It is conjectured that Z_2 is not prime and, in fact, is a counter-example to the hyperplane problem for Banach spaces. That is, there is no known example of a Banach space which is not isomorphic to its hyperplanes. It is believed that Z_2 is just such a space. It is not known (but should be easier to prove if it is true) if Z_2 has at most two non-isomorphic complemented subspaces (namely, the whole space and its hyperplanes). If Z_2 is not isomorphic to its hyperplane Y then since $Y^2 \approx Z_2$, it would follow that $Y \not\approx Y^2$ but $Y \approx Y^3$, answering one of our earlier questions. This example certainly contains a wealth of interesting properties which should be gleaned out one day. For now, it seems to be "sitting on the shelf" waiting for someone to take up the challenge of understanding its true nature. And the SBP property waits with it.

7. THE SPACE ℓ_∞/C_0.

In a recent paper, Drewnowski and Roberts [17] have shown that ℓ_∞/C_0 is primary under the assumption of the continuum hypothesis (CH). But what they really showed was that ℓ_∞/C_0 has SBP if we assume (CH). This raises the interesting possibility that SBP fails in ZF but holds if we assume (CH). In particular, ℓ_∞/C_0 is a candidate for a space failing SBP.

BIBLIOGRAPHY

1. Alspach, D., Bengamini, Y., *Primariness of spaces of continuous functions on ordinals,* Israel J. Math. **27**, pp. 65–92, (1977).
2. Alspach, D., Enflo, P., Odell, E., *On the structure of separable L_p spaces* $(1 < p < \infty)$, Studia Math. **60**, pp. 79–90, (1977).
3. Altshuler, Z., Casazza, P.G., Lin, B.L., *On symmetric basic sequences in Lorentz sequence spaces*, Israel J. Math. **15**, pp. 140–155.
4. Andrews, A.D., *the Banach space JT is primary*, Pacific Journal of Math. **108**, pp. 9–17, (1983).
5. Arazy, J., *On large subspaces of the Schatten p-classes*, Composito Math. **41**, pp. 297–336, (1980).
6. Bourgain, J., *On the primarity of H^∞ - spaces*, (preprint).
7. Bourgain, J., Casazza, P.G., Lindenstrauss, J., Tzafriri, L., *Banach spaces with a unique unconditional basis, up to permutations*, Memoirs of A.M.S. (1985),No **322**.
8. Capon, M., *Sur la primarite de certains espaces de Benach*, (preprint).
9. Capon, M., *Primarite de $L^p(X)$, lorsque X ne Verifie pas P_p*, (preprint).
10. Casazza, P. G., *James' quasi-reflexive space is primary*, Israel J. Math. **26**, pp. 294–305, (1977).
11. Casazza, P. G., Johnson, W. B., Tzafriri, L., *On Tsirelsons space*, Israel J. Math. **47**, pp. 81–98, (1984).
12. Casazza, P. G., Kottman, C., Lin, B. L., *On primary Banach spaces*, Canad. J. Math. **29**, pp. 856–873, (1977).
13. Casazza, P., Lin, B. L., *Projections on Banach spaces with symmetric bases*, Studia Math. **52**, pp. 189–193, (1974).
14. Casazza, P. G., Lin, B. L., *On symmetric basic sequences in Lorentz sequence spaces II*, Israel J. Math. **17**, pp. 191–218, (1974).

15. Casazza, P. G., Odell, E., *Tsirelsons space and minimal subspaces*, Longhorn Lecture Notes, (University of Texas at Austin) pp. 61–73, (1982–83).

16. Casazza, P. G., Shura, T., *On Tsirelson's space*, (Lecture Notes - To appear).

17. Drewnowski, L. and J. Roberts, *On the Primariness of the Banach Space* ℓ_∞/C_0, (Preprint).

18. Edelstein, I.S., Wojtaszczyk, P., *On projections and unconditional bases in direct sums of Banach spaces*, Studia Math **56**, pp. 263–276, (1976).

19. Figiel, T., *An example of an infinite dimensional Banach space non-isomorphic to its Carterian square*, Studia Math. **42**, pp. 295–306, (1972).

20. Figiel, T., Johnson, W. B., *A uniformly convex Banach space which contains no* ℓ_p, Composito Math. **29**, pp. 179–190, (1974).

21. Johnson, W. B., *Banach spaces all of whose subspaces have the approximation property*, (preprint).

22. Johnson, W. B., Lindenstrauss, J., Schechtman, G., *On the relationship between several notions of unconditional structure*, Israel J. Math. **37** pp. 120–129, (1980).

23. Kalton, N. J., Peck, T., *Twisted sums of sequence spaces and the three space problem*, Trans. A.M.S. **255**, pp. 1–30, (1979).

24. Lindenstrauss, J., *On complemented subspaces of m.*, Israel J. Math. **5**, pp. 153–1566, (1967).

25. Lindenstraus, J., Pelczynski, A., *Contributions to the theory of classical Banach spaces*, J. Funcl. Anal. **8**, pp. 225–249, (1971).

26. Lindenstrauss, J. Tzafriri, l., *Classical Banach spaces I, sequence spaces*, Springer-Verlag, N. Y., (1977).

27. Lindenstrauss, J., Tzafriri, L., *Classical Banach spaces II, Function Spaces*, Springer-Verlag, N. Y., (1979).

28. Lindenstrauss, J., Tzafriri, L., *On Orlicz sequence spaces*, Israel J. Math. **9**, pp. 263–269, (1971).

29. Lindenstrauss, J., Tzafriri, L., *On Orlicz sequence spaces II*, Israel J. Math. **11**, pp. 255–379, (1972).

30. Lindenstrauss, J., Tzafriri, L., *On Orlicz sequence spaces III*, Israel J. Math. **14**, pp. 368–389, (1973).

31. Pelczynski, A., *Projections in certain Banach spaces*, Studia Math. **19**, pp. 209–227, (1960).

32. Pelczynski, A., *On C(s) - subspaces of separable Banach spaces*, Studia Math. **31**, pp. 513–522, (1968).

33. Pelczynski, A., *Universal bases*, Studia Math. **32**, pp 247–268, (1969).

34. Scheetman, G., *On Pelczynski's paper "Universal Bases"*,Israel J. Math. **22**, pp. 181–184, (1975).

35. Sobczyk, A., *Projection of the space m onto its subspace* C_0, Bull. Amer. Soc. **47**, pp. 938–947, (1941).

36. Tsirelson, B. S., *Not every Banach space contains* ℓ_p *or* C_0, Functional Anal. & Appl. **8**, pp. 138–141, (1974). [Translated from Russian].

37. Wojtaszczk, P., *On projections and unconditional bases in Direct sums of Banach spaces II*, Studia Math. **62**, pp. 193–201, (1978).

38. Wojtaszczyk, P., *Decompositions of* H^p *spaces*, Duke Math J. **46**, pp. 635–644, (1979).

Department of Mathematics

University of Missouri

Columbia, Missouri 65211

Contemporary Mathematics
Volume **85**, 1989

Rotund reflexive Orlicz spaces are fully convex

Shutao Chen*, Bor–Luh Lin and Xintai Yu*

ABSTRACT: A reflexive Orlicz space is fully convex if and only if it is rotund.

In 1955, two generalizations of the class of uniformly convex Banach spaces were introduced. Lovaglia [L] introduced the class of locally uniformly rotund Banach spaces. A Banach space X is said to be locally uniformly rotund if for any sequence $\{x_n\}$ and x in X such that $\|x_n\| \le 1$, $\|x\| = 1$ and $\lim_n \|x+x_n\| = 2$, then $\lim_n \|x_n - x\| = 0$. Fan and Glicksberg [FG1] introduced the class of fully convex normed linear spaces. Let $k \geq 2$ be an integer. A normed space X is said to be fully k–convex (kR) if for every sequence $\{x_n\}$ in X, $\lim_{n_1,\ldots,n_k\to\infty} \frac{1}{k}\|x_{n_1}+\cdots+x_{n_k}\| = 1$ implies that $\{x_n\}$ is a Cauchy sequence in X. It is proved in [FG1] that every uniformly convex space is 2R, kR space is (k+1)R space, and every kR Banach space is reflexive. The converse of these statements are false (see [FG1] and [LY]).

An Orlicz function M is a continuous, non–decreasing and convex function defined on $\mathbb{R}^+$ such that $M(0) = 0$ and $\lim_{u\to\infty} M(u) = \infty$. We shall assume that $M(u) > 0$ for all

---

*Supported by the National Science Fund of China.

1980 Mathematics Subject Classification (1985 Revision): Primary 46B20, 46B30.

Key words and phrases: Orlicz spaces, fully convex spaces, locally uniformly rotund spaces, rotund spaces.

$u > 0$. M is said to satisfy the Δ_2–condition at infinite (respectively, zero) if $\limsup_{u\to\infty} M(2u)/M(u) < \infty$ (resp. $\limsup_{u\to 0^+} M(2u)/M(u) < \infty$). For an Orlicz function M, let M^* denote the complementary function of M. Let (Ω,Σ,μ) be a measure space. We shall consider only the case that $\mu(\Omega)$ is finite and that Ω is atomless. For a given measurable function x on Ω (resp. a sequence $x = (a_1, a_2, \cdots)$), we define the module of x to be $\rho(x) = \int_\Omega M(|x|)d\mu$ (resp. $\rho(x) = \sum_{n=1}^{\infty} M(|a_n|)$). The Orlicz space L_M (resp. Orlicz sequence space ℓ_M) is the Banach space of all measurable functions x (resp. sequences of scalar x) such that $\rho(x/\lambda) < \infty$ for some $\lambda > 0$ with the norm $\|x\| = \inf\{\lambda\colon\ \lambda > 0,\ \rho(x/\lambda) < 1\}$. In [K], it is proved that if the measure space (Ω,Σ,μ) is finite and atomless, then an Orlicz space L_M is locally uniformly rotund if and only if it is rotund, and it is equivalent to the conditions that M satisfies the Δ_2–condition at infinite and M is strictly convex, i.e. $M(\frac{u+v}{2}) < \frac{M(u)+M(v)}{2}$ for all $u \neq v$ in $\mathbb{R}^+$. In this paper, we prove that a reflexive Orlicz space L_M (resp. Orlicz sequence space ℓ_M) is 2R if and only if it is rotund. Hence in the class of reflexive Orlicz spaces (resp. Orlicz sequence spaces), the concepts of locally uniformly rotundity and the fully convexity are the same.

<u>Theorem 1</u>. Let (Ω,Σ,μ) be a finite measure space which is atomless. Then an Orlicz space L_M is 2R if and only if M is strictly convex and both M and M^* satisfy the Δ_2–condition at infinite.

<u>Proof</u>. It is well–known that L_M is reflexive if and only if M and M^* satisfy the Δ_2–condition at infinite. since every kR Banach space is reflexive and rotund, it follows that M and M^* satisfy the Δ_2–condition at infinite and furthermore M is strictly convex [K].

To prove the converse. Let $\{x_n\}$ be a sequence in X such that $\lim_{n,m} \|x_n+x_m\| = 2$. We may assume that $\|x_n\| = 1$, $n = 1,2,\cdots$. We first show that $\{x_n\}$ converges in measure.

Given $\epsilon > 0$ and $\delta > 0$, since M is strictly convex, there exists $\sigma > 0$ such that if $M(u) \le 4/\epsilon$, $M(v) \le 4/\epsilon$ and $|u-v| \ge \delta$, then

$$M(\tfrac{u+v}{2}) \le (1-\sigma)\,\frac{M(u)+M(v)}{2}\,. \tag{1}$$

Let $E(m,n) = \{\omega \in \Omega : |x_n(\omega)-x_m(\omega)| \ge \delta\}$ and $F(m,n) = \{\omega \in \Omega : M(|x_n(\omega)|) > 4/\epsilon$ or $M(|x_m(\omega)|) > 4/\epsilon\}$. Then $2 \ge \rho(x_n) + \rho(x_m) \ge 4/\epsilon\ \mu(F(m,n))$. Hence $\mu(F(m,n)) < \epsilon/2$. Since $\lim_{n,m} \|x_n+x_m\| = 2$ and M satisfies the Δ_2–condition at infinite, there is an integer N such that for all $n,m \ge N$,

$$1 - \rho\left[\frac{x_n+x_m}{2}\right] \le \frac{1}{2}\,\epsilon\sigma M(\tfrac{\delta}{2}). \tag{2}$$

Let $n,m \ge N$. Then, by (1) and (2),

$$\rho\left[\frac{x_n+x_m}{2}\right] \le (1-\sigma)\int_{E(m,n)\setminus F(m,n)} \frac{M(|x_n|)+M(|x_m|)}{2}\,d\mu$$

$$+\int_{\Omega\setminus(E(m,n)\setminus F(m,n))} \frac{M(|x_n|)+M(|x_m|)}{2}\,d\mu$$

$$= \frac{\rho(x_n)+\rho(x_m)}{2} - \sigma \int_{E(m,n)\backslash F(m,n)} \frac{M(|x_n|+M(|x_m|)}{2} d\mu$$

$$\leq 1 - \sigma \int_{E(m,n)\backslash F(m,n)} M\left[\frac{|x_n-x_m|}{2}\right] d\mu$$

$$\leq 1 - \sigma M(\tfrac{\delta}{2})\ \mu(E(m,n)\backslash F(m,n)).$$

Hence $\sigma M(\frac{\delta}{2})\mu(E(m,n)\backslash F(m,n)) \leq 1 - \rho\left[\frac{x_n+x_m}{2}\right] \leq \frac{\epsilon\sigma}{2} M(\frac{\delta}{2})$. Thus $\mu(E(m,n)\backslash F(m,n)) \leq \frac{\epsilon}{2}$. It follows that $\mu(E(m,n)) < \epsilon$ for all $n,m \geq N$ and this completes the proof that $\{x_n\}$ converges in measure.

Let x_0 be a measurable function such that $\{x_n\}$ converges in measure to x_0. By Faton's lemma, $\rho(x_o) \leq 1$. If $\rho(x_0) = 1$ then $\lim_n \rho(x_n) = \rho(x_0)$, $\{x_n\}$ converges in measure to x_0 and M satisfies the Δ_2-condition at infinite. We conclude that $\lim_n ||x_n-x|| = 0$ and so L_M is 2R.

Suppose in contrary. Let $a = 1 - \rho(x_0) > 0$. Choose $u_0 \in \mathbb{R}^+$ such that

$$M(u_0) = \frac{a}{5\mu(\Omega)}\ . \tag{3}$$

Since M^* satisfies the Δ_2-condition at infinite, there exists $b > 0$ such that for all $u \geq u_0$,

$$M(\tfrac{u}{2})) \leq (\tfrac{1}{2} - b)M(u). \tag{4}$$

Now, choose N_0 such that for all $n,m \geq N_0$,

$$1 - \rho\left[\frac{x_n + x_m}{2}\right] < \frac{ab}{5} . \tag{5}$$

Since M satisfies the Δ_2–condition at infinite, there exists $\epsilon_0 > 0$ (see, e.g. Lemma 0,2 in [K]) such that for all x,y satisfying $\rho(x) \leq 1$ and $\rho(y) \leq \epsilon_0$,

$$|\rho(x+y) - \rho(x)| < \frac{ab}{5} . \tag{6}$$

Again, use the fact that M satisfies the Δ_2–condition at infinite, for $\epsilon_0 > 0$, there is $\delta_0 > 0$ such that for all $E \in \Sigma$, $\mu(E) < \delta_0$, then

$$\rho(x_0 \chi_E) < \epsilon_0 . \tag{7}$$

Since $\{x_n\}$ converges in measure to x_0, for $\delta_0 > 0$, by choosing a subsequence if necessary, we may assume that there is a set $E_0 \in \Sigma$, $\mu(E_0) < \delta_0$ and $\{x_n\}$ converges uniformly to x_0 on $\Omega \backslash E_0$. Now,

$$\lim_n \rho(x_n \chi_{E_0}) = \lim_n (\rho(x_n) - \rho(x_n \chi_{\Omega\backslash E_0})) = 1 - \rho(x_o \chi_{\Omega\backslash E_0})$$

$$= 1 - \rho(x_0) + \rho(x_0 \chi_{E_0}) = a + \rho(x_0 \chi_{E_0}).$$

Choose $k \geq N_0$ such that $\rho(x_k \chi_{E_0}) > \frac{4a}{5}$. If $E_k = \{\omega : \omega \in E_0, x_k(\omega) \geq u_0\}$, then, by (3),

$$\rho(x_k\chi_{E_k}) = \rho(x_k\chi_{E_0}) - \rho(x_k\chi_{E_0\setminus E_k})$$

$$\geq \frac{4a}{5} - M(u_0)\mu(E_0\setminus E_k) \geq \frac{4a}{5} - M(u_0)\mu(\Omega) = \frac{3a}{5}. \tag{8}$$

For x_k, choose $\delta_k > 0$ such that $\rho(x_k\chi_E) < \frac{a}{5}$ for all $E \in \Sigma$, $\mu(E) < \delta_k$. Finally, choose $E \in \Sigma$ with $\mu(E) < \min(\delta_0,\delta_k)$ such that $\{x_n\}$ converges uniformly to x_0 on $\Omega\setminus E$. Then, by (7), $\lim_n \rho(x_n\chi_{E_k\setminus E}) = \rho(x_0\chi_{E_k\setminus E}) \leq \rho(x_0\chi_{E_0}) < \epsilon_0$. Hence there is $j \geq N_0$ such that $\rho(x_j\chi_{E_k\setminus E}) < \epsilon_0$. By (4), (6), (7) and (8),

$$\rho\left[\frac{x_k+x_j}{2}\chi_{E_k\setminus E}\right] < \rho\left[\frac{x_k}{2}\chi_{E_k\setminus E}\right] + \frac{ab}{5} \leq (\tfrac{1}{2} - b)\,\rho(x_k\chi_{E_k\setminus E}) + \frac{ab}{5}$$

$$< \tfrac{1}{2}[\rho(x_k\chi_{E_k\setminus E}) + \rho(x_j\chi_{E_k\setminus E})] - b(\tfrac{3a}{5} - \tfrac{a}{5}) + \frac{ab}{5}$$

$$= \tfrac{1}{2}[\rho(x_k\chi_{E_k\setminus E}) + \rho(x_j\chi_{E_k\setminus E})] - \frac{ab}{5}.$$

By (5), we have

$$1 - \frac{ab}{5} < \rho\left[\frac{x_k+x_j}{2}\right] \leq \rho\left[\frac{x_k+x_j}{2}\chi_{\Omega\setminus(E_k\setminus E)}\right] +$$

$$\tfrac{1}{2}[\rho(x_k\chi_{E_k\setminus E}) + \rho(x_j\chi_{E_k\setminus E})] - \frac{ab}{5}$$

$$\leq \tfrac{1}{2}[\rho(x_k) + \rho(x_j)] - \frac{ab}{5} \leq 1 - \frac{ab}{5}.$$

Which is a contradiction. □

Using the similar (and simpler) argument as in Theorem 1, the following theorem can be proved. We shall omit the detail.

Theorem 2. The following are equivalent.

(i) ℓ_M is 2R,

(ii) ℓ_M is kR for some $k \geq 2$

(iii) ℓ_M is reflexive and rotund

(iv) M and M^* satisfy the Δ_2–condition at zero and M is strictly convex.

Remark.

Let X be a Banach space. It is known [FG2] that if X is kR then X has property (G), that is, every point of the unit sphere of X is a denting point of the unit ball of X. Furthermore, if ℓ_1 is not isomorphic to a subspace of X, then X having the property (G) is equivalent to X is rotund and X has the Kadec property (K), that is, the weak topology and norm topology coincide on the unit sphere of X. Those every rotund reflexive Orlicz space L_M (or ℓ_M) has the properties (G) and (K). [LLT].

References

[FG1] Ky Fan and I. Glicksberg, Fully convex normed linear spaces, *Proc. Nat. Acad. Sci.*, U.S.A. 4(1955), 947–953.

[FG2] Ky Fan and I. Glicksberg, Some Geometric properties of the spheres in a normed linear space, *Duke Math. J.* 25(1958), 553–568.

[K] A. Kaminska, The criteria for local uniform rotundity of Orlicz spaces, *Studia Math.* 79(1984), 201–215.

[L] A.R. Lovaglia, Locally uniformly convex Banach space, *Trans. Amer. Math. Soc.* 78(1955), 225–238.

[LLT] Bor–Luh Lin, Pei–kee Lin and S.L. Troyanski, Some geometric and topological properties of the unit sphere in a Banach space, *Math. Annalen* 274(1986), 613–616.

[LY] Bor–Luh Lin and Yu, Xintai, On the k–uniform rotund and the fully convex Banach spaces, *J. Math. Anal. Appl.* 110(1985), 407–410.

[WWCW] Congxin Wu, Tingfu Wang, Shutao Chen and Yuwen Wang, Geometric theory in Orlicz spaces (Chinese), Harbin Institute of Technology, 1986.

Shutao Chen
Department of Mathematics
Harbin Normal University
Harbin, China

Bor–Luh Lin
Department of Mathematics
University of Iowa
Iowa City, Iowa 52242
U.S.A.

Xintai Yu
Department of Mathematics
East China Normal University
Shanghai, China

Contemporary Mathematics
Volume **85**, 1989

BIORTHOGONAL SYSTEMS AND BIG QUOTIENT SPACES

Catherine Finet and Gilles Godefroy

Abstract: Any Banach space which has a quotient space which does not linearly inject into $\ell^\infty(A)$ contains a biorthogonal system of cardinality $K > |A|$. The converse holds if X has $(\mathscr{C})$. Weak–star analytic subspaces of $\ell^\infty(\mathbb{N})$ are classified by their biorthogonal systems. Non ω^*–separable dual unit balls are constructed. If Y is separable and does not contain $\ell^1(\mathbb{N})$, every operator $T:Y^* \longrightarrow Y^*$ with non–separable range fixes a biorthogonal system with c elements. The space $\mathscr{C}(K)$ of Kunen is a non–separable Banach space in which every closed subspace is a countable intersection of closed hyperplanes.

I. Introduction.

It is well known that a quotient space $X/_Y$ of a Banach space X may be "bigger", in some sense, than the space X. A classical example of this situation is $X = \ell^\infty(\mathbb{N})$,

---

1980 Mathematics Subject Classification (1985 Revision): Primary 46B20

Key words and phrases: Analytic sets, analytic compact sets, representable spaces, Markushevich basis.

$Y = c_0(\mathbb{N})$: the quotient space ℓ^∞/c_0 cannot be separated by a countable number of continuous linear forms. In other words, there is no continuous linear injection from ℓ^∞/c_0 into $\ell^\infty(\mathbb{N})$.

The aim of the present work is to connect this property with another important feature of non–separable Banach spaces: biorthogonal systems. The main gist of our results is that the cardinality of the biorthogonal systems which live in X is closely related to the ω^*–density character of the ω^*–closed subspaces of X^*. We will also investigate in detail the structure of the biorthogonal systems into "natural" subspaces of $\ell^\infty(\mathbb{N})$; and apply our result to renormings. The assumption of "representability" (in the sense of [9]) will be shown to discard several pathologies.

Most of our results can be translated, by simple duality arguments, into results about the ω^*–topology of the dual space; we did not formulate the corresponding statements in this paper. However, let us mention that what we investigate is – dually – the classical fact that subsets of separable non–metrizable topological spaces are not separable in general; what we actually show is that metrizability is very often a necessary condition for the separability of subsets. The crucial example of Kunen ([12]; see [16]) stresses the central role played by conditions of hereditary separability.

<u>Notation</u>. The closed unit ball of a Banach space X is denoted by X_1. "Operator" means "bounded linear operator". X "*linearly injects*" into Y if there is a one–to–one operator T from X into Y; this operator T is *not* supposed to induce an isomorphism between X and T(X). A Banach space X has the *property* $(\mathscr{C})$ of Corson if every family of convex closed bounded sets with empty intersection contains a countable subfamily with empty intersection (see [21]). A *biorthogonal system* – in short b.s. – is a subset (x_α, x_α^*) $(\alpha \in I)$ of $X \times X^*$ which satisfies

$$\begin{cases} (1)\ x^*_\beta(x_\alpha) = 0 & \forall\, \alpha \neq \beta \\ (2)\ x^*_\alpha(x_\alpha) = 1 & \forall\, \alpha \in I \\ (3)\ \sup \{\|x^*_\alpha\|\ \|x_\alpha\| \mid \alpha \in I\} < \infty \end{cases}$$

Observe that the boundedness condition (3) is not always figuring in the classical definitions of a b.s. We will sometimes call the set $\{x_\alpha \mid \alpha \in I\}$ itself a b.s. If a b.s. (x_α, x^*_α) $(\alpha \in I)$ in $X \times X^*$ satisfies also

$$\begin{cases} (4)\ \overline{\text{span}}^{\|\ \|}(x_\alpha) = X \\ (5)\ \cap\{\text{Ker}(x^*_\alpha) \mid \alpha \in I\} = \{0\} \end{cases}$$

then it is called a *Markushevich basis* – in short a M–basis. Every separable space has a M–basis ([18], [19]); actually, every space which contains a set satisfying (1)–(2)–(4)–(5) has a M–basis [20]. Note that the conditions (4) and (5) are mutually independent (see e.g. the examples II.5 below). If a b.s. (x_α, x^*_α) is an M–basis of $\overline{\text{span}}^{\|\ \|}(x_\alpha)$ – when we consider the restrictions of (x_α^*) to this space – then it is called a *Markushevich basic set*; in short an M–basic set. A topological space is *analytic* if it is the continuous image of $\mathbb{N}^{\mathbb{N}}$.

The present work is a revised version of the preprint [5] of the same authors.

Acknowledgements. This work was completed while the authors were visiting the University of Missouri at Columbia. It is their pleasure to thank the Department of Mathematics of U.M.C. and in particular Nigel Kalton, Paula and Elias Saab.

II. Biorthogonal systems and big quotient spaces.

Our first result describes the tight connection which relates the existence of "big" quotient spaces of X and the presence of "large" biorthogonal systems in X.

Theorem II.1. (i) Let X be a banach space, and A be a set. If there is a quotient space Y of X which does not linearly inject into $\ell^\infty(A)$, then X contains a biorthogonal system of cardinality $K > |A|$.

(ii) Let X be a Banach space with the property ($\mathscr{C}$) of Corson. If (x_α, x_α^*) $(\alpha \in I, |I| > \aleph_0)$ is an uncountable biorthogonal system and if we let $Y = \cap\{\text{Ker}(x_\alpha^*) \mid \alpha \in I\}$, then $X_{/Y}$ does not linearly inject into $\ell^\infty(A)$ with $|A| < K = |I|$.

Proof. The case A finite is trivial. We assume now $|A| \geq \aleph_0$.

Let us recall first the obvious fact that a Banach space Z linearly inject into $\ell^\infty(A)$ if and only if $\omega^*\text{-dens}(Z^*) \leq |A|$.

(i) If $V = X_{/Y}$ contains a biorthogonal system (v_α, v_α^*) of cardinality K, then X contains such a system; indeed, we pick for every α a point $x_\alpha \in X$ such that $Q(x_\alpha) = v_\alpha$ and $\|x_\alpha\| < 2\|v_\alpha\|$ – where Q denotes the canonical quotient map from X onto V; it is clear that $(x_\alpha, Q^*(v_\alpha^*))$ is a biorthogonal system in X.

Hence for proving (i), it suffices to show that a Banach space V which does not linearly inject into $\ell^\infty(A)$ contains a b.s. of cardinality $K > |A|$. For this, we consider the family B of b.s. in $V \times V^*$ which satisfy the additional condition

$$\sup_\alpha \|v_\alpha\| \cdot \|v_\alpha^*\| \leq 2$$

B is clearly inductive if we order it by inclusion; hence, Zorn's lemma will prove (i) if we can show that a biorthogonal system (v_α, v_α^*) $(\alpha \in I)$ in B cannot be maximal if $|I| \leq |A|$.

For showing this we let

$$W = \overline{\text{span}}'' \; '' \{x_\alpha | \alpha \in I\}$$

and

$$Z = \cap \{\mathrm{Ker}(x_\alpha^*) \mid \alpha \in I\}$$

We clearly have

$$\|\,\| - \mathrm{dens}(W) = \omega^*\!\!-\mathrm{dens}(Z^\perp) = |I| \leq |A|$$

and since one always have

$$\omega^*\!\!-\mathrm{dens}(W^*) \leq |A|$$

We claim that $Z \not\subset W$; indeed $Z \subset W$ implies

$$\omega^*\!\!-\mathrm{dens}(V^*/Z^\perp) = \omega^*\!\!-\mathrm{dens}(Z^*) \leq \omega^*\!\!-\mathrm{dens}(W^*) \leq |A|$$

but it is easily seen that

$$\left.\begin{array}{l} \omega^*\!\!-\mathrm{dens}(V^*/Z^\perp) \leq |A| \\ \omega^*\!\!-\mathrm{dens}(Z^\perp) \leq |A| \end{array}\right\} \Rightarrow \omega^*\!\!-\mathrm{dens}(V^*) \leq |A|$$

and this is impossible since by assumption V does not linearly inject into $\ell^\infty(A)$. Now, since $Z \not\subset W$, we can pick $v \in Z$, with

$$\begin{cases} \|v\| = 1 \\ \mathrm{dist}(v, W) > 1/2 \end{cases}$$

By Hahn–Banach, there exists $v^* \in W^\perp$ $\|v^*\| < 2$ and $v^*(v) = 1$. It is clear that $(v_\alpha, v;\ v_\alpha^*, v^*)$ is in B; and this concludes the proof of (i).

(ii) By [21], the property ($\mathscr{C}$) is equivalent to the following condition $(*)$:

$(*)$ For any subset D of X_1^* and any $y \in \bar{D}^*$, there is a sequence $S = \{a_n | n \geq 1\}$ in D such that $y = \overline{\mathrm{conv}}^*(S)$.

It follows from the Banach–Dieudonné theorem (see [21]) that $(*)$ implies: if Z is a subspace of X^* and $y \in \bar{Z}^*$, then there is a sequence $S' = \{z_n | n \geq 1\}$ in Z such that $y \in \overline{\mathrm{conv}}^*(S')$.

Let now (x_α, x_α^*) $(\alpha \in 1,\ |I| > \aleph_0)$ be an uncountable b.s. in X and let

$$Y = \cap\{\mathrm{Ker}(x_\alpha^* | \alpha \in 1\}$$

we have to show that

$$\omega^*\text{–dens}(Y^\perp) = \omega^*\text{–dens}[\ \overline{\mathrm{span}}^*(x_\alpha^*)] \geq |I|$$

For this we consider $\{y_\beta^*\} \mid \beta \in J\}$ a ω^*–dense subset of $Y^\perp$; $(*)$ implies that for every $\beta \in J$, there is a countable subset (I_β) of I such that

$$y_\beta \in \overline{\text{span}}^*\{x^*_\alpha \mid \alpha \in I_\beta\}$$

If $|J| < |I|$, we have then

$$\left| \cup\{I_\beta | \beta \in J\} \right| < |I|$$

since $|I_\beta| \le \aleph_0$ for all β and $|I| > \aleph_0$; in particular, there is $\alpha_0 \in I \backslash \cup\{I_\beta \mid \beta \in J\}$. And then

$$\begin{cases} y^*_\beta(x_{\alpha_0}) = 0 & \forall \beta \in J \\ x^*_{\alpha_0}(x_{\alpha_0}) = 1 \end{cases}$$

and this contradicts the fact that a $\{y^*_\beta | \beta \in J\}$ is ω^*–dense in $Y^\perp$.

■

This obvious corollary of theorem II.1 covers the more natural applications.

<u>Corollary II.2</u>: (i) Every Banach space X which has a quotient space which does not linearly inject into $\ell^\infty(\mathbb{N})$ contains an uncountable biorthogonal system.

(ii) If X has the property $(\mathscr{C})$ of Corson and if (x_α, x^*_α) $(\alpha \in I, |I| > \aleph_0)$ is an uncountable biorthogonal system, then if we let $Y = \cap\{\text{Ker}(x^*_\alpha) \mid \alpha \in I\}$, the space X/Y does not linearly inject into $\ell^\infty(\mathbb{N})$.

<u>Example II.3</u>. 1) K. Kunen has constructed – assuming the continuum hypothesis (CH) – a non–separable Banach space X, of the form $X = \mathscr{C}(K)$, where K is a separable non–metrizable scattered compact space, which satisfies: For any uncountable subset A of X, there is $a \in A$ such that $a \in \overline{\text{conv}}^{\|\ \|}(A\backslash\{a\})$ (see [16]). Of course, the

space X contains no uncountable biorthogonal system; hence by Corollary II.2, every quotient space of X injects into $\ell^\infty(\mathbb{N})$. In other words, Kunen's space is an example of a non–separable Banach space in which any closed subspace is a *countable* intersection of closed hyperplanes.

2) The example of $X = \ell^1(\aleph_1)$ shows that Corollary II.2(i) is optimal. Indeed, $c_0(\aleph_1)$ is a quotient of $\ell^1(\aleph_1)$ and $c_0(\aleph_1)$ does not inject into $\ell^\infty(\mathbb{N})$; on the other hand, the density character of $\ell^1(\aleph_1)$ is $\aleph_1$, and thus a b.s. in $\ell^{\hat{}}(\aleph_1)$ has at most $\aleph_1$ elements. Classical techniques permit to construct a $\mathscr{C}(K)$–space with similar properties (see [5]).

Our next corollary shows that the spaces which enjoy $(\mathscr{C})$ do not contain "big" Markushevich bases.

Corollary II.4. Let X be a Banach space which has the property $(\mathscr{C})$ of Corson. If X linearly injects into $\ell^\infty(A)$, then X contains no M–basic set (x_α) $(\alpha \in 1)$ with $|I| > |A|$.

Proof. If A is finite, the result is trivial; hence we may assume $|A| \geq \aleph_0$. Consider (x_α, x_α^*) $(\alpha \in I)$ a biorthogonal system. We let

$$Y = \overline{\text{span}}'' "\{x_\alpha \,|\, \alpha \in I\}$$

Since the property $(\mathscr{C})$ is hereditary, Y has $(\mathscr{C})$.

We call (y_α^*) the restriction of (x_α^*) to Y, and we define

$$Z = \cap\{\operatorname{Ker}(y_\alpha^*) \mid \alpha \in I\}$$

Now if $|I| > |A| \geq \aleph_0$, Theorem II.1 (ii) shows that $Y_{/Z}$ does not inject into $\ell^\infty(A)$; since Y injects into $\ell^\infty(A)$, we must have $Z \neq \{0\}$; this means that (x_α, x^*_α) is not an M–basic set.

∎

Of course, Corollary II.4 implies that any space with $(\mathscr{C})$ which injects into $\ell^\infty(\mathbb{N})$ contains only countable M–basic sets. This special case is due to Suarez ([25]) and suffices for the following examples.

Examples II.5: 1) Recall that a compact topological space K is called a *Rosenthal compact* if K is homeomorphic to a pointwise compact set of first Baire class functions on a Polish space (see [1], [8]). By [8], we may apply Corollary 4 to this class, and thus we have: let K be a separable Rosenthal compact. Then every Markushevich basic set in $\mathscr{C}(K)$ is countable; a fortiori, $\mathscr{C}(K)$ has no Markushevich basis if K is not metrizable.

The special case of $K_0 = [0,1] \times \{0,1\}$ equipped with the lexicographical order topology was observed in [25]. Actually, $X = \mathscr{C}(K_0)$ provides us with a natural example of the situation described in Theorem II.1 (ii): X is the space of functions on $[0,1]$ which satisfy:

(α) $\forall\, x \in [0,1]$, $f(x^-)$ and $f(x^+)$ exist.

(β) $\forall\, x \in [0,1]$, $f(x^-) = f(x)$

Consider now for every $x \in [0,1)$

$$\begin{cases} f_x = \mathbf{1}_{(x,1]} \\ \mu_x(f) = \lim\limits_{t \to x^+} f(t) - f(x) \end{cases}$$

It is clear that $\{f_x, \mu_x) \mid x \in [0,1)\}$ is a biorthogonal system in $X \times X^*$. If we consider now

$$Y = \cap\{\mathrm{Ker}(\mu_x) \mid x \in [0,1)\}$$

then $Y = \mathscr{C}([0,1])$ and $X/_Y \simeq c_0([0,1])$; clearly, $c_0([0,1])$ does not inject into $\ell^\infty(\mathbb{N})$. This example motivated Theorem II.1(ii); moreover, it provides us with a natural example of a b.s. such that

$$\begin{cases} \overline{\mathrm{span}}^{\|\ \|}(f_x) = X \\ \cap\ \mathrm{Ker}(f_x^*) = \cap\ \mathrm{Ker}(\mu_x) \neq \{0\} \end{cases}$$

2) Let X be a separable Banach space not containing $\ell^1(\mathbb{N})$. By [17] and [21], X^* has the property ($\mathscr{C}$) and thus, by Corollary II.4., X^* contains no uncountable M–basic set. A fortiori, if X^* is not separable, then X^* has no Markushevich basis. This result should be compared to the following facts:

(i) Under the same assumptions, X^* contains a separable subspace Y such that $X^*/_Y$ does not linearly inject into $\ell^\infty(\mathbb{N})$; a fortiori, Y is not complemented into X^* ([6], Th. VII. 6; see also [7], Th. I.8.). Let us mention that the techniques of [6], [7] and of the present work are completely different.

(ii) On the other hand, any dual space with the Radon–Nikodym property has a Markushevich basis ([4]).

(iii) Consider any separable space X, such as $X = JT$ ([11]), such that $X \not\supset \ell^1(\mathbb{N})$ and X^* is non separable. By [18], X has a Markushevich basis $\{(x_n, x_n^*) \mid n \geq 1\}$. If we consider now $(x_n^*, x_n^{**} = x_n)$ as a b.s. in X^*, we see that the assumption $|I| > \aleph_0$ is indeed necessary in Theorem II.1(ii); this b.s. provides us also with a funny fact: if $|I| > \aleph_0$, we *cannot* have

$$\cap\{\text{Ker}\,(x^*_\alpha)\ |\ \alpha \in I\} = \{0\}$$

although it may happen when $|I| = \aleph_0$! We invite the reader to figure out why this is not so surprising after all. Finally, this example shows that there is no hope to apply maximality arguments to Markushevich basis.

(iv) By the above, if X is separable and X^* contains a non–separable subspace Y which has a Markushevich basis, then X contains $\ell^1(\mathbb{N})$. This extends the classical result where $Y = \ell^1(c)$ (see [17], [10]).

Let us observe that the two examples above and both "representable" Banach spaces, that is, are both isomorphic to ω^*–analytic subspaces of $\ell^\infty(\mathbb{N})$ – see [9] where this notion is introduced. Within this frame, we will be able to give a quite complete description of the b.s. and their relation with the structure of the space.

Theorem II.6. Let X be a Banach space which is isomorphic to a ω^*–analytic subspace of $\ell^\infty(\mathbb{N})$. Then one has:

(i) X is separable if and only if X contains no biorthogonal system ([9]).

(ii) X does not contains $\ell^1(c)$ if and only if X contains no uncountable M–basic set.

(iii) X contains $\ell^1(c)$ if and only if X contains an uncountable biorthogonal system (x_α, x^*_α) such that $\cap\{\text{Ker}\,(x^*_\alpha)\ |\ \alpha \in I\} = \{0\}$.

Before showing this result, let us mention that [9] permits to complete this theorem with much more equivalences.

Proof. (i) is ([9], Theorem 5).

(ii) If $X \subseteq \ell^\infty(\mathbb{N})$ is ω^*–analytic and does not contain $\ell^1(c)$ then by ([9], Théorème 6) the dual unit ball (X_1^*, ω^*) is angelic and thus X has the property $(\mathscr{C})$; now Corollary II.4 implies that X contains no uncountable M–basic set; the converse is clear since $\ell^1(c)$ has an obvious Markushevich basis.

(iii) If a subspace X of $\ell^\infty(\mathbb{N})$ contains $\ell^1(c)$, then by ([3], p. 178 and [23]), X contains a b.s. (x_α, x_α^*) of cardinality c such that $\cap \{\mathrm{Ker}(x_\alpha^*) \mid \alpha \in I\} = \{0\}$. Conversely, if such a b.s. exists, we let

$$V = \overline{\mathrm{span}}''(x_\alpha)$$

and y_α^* be the restriction of x_α^* to V; clearly (x_α, y_α^*) is a M–basis of V, and since $V \subseteq X$, (ii) shows that $\ell^1(c) \subseteq X$.

■

Example II.7. The dual of any separable Banach space is isomorphic to a ω^*–K_σ subspace of $\ell^\infty(\mathbb{N})$. Hence Theorem 6 applies to such spaces; the classification (i), (ii), (iii) corresponds, if $X = Y^*$ with Y separable, to: Y^* separable – $Y \not\supset \ell^1(\mathbb{N})$ and Y^* nonseparable – $Y \supset \ell^1(\mathbb{N})$.

Theorem II.6 shows that an assumption of ω^*–analyticity permits to discard certain pathologies which may occur for an arbitrary Banach space. We pursue this idea in the next two results; see also [10].

Proposition II.8. Let X be a norm closed subspace of $\ell^\infty(\mathbb{N})$. If X contains a ω^*–analytic not norm–separable subset A, then there is a subspace Y of X such that

$X/_Y$ does not linearly inject into $\ell^\infty(\mathbb{N})$.

Proof. We first observe that the space $V = \overline{\text{span}}^{w^*}(A)$ is ω^*–analytic; indeed, span(A) is a continuous image of

$$\mathscr{A} = \bigcup_{n\geq 1} (A^n \times \mathbb{R}^n)$$

hence span(A) is ω^*–analytic; we write now $\overline{\text{span}}^{w^*}(A)$ as

$$\overline{\text{span}}^{w^*}(A) = \bigcap_{k\geq 1} (\text{span}(A) + [-k^{-1}, k^{-1}]^{\mathbb{N}})$$

and this shows that $V = \overline{\text{span}}^{w^*}(A)$ is ω^*–analytic; note also that V is not norm–separable since it contains A.

Now, two situations may occur: if $V \not\supset \ell^1(c)$, V has $(\mathscr{C})$ [9]; but V contains a b.s. of cardinality c, and thus theorem 1(ii) provides us with a subspace W of V such that $V/_W \hookrightarrow \ell^\infty(\mathbb{N})$. If $V \supset \ell^1(c)$, then $\ell^\infty(\mathbb{N})$ is a quotient space of V, and so is ℓ^∞/c_0.

If both cases, V has a quotient space $V/_W$ which doe not linearly inject into $\ell^\infty(\mathbb{N})$; a fortiori, $X/_W$ does not linearly inject into $\ell^\infty(\mathbb{N})$.

■

Remark II.9. In a joint paper of A. Louveau and the second–named author [10], sufficient conditions for the existence of a ω^*–analytic set A like in Proposition II.8 are investigated. It is shown for instance that under a suitable determinacy axiom, every non–separable space X which belongs to the projective hierarchy (for the ω^*–topology) contains such a set. This confirms the highly pathological character of Kunen's space and of any similar counterexample.

In the next section (§ III), we will apply the existence of a big quotient space to the construction of certain equivalent norms.

The last result of this section shares a common feature with several classical results about operators, that is: if $T: X \longrightarrow Y$ has a "big" range, then T restricted to a "big" subset of X is an isomorphism.

Theorem II.10. Let X be a separable Banach space not containing $\ell^1(\mathbb{N})$. Let $T: X^* \longrightarrow X^*$ be a bounded linear operator. Then the following are equivalent:

(i) $T(X^*)$ is not norm–separable.

(ii) X^* contains a b.s $(x^*_\alpha, x^{**}_\alpha)$ of cardinality c such that $T(x^*_\alpha)$ is also a biorthogonal system (in other words, T "fixes" a b.s. of cardinality c).

Proof. (ii) $\Longrightarrow$ (i) is obvious, since any b.s. is norm ϵ–separated for some $\epsilon > 0$. (i) $\Longrightarrow$ (ii): Let $T: X^* \longrightarrow X^*$ be an operator with non–separable range; without loss of generality, we may assume that $\|T\| \leq 1$. Let $K = (X^*_1, \omega^*)$ be the dual unit ball. We claim that $T: K \longrightarrow K$ is a Borel map; indeed, if

$$V = K \cap \bigcap_{j=1}^{n} \{x_j^{**-1}(a_j, b_j)\}$$

is an elementary ω^*–open set in K, we have

$$T^{-1}(V) \cap K = K \cap \bigcap_{j=1}^{n} \{T^*(x_j^{**})^{-1}(a_j, b_j)\}$$

and by [17], $T^*(x_j^{**})$ is (ω^*) of the first Baire class; hence $T^{-1}(V) \cap K$ is ω^*–Borel in K. Now, if $\varphi: K \longrightarrow K$ is a Borel map from a metrizable compact K into itself, $\varphi(K)$ is an analytic set (see [15]); hence by the above, the space $T(X^*)$ is ω^*–analytic in X^*.

By assumption, $T(X^*)$ is not norm–separable. Now a glance at the proof of the key lemma 4 of [9] shows that it permits to construct in $T(X^*)$ a biorthogonal system $(y_\alpha^*, y_\alpha^{**})$ $(\alpha \in 1)$ with $|I| = c$. For each α, we pick $x_\alpha^* \in X^*$ such that $T(x_\alpha^*) = y_\alpha$; and for every $n \geq 1$, we let

$$I_n = \{\alpha \in I \mid \|x_\alpha^*\| \leq n\}$$

We clearly have $I = \cup\{I_n \mid n \geq 1\}$. Since c is not of countable cofinality, there exists $n_0 \in \mathbb{N}$ such that $|I_{n_0}| = c$. Now it is straightforward to check that the set

$$\{(x_\alpha^*, T^*(y_\alpha^{**})) \mid \alpha \in I_{n_0}\}$$

satisfies the conditions requested in (ii).

∎

Along the same lines, let us mention that the results of [9] can be extended without much effort to the more general frame of "representable operators" (see [5]); the interesting point in theorem II.10 is that every such operator T is "representable".

III. An application to renormings.

If X is a separable Banach space, then the compact (X_1^*, ω^*) is metrizable and hence separable; if (y_n) is a ω^*–dense sequence in X_1^*, then the operator

$$T:X \longrightarrow \ell^\infty(\mathbb{N})$$
$$x \longmapsto (y_n(x))$$

is an isometric embedding from X into $\ell^\infty(\mathbb{N})$. Actually, since the unit ball of $\ell^{\infty*}$ is ω^*–separable, X embeds isometrically into ℓ^∞ if and only if (X_1^*,ω^*) is separable. This condition clearly holds for every equivalent norm if X is separable and it seems to be an open problem whether or not the converse holds in full generality. We will show that this is indeed the case for "reasonable" Banach spaces (by II. 8).

We start with a simple lemma. Let us define for a Banach space X the quantity

$$d_\infty(X) = \inf\{d(X,Y) \mid Y \text{ subspace of } \ell^\infty(\mathbb{N})\}$$

where $d(X,Y)$ is the Banach–Mazur distance

$$d(X,Y) = \inf\{||T|| \cdot ||T^{-1}|| \mid T: X \longrightarrow Y \text{ isomorphism}\}$$

obviously $d_\infty(X) \in [1,+\infty]$ and $d_\infty(X) < \infty$ if and only if X is isomorphic to a subspace of $\ell^\infty(\mathbb{N})$. Moreover, one has

Lemma III.1. Let X be a Banach space. Then

$$d_\infty(X) = \inf[\sup\{r \geq 0 \mid (rX_1^*) \subseteq \overline{\text{conv}}^*(y_n)\}]^{-1}$$

where the infimum is taken over all the sequences $(y_n)_{n\geq 1}$ in X_1^*. There exists an isomorphism T between X and a subspace Y of $\ell^\infty(\mathbb{N})$ such that $d_\infty(X) = \|T\| \cdot \|T^{-1}\|$. In particular, $d_\infty(X) = 1$ if and only if X is isometric to a subspace of $\ell^\infty(\mathbb{N})$.

Proof. Let $\alpha > d_\infty(X)$; there exists $T: X \longrightarrow Y \subseteq \ell^\infty(\mathbb{N})$ such that $\|T\| \cdot \|T^{-1}\| < \alpha$; we can assume that $\|T\| = 1$ and then $\|T^{-1}\| < \alpha$. Let (e_n) be the canonical basis of $\ell^1(\mathbb{N})$ and let $y_n = T^*(e_n)$. We have $\|y_n\| \le 1$ since $\|T^*\| = 1$; moreover,

$$\begin{aligned}\|T^{-1}\| < \alpha &\Longleftrightarrow \forall x \in X,\ \|T(x)\| \ge \alpha^{-1}\|x\| \\ &\Longleftrightarrow \forall x \in X,\ \sup_{n\ge1} \{|\langle e_n, T(x)\rangle|\} \ge \alpha^{-1}\|x\|\end{aligned}$$

that is,

$$\forall x \in X,\ \sup_{n\ge1} \{|\langle x, y_n\rangle|\} \ge \alpha^{-1}\|x\|$$

and then Hahn–Banach shows that

$$\sup\{r \ge 0 \mid (rX_1^*) \subseteq \overline{\mathrm{conv}}^*(y_n)\} \ge \alpha^{-1}$$

hence we have the inequality $\ge$ since $\alpha > d_\infty(X)$ is arbitrary. Conversely, for any $k \ge 1$, there is a sequence (y_n^k) in X_1^* such that

$$\overline{\mathrm{conv}}^*(y_n^k) \supseteq r_k X_1^*$$

for some r_k such that

$$r_k^{-1} \langle [\sup\{r \geq 0 \mid (rX_1^*) \subseteq \overline{\text{conv}}^*(y_n^k)\}]^{-1} + k^{-1}$$

and it is standard to check that the operator

$$\begin{aligned} T_k\colon X &\longrightarrow \ell^\infty(\mathbb{N}) \\ x &\longmapsto y_n^k(x) \end{aligned}$$

is an isomorphism between X and $T_k(X)$ which satisfies $\|T_k\| \cdot \|T_k^{-1}\| \leq r_k^{-1}$, hence we have the reverse inequality and this establishes the formula.

For constructing an operator T such that $\|T\| \cdot \|T^{-1}\| = d_\infty(X)$, it suffices, with the above notation, to consider the operator

$$\begin{aligned} T\colon X &\longrightarrow \ell^\infty(\mathbb{N}^2) \\ x &\longmapsto (y_n^k(x))_{n,k\geq 1} \end{aligned}$$

and the last assertion is obvious.

■

It is clear that $d_\infty(X)$ depends upon the norm of X; the corresponding isomorphic invariant is

$$\mu(X) = \sup\{d_\infty(X, |\cdot|)\}$$

where the supremum is computed over the set of equivalent norms on X. Very little seems to be known about $\mu(X)$; for instance, $\mu(X) = 1$ if X is separable but the converse appears to be open; it is also not known if $\mu(X) = \infty$ as soon as $\mu(X) > 1$. Concerning subspaces and quotients, we have

<u>Theorem III.2</u>. Let X be a Banach space, and Y a subspace of X. Then $\mu(X) \geq \mu(Y)$ and $\mu(X) \geq \mu(X_{/Y})$.

<u>Proof</u>. The relation $\mu(X) \geq \mu(Y)$ is an immediate consequence of the well-known fact that any equivalent norm on Y is the restriction of an equivalent norm on X.

Let us prove now that $\mu(X) \geq \mu(X_{/Y})$. Pick $\alpha < \mu(X_{/Y})$; there exists a norm $|\cdot|$ on $X_{/Y}$ such that $d(X_{/Y}, |\cdot|) > \alpha$. It is classical, and easily seen, that there is a norm $\|\cdot\|$ on X whose quotient norm is Y; we also note $\|\cdot\|$ the dual norm on X^*, and $X_1^* = \{y \in X^* \mid \|y\| \leq 1\}$.

The space $(X_{/Y}, |\cdot|)^*$ is isometric to $(Y^\perp, \|\cdot\|)$; hence by III.1, for every countable subset (y_n) in $Y^\perp$ with $\|y_n\| \leq 1$, there is $y \in Y^\perp$ with $\|y\| < \alpha^{-1}$ and $y \notin \overline{\text{conv}}^*((y_n))$. We take now $\epsilon \in (0,1)$ and we consider

$$K_\epsilon = \text{conv}(\{y \in Y^\perp \mid \|y\| \leq 1\} \cup \epsilon X_1^*).$$

The set K_ϵ is convex, balanced and ω^*-closed, and it contains an open ball; hence K_ϵ is the unit ball of an equivalent dual norm $\|\cdot\|_\epsilon$ on X^*. Take now any sequence (y_n) in K_ϵ; every (y_n) may be written

$$y_n = \lambda_n u_n + (1-\lambda_n) v_n$$

for some $\lambda_n \in [0,1]$, with $u_n \in Y^\perp$, $v_n \in X^*$ such that $||u_n|| \leq 1$ and $||v_n|| \leq \epsilon$. It is clear that we have

$$\overline{\text{conv}}^*((y_n)) \subseteq \overline{\text{conv}}^*((u_n)) + \epsilon X_1^*.$$

But we know that there exists $y \in Y^\perp$ with $||y|| < \alpha^{-1}$ and $y \notin \overline{\text{conv}}^*((y_n))$; therefore we have

$$y' = (1+\epsilon)y \notin \overline{\text{conv}}^*((u_n))$$

clearly, $||y'|| < \alpha^{-1} + \epsilon$; and on $Y^\perp$, the norms $||\cdot||$ and $||\cdot||_\epsilon$ coincide. Since $y' \in Y^\perp$, we deduce that $||y'||_\epsilon < \alpha^{-1} + \epsilon$. Now, lemma III.1 shows that $d_\infty(X,||\cdot||_\epsilon \geq (\alpha^{-1}+\epsilon)^{-1}$.

Since $\epsilon \in (0,1)$ was arbitrary, this implies $\mu(X) \geq \alpha$; and this is true for any $\alpha < \mu(X/Y)$, thus we have $\mu(X) \geq \mu(X/Y)$.

■

Not surprisingly, the technique we used for showing III.2 is very similar to the interpolation technique of [2]. An immediate consequence of III.2 is the

Corollary III.3. Let X be a Banach space which has a closed subspace Y such that X/Y does not inject into $\ell^\infty(\mathbb{N})$. Then $\mu(X) = \infty$.

IV. Remarks – Questions.

IV.1: If X is a Banach space, let us consider the following cardinals:

$$\lambda_1(X) = \sup\{K \mid X \text{ contains a b.s. of cardinality } K\}$$

$$\lambda_2(X) = \inf\{K \mid \forall Y \subseteq X,\ X/Y \hookrightarrow \ell^\infty(K)\}$$

Theorem II.1 shows in particular that if X has the property ($\mathscr{C}$), $\lambda_1(X) = \lambda_2(X)$; indeed II.1(i) shows that $\lambda_1(X) \geq \lambda_2(X)$ is always true. The equation $\lambda_1(X) = \lambda_2(X)$ is easily checked if $\lambda_1(X) \leq \aleph_0$; and if $\lambda_1(X) > \aleph_0$, $\lambda_1(X) = \lambda_2(X)$ follows from II.1(ii). It is not clear to us whether the equation always holds:

Question IV.1: Is it true that $\lambda_1(X) = \lambda_2(X)$ for any Banach space X?

It follows from the results of §II that if a counterexample exists within the subspaces of $\ell^\infty(\mathbb{N})$, this space must be very pathological.

IV.2: If a Banach space X contains an uncountable b.s (x_α, x_α^*) $(\alpha \in I)$ then X contains a closed bounded convex set C – namely, $C = \overline{\text{conv}}^{\|\ \|}\{x_\alpha \mid \alpha \in I\}$ such that any point is a support point; that is, for every $x \in C$, there exists $x^* \in X$ such that $x^*(x) = \sup_C (x) > \inf_C(x)$ ([14], [13]). Hence by ([24], [4]), such a set exists in any non–separable dual space. It is not known whether any non–separable Banach space X contains such a set. What if X is the $\mathscr{C}(K)$–space of Kunen [12]?

IV.3: It is an instructive exercise to show that if X is the $\mathscr{C}(K)$–space of Kunen and Γ is any set, then any operator from X into $Y = c_0(\Gamma)$ or from $Y = c_0(\Gamma)$ into X has a separable image.

Question IV.4: What are the spaces which satisfy the equivalence of Theorem II.10? Let us observe that by [22], $\ell^\infty(\mathbb{N})$ satisfies II.10.

Question IV.5: Does there exist a non–separable Banach space X such that every quotient of X is isomorphic to a subspace of $\ell^\infty(\mathbb{N})$? We do not know whether or not Kunen's space satisfies this stronger property.

Question IV.6: If X is a non–separable ω^*–analytic subspace of $\ell^\infty(\mathbb{N})$, does there exist a separable subspace Y of X such that $X/_Y$ does not inject into $\ell^\infty(\mathbb{N})$? This is true by ([6],§VII) and ([7], Theorem I.8) if X is the dual of a space not containing $\ell^1(\mathbb{N})$; this is also true of course if $X = \ell^\infty(\mathbb{N})$ – take $Y = c_0(\mathbb{N})$.

Question IV.7: If X is a ω^*–analytic subspace of $\ell^\infty(\mathbb{N})$ and $X \not\supset \ell^1(c)$, does X contain a b.s. (n_α, n_α^*) $(\alpha \in I)$ such that $X = \overline{\mathrm{span}}'' \; '' \{x_\alpha \,|\, \alpha \in I\}$?

REFERENCES

[1] J. BOURGAIN, D.H. FREMLIN, M. TALAGRAND, Pointwise compact sets of Baire–measurable functions, Amer. J. Math., 100 (1978), 845–886.

[2] W.B. DAVIS, T. FIGIEL, W.B. JOHNSON, A. PELCZYNSKI, Factoring weakly compact operators, J. Funct. Analysis, 17 (1974), 311–327.

[3] W.B. DAVIS, W.B. JOHNSON, On the existence of fundamental and total biorthogonal systems in Banach spaces, Studia Math., 45 (1973), 173–179.

[4] M. FABIAN, G. GODEFROY, The dual of every Asplund space admits a projectional resolution of the identity, Studia Math., 91, 2 (to appear).

[5] C. FINET, G. GODEFROY, Representable operators, big quotient spaces and applications, preprint (1988).

[6] N. GHOUSSOUB, G. GODEFROY, B. MAUREY, W. SCHACHERMAYER, Some topological and geometrical structures in Banach spaces, Memoirs of the A.M.S. 378 (1987).

[7] N. GHOUSSOUB, B. MAUREY, H_δ–embeddings in Hilbert spaces and optimization of G_δ–sets, Memoirs of the A.M.S. 349 (1986).

[8] G. GODEFROY, Compact de Rosenthal, Pacific J. Maths, 91, 2 (1980), 293–306.

[9] G. GODEFROY, M. TALAGRAND, Espaces de Banach representables, Israel J. Maths, 41, 4 (1982), 321–330.

[10] G. GODEFROY, A. LOUVEAU, Axioms of determinacy and biorthogonal systems, to appear.

[11] R.C. JAMES, A separable somewhat reflexive Banach space with non–separable dual, Bull. Amer. Math. Soc. 80 (1974), 738–743.

[12] K. KUNEN, On hereditarily Lindelöf Banach spaces, Manuscript (1980).

[13] D. KURZAROVA, Convex sets containing only support points in Banach spaces with an uncountable minimal system, C.R. de l' Acad. Bulg. Sciences, 39–12 (1986), 13–14.

[14] V. MONTESINOS, On a problem of S. Rolewicz, Studia Math. 81, 1 (1985), 65–69.

[15] Y.N. MOSCHOVAKIS, Descriptive set theory, Studies in Logic, 100, North–Holland (1980).

[16] S. NEGREPONTIS, Banach spaces and topology, in Handbook of set–theoretic topology (ed. K. Kunen, J.E. Vaughan), North–Holland (1984).

[17] E. ODELL, H.P. ROSENTHAL, A double dual characterization of separable Banach spaces containing ℓ^1, Israel J. Maths, 20 (1975), 375–384.

[18] R.I. OVSEPIAN, A. PELCZYNSKI, The existence in every separable Banach space of a fundamental total and bounded biorthogonal sequence and related constructions of uniformly bounded orthonormal systems in L^2, Studia Math. 54 (1975), 149–159.

[19] A. PELCZYNSKI, All separable Banach spaces admit for every $\epsilon > 0$ fundamental and total biorthogonal sequences bounded by $1+\epsilon$, Studia Math., 55 (1976), 295–304.

[20] A.N. PLICHKO, On projectional resolutions, Markushevich basis, and equivalent norms (Russian), Mat. Zametki, 34 (1983), 719–726.

[21] R. POL, On a question of H.H. Corson and some related problems, Fund. Math., 109, 2 (1980), 143–154.

[22] H.P. ROSENTHAL, On relatively disjoint families of measures, with some applications to Banach space theory, Studia Math., 37 (1970), 13–36.

[23] I. SINGER, On biorthogonal systems and total sequences of functionnals II, Math. Ann., 201 (1973), 1–8.

[24] C. STEGALL, The Radon–Nikodym property in conjugate Banach space, Transactions of the A.M.S., 206 (1975), 213–223.

[25] G.A. SUAREZ, Some uncountable structures and the Choquet–Edgar property in non–separable Banach spaces, Proceedings of the 10th Spanish–Portuguese Conf. in Math. III, Murcia (1985), 397–406.

Catherine Finet
Université de l'Etat à Mons
Institut de Mathématiques
15, Avenue Maistriau
B–7000 MONS
BELGIUM

Gilles Godefroy
Université Paris VI
Equipe d'Analyse Tour 46–0
4, Place Jussieu
75252 PARIS CEDEX 05
FRANCE

Contemporary Mathematics
Volume **85**, 1989

Analytic Martingales and Plurisubharmonic Barriers in Complex Banach spaces

N. Ghoussoub J. Lindenstrauss B. Maurey

Abstract

We describe the geometrical structure on a complex Banach space X that is necessary and sufficient for the existence of radial limits for bounded X-valued analytic functions on the open unit disc in the complex plane. It is shown that in such spaces, closed bounded subsets have many plurisubharmonic barriers. The proofs rely on (analytic) martingale techniques.

1980 Mathematics Subject Classification: primary 46B20, 46E40, 60G46

Key words and phrases: Analytic functions and martingales - plurisubharmonic barrier points and functions.

1. Introduction

A central place in functional analysis is occupied by the study of convex sets and their extremal structure. The basic theorems in this direction are the separation theorem by linear functionals (Hahn-Banach theorem) and the Krein-Milman theorem. The original setting of the Krein-Milman theorem was that of compact convex sets. In the past decades there was a considerable effort to study versions of the Krein-Milman theorem in non compact settings. The main outgrowth of this research work is the theory of the so-called *Radon-Nikodym property* (R.N.P). It turned out that there is a remarkable class of Banach spaces, spaces having R.N.P, which can be characterized in several completely different ways. These spaces X can be characterized e.g. by the validity of the Radon-Nikodym theorem for X-valued measures, by the convergence of X-valued bounded martingales and by the validity of versions of the Krein-Milman theorem for (non compact) closed bounded convex sets in X. There are also structural results describing the spaces which have this property. The R.N.P theory is considered in detail in the book [D-U], the lecture notes [Bour] and the Memoir [G-M1].

It has been known for a long time that the theory of convexity has a natural non-linear analogue; the so-called *complex convexity* in which linear functionals are replaced by harmonic functions, convex functions by subharmonic functions etc. Also, the notions related to the Krein-Milman theorem (e.g. extreme points, exposed points) have natural generalizations in the setting of complex convexity. This circle of ideas is exposed e.g. in [Gam]. In section 2 below we shall recall the relevant definitions.

In [B-D] a study was started of the class of Banach spaces which satisfy the natural complex analogue of one of the properties which characterize R.N.P spaces. The basic definition is the following:

Definition. A complex Banach space X is said to have the *analytic Radon-Nikodym property* (A.R.N.P) if for every bounded analytic function f from the open unit disc D into $X, \lim_{r \uparrow 1} f(re^{i\theta})$ exists (in norm) for almost all θ.

It is easily seen that in the definition above we can replace the Hardy class $H^\infty(X)$ by any of the Hardy spaces $H^p(X), p > 0$. Another easy and well known remark is that if in the definition above the word "analytic" is replaced by "harmonic" we get a characterization of spaces having the R.N.P. As observed by Dowling [D], a space X has the A.N.R.P iff the vector valued version of the theorem of the brothers Riesz holds, i.e. every finite X-valued measure μ on [0,1] such that $\int_0^1 e^{2\pi i n x} d\mu(x) = 0$ for $n > 0$ has a Radon-Nikodym derivative with respect to the Lesbegue measure on [0,1].

Some examples will give an orientation on the class of spaces under discussion. Clearly any space with R.N.P (in particular any reflexive or separable conjugate space) has the A.R.N.P. Of the two main examples of spaces which fail to have the R.N.P (namely c_0 and $L_1(0,1)$) the first fails to have also the A.R.N.P (consider the function $f\colon D \to c_0$ defined by $f(z) = (1, z, z^2, \ldots, z^n, \ldots)$) while the second is known to have the A.R.N.P. As a matter of fact, a Banach lattice has the A.R.N.P iff it has no subspace isomorphic to c_0 [B-D]. The most important class of examples of spaces having the A.R.N.P but not the R.N.P are the non-commutative generalizations of L_1 namely the preduals of W^* algebras [H-P]. Some other known facts: Besides c_0 the space L_1/H^1 plays the role of the main example of a space failing A.R.N.P (taking the rôle L_1 had in the R.N.P theory). The space J_*T, the predual of the James tree space, is another example of a space having A.R.N.P but not R.N.P (cf. [G-M-S]).

Edgar [E1], [E2], initiated the study of the structure of spaces having A.R.N.P. His main achievement in those papers was to characterize A.R.N.P spaces X in terms of the convergence of some X-valued bounded martingales (so-called analytic martingales). The statement of his result is analogous to the characterization of R.N.P spaces by regular martingales but the proof in the A.R.N.P case is more involved. Subsequently other versions of martingale characterizations of A.R.N.P spaces were obtained (cf. [Gar] and in particular [B-S]). In the papers mentioned above Edgar also considered possible geometric characterizations of A.R.N.P spaces related to complex convexity versions of the Krein-

Milman theorem. In the present paper we obtain a geometric characterization of A.R.N.P spaces which parallels known results on R.N.P. In particular we answer some questions raised by Edgar. Our main result is the following.

Theorem (A). *Let X be a complex Banach space. The following properties are equivalent:*

1. X has the A.R.N.P.

2. Every closed and bounded subset of X is contained in the plurisubharmonic convex hull of its strong barrier points.

3. Every non-empty bounded subset of X has non empty plurisubharmonic slices with arbitrary small diameter.

4. Every bounded upper semicontinuous f on a closed bounded subset A of X can be perturbed by a plurisubharmonic function φ on X with arbitrary small maximum norm so that $f + \varphi$ strongly exposes A (and thus in particular attains its maximum on A).

As mentioned above the definitions of the relevant notions which are not self-evident will be given in section 2. The proof of the theorem will be given in Section 3. In its present formulation the theorem incorporates also a result of Bu [Bu] (in our original formulation, assertion 3 was relaced by another - slightly stronger - property, see section 3).

Since we deal here with non-linear and non-convex functions, Banach spaces are not really the natural setting for our study. The natural setting seems to be the larger class of linear spaces which admit a plurisubharmonic quasi-norm (the so-called A-convex spaces introduced by Kalton [K]). A detailed study of that setting will be carried out in [G-M2]. For the sake of brevity, we limit ourselves in this paper to the Banach space situation.

2. Definitions and preliminary remarks.

In this section we present the notions which are relevant to our study and make some remarks concerning these notions.

The basic object in our discussion is that of a *plurisubharmonic function*. Let X be a complex Banach space. A real-valued function f on X is said to belong to $PSH^1(X)$ if

$$f(x) \leq \frac{1}{2\pi}\int_0^{2\pi} f(x+e^{i\theta}y)d\theta \qquad x,y \in X \tag{2.1}$$

and

$$|f(x)-f(y)| \leq \|x-y\| \qquad x,y \in X. \tag{2.2}$$

The main condition is of course (2.1). The word plurisubharmonic function is used in the literature for functions f which satisfy (2.1) and are upper semi continuous. We find it more convenient to work with functions which satisfy a Lipschitz condition and in order to be specific we added to the definition of $PSH^1(X)$ the normalizing condition (2.2).

Note that the real part of every element in the unit ball of X^* belongs to $PSH^1(X)$ and so is the function $f(x) = \|x - x_0\|$ for every $x_0 \in X$. If $f, g \in PSH^1(X)$ so do $f \vee g$ and $(f+g)/2$.

There is a well established procedure for associating to a real-valued function f on X its plurisubharmonic envelope $\hat{f}$, i.e. the largest plurisubharmonic function which is bounded above by f (cf. e.g. [E1]). The procedure is the following: A sequence $\{f_n\}_{n=0}^{\infty}$ is defined inductively by putting $f_0(x) = f(x)$ and

$$f_{n+1}(x) = \inf_{y\in X} \frac{1}{2\pi}\int_0^{2\pi} f_n(x+e^{i\theta}y)d\theta \qquad n \geq 0. \tag{2.3}$$

The sequence $\{f_n\}_{n=0}^{\infty}$ is by definition a decreasing sequence of functions and its limit is $\hat{f}(\hat{f}$ and even f_n for $n > 0$ may take the value $-\infty$ at some points. However if f is bounded below by a constant, or more generally, by a plurisubharmonic function this cannot happen). In the situations we consider here $\hat{f}$ will always be finite. If f satisfies (2.2) then a trivial induction shows that all the $\{f_n\}_{n=0}^{\infty}$ and thus also $\hat{f}$ will satisfy (2.2): that is $\hat{f}$ belongs to $PHS(X)$.

If A is a closed and bounded subset of X we denote by $PSH^1(A)$ the functions which are restrictions to A of functions belonging to $PSH^1(X)$. We consider $PSH^1(A)$ as a subset of $C(A)$ the space of bounded continuous real-valued functions on A with the supremum norm.

Remark 1. *$PSH^1(A)$ is a closed subset of $C(A)$.*

Proof: We define a non-linear extension map $\psi: PSH^1(A) \to PSH^1(X)$ (i.e. $\psi(f)(x) = f(x)$ for $x \in A$) which satisfies

$$(2.4) \qquad |\psi(f)(x) - \psi(g)(x)| \leq \|f - g\|_{C(A)}$$

for all $f, g \in PSH^1(A)$ and $x \in X$. This, of course, will suffice since if $\{g_n\}_{n=1}^{\infty}$ is a Cauchy sequence in $PSH^1(A)$, the sequence $\psi(g_n)$ will converge pointwise on X to a function in $PSH^1(X)$ whose restriction to A is the limit of $\{g_n\}_{n=1}^{\infty}$.

To construct ψ consider a function $f \in PSH^1(X)$ and consider $\tilde{f}(x) = \inf_{y \in A}\{\|x - y\| + f(y)\}, x \in X$. The function $\tilde{f}$ agrees with f on A and satisfies (2.2). Moreover it is the largest function having these properties and in particular $\tilde{f}(x) \geq f(x)$ all $x \in X$. It is also evident that $|\tilde{f}(x) - \tilde{g}(x)| \leq \|f|_A - g|_A\|_{C(A)}$ for every $f, g \in PSH^1(X)$. The map ψ is defined by

$$\psi(f|_A) = \hat{\tilde{f}} \qquad \text{for } f \in PSH^1(X).$$

From the remarks above as well as the procedure which defines the envelope it is clear that (2.4) holds and that for $x \in X, \tilde{f}(x) \geq \hat{\tilde{f}}(x) \geq f(x)$. In particular $\hat{\tilde{f}}(x) = f(x)$ for $x \in A$.

□

With X and A as above we define *the set* $PSH(A)$ to be the closure in $C(A)$ of $\bigcup_{n=1}^{\infty} nPSH^1(A)$, i.e. those functions on A which can be approximated uniformly on A by functions which are restrictions on A of plurisubharmonic Lipschitz functions on X. The functions in $PSH(A)$ clearly satisfy (2.1) whenever x and $x + e^{i\theta}y, 0 \leq \theta \leq 2\pi$, belong to A. The set $PSH(A)$ is a closed cone in $C(A)$.

A Radon probability measure μ on X is said to be a *Jensen measure representing a point* $x_0 \in X$ if

$$f(x_0) \leq \int f(x) d\mu(x) \qquad f \in PSH^1(X). \tag{2.5}$$

It follows from (2.5) that $f(x_0) = \int f(x) d\mu(x)$ for every $f \in X^*$ and thus x_0 is the barycenter of the measure μ. The normalized Haar measure on the circle $x_0 + e^{i\theta}y, 0 \leq \theta \leq 2\pi$ is an obvious example of a Jensen measure representing x_0.

Let A be a bounded set in X. We define *the plurisubharmonic convex hull* $\hat{A}$ of A by

$$\hat{A} = \{x \in X, \hat{d}_A(x) = 0\} \tag{2.6}$$

where

$$d_A(y) = d(y, A) = \inf_{z \in A} \|y - z\|. \tag{2.7}$$

Obviously $d_A = d_{\bar{A}}$ and thus A and its closure have the same plurisubharmonic convex hull. The function d_A is the largest function which vanishes on A and satisfies (2.2). Hence it is possible to replace (2.6) by an equivalent definition which does not single out the particular function d_A

$$(2.6)' \hat{A} = \{x \in X, f(x) \leq 0 \text{ whenever } f \in PSH^1(X) \text{ and vanishes on } A\}.$$

Clearly the barycenter of every Jensen measure supported on A belongs to $\hat{A}$. For compact sets A the converse is true.

Remark 2. *Let A be compact set in a complex Banach space X. Then $x \in \hat{A} \Leftrightarrow x$ is the barycenter of a Jensen measure supported on A.*

Proof: Assume $x_0 \in \hat{A}$ and consider the convex cone $\mathcal{U}$ in $C(A)$ defined by

$$\mathcal{U} = \{g \in C(A); \exists f \in PSH^1(X), \exists \lambda > 0, f(x_0) = 0, g(y) \geq \lambda f(y) \quad \forall y \in A\}$$

By (2.6)', the function $-1 \notin \overline{\mathcal{U}}$. By the Hahn-Banach separation theorem and the Riesz representation theorem there is a Radon probability measure μ on A so that $\int_A f(x)d\mu(x) \geq 0$ for every $f \in PSH^1(X)$ which vanishes at x_0. This measure μ is a Jensen measure representing x_0.

□

A set A in a Banach space X is called *PSH-convex if* $A = \hat{A}$. Every closed convex bounded set is PSH convex.

Let again A be a closed and bounded subset of a complex Banach space X.

(i) A point x_0 is a *complex extreme point* of A if there is no non-zero vector $y \in X$ with

$$\{x_0 + re^{i\theta}y, 0 \leq r \leq 1, 0 \leq \theta \leq 2\pi\} \subset A.$$

(ii) A point x_0 is a *Jensen boundary point* of A if the dirac measure δ_{x_0} is the only Jensen measure supported on A which represents x_0.

(iii) The point x_0 is called a *PSH-denting point of A* if for every $\varepsilon > 0, x_0 \notin \hat{A}_\varepsilon$ where $A_\varepsilon = A \backslash B(x_0, \varepsilon)$ and $B(x_0, \varepsilon) = \{y \in X; \|y - x_0\| \leq \varepsilon\}$. Alternatively, x_0 is a PSH-denting point of A if there are PSH-slices of A containing x_0 and of arbitrary small diameter. *A PSH-slice S of A* being a non-empty set of the form $S = A \cap \{x; f(x) > 0\}$ with $f \in PSH^1(X)$.

(iv) A point x_0 is said to be a *barrier* (resp. *strong barrier*) of A if there is an $f \in PSH^1(X)$ which *exposes* (resp. *strongly exposes*) A at x_0, i.e.

(i) $f(x_0) = \sup_{x \in A} f(x)$

(ii) $f(x) < f(x_0)$ for every $x \neq x_0$ in A (resp. $\sup_{x \in A_\varepsilon} f(x) < f(x_0)$ for every $\varepsilon > 0$).

Such a function f will be called a *PSH-barrier* (resp. *PSH-strong barrier*) for A.

Note that any real extreme point of a set is also a complex extreme point. The unit ball of $L_1(0,1)$ is an example of a set which fails to have real extreme points but all its boundary points are complex extreme. It is clear that a barrier point is a Jensen boundary point which in turn is a complex extreme point. Every strong barrier point is a PSH denting point.

The same standard argument as that used to prove remark 2 proves also the following well known

Remark 3. *Let A be a compact set in a complex Banach space and let x_0 be a Jensen boundary point of A. Then for any $g \in C(A)$.*

$$g(x_0) = \sup\{\lambda f(x), f \in PSH^1(X), \lambda > 0, g(y) \geq \lambda f(y) \quad \forall y \in A\}. \tag{2.8}$$

Proof: We may assume that $g(x_0) = 0$. If (2.8) fails then g does not belong to the closure of the convex cone

$$\mathcal{U} = \{h \in C(A), \exists f \in PSH^1(X), f(x_0) = 0, \exists \lambda > 0, h(y) \geq \lambda f(y) \quad \forall y \in A\}.$$

By the separation theorem there is a Radon probability measure μ on A so that $\int f(x)d\mu(x) \geq 0$ for all $f \in PSH^1(X)$ which vanish at x_0 and $\int g(x)d\mu(x) < 0$. This in turn implies that μ is a Jensen representing measure of x_0, i.e. $\mu = \delta_{x_0}$ and this contradicts the fact that $g(x_0) = 0$. □

The last notion we need is that of an analytic martingale introduced in [D-G-T]. An *analytic martingale* is a sequence $\{M_n\}_{n=0}^{\infty}$ of X-valued random variables defined on $\Omega = [0, 2\pi]^N$ - with the usual normalized measure - which are of the form

$$M_n(\theta) = h_0 + \sum_{k=1}^{n} h_k(\theta_1, \theta_2, \ldots, \theta_{k-1})e^{i\theta_k} \qquad n = 0, 1, 2, \ldots \tag{2.9}$$

where $\theta = (\theta_1, \theta_2, \ldots) \in \Omega, h_0 \in X, h_k : [0, 2\pi]^{k-1} \to X$ are bounded random variables. We shall also consider finite analytic martingales $\{M_n\}_{n=0}^{m}$ which are defined in an obvious way.

The theorem of Edgar [E2] mentioned in the introduction states the following: *A complex Banach space X has the A.R.N.P if and only if every X-valued analytic martingale $\{M_n\}_{n=0}^{\infty}$ such that $\sup_n \|M_n\|_{L_1(\Omega, X)} < \infty$ converges a.e. on Ω.*

Finally we need the following slight variant of a Lemma in [E1],

Remark 4. *Let f be a real-valued function on X satisfying (2.2) and let $\{f_n\}_{n=0}^{\infty}$ be defined by (2.3). Then for every n and $x \in X$*

$$f_n(x) = \inf Ef(M_n) \tag{2.10}$$

where E denotes the expectation and the inf is over all M_n of the form (2.9) with $h_0 = x$ and all the functions h_k are simple (i.e. finitely valued).

Proof: It is clear that for every M_n as above $f_n(x) \le Ef(M_n)$. To prove the converse, take any $\varepsilon > 0$ and let y be such that $f_n(x) > \frac{1}{2\pi}\int_0^{2\pi} f_{n-1}(x + e^{i\theta}y)d\theta - \varepsilon/2$. We put $h_1 = y$. Since all the f_k's satisfy (2.2), it follows that there is a simple function $h_2(\theta_1) : [0, 2\pi] \to X$ so that for all $\theta_1 \in [0, 2\pi]$

$$f_{n-1}(x + e^{i\theta_1}y) > \frac{1}{2\pi}\int_0^{2\pi} f_{n-2}(x + e^{i\theta_1}y + e^{i\theta_2}h_2(\theta_1))d\theta_2 - \varepsilon/4.$$

We continue in an obvious manner and find simple $\{h_k\}_{k=1}^{n}$ so that the M_n they define via (2.9) satisfies $Ef(M_n) \le f_n(x) + \varepsilon$. □

3. Proof of the Theorem

In this section we prove the theorem stated in the introduction. We shall first however consider the case of compact sets. In this case, the result holds in general Banach spaces and two of its three assertions are already well known. It is of interest to compare the situation for compact sets with that of closed bounded sets (where the theorem provides characterizations of A.R.N.P spaces).

Proposition (B). *Let K be a compact set in a complex Banach space X. Then*

a. *Every function in $PSH(K)$ attains its maximum on K at a Jensen boundary point of K.*

b. *Every Jensen boundary point of K is exposed by a function in $PSH(K)$.*

c. *K is contained in the plurisubharmonic convex hull of its barrier points.*

Proof: Part a) is proved by the usual argument used in the proof of the Krein-Milman theorem. A closed subset F of a set A is called a J-face of A if every Jensen measure supported on A which represents a point in F must be supported already on F. One verifies easily that if $f \in PSH(K)$ then the set where f attains its maximum on K is a J-face of K, every J-face of K contains a minimal J-face and that every minimal J-face is a single point i.e. a Jensen boundary point of K.

Part b) follows by an argument due essentially to E. Bishop. The details are presented in ([Gam] pages 14-16). If x_0 is a Jensen boundary point of K then by remark 2 in section 2, for every closed set E of K such that $x_0 \notin E$ we have also $x_0 \notin \widehat{E}$. Using this fact and remark 3 the argument in [Gam] produces for every $h \in C(K)$ an $f \in PSH(K)$ such that $f(x_0) = h(x_0)$ and $f(x) \leq h(x)$ for all $x \in K$. If h exposes K at x_0 the same will be true for f.

Proof of c). Since $\hat{d}_A \in PSH^1(K)$, it is clearly enough to show that the barrier functions are dense in $PSH^1(K)$. Since $PSH^1(K)$ is a complete metric space (remark 1) it suffices, by Baire's theorem, to show that for every $\varepsilon > 0$

$$G_\varepsilon = \{g \in PSH^1(K), diam\ S(K,g,\tau) \leq \varepsilon \text{ some some } \tau > 0\}$$

is open and dense in $PSH^1(K)$ where

$$S(K,g,\tau) = \{x \in K, g(x) > \sup_{y \in K} g(y) - \tau\}.$$

It is evident that G_ε is open. We only have to verify that it is dense. Let O be a non-empty open set in $PSH^1(K)$. We cover K by a finite union of balls of radius $\varepsilon/4$. For at least one of these balls say $B = B(x_1, \varepsilon/4)$ it is true that the set of function in O which attain their maximum on K at a point of B has a non empty interior. Let g_0 be such an interior point, i.e. for some $\alpha > 0$ every $f \in PSH^1(K)$ with $\|f - g_0\|_{C(K)} < \alpha$ belongs to O and attains its maximum on K at a point of B. We claim that $g_0 \in G_\varepsilon$. Let $\tau > 0$ and consider the slice $S(K, g_0, \tau)$. If this slice has a diameter larger than ε it must contain a point $x_2 \in K$ with $\|x_2 - x_1\| \geq \varepsilon/2$. The function $h(x) = (\|x - x_1\| - \varepsilon/4) \vee 0$ belongs to $PSH^1(X)$, vanishes on B and satisfies $h(x_2) \geq \varepsilon/4$. The function $(1 + 8\tau/\varepsilon)^{-1}(g_0 + (8\tau/\varepsilon)h)$ belongs to $PSH^1(X)$. If τ is small enough its distance from g_0 (in the norm induced by $C(K)$) is smaller than α and thus this function attains its maximum on K at a point in B. However

$$\max_{y \in B \cap K}(g_0(y) + 8\tau h(y)/\varepsilon) = \max_{y \in B \cap K} g_0(y) = \max_{y \in K} g_0(y)$$

while

$$g_0(x_2) + 8\tau h(x_2)/\varepsilon \geq \max_{y \in K} g_0(y) - \tau + 2\tau = \max_{y \in K} g_0(y) + \tau$$

and we are led to a contradiction. □

We pass now to our main theorem. We state it again mainly in order to add to it still another equivalent condition.

Theorem (A'). *Let X be a complex Banach space. The following are equivalent.*

1. *X has the A.R.N.P.*

2. *Every closed and bounded subset of X is contained in the plurisubharmonic convex hull of its strong barrier points.*

3. *Every non empty bounded subset of X has PSH-slices of arbitrary small diameter.*

4. *For every bounded upper semi continuous real-valued function f on a closed and bounded subset A of X and for every $\varepsilon > 0$ there is a $g \in PSH^1(X)$ with $\|g\|_{C(A)} < \varepsilon$ so that $f + g$ strongly exposes A.*

5. *Every closed and bounded subset of X is contained in the plurisubharmonic convex hull of its PSH-denting points.*

Proof: It is evident that (2) $\Rightarrow$ (5) $\Rightarrow$ (3). Also, by applying (4) to f with $f \in \lambda PSH^1(A)$ for some $\lambda > 0$, it follows that the strong barriers are dense in $PSH^1(A)$ and hence (4) $\Rightarrow$ (2). It remains to prove that (1) $\Rightarrow$ (4) and that (3) $\Rightarrow$ (1).

The assertion that (3) $\Rightarrow$ (1) is due to Bu [Bu] and we shall not reproduce its proof here. Instead we present here our original proof of the formally weaker implication (5) $\Rightarrow$ (1). In order to do this we formulate an additional property.

(*) *For every $\varepsilon > 0$ and every bounded set A in X not contained in the ball $B(0, \varepsilon/2) = \{x; \|x\| \le \varepsilon/2\}$ there is a PSH-slice S of A of diameter $\le \varepsilon$ so that $S \cap B(0, \varepsilon/2) = \emptyset$.*

It is easy to see that (5) $\Rightarrow$ (*). Indeed, if $A \not\subset B(0, \varepsilon/2)$ and (5) holds, then since $B(0, \varepsilon/2)$ is PSH-convex, $\overline{A}$ must have a PSH-denting point x_0 with $\|x_0\| > \varepsilon/2$. If $\eta < \min(\varepsilon, \|x_0\| - \varepsilon/2)$ any non empty PSH-slice S of A containing x_0 and of diameter $\le \eta$ will satisfy the requirements of (*).

We show now that (*) $\Rightarrow$ (1). Assume that (*) holds and that $F: D \to X$ is an analytic function such that $\|F(z)\| \le 1$ for every $z \in D$. By a direct exhaustion argument it is clear that in order to show that $\lim\limits_{r \uparrow 1} F(re^{i\theta})$ exists a.e. on $[0, 2\pi]$ it is enough to show that for every $\varepsilon > 0$ and every measurable subset Δ of $[0, 2\pi]$ of positive measure there is a subset $\Delta' \subset \Delta$ of positive measure so that

$$\limsup_{r,r'\uparrow 1} \|F(re^{i\theta}) - F(r'e^{i\theta})\| \le \varepsilon \qquad \text{a.e. on } \Delta'. \tag{3.1}$$

Let $\varepsilon > 0$ and Δ be given. Let $H(z)$ be a complex valued outer function on $\overline{D}$ so that $|H(e^{i\theta})| = 1$ a.e. on Δ and $|H(e^{i\theta})| = \varepsilon/2$ a.e. on $[0, 2\pi]\backslash\Delta$ (cf. e.g. [Ko] for the existence of such a function). Let

$$A = H \cdot F(D) \subset X.$$

If $A \subset B(0, \varepsilon/2)$ then $\limsup_{r\uparrow 1} \|F(re^{i\theta})\| \le \varepsilon/2$ for almost all $\theta \in \Delta$ and (3.1) holds with $\Delta' = \Delta$. If $A \not\subset B(0, \varepsilon/2)$ then by $(*)$ there is a $g \in PSH^1(X)$ so that $S = A \cap \{x; g(x) > 0\}$ is non empty, does not meet $B(0, \varepsilon/2)$ and is of diameter $\le \varepsilon$. The real-valued subharmonic function $g \circ (H \cdot F)$ on D has a radial limit $\psi(\theta) = \lim_{r\uparrow 1} g(H \cdot F(re^{i\theta}))$ a.e. on $[0, 2\pi]$. Since S does not meet $B(0, \varepsilon/2)$ and $\limsup_{r\uparrow 1} \|H \cdot F(re^{i\theta})\| \le \varepsilon/2$ a.e. on $[0, 2\pi]\backslash\Delta$ it follows that $\psi(\theta) \le 0$ a.e on $[0, 2\pi]\backslash\Delta$. Since S is non empty $g \circ (H \cdot F)$ takes positive values on D and thus since it is subharmonic and bounded the set $\Delta' = \{\theta; \psi(\theta) > 0\}$ is of positive measure. Finally, since the diameter of S is $\le \varepsilon$, (3.1) holds for every $\theta \in \Delta'$. This concludes the proof of $(*)$ and thus (5) $\Rightarrow$ (1).

We pass to the main assertion of the theorem namely that (1) $\Rightarrow$ (4). The proof of this implication uses some ideas in Bourgain's proof of the optimization theorem for R.N.P sets (see [Bour]). It is based on the following:

Lemma (5). *Assume the space X has the A.R.N.P, A is a closed bounded set in X and f a bounded function on A. For every $g \in PSH^1(A)$ and $\tau > 0$ put*

$$S(A, f + g, \tau) = \{x \in A; f(x) + g(x) \ge \sup_{y\in A}(f(y) + g(y)) - \tau\}. \tag{3.2}$$

For $\eta > 0$, put further

$$\delta_f(\eta) = \inf\{diam\ S(A, f + g, \tau), \tau > 0, g \in PSH^1(A), \|g\|_{C(A)} \le \eta\}. \tag{3.3}$$

Then $\delta_f(\eta) = 0$ *for all* $\eta > 0$.

Once the Lemma is proved the proof of (1) $\Rightarrow$ (4) is concluded as follows. For $\eta > 0$ the set

$$V_\eta = \{g \in PSH^1(A), \|g\|_{C(A)} \leq \eta\}$$

is a complete metric space (by Remark 1). By Lemma (5) the set

$$O_n = \{g \in V_\eta, \exists \tau > 0 \text{ so that } diam\ S(A, f+g) < n^{-1}\}$$

is non empty. O_n is clearly an open subset of V_η. It is also a dense set. Indeed if $g_0 \in V_\eta$ and $\varepsilon > 0$ we apply the Lemma to the function $\varepsilon^{-1}(f + (1-\varepsilon)g_0)$ to find a $g \in V_\eta$ and a $\tau > 0$ so that

$$diam\ S(A, \varepsilon^{-1}(f + (1-\varepsilon)g_0) + g, \tau) = diam\ S(A, f + (1-\varepsilon)g_0 + \varepsilon g, \varepsilon\tau) < n^{-1}.$$

The function $(1-\varepsilon)g_0 + \varepsilon g$ belongs thus to O_n and this proves the denseness of O_n. By the Baire category theorem $\bigcap_{n=1}^{\infty} O_n$ is a dense G_δ subset of V_η and in particular is non empty. Since f is upper semi continuous the sets $S(A, f+g, \tau)$ are all closed and thus for every $g \in \bigcap_{n=1}^{\infty} O_n$, $f+g$ strongly exposes A.

Proof of Lemma (5): We assume as we clearly may that $A \subset B(0,1)$. If the Lemma were false there would exist some $0 < \eta_0 \leq 1/2$ so that $\delta_f(\eta) \geq 4r$ for some $r > 0$ and all $0 < \eta \leq \eta_0$.

For $\tau > 0, \eta > 0, \lambda > 0$ put

$$D_f(\tau, \eta, \lambda) = \cup\{S(A, f+g, \tau), \|g\|_{C(A)} \leq \eta, g \in \lambda PSH^1(A)\}. \tag{3.4}$$

We claim first that if $\tau, \eta, \lambda, \gamma$ and α are all positive such that $\eta + 2\alpha \leq \eta_0, \lambda + \alpha \leq 1$ then

$$\hat{d}_E(x) \leq (\tau+\gamma)/\alpha \qquad \text{for} \qquad x \in D_f(\tau,\eta,\lambda) \tag{3.5}$$

where

$$E = D_f(\gamma, \eta+2\alpha, \lambda+\alpha)\backslash B(x,r).$$

Note that since $\eta+2\alpha \leq \eta_0$ and $\lambda+\alpha \leq 1$ it follows from (3.3) and the definition of r that each summand in $D_f(\gamma,\eta+2\alpha,\lambda+\alpha)$ is a set of diameter $\geq 4r$ and thus in particular E is non empty.

Recall that $\hat{d}_E \in PSH^1(X)$ and vanishes on E. Since $A \subset B(0,1)$ it follows that $\|\hat{d}_E\|_{C(A)} \leq 2$. Let $x \in S(A, f+g, \tau)$ for some $g \in \lambda PSH^1(A)$ with $\|g\|_{C(A)} \leq \eta$. Then $g + \alpha\hat{d}_E \in (\lambda+\alpha)PSH^1(A)$ and $\|g+\alpha\hat{d}_E\|_{C(A)} \leq \eta+2\alpha$. As observed above we can find a

$$y \in S(A, f+g+\alpha\hat{d}_E, \gamma)\backslash B(x,r) \subset E.$$

Then

$$\begin{aligned}\alpha\hat{d}_E(x) + \sup_{u\in A}(f(u)+g(u)) - \tau &\leq \alpha\hat{d}_E(x) + f(x) + g(x) \leq \\ &\leq \sup_{u\in A}(f(u)+g(u)+\alpha\hat{d}_E(u)) \leq f(y)+g(y)+\alpha\hat{d}_E(y)+\gamma \\ &= f(y)+g(y)+\gamma \leq \gamma + \sup_{u\in A}(f(u)+g(u))\end{aligned}$$

and this implies (3.5).

We let now $\{\varepsilon_k\}_{k=1}^{\infty}$ be a decreasing sequence of positive numbers so that $\sum_{k=1}^{\infty} \varepsilon_k = \eta_0/2$ and consider the subsets

$$D_n = D_f(\varepsilon_{n+1}^2, 2\sum_{k=1}^{n}\varepsilon_k, \sum_{k=1}^{n}\varepsilon_k) \qquad n = 1,2,\ldots$$

of A. By (3.5)

$$\hat{d}_{D_{n+1}\setminus B(x,n)}(x) \leq 2\varepsilon_n \quad \text{for} \quad x \in D_n, n = 1, 2, \dots \tag{3.6}$$

Using (3.6) we shall constuct an L_1-bounded X-valued analytic martingale which does not converge and thus arrive at a contradiction.

We start with some $x_0 \in D_1$. By Remark 4 there is an integer m_1 and simple functions $\{h_k(\theta_1, \theta_2, \dots, \theta_{k-1})\}_{k=0}^{m_1}$ (with $h_0 \equiv x_0$) so that if $\{M_n\}_{n=0}^{m_1}$ are defined by (2.9) then

$$E\{d_{D_2 \setminus B(x_0,r)}(M_{m_1})\} < 3\varepsilon. \tag{3.7}$$

(3.7) means that there is a simple function $Z_1: [0, 2\pi]^{m_1} \to D_2$ so that $\|Z_1(\theta) - x_0\| \geq r$ for every θ and so that $\|Z_1(\theta) - M_{m_1}(\theta)\|_{L_1(\Omega)} \leq 3\varepsilon_1$. By using again (3.6) and Remark 4 we can find an $m_2 > m_1$ and simple functions $\{h_k\}_{k=m_1+1}^{m_2}$ so that if M_{m_2} is defined by (2.9), then

$$E\{d_{D_3 \setminus B(Z_1,r)}(M_{m_2})\} < 3\varepsilon_2.$$

Continuing in an obvious manner we construct an analytic X-valued martingale $\{M_n\}_{n=0}^{\infty}$, a sequence of integers $\{m_j\}_{j=1}^{\infty}$ and simple A-valued random variables Z_j on $\Omega = [0, 2\pi]^N$ so that

$$\|Z_j(\theta) - Z_{j-1}(\theta)\| \geq r, \qquad \text{for all} \quad j \text{ and } \theta \tag{3.8}$$

and

$$\|Z_{j+1} - Z_j - (M_{m_{j+1}} - M_{m_j})\|_{L_1(\Omega)} \leq 3\varepsilon_j, \ j = 1, 2, \dots . \tag{3.9}$$

Since $\|Z_j(\theta)\|_{L_1(\Omega)} \leq 1$ for all j and $\sum_{j=1}^{\infty} \varepsilon_j < \infty$ it follows from (3.9) that

$$\sup_m \|M_m\|_{L_1(\Omega)} = \sup_j \|M_{m_j}\|_{L_1(\Omega)} < \infty.$$

On the other hand it follows from (3.8) and (3.9) that $\{M_m(\theta)\}_{m=1}^{\infty}$ fails to converge for almost every $\theta \in \Omega$. □

Final remark: It is natural to ask whether in the setting of Theorem (A), one can actually obtain arbitrarily small *pluriharmonic* - as opposed to plurisubharmonic - slices of the closed bounded subsets of the Banach space X. Unfortunately, the example of the space L^1 gives a negative answer. Indeed, as mentioned above, L^1 has the A.R.N.P. On the other hand, every pluriharmonic function φ (i.e. φ and $-\varphi$ are in $PSH(X)$) on L^1 is the real part of a holomorphic function (see [C] p. 229) and, hence it is necessarily weakly continuous. To see that, it is enough to notice that *monomials* on L^1 are of the form:

$$p(f_1, f_2, \ldots, f_n) = \int\int\cdots\int f_1(x_1)f_2(x_2)\ldots f_n(x_n)K(x_1, x_2, \ldots, x_n)dx_1dx_2\ldots dx_n$$

where $K(x_1, x_2, \ldots, x_n)$ is an L^∞-bounded kernel.

It follows that slices determined by pluriharmonic functions give rise to weak neighborhoods in L^1. Since the unit ball of L^1 has no *"points of weak to norm continuity"*, one cannot expect to find pluriharmonic slices of arbitrarily small norm diameter. The question of when the slices can be determined by weakly continuous plurisubharmonic or pluriharmonic functions will be studied in [G-M-S].

REFERENCES:

[Bour] J. Bourgain: *La propriété de Radon-Nikodym,* Cours de 3° cycle, Université Paris VI, France (1979).

[B-S] S. Bu, W. Schachermayer: In preparation (1988).

[Bu] S. Bu: *Quelgues remarques sur la propriété de Radon-Nikodym analytique.* C.R. Acad. Sci. Paris 306 (1988) 757-760.

[B-D] A.V. Bukhvalov and A.A. Danilevich: *Boundary properties of analytic and harmonic functions iwth values in Banach spaces.* Math Zametki 31 (1982) 203-214; English translation: Math Notes 31 (1982) 104-110.

[C] S.B. Chae: *Holomorphy and Calculus in normed spaces,* Pure and Applied Mathematics, Marcel Dekker, Inc. - New York and Basel (1985).

[D-G-J] W.J. Davis, D.J.H. Garling, N. Tomczak-Yaegermann: *The complex convexity of complex quasi-normed linear spaces.* J. Func. Anal. **55** (1984) 110-150.

[D-U] J. Diestel, J.J. Uhl: *Vector measures.* Math Surveys, **15** A.M.S. (1977).

[D] P.M. Dowling: *Representable operators and the analytic Radon-Nikodym property in Banach spaces.* Proc. Royal. Irish. Acad. 85A (1985), pp. 143-150.

[Du] R. Durett: *Brownian motion and Martingales in Analysis,* Wadsworth (1984).

[E1] G.A. Edgar: *Complex martingale convergence,* Springer-Verlag, Lecture Notes (1116) pp. 38-59 (1985).

[E2] G.A. Edgar: *Analytic martingale convergence,* J. Funct. Analysis, **69**, No. 2 (1986), pp. 268-280.

[Gam] T.W. Gamelin: *Uniform algebras and Jensen measures,* Cambridge University Press - Lecture notes series (32), (1978).

[Gar] D.J.H. Garling: *"On martingales with values in a complex Banach space",* (Preprint) (1987).

[G-M1] N. Ghoussoub, B. Maurey: *"H_δ-embedding in Hilbert space and Optimization on G_δ-sets".* Memoirs of A.M.S. **62**, No. 349 (1986).

[G-M2] N. Ghoussoub, B. Maurey: *Plurisubharmonic Martingales and Barriers in Complex Quasi-Banach Spaces.* (To appear) (1988).

[G-M-S] N. Ghoussoub, B. Maurey, W. Schachermayer: *"Pluriharmonically dentable complex Banach spaces".* (To appear) (1988).

[H-P] U. Haagerup, G. Pisier: *Factorization of analytic functions iwth values in non-commutative L^1-spaces and applications* - Preprint (1988).

[K] N.J. Kalton: *Plurisubharmonic functions on quasi-Banach spaces,* Studia Mathematica, TLXXXIV (1986) pp. 297-324.

[Ko] P. Koosis: *Lectures on H^p-spaces.* LMS Lecture notes. Cambridge University Press, Cambridge (1980).

N. Ghoussoub
Department of Mathematics
University of British
Columbia, Vancouver, B.C.,
Canada, V6T 1Y4

J. Lindenstrauss
Department of Mathematics
Hebrew University of
Jerusalem
Jerusalem, Israel

B. Maurey
U.F.R. de Mathématiques
Université Paris VII
Paris, France

Contemporary Mathematics
Volume **85**, 1989

EXISTENCE AND UNIQUENESS OF ISOMETRIC PREDUALS: A SURVEY

Gilles Godefroy

Equipe d' Analyse
Université Paris VI
Tour 46–0.4éme étage
4, Place Jussieu
75252 PARIS CEDEX 05
FRANCE

University of Missouri – Columbia
Department of Mathematics
Columbia, MO 65211
U.S.A.

TABLE OF CONTENTS

1980 Mathematics Subject Classification (1985 Revision): Primary 46B20

Key Words and phrases: Hahn–Banach theorem, existence and uniqueness of preduals, the ball topology, ω^*–continuity.

0. Introduction – Notation.

In the mid 1950's, A. Grothendieck [33] showed the following results:

(1) If Y is an L^1–space and if Z is a Banach space such that Z^* is isometric to Y^*, then Z is isometric to Y.

(2) If X is isometric to a $\mathscr{C}(K)$ space, then X is isometric to a dual space if and only if K is hyperstonean.

(2) is a converse of an earlier result of Dixmier [14]: K hyperstonean $\Rightarrow \mathscr{C}(K)$ dual space, the predual being the L^1 space of normal Radon measures on K.

In this paper we will say that Y is *unique predual* if it satisfies: Y^* isometric to $Z^* \Rightarrow Y$ isometric to Z. We will say that X is a *dual space* if there is a Banach space Y such that Y^* is *isometric* to X. In the spirit of the early results of Grothendieck, we will seek conditions that imply that a Banach space Y is unique predual, or non trivial conditions that imply that a Banach space X is a dual space.

Not much was done along these lines during the twenty years that followed Grothendieck's discovery, with the notable exception of the extension of his results of the non–commutative case: a predual W_* of a von Neumann algebra W is unique predual; a $\mathbb{C}^*$–algebra is a dual space if and only if it is a von Neumann algebra (see [15]). In the mid 1970's, important discoveries were made about the duality between Banach spaces, such as the duality Asplund vs. Radon–Nikodym (see [49]) or the theorem of Odell–Rosenthal [43]. Not surprisingly, these discoveries were followed by some progresses on isometric duality; we will display in this survey most of the results which were obtained in the last decade.

The reader will see that some non–trivial positive results are now available, although natural and important questions remain open. My impression is that the existing techniques are now almost exhausted, and that new ideas are needed for going further; hence this seems to be an appropriate time for writing a survey.

Most of the results presented here have already been published – frequently in French – although some results and many remarks are published here for the first time; this is in particular the case for most of the section VII and VIII. This survey is of course not comprehensive; moreover some of the results we are presenting are not the optimal ones; however they cover the natural applications and have proofs which are sometimes much easier than those of the optimal results. The reader will find in this survey most of the existing techniques. Let me mention that these techniques are qualitative; each time this will be possible, we will use a few words of explanation instead of mathematical formulas.

We deal in this article with isometric unique preduals. The corresponding isomorphic question: "when is it true that Y^* isomorphic to Z^* implies Y isomorphic to Z?" uses completely different techniques and there are only a few results ([7]) which we will not discuss here. Of course it will frequently happen that a space is unique isometric predual for every equivalent norm, but this is a distinct question.

Notation: If Y is a subspace of the dual space X^*, $\sigma(X,Y)$ is the topology on X of the pointwise convergence on Y. We denote by ω the weak topology $\sigma(X,X^*)$ and by ω^* the weak* topology $\sigma(X^*,X)$. The balls we consider are assumed to be norm–closed; the unit ball of E is denoted by E_1. A space is identified without special notation to a subspace of its bidual; hence the canonical decomposition of the third dual Y^{***} is denoted by $Y^{***} = Y^* \oplus Y^{\perp}$. "Operator" means "bounded linear map", and "subspace" means "norm closed vector subspace". If (y_α) is a net in a dual space Y^*, we denote by $\omega^*\text{–}\lim(y_\alpha)$ the limit of this net in (Y^*,ω^*) when it converges; the closure of a subset C of Y^* into (Y^*,ω^*) is $\bar{C}^*$.

We denote by $x,x'\dots$ – respectively, $y,y',\dots$, $z,z',\dots$, $t,t',\dots$ – the elements of a Banach space Y – respectively of Y^*, Y^{**}, Y^{***}. When we investigate whether a space has a predual – resp. is unique predual – we call this space X – resp. Y.

A space Y is said to be *strongly unique predual* if there is a unique projection π of norm one from Y^{***} onto Y^* with $\mathrm{Ker}(\pi)$ ω^*–closed (namely, the projection of kernel $Y^\perp$). Clearly strongly unique predual implies unique predual (with a "unique position" of the predual) and the converse is an open questions (§X(2)). Each time we show that a space is unique predual, what is actually shown is that it is strongly unique predual; we will not repeat it in the statements.

Acknowledgements: The idea of writing this survey was conceived while I was visiting the University of Iowa at Iowa City in April 1988. It is my pleasure to thank the University of Iowa, and in particular Professor Bor–luh Lin, for this invitation and for the excellent typing of this article.

I. Applications of the Hahn–Banach theorem: some basic lemmas.

We start this section by gathering a few lemmas which we will find useful. Our main tool for showing them will be the Hahn–Banach theorem.

Lemma I.1: Let X be a Banach space and Y be a subspace of X^*. the following are equivalent:

(i) $\forall x \in X,\ \|x\| = \sup\{y(x) \mid y \in Y,\ \|y\| \le 1\}$

(ii) Y_1 is dense in (X_1^*, ω^*)

(iii) $Y^{\perp}$ is contained in the set

$$0_X = \{z \in X^{**} \mid ||z-x|| \geq ||x|| \ \forall x \in X\}$$

(iv) Every closed ball of X is $\sigma(X,Y)$–closed.

A space Y which satisfies the above conditions is called a *norming* subspace of X^*.

Proof.

(i) ⟹ (ii): If $\overline{Y_1}^{*} = K$ is different from X_1^*, there is $x \in X$ such that

$$\sup_{Y_1} (x) < \sup_{X_1^*} (x) = ||x||$$

and this contradicts (i).

(ii) ⟹ (iii): Let z be in $Y^{\perp}$, and n in X. For any $\varepsilon > 0$, there exists $0 \subseteq X_1^*$ ω^*–open such that

$$-x(y) > ||x|| - \varepsilon \qquad \forall y \in 0$$

By (ii), there is $y \in Y \cap 0$, and then

$$||z-x|| \geq z-x(y) = -x(y) > ||x|| - \varepsilon$$

this shows (iii).

(iii) ⟹ (i): For any subspace Y of X^*, we let i_{Y^*} be the canonical quotient map from X^{**} onto $Y^{**} = X^{**}/_{Y^{\perp}}$. By definition of the norm of $X^{**}/_{Y^{\perp}}$, one has

$$\sup_{Y_1}(x) = \|i_{Y^*}(x)\| = \text{dist}(x, Y^\perp)$$

hence if $Y^\perp \subseteq 0_X$,

$$\sup_{Y_1}(x) \geq \text{dist}(x, 0_X) = \|x\|$$

which shows (i).

(i) $\Rightarrow$ (iv): Of course (iv) is equivalent to: X_1 is $\sigma(X,Y)$–closed. If $x \notin X_1$ and (i) is satisfied, there is $y \in Y$ such that $y(x) > 1 \geq \|y\|$, which shows that X_1 is $\sigma(X,Y)$–closed.

Conversely, assume (iv) and pick $x \in X \backslash \{0\}$. For any $\alpha < \|x\|$, there is $y \in Y$ such that

$$y(x) > \sup\{y(x') \mid \|x'\| \leq \alpha\} = \alpha\|y\|$$

thus if $y' = \|y\|^{-1} y$

$$y'(x) > \alpha$$

and this shows (i). ■

We denote by $\mathcal{N}_X$ – or sometimes simply by $\mathcal{N}$ – the family of norming subspaces of X^*. We say that X^* contains a smallest norming subspace if the space $N = \cap\{Y\}\ Y \in \mathcal{N}\}$ is itself norming.

<u>Lemma I.2</u>: *Let* X *be a Banach space,* $\mathcal{N}$ *the family of norming subspaces of* X^*,

and

$$0_X = \{z \in X^{**} \mid \|z-x\| \geq \|x\| \ \forall x \in X\}.$$

Then there is a smallest norming subspace — namely, $N = 0_X^{\top}$ *— if and only if* 0_X *is a* ω^**—closed vector subspace of* X^{**}.

Proof: By I.1, $Y \in \mathcal{N}$ if and only if $Y^{\perp} \subseteq 0_X$. Hence if $N = \cap\{Y \mid Y \in \mathcal{N}\}$ belongs to $\mathcal{N}$, we clearly have $N^{\perp} \subseteq 0_X$; and conversely, if $z \in 0_X$, $\ker(z) \in \mathcal{N}$, hence $\mathrm{Ker}(z) \subseteq N$ and $z \in N^{\perp}$; this shows $0_X = N^{\perp}$.

If we assume now that 0_X is a ω^*—closed subspace, let us write $0_X = N^{\perp}$; by I.1. $N \in \mathcal{N}$ and for every $Y \in \mathcal{N}$, $Y^{\perp} \subseteq N^{\perp}$ and thus $N \subseteq Y$.

■

Remark I.3: By using a much deeper technique, it can be shown ([20], Theorem V.6) that is X is separable, 0_X is ω^*—closed as soon as 0_X is a vector space. In other words, if X is separable, there is a smallest norming subspace as soon as the intersection of any pair of norming hyperplanes is norming; in particular, if X is separable, there is a smallest norming subspace if and only if $\mathcal{N}$ is stable under intersection.

Lemma I.1 provides us with a practical tool for checking that $z \in X^{**}$ belongs to 0_X.

Lemma I.3: *Let* X *be a Banach space, and* $z \in X^{**}$. *The following are equivalent.*

1) $z \in 0_X$.

2) $\mathrm{Ker}(z) \cap X_1^*$ is ω^**—dense in* X_1^*.

Proof: Apply I.1 to $Y = \mathrm{Ker}(z)$.

■

Our next lemma will be quite useful. Recall that we denote by $\mathscr{C}_*(z)$ the set of points of ω^*–continuity of z on X_1^*. With this notation, one has:

Lemma I.4. *Let* X *be a Banach space, and* $z \in X^{**}$. *If the subset* $\mathscr{C}_*(z)$ *of* X_1^* *separates* X, *the set*

$$P(z) = \cap \{B(x, \|z-x\|) \mid x \in X\}$$

contains at most one point.

Proof: Let x_1 and x_2 be in $P(z)$; we have

$$\|x-x_i\| \leq \|z-x\| \ \forall x \in X, \ \forall \ i = 1,2$$

hence if we let $x' = x - x_i$,

$$\|x'\| \leq \|(z-x_i) - x'\| \quad \forall x' \in X, \ \forall i = 1,2$$

that is, $(z-x_1)$ and $(z-x_2)$ belong to 0_X.

By lemma I.3 this implies that

$$(*) \qquad [\mathrm{Ker}(z-x_i) \cap X_1^*]^{-*} = X_1^* \qquad (i = 1,2).$$

Now if $y \in \mathscr{C}_*(z)$, y is also a point of ω^*–continuity of $(z-x_i)$ $(i=1,2)$, hence by

(1), we have

$$\langle z-x_1,y\rangle = \langle z-x_2,y\rangle = 0$$

and thus

$$\forall y \in \mathcal{C}_*(z),\ \ x_1(y) = x_2(y) = z(y).$$

Now since $\mathcal{C}_*(z)$ is separating X by assumption, this implies $x_1 = x_2$.

■

Lemma I.4 has an easy analog when the sets $\mathcal{C}_*(z)$ are "independent" of z; we denote by $\mathcal{C}(\omega^*,\omega)$ the set

$$\mathcal{C}(\omega^*,\omega) = \cap \{\mathcal{C}_*(z) \mid z \in X^{**}\}$$

Clearly, $\mathcal{C}(\omega^*,\omega)$ is also the set of points of continuity of

$$\mathrm{Id}\colon\ (X_1^*,\omega^*) \longrightarrow (X_1^*,\omega)$$

With this notation, one has:

<u>Lemma I.5</u>: *Let* X *be a Banach space such that the space* $N = \overline{\mathrm{span}}(\mathcal{C}(\omega^*,\omega))$ *is norming. Then* N *is the smallest norming subspace of* X^* *and* $0_X = N^\perp$.

<u>Proof</u>: By assumption, $N \in \mathcal{N}_X$. Now if $Y \in \mathcal{N}$, its unit ball Y_1 is ω^*–dense and weakly closed in X_1^*; hence $\mathcal{C}(\omega^*,\omega) \subseteq Y_1$ and $N \subseteq Y$. Finally $0_X = N^\perp$ follows

from I.2.

■

In practice, the condition "$\overline{\text{span}}(\mathscr{C}(\omega^*,\omega))$ norming" will often be checked through a stronger condition, namely

$$X_1^* = \overline{\text{conv}}^*(\mathscr{C}(\omega^*,\omega)).$$

Remark I.6: It is possible to complete a deeper study of the "metric vs. topological" properties of the elements of the bidual; this can be done by using the u.s.c. and l.s.c. envelopes $\hat{z}$ and $\check{z}$ of z on (X_1^*,ω^*). this will not be necessary, however, for the applications we have in mind in the present survey.

II. Existence and uniqueness of preduals for a first family of "smooth" spaces.

When 0_X is a "nice" subset of X^{**} – that is, a ω^*–closed vector subspace – we can easily obtain conditions for the existence of an isometric predual of X, and uniqueness of that predual. Let us state:

Theorem II.1: *Let* X *be a Banach space such that* 0_X *is a* ω^**–closed vector subspace of* X^{**}. *The following is equivalent.*

(i) X *is isometric to a dual space.*

(ii) $X^{**} = X \oplus 0_X$.

(iii) *For every family* $(B_\alpha)_{\alpha\in I}$ *of balls of* X *with* $\bigcap_{\alpha\in I} B_\alpha = \emptyset$, *there is a finite subset* F *of* I *such that* $\bigcap_{\alpha\in F} B_\alpha = \emptyset$ *– in other words,* X *has the* $I.P_{f,\infty}$.

(iv) *There exists a projection* $\pi\colon X^{**} \longrightarrow X$ *of norm* 1.

Moreover, if the above conditions are satisfied, there is a unique projection of norm 1 from X^{**} *onto* X, *and the isometric predual of* X (*namely,* $N = 0_X^{\top}$) *is unique.*

Proof: We write $0_X = N^{\perp}$ like in §I. By lemma I.2 our assumption means that there is a smallest norming subspace N of X^*.

(i) ⟹ (iv) is always true: if X_* is a predual of X, one has $X^{**} = X \oplus (X_*)^{\perp}$ and the projection "restriction to X_*" from X^{**} onto X has norm 1.

(iv) ⟹ (iii) also is always true: let $(B_\alpha)_{\alpha\in I}$ be a family of balls in X such that

$$\forall F \subseteq I \text{ finite}, \bigcap_{\alpha\in F} B_\alpha \neq \emptyset$$

then if $\overline{B}^*_\alpha$ is the closure of B_α in (X^{**}, ω^*)

$$\forall F \subseteq I \text{ finite}, \bigcap_{\alpha\in F} \overline{B}^* \neq \emptyset$$

hence by ω^*–compactness

$$\bigcap_{\alpha\in I} \overline{B}^*_\alpha \neq \emptyset$$

now if $z \in \bigcap_{\alpha\in I} \overline{B}^*_\alpha$, $\pi(z) \in \bigcap_{\alpha\in I} B_\alpha$ since $\|\pi\| = 1$.

(iii) ⟹ (ii): obviously $X \cap 0_X = \{0\}$. What we have to prove is $X^{**} = X + 0_X$. This means that for every $z \in X^{**}$, the set

$$P(z) = \cap\{B(x, \|z-x\|) \mid x \in X\}$$

is non–empty. We consider the following family of balls

$$B = \{B(x,r) \mid x \in X,\ r > \|z-x\|\}.$$

The local reflexivity principle ([41]) implies that the intersection of every finite subfamily of B is non–empty; by (iii), $\cap B \neq \emptyset$, and clearly $\cap B \subseteq P(z)$.

(ii) $\Rightarrow$ (i): We have $0_X = N^\perp$ and the projection $\pi\colon X^{**} \longrightarrow X$ with kernel $0_X = N^\perp$ has norm one since $\|z-x\| \geq \|x\|$ for every $x \in X$ and $z \in 0_X$. Since $\|\pi\| = 1$, $X^{**}/_{\mathrm{Ker}(\pi)} = X^{**}/_{N^\perp}$ is isometric to $\mathrm{Im}(\pi) = X$; on the other hand, $X^{**}/_{N^\perp}$ is isometric to N^*; hence X is isometric to N^*.

Every isometric predual of X is a minimal norming subspace of X^*, that is a minimal element of $\mathcal{N}_X$; but $\mathcal{N}_X$ has a smallest element N, hence N is unique predual of X (if there is a predual, i.e. if (i)–(iv) are satisfied). Finally, if $\pi\colon X^{**} \longrightarrow X$ is a projection with $\|\pi\| = 1$, we have

$$\|z-x\| \geq \|x\| \quad \forall x \in X,\ \forall z \in \mathrm{Ker}(\pi)$$

hence $0_X \subseteq \mathrm{Ker}\,\pi$; but since $X^{**} = X \oplus 0_X$, this forces π to be the projection of kernel 0_X.

■

The examples of spaces to which II.1 applies are of two kinds: on one hand, spaces which satisfy II.1 for one given norm ("isometric" examples II.2); on the other hand, space which satisfy II.1 for every equivalent norm ("isomorphic" situation II.3). It turns out that this latter class of spaces is *exactly* the class of Banach spaces not containing $\ell^1(\mathbb{N})$.

The following examples clearly indicate that the assumption of II.1. is a smoothness assumption; a moment of reflexion (and a few pictures) should be even more convincing.

Examples II.2: 1) If the norm of X is Frechet–smooth on a dense set, then by Smulyan's lemma (see [13], p. 24), X_1^* is the ω^*–closed convex hull of its ω^*–strongly exposed points; a fortiori, one has

$$X_1^* = \overline{\text{conv}}^*(\mathscr{C}(\omega^*,\omega))$$

and then lemma I.5 applies. In that case, it is possible to give an amusing description of the space N, which is the "tangent space" of X, that is, the norm closed linear span of the derivatives of the norm at its points of Frechet–smoothness; by II.1 this "tangent space" is the only possible predual of X.

A very simple example of this situation is $X = \ell^\infty(I)$.

2) It is easily seen that the condition $X_1^* = \overline{\text{conv}}^*(\mathscr{C}(\omega^*,\omega))$ still holds if X meets the set where the norm of X^{**} is Gateaux–smooth on a dense set. Actually, an easy compactness argument [21] shows that if $y \in S_1(X^*)$,

$$y \in \mathscr{C}(\omega^*,\omega) \Longleftrightarrow \exists\,!\, z \in S_1(X^{***}) \text{ s.t. } z_{|X} = y.$$

3) If now Y is a Banach space which satisfies

$$Y_1 = \overline{\text{conv}}(\mathscr{C}(\omega,\|\ \|))$$

where $\mathscr{C}(\omega,\|\ \|)$ denotes the set of points of continuity of the map

$$\text{Id} = (Y_1,\omega) \longrightarrow (Y_1,\|\ \|)$$

then $X = Y^*$ satisfies the assumption of II.1; in particular, there is a unique projection of norm 1 from $X^{**} = Y^{***}$ onto $X = Y^*$ and Y is unique predual of $Y^* = X$.

For checking II.1 it suffices to observe that if $y \in \mathscr{C}(\omega,\|\ \|)$ then the canonical image of y in $Y^{**} = X^*$ is a point of continuity of $\text{Id} = (X_1^*,\omega^*) \longrightarrow (X_1^*,\|\ \|)$ and then to apply I.5.

Hence we obtain that the following spaces *are* unique preduals:

(a) spaces with a locally uniformly convex norm – or more generally a "Kadeč norm", that is, a norm for which the norm and weak topologies coincide on the sphere.

(b) spaces with the Radon–Nikodym property, for every equivalent norm. Indeed, the unit ball of such spaces is the norm–closed convex hull of its strongly exposed points (see [13]).

(c) Separable spaces Y such that (Y_1,ω) is a Baire topological space; more generally, Banach spaces Y for which (Y_1,ω) is a Namioka space (by [12]).

(d) Projective tensor products of spaces with the R.N.P. equipped with the tensor norm. From this it follows that for any reflexive space R, the space $L(R)$ of bounded operators on R has a unique predual [22].

We turn now to the spaces for which the assumption of II.1 is satisfied for every equivalent norm. For these spaces, we have both the uniqueness of the predual if it exists and weak sufficient conditions – such as II.1 (iv) – for its existence. It turns out that this class is very simple to describe:

Theorem II.3: *Let* X *be a Banach space. The following are equivalent:*

(i) *For every equivalent norm on* X, *the set* 0_X *is a* ω^**–closed vector subspace of* X^{**}.

(ii) X *does not contain* $\ell^1(\mathbb{N})$.

Proof:

(ii) ⟹ (i): We make the proof in the case X separable; the result is also true in the non–separable situation, but the proof – given in [20] – is much more difficult.

If X does not contain $\ell^1(\mathbb{N})$, then by Ódell–Rosenthal's theorem [43], every $z \in X^{**}$ is of the first Baire class on (X_1^*, ω^*); hence by Baire's theorem, $\mathscr{C}_*(z)$ is a ω^*–dense G_δ of X_1^* for every $z \in X^{**}$. Hence we have by I.3:

$$z \in 0_X \Longleftrightarrow [\mathrm{Ker}(z) \cap X_1^*]^{-*} = X_1^*$$

$$\Longleftrightarrow z(y) = 0 \quad \forall y \in \mathscr{C}_*(z).$$

But for any $z_1, z_2 \in 0_X$, $\lambda_1, \lambda_2 \in \mathbb{R}$,

$$\mathscr{C}_*(z_1) \cap \mathscr{C}_*(z_2) \subseteq \mathscr{C}_*(\lambda_1 z_1 + \lambda_2 z_2)$$

and since the intersection of two ω^*–dense G_δ is ω^*–dense, it follows that $(\lambda_1 z_1 + \lambda_2 z_2) \in 0_X$, and 0_X is a vector space.

By Banach–Dieudonné, 0_X is ω^*–closed if and only if $(0_X \cap X_1^{**})$ is ω^*–closed; since X_1^{**} is a compact of first Baire class functions, it suffices ([6]) to check that

$(0_X \cap X_1^{**})$ is sequentially closed; and if $z = \lim_{n\to\infty} z_n$ in (X^{**},ω^*), we have

$$\bigcap_{n\geq 1} \mathrm{Ker}(z_n) \subseteq \mathrm{Ker}(z)$$

and since $\mathrm{Ker}(z_n) \cap X_1^*$ is a ω^*–dense G_δ of (X_1^*,ω^*) for every $n \geq 1$, Baire's theorem shows that $\mathrm{Ker}(z) \cap X_1^*$ is also ω^*–dense in X_1^*, hence $z \in 0_X$.

(i) ⟹ (ii): We assume now that X contains $\ell^1(\mathbb{N})$, and we have to construct a norm such that 0_X is not a ω^*–closed vector space. We will actually construct a norm which satisfies the much stronger property span $(0_X) = X^{**}$.

It is standard to deduce from the Hahn–Banach theorem that exists an equivalent norm $\| \ \|$ on X such that $(X,\| \ \|)$ contains a subspace Z isometric to $(\ell^1(\mathbb{N}),\| \ \|_1)$. Let $Q\colon X^* \longrightarrow X^*/_{Z^\perp} = Z^*$ be the canonical quotient map; Z^* is isometric to $\ell^\infty(\mathbb{N})$.

The set $K = \mathrm{Ext}(Z_1^*)$ is ω^*–homeomorphic to $\{-1,1\}^{\mathbb{N}}$; by Zorn's lemma, there is a minimal ω^*–compact subset K_0 of X_1^* such that $Q(K_0) = K$.

We pick a non–trivial ultrafilter U on $\mathbb{N}$, which we consider as an element z of $Z^{**} = Z^{\perp\perp}$. The sets

$$A^+ = \{y \in K \mid z(y) = 1\}$$
$$A^- = \{y \in K \mid z(y) = -1\}$$

are both ω^*–dense in K, and $K = A^+ \cup A^-$; since K_0 is minimal, the sets $K_0 \cap Q^{-1}(A^+)$ and $K_0 \cap Q^{-1}(A^-)$ are both ω^*–dense in K_0.

We let now $K_1 = K_0 \cup (-K_0)$; clearly $|z| = 1$ on K_1 are the sets $K_1 \cap z^{-1}(\pm 1)$

are both ω^*–dense in K_1. We finally define

$$B = \overline{\mathrm{conv}}^*(K_1 + \{y \in Z^\perp \mid \|y\| \le 2\})$$

We have $Q(B) = \overline{\mathrm{conv}}^*(K_0) = Z_1^*$; for every $y \in Z^*$ with $\|y\| \le 1$ there exists $y_0 \in \overline{\mathrm{conv}}^*(K_0)$ such that $Q(y) = Q(y_0)$, and thus $(y-y_0) \in Z^\perp$ and $\|y_0-y\| \le \|y_0\| + \|y\| \le 2$; this shows that $X_1^* \subseteq B$, and it follows that B is the unit ball of an equivalent dual norm on X^*; we denote by $|||\ |||$ the predual norm on X.

We have $Q(B) = Z_1^*$ and thus $|||z||| = 1$. The set $K_2 = K_1 + \{y \in Z^\perp \mid \|y\| \le 2\}$ is ω^*–compact and thus it contains $\mathrm{Ext}(B)$; but the sets $K_2 \cap z^{-1}(\pm 1)$ are both ω^*–dense in K_2, hence the Krein–Milman theorem shows that the sets $z^{-1}(\pm 1) \cap B$ are both ω^*–dense in B.

Pick now $z' \in X^{**}$, and let V be any ω^*–open convex subset of X_1^*; the affine function $(z' + 2\,|||x|||\,z)$ satisfies

$$\begin{cases} \sup\limits_V\, (z' + 2\,|||x|||\,z) \ge |||x||| \\ \inf\limits_V\, (z' + 2\,|||x|||\,z) \le -\,|||x||| \end{cases}$$

and since V is convex, this implies that

$$V \cap \mathrm{Ker}(z' + 2\,|||x|||\,z) \neq \emptyset$$

and it follows that $[\mathrm{Ker}(z' + 2\,|||x|||\,z) \cap B]$ is ω^*–dense in B, hence by I.3

$$(z' + 2|||x|||z) \in 0_X$$

when X is equipped with the norm $|||\ \ |||$, and the same argument shows that

$$(z' - 2\,|||x|||\,z) \in 0_X$$

from which follows that $\text{span}(0_X) = X^{**}$.

■

What we constructed in the proof of II.3 (i) ⟹ (ii) is an "everywhere octahedral norm"; for this notion we refer to [23] and [24].

Let us observe that II.3 describes a very strong dichotomy about the structure of norming subspaces: if $X \not\supset \ell^1(\mathbb{N})$, the set of norming subspaces has always a smallest element; if $X \supset \ell^1(\mathbb{N})$, there is a norm on X such that every hyperplane of X^* contains the intersection of two norming hyperplanes; for this norm, the set $\mathscr{N}$ is of course not stable under intersection.

Corollary II.4: *Let* X *be a Banach space which does not contain* $\ell^1(\mathbb{N})$. *The following are equivalent:*

(i) X *is isometric to a dual space.*

(ii) X *has the* $I.P_{f,\infty}$.

(iii) *There is a projection* $\pi = X^{**} \to X$ *with* $\|\pi\| = 1$.

If the conditions (i), (ii), (iii) *are satisfied, the norm–one projection from* X^{**} *onto* X *and the predual of* X *are unique.*

Proof: Apply II.1 and II.3 (ii) ⟹ (i).

■

Note that II.3 (i) ⟹ (ii) means that II.4 is somehow optimal, as far as the equivalent conditions to the existence of the predual are concerned. But the uniqueness of the norm–one projection and, a fortiori, the uniqueness of the isometric predual under isomorphic conditions can be widely extended. For this we will need different techniques which are explained in the section V.

III. An alternative approach: the ball topology.

Let X be a Banach space equipped with a given norm $\| \ \|$; we denote by b_x the *ball topology* on X, that is, the coarsest topology for which the closed balls of X are closed. If $x \in X$, a typical b_x–neighborhood of x is a set V of the form

$$V = X \backslash \bigcup_{i=1}^{n} B(x_i; \ell_i)$$

where $\ell_i < \|x - x_i\|$ for $i = 1, 2, ..., n$. This topology is defined and extensively studied in [20].

The topology b_x looks bad at first sight; for instance, it is never Hausdorff if $X \neq \{0\}$. The interesting point is that it can be very nice when restricted to bounded subsets of X. Let us illustrate this fact by the following simple lemma.

Lemma III.1: *Let $x \in X$ be a point of Frechet–smoothness of the norm of X. Then the derivative* $f_x \in X^{**}$ *is* b_x*–continuous on* x_1.

Proof: We have for every $x' \in X$

$$\lim_{t\to 0} t^{-1}[\,\|x + tx'\| - \|x\|\,] = f_x(x')$$

and the limit is uniform in $x' \in X_1$; now for any $t \in \mathbb{R}$, the function

$$\Phi_t(x') = t^{-1}[\,\|x + tx'\| - \|x\|\,]$$

is b_X–l.s.c. since the balls of X are b_X–closed. Hence f_x is a uniform limit on X_1 of b_X–l.s.c. functions and thus f_x itself is b_X–l.s.c. on X_1. But since $f_x(-x') = -f(x')$ and multiplication by (-1) is a b_X–homeomorphism of X_1, f_x is also b_X–u.s.c., hence it is b_X–continuous.

∎

From this lemma follows immediately that if the norm of X is Frechet–smooth on a dense set, then (X_1,b_X) is Hausdorff – actually, a much stronger statement is true. If now $X = Y^*$ is a dual space, b_X is coarser than ω^*; hence by compactness, ω^* coincide with b_X on $X_1 = Y_1^*$ as soon as (X_1,b_X) is Hausdorff. A very simple example of this situation is $X = \ell^\infty(I)$.

Now let us observe that b_X is obviously independent of the predual of X; hence if (X_1,b_X) is Hausdorff, there is at most one ω^*–topology on X_1 – namely, b_X – and the isometric predual is unique. We even have much better results, as we will see right now.

The comparison of I.5 and III.1 suggests that there is a close connection between the questions we considered in §I and II, and the ball topology; this is indeed the case.

We can summarize the main applications of the ball topology to duality in the following theorem, which deals with both the existence of the uniqueness of the predual.

Theorem III.2: *Let X be a Banach space such that (X_1,b_X) is Hausdorff. Then the following are equivalent:*

(i) X *is isometric to a dual space.*

(ii) X *has the* $I.P_{f,\infty}$.

(iii) (X_1, b_X) *is compact.*

Moreover, if (i), (ii), (iii) *are satisfied, then there is a unique projection of norm* 1 *from* X^{**} *onto* X, *and* b_X *is the unique compact Hausdorff topology on* X_1 *such that for every* $\lambda \in [0,1]$, *the map* $g_\lambda(x,y) = \lambda x + (1-\lambda)y$ *from* $(X_1 \times X_1)$ *to* X_1 *is separately continuous.*

Theorem III.2 means that if (X_1, b_X) is Hausdorff, the geometry of X_1 determines uniquely the compact topology on X_1, if this topology has some connection with the affine structure; the separate continuity of the g_λ's is essentially the weakest assumption one can think of.

Theorem III.2 is a rewriting of ([20], Theorem VI.3). The proof is quite difficult, and too long for being presented here.

By III.1, Theorem III.2 applies for instance to spaces whose norm is Frechet–smooth on a dense set; and in this case III.2 is an improvement of II.1. But the most interesting situations where III.2 applies arise in the same context as Theorem II.3. Indeed ([20], Theorem V.4) reads:

Theorem III.3: *Let* E *be a Banach space such that* E^* *contains a norming subspace* Y *which does not contain* $\ell^1(\mathbb{N})$. *Then for every space* X *such that* $Y \subseteq X \subseteq E^*$, (X_1, b_X) *is Hausdorff and thus* III.2 *applies.*

A special case of this result is when Y is a predual of E:

Corollary III.4: *Let* Y *be a Banach space which does not contain* $\ell^1(\mathbb{N})$. *Let* X *be a space such that* $Y \subseteq X \subseteq Y^{**}$. *Then* (X_1, b_X) *is Hausdorff and thus* III.2 *applies.*

Remarks III.5: 1) What is actually shown in [20] is that the spaces X like in III.3. and III.4 satisfy: 0_X ω^*–closed vector space. Hence III.4 is an improvement of II.3 (ii) ⟹ (i). The proof of III.4 actually uses the arguments of II.3, plus several other techniques. It is instructive to understand why this is not in contradiction with II.3. (i) ⟹ (ii); indeed the spaces X of III.4 can of course contain $\ell^1(\mathbb{N})$ – take $Y = c_0(\mathbb{N})$; and they will satisfy "0_X ω^*–closed vector subspace" for any equivalent norm on X which is induced by the bidual norm of an equivalent norm on Y – but *not* by every equivalent norm on X.

2) It follows from III.4 that the dual $Z = Y^*$ of a Banach space Y which does not contain $\ell^1(\mathbb{N})$ is unique predual (take $X = Y^{**}$ in III.4). In other words, dual spaces with the weak R.N.P. are unique preduals. By using a different method, namely, "small convex combinations of slices" (see [18]), this result is obtained in [25] for dual spaces with the weak R.N.P. and for their subspaces.

We will conclude here this brief account of the applications of the ball topology to duality questions. I would like to mention however that the few results presented here are far from exhausting these applications and moreover that the ball topology is a versatile tool, which permits for instance to show ([20], Theorem VIII.2).

Theorem III.6: *Let* X *be any Banach space, and* W *be any weakly compact subset of* X. *Then* W *is an intersection of finite unions of balls.*

Theorem III.6 is a sharp extension of [10] where X was supposed to be reflexive.

The proof of III.6 is difficult; a simple proof would be welcome, although it is not clear to me that such a proof exists.

IV. A critical example: the space L^1.

In the examples of unique preduals that we have considered so far, a stronger property was available: namely, the uniqueness of the projection of norm one from X^{***} to X^*. We will see now that this property is strictly stronger than "X unique predual". An example will be provided by the oldest non–trivial example of unique predual: the space L^1([33]). In what follows, we denote by L^1 the space $L^1([0,1]; dx)$.

Proposition IV.1: *There exist infinitely many linear projections of norm one from* L^{1***} *onto* $L^{1*} = L^\infty$.

Proof: Let Ω be the spectrum of the Banach algebra L^∞; Ω is compact and L^∞ is isometric to $\mathscr{C}(\Omega)$, hence $L^{\infty*}$ is isometric to the space $\mathscr{M}(\Omega)$ of Radon measures on Ω.

Let $\pi\colon \mathscr{M}(\Omega) \longrightarrow L^1$ be the Radon–Nikodym projection. We consider the function $\mathbf{1} = \mathbf{1}_{[0,1]}$ as an element of $L^\infty = L^{1*}$ and we let $t = \pi^*(\mathbf{1})$; t belongs to $L^{\infty**}$.

Every Dirac measure $\epsilon_z (z \in \Omega)$ belongs to $\operatorname{Ker} \pi$; hence $t(\epsilon_z) = 0$ for every $z \in \Omega$ and thus, by the Krein–Milman theorem, $\operatorname{Ker}(t)$ is a norming subspace of $L^{\infty*}$. Now I.1 shows that $t \in 0_{L^\infty}$; hence the projection

$$p = L^\infty \oplus \mathbb{R}t \longrightarrow L^\infty$$

of kernel $\{\mathbb{R}t\}$ has norm 1. Since L^∞ is an injective Banach space, p extends to a norm one projection P_0: $L^{\infty**} \longrightarrow L^\infty$.

On the other hand, we may write $L^{\infty**} = L^\infty \oplus (L^1)^\perp$ and the projection π of kernel $(L^1)^\perp$ has norm one; and clearly $t \notin (L^1)^\perp$, hence P_1 and π are distinct.

If we let now, for $\lambda \in [0,1]$, $P_\lambda = \lambda\pi + (1-\lambda)P_0$, it is routine to check that the P_λ's are distinct projections of norm one from $L^{\infty**}$ onto L^∞.

■

If we want instead to ge the condition "span $0_{X^*} = X^{***}$", we can apply the following general result:

<u>Proposition IV.2</u>: *Let* X *be a Banach space such that:*

(i) *There exists a projection* π *from* X^{**} *onto* X *such that*

$$\forall\, z \in X^{**},\ \ ||z|| = ||\pi(z)|| + ||z-\pi(z)||$$

(ii) The unit ball X_1 of X has no extreme points. Then one has: span $0_{X^*} = X^{***}$.

<u>Proof</u>: If (i) is satisfied, then clearly

$$\mathrm{Ext}(X_1^{**}) = \mathrm{Ext}(X_1) \cup \mathrm{Ext}((\mathrm{Ker}\ \pi)_1).$$

Hence by (ii) we have $\mathrm{Ext}(X_1^{**}) = \mathrm{Ext}((\mathrm{Ker}(\pi))_1)$. Now the Krein–Milman theorem shows that Ker π is a norming subspace of X_1^{**}; hence $(\mathrm{Ker}\ \pi)^\perp \subseteq 0_{X^*}$. On the other hand, X is obviously a norming subspace of X^{**}, hence $X^{\perp} \subseteq 0_{X^*}$. And since $X^{**} = X \oplus \mathrm{Ker}(\pi)$, we have $X^{***} = X^{\perp} \oplus (\mathrm{Ker}\ \pi)^{\perp}$, and a fortiori $\mathrm{span}(0_{X^*}) = X^{***}$.

∎

It turns out that IV.2 is satisfied not only by L^1, but by a bunch of spaces which are unique preduals (see §V). For such spaces, the techniques of §II and III fail completely. It turns out that a different technique, which relies on Lemma 1.4, lead to more positive results. This will be explained in the next chapter.

However, we should point out a major difference between the two family of examples: so far, we have been dealing with spaces Y which were the smallest norming subspaces of Y^{**}. From now on, we will consider spaces – such as L^1 – which will be unique preduals but will not satisfy this stronger condition. In particular, we will not be able to obtain non–trivial results of *existence* of the predual for the dual class; indeed if there is no smallest norming subspace, there is no natural "candidate" for being a predual.

V. Uniqueness of preduals for a second family of spaces.

The following theorem should be understood as an abstract Radon–Nikodym theorem. It roughly means that if one has an "easy way" to recognize which elements $z \in Y^{**}$ belong to Y, then Y is unique isometric predual of Y^*.

Theorem V.1: *Let* Y *be a Banach space. If the following condition holds:*

(∗): *for every* $z \in Y^{**}$, *one has:*

$z \in Y \Longleftrightarrow$ *for every* ω*–Cauchy sequence* (y_n) *in* Y^*, $z(\omega^*\text{–}\lim(y_n)) = \lim z(y_n)$.

Then Y *is the unique isometric predual of* Y^{**}.

Proof: It is enough to show that there exists a unique projection $\pi\colon Y^{***} \to Y^*$ of norm one such that $\mathrm{Ker}(\pi)$ is ω^*–closed; indeed, to any isometric predual Z of Y^* corresponds such a projection (of kernel $Z^\perp$).

Let S be the ω^*–*sequential* closure of Y^* into (Y^{***},ω^*), and let $S_0 = S \cap Y^\perp$. For every $t \in S_0$, one has

$$P(t) = \cap\{B(y,\|t-y\|) \mid y \in Y^*\} = \{0\}$$

Indeed $0 \in P(t)$ since the projection from Y^{***} onto Y^* of kernel $Y^\perp$ has norm one; on the other hand, every $t \in S$ is of the first Baire class on (Y_1^{**},ω^*), hence by Baire's theorem $\mathscr{C}_*(t)$ is a ω^*–dense G_δ of Y_1^{**}; thus $\mathscr{C}_*(t)$ separates Y^* and we may apply Lemma I.4 which shows that $P(t)$ contains at most one point; hence we have $P(t) = \{0\}$.

Now if $P(t) = \{0\}$, t belongs to the kernel of any projection π of norm one from Y^{***} onto Y^*, hence $\mathrm{Ker}(\pi) \supseteq S_0$; if $\mathrm{Ker}(\pi)$ is ω^*–closed, it will contain $\overline{S_0}^* = \overline{S \cap Y^\perp}^*$. For showing that $\mathrm{Ker}\,\pi = Y^\perp$, it is therefore enough to show that

$$Y^\perp = \overline{S \cap Y^\perp}^*$$

We prove it by contradiction: if $Y^\perp \neq [S \cap Y^\perp]^{-*}$, there is $z \in Y^{**}\backslash Y$ such that $z = 0$ on $(S \cap Y^\perp)$; but if (y_n) is any ω–Cauchy sequence in Y^* and if we call y' the limit of (y_n) in (Y^*,ω^*) and t' the limit of (y_n) in (Y^{***},ω^*) we have $(y'-t') \in (S \cap Y^\perp)$ and thus $z(y'-t') = 0$; since $z(t') = \lim z(y_n)$, this means that

$$z(\omega^*\text{-}\lim(y_n)) = \lim z(y_n)$$

for any ω–Cauchy sequence (y_n) in Y^*, and by $(*)$ this implies that $z \in Y$; a contradiction.

■

Remark V.2: The key point of the proof is the fact that if $z \in X^{**}$ belongs the ω^*–sequential closure of X then by I.4

$$P(z) = \cap\{B(x, \|z-x\|) \mid x \in X\}$$

contains at most one point; this is true for every equivalent norm on X since the assumption on z is of isomorphic nature. It turns out that under this form, the result is sharp – indeed, it is shown in [57] that: X Banach space, $z \in X^{**}$. Then the following are equivalent:

(i) $\forall K \subset X^*$ ω^*–compact, $z_{|K}$ has a point of ω^*–continuity.

(ii) For every equivalent norm on X, P(z) contains at most one point.

This is a "pointwise" analog of II.3 (i) $\Rightarrow$ (ii); the proof is completed by similar methods.

On the other hand, if we compare V.1. with II.3. (ii) $\Rightarrow$ (i), we see that we can work with ω–Cauchy sequences in Y^* instead of working with the unit ball Y_1^* itself; Y is still unique predual; what we loose (by §IV) is the uniqueness of the norm one projection from Y^{***} onto Y^*.

It turns out that in many important examples, a stronger property than $(*)$ (of V.1) is actually satisfied. Before stating our next result, let us recall that a sequence (x_n) in a Banach space X is said to be a weakly unconditionally convergent (in short, w.u.c.) series if one has

$$\forall\, y \in X^*,\ \sum_{n=1}^{\infty} |y(x_n)| < \infty$$

Obviously, if (x_n) is a w.u.c. series, the sequence $(s_n = \sum_{i=1}^{n} x_i)$ is a ω–Cauchy sequence, In the case where (x_n) is a subset of a dual space X^*, we denote by

$$x' = \Sigma^* x_n$$

the limit of (s_n) in (X^*,ω^*); it turns out that $\Sigma^* x_n$ does *not* depend upon the isometric predual of X^*(by I.4; see §VII below).

The assumption of our next result was called "property (X)" in [26], where this theorem is proved.

Theorem V.3: *Let* Y *be a Banach space. If the following condition holds:*

(∗∗) = *for every* $z \in Y^{**}$, *one has:*

$z \in Y \Longleftrightarrow$ *for every w.u.c. series* (y_n) *in* Y^*,

$$z(\Sigma^* y_n) = \Sigma\, z(y_n)$$

Then:

(i) Y *is the unique isometric predual of* Y^{**}.

(ii) *Every* $z \in Y^{**}$ *which is strongly Baire measurable on* (Y^*,ω^*) – *in particular, every* $z \in Y^{**}$ *which is Borel on* (Y^*,ω^*) – *belongs to* Y.

Before proceeding to the proof, let us observe that (ii) trivially implies that Y is weakly sequentially complete – replace "Borel" by "first Baire class".

Proof:

(i) is an immediate application of V.1. since the partial sums of a w.u.c. series form a ω–Cauchy sequence.

(ii) Let us recall that $z \in Y^{**}$ is strongly Baire measurable on (Y^*,ω^*) if for every ω^*–compact subset K of Y^* and any $\alpha < \beta$ in $\mathbb{R}$, there is a ω^*–open subset 0 of K such that the set $(z^{-1}((\alpha,\beta)) \cap K) \Delta 0$ is meager in (K,ω^*); the class of strongly Baire measurable functions contains the Borel functions (and much more).

We denote by (e_n^*) $(n \geq 1)$ the coordinate linear forms on $\ell^1(\mathbb{N})$. If $T: Y \longrightarrow \ell^1(\mathbb{N})$ is a bounded linear operator, then clearly $(y_n = T^*(e_n))$ is a w.u.c. series in Y^*; conversely, any w.u.c. series in Y^* defines an operator $T:Y \longrightarrow \ell^1(\mathbb{N})$ by the formula $T(x) = (y_n(x))$.

If $z \in Y^{**}$ is Borel on (Y^*,ω^*), then for every operator $T: Y \longrightarrow \ell^1(\mathbb{N})$, $T^{**}(z)$ is Borel on (ℓ^∞,ω^*), hence by [9] $T^{**}(z)$ belongs to $\ell^1(\mathbb{N})$. This implies that $z(\Sigma^* y_n) = \Sigma z(y_n)$ for every w.u.c. series in Y^*, hence by property (**) that $z \in Y$. If more generally, z is strongly Baire measurable on (Y^*,ω^*), the technique of [51] shows that $T^{**}(z)$ has the strong Baire property on (ℓ^∞,ω^*) and again [9] concludes the proof.

■

Examples V.4: Let us observe first that the classes of spaces which satisfies the assumption of V.1 or V.3 are easily seen to be hereditary; hence any subspace of the following examples will also be unique predual. Moreover, our assumptions are of isomorphic nature, and thus these spaces will be unique isometric predual for every equivalent norm. We will check that every space Y of the following list satisfies the condition (**) of V.3.

(1) separable weakly sequentially complete Banach lattices.

Proof: If Y is a w.s.c. separable Banach lattice, then the abstract version of the Radon–Nikodym theorem reads: if $z \in Y^{**}$, then $z \in Y$ if and only if for every decreasing sequence (y_n) in Y^*_+ with $\inf(y_n) = 0$, one has $\lim z(y_n) = 0$ (see [45]). If we let $x_1 = y_1$ and $x_n = y_n - y_{n-1}$ for $n \geq 2$, then the sequence (x_n) is a w.u.c. series; indeed for every $n \geq 1$,

$$\begin{aligned}\sum_{i=1}^{n} |z(x_i)| &\leq \sum_{i=1}^{n} \langle |z|, |x_i| \rangle \\ &= \langle |z|, y_1 \rangle + \sum_{i=2}^{n} \langle |z|, y_{i-1} - y_i \rangle \\ &= \langle |z|, y_1 \rangle + \langle |z|, y_1 - y_n \rangle \\ &\leq 2 \langle |z|, y_1 \rangle\end{aligned}$$

Moreover, since $\sum_{i=1}^{n} x_i = y_n$, we have $\Sigma^* x_i = 0$; hence we can check the condition (**) on these particular w.u.c. series and V.3 applies.

A special case of the above is of course the example $Y = L^1$ of Grothendieck, which is therefore unique predual for every equivalent norm; the original argument of Grothendieck [33] was using the clever observation that the map $T(f) = (1+f)/2$ transforms the unit ball of L^∞ into the positive unit ball; this argument does not extend to equivalent norms.

The above argument can be refined for showing that a w.s.c. subspace of a Banach lattice with an order continuous norm satisfies also (**) [26]; this permits to apply V.3 to:

(2) separable w.s.c. spaces which are complemented into a Banach lattice

(3) separable spaces with local unconditional structure which do not contain ℓ_n^∞ uniformly.

The non–commutative analog is also valid.

(4) Preduals of von Neumann algebras.

Proof: We need of course a non–commutative Radon–Nikodym theorem, which reads ([50]): If W is a von Neumann algebra and $z \in W^*$, then z belongs to the predual W_* if and only if

$$z(\Sigma P_n) = z(Id_H)$$

for every orthogonal family of projections (P_n) in W such that $\Sigma P_n = Id_H$ for the weak operator topology (that is, $\Sigma^* P_n = Id_H$). It is easily seen that (P_n) is a w.u.c. series.

By [36], the result extends to preduals of JB^*–triples.

(5) $L^1(\mathbf{T})/_{H^1(D)}$.

We have this time a non self–adjoint analog; and now our "Radon–Nikodym" theorem is provided by Ando, who used it in [2] for showing that $L^1(\mathbf{T})/_{H^1(D)}$ equipped with its canonical norm is unique predual. The result reads: let Ω be the spectrum of $L^\infty(\mathbf{T};d\theta)$ and $\hat{m} \in \mathcal{M}(\Omega)$ the measure which corresponds to $d\theta$. Then $\nu \perp \hat{m} \Longleftrightarrow$ there is $A \subseteq \Omega$ compact and a w.u.c. series (g_n) in $H^\infty(D)$ such that

(i) $|\nu| (\Omega \backslash A) = 0$

(ii) $\sum_{n=1}^{\infty} \hat{g}_n(\omega) = \mathbf{1}_A(\omega)\ \ \forall \omega \in \Omega$

(iii) $\sum_{n=1}^{\infty} g_n(\zeta) = 0$ for almost every $\zeta \in \mathbf{T}$

(**) follows easily from this fact and the identification of the dual of $L^1(\mathbf{T})/_{H^1(D)}$ with $H_0^\infty(D)$.

This example (5) can be extended into several directions; the proof shows that every ω^*–closed subspace X of $L^\infty(\mathbf{T})$ containing $H^\infty(D)$ and which is an H^∞–module has a unique predual $L^1/_{X^\perp}$ which satisfies (**); under the same assumption, it can be shown that this predual is L–complemented in X^*. On the other hand, the result can be extended to spaces of the type $L^1(m)/_{H^1(m)}$, where m is a unique representing measure for a uniform algebra A (in the classical situation, A = disk algebra); this is done in [27].

(6) If X is separable, $X \not\supset \ell^1(\mathbb{N})$ and X has the property (u) of Pelczynski, then $Y = X^*$ satisfies (**) of V.3. Indeed, since $X^{***} = X^* \oplus X^\perp$, it suffices to show that if $t \in X^\perp$ and

$$t(\Sigma^* z_n) = \Sigma\, t(z_n)$$

for any w.u.c. series (z_n) in X^{**}, then $t = 0$. Pick such a t; the assumption made on X means that every $z \in X^{**}$ can be written $z = \Sigma^* x_n$, where (x_n) is a w.u.c. series in X; now since $t \in X^\perp$, $t(n_n) = 0$ for every n, and thus $t(z) = t(\Sigma^* x_n) = \Sigma t(x_n) = 0$.

An interesting special case of (6) is when X is an M–ideal in its bidual, since it has been very recently shown that such spaces always have property (u).

Let us mention that the property "X^* unique predual" under the assumption (6) is also a consequence of III.4. Finally, it has been recently shown [38] that the assumptions of (6) do not imply X^* separable.

Remark V.5: The "property (X)" – that is, property (**) of V.3 – looks quite similar to the classical property (V^*); actually, one has

a) Y has (X) $\Longleftrightarrow \forall z \in Y^{**} \backslash Y$, $\exists T: Y \rightarrow \ell^1(\mathbb{N})$ such that $T^{**}(z) \notin \ell^1$.

b) Y has (V^*) $\Longleftrightarrow$ For every bounded not weakly compact subset C of Y, $\exists T: Y \rightarrow \ell^1(\mathbb{N})$ such that $T^{**}(\overline{C}^*) \not\subset \ell^1(\mathbb{N})$.

Hence (X) implies (V^*); it turns out that the converse is false, by an example of M. Talagrand [52]; the same example shows that (V^*) fails to imply the condition (ii) of Theorem V.3; there is [52], in the bidual of Talagrand's space, of element z which belongs to the ω^*–second Baire class and satisfies the barycentric calculus, but is not ω^*–continuous (although the space has (V^*) and thus is w.s.c.).

VI. Stability properties of the class of unique preduals.

In this section we will investigate the stability of the class of unique preduals under the standard operations; for sake of shortness, we will content ourselves with sketchy proofs. We will also meet several open questions which will be recalled in our final section (§X). Let us start our list now:

(1) The class of unique preduals is stable under ℓ^1–sums; this is a consequence of the fact that M–projections (i.e. ℓ^∞–projections) in a dual space are always ω^*–continuous (a result of [11]; see §VII).

(2) It is unknown whether $L^1(X)$ is always unique predual when X is; several positive results along these lines may be found in [8]. Let us mention however that if (P) is the

property: "X is unique predual for every equivalent norm", then X and Y have (P) $\not\Rightarrow$ $X \otimes_\pi Y$ has (P). Indeed there is an example where X and Y have R.N.P. but $X \otimes_\pi Y$ contains $c_0(\mathbb{N})$, and a separable space which contain $c_0(\mathbb{N})$ does not have (P) – see §X.

(3) The class of unique preduals is not stable under equivalent renorming: for instance, if $c_0(\mathbb{N})$ is equipped with an equivalent l.u.r. norm $|||\ \ |||$, then $(c_0(\mathbb{N}), |||\ \ |||)$ is unique predual by II.2. (3); but clearly $c_0(\mathbb{N})$ is not unique predual for its natural norm (see §IX).

(4) Although many families of unique preduals (such as the spaces which satisfy (∗) in V.1) are hereditary, the class of unique preduals is not hereditary: for instance, it is possible to construct an equivalent norm $\|\ \ \|$ on $c_0(\mathbb{N})$ such that the unit ball of $\|\ \ \|$ is the norm–closed convex hull of $\mathscr{C}(\omega, \|\ \ \|)$ – hence $(c_0, \|\ \ \|)$ is unique predual by II.2. (3) – but $(c_0, \|\ \ \|)$ contains an isometric copy of c_0 equipped with the canonical norm.

(5) It is not known whether the bidual Y^{**} of a unique predual Y is itself unique predual; I would think that it is not the case. Investigating the question requires a look at the fifth dual Y^{*****} of Y – an exciting task.

The trouble here comes of course from the fact that being unique predual is not a local property, and hence the bidual can behave quite differently from the space; let us observe however that by II.4 and the local reflexivity principle (plus a classical lifting argument), if X does not contain ℓ^1_n uniformly, then any dual of X is unique predual; indeed, no dual of X can contain $\ell^1(\mathbb{N})$.

(6) The class of unique preduals is obviously not stable under quotient, since ℓ^1 is unique predual; however, some positive results hold true. The proof of the following claim is formal and will be omitted (see [28]):

<u>Claim</u>: Let Y be a Banach space, and X be a subspace of Y. Then X is reflexive if and only if $X^{\perp\perp\perp} = X^{\perp} \oplus Y^{\perp}$ — where $X^{\perp}$ is the orthogonal of X in Y^* and $Y^{\perp}$ is the orthogonal of Y in Y^{***}.

From this claim it is easily seen that if X is a reflexive subspace of Y, then Y satisfies the assumption (*) of V.1. if and only if $Y_{/X}$ does. Hence, for instance, if $X \subseteq Y \subseteq Z$ with X reflexive and $Z = W_*$ a predual of a von Neumann algebra, then $Y_{/X}$ is unique predual for every equivalent norm.

It is not clear whether, more generally, the assumption (*) of V.1, or (**) of V.3, have the "3–space property"; that is, if X and $Y_{/X}$ have (*) − or (**) − is it also true for Y?

(7) Recall that the Edgar ordering of Banach spaces is defined as follows: $X \prec Y$ if and only if for every $z \in X^{**}\backslash X$, there is $T: X \longrightarrow Y$ such that $T(z) \notin Y$ [16]. For instance, X satisfies (**) of V.3. if and only if $X \prec \ell^1(\mathbb{N})$ [16]. It is now easily seen that (**) is stable under "smaller" subspaces for $\prec$; it can be checked that the natural classes of unique preduals for every equivalent norms are stable under $\prec$ ([29]).

(8) If Y is "strongly regular" (see [18]) then Y is unique predual ([18], Th. VII.2) and this class is stable under G_δ–embedding [19]. Is it also true of the spaces which have (*) or (**)? What if there is a G_δ–embedding from Y into L^1?

We conclude this section by mentioning that "almost every" separable Banach space is unique predual [25], with two different interpretations: if $G(\ell^1)$ is the set of closed subspaces of $\ell^1(\mathbb{N})$ equipped with its natural topology, there is a dense G_δ Ω of $G(\ell^1)$ such that for $X \in \Omega$, $\ell^1_{/X}$ is unique predual. On the other hand, if we call $N(Y)$ the set of equivalent norms on a Banach space Y, equipped with the topology of uniform convergence on bounded sets, there is a G_δ–subset Ω' of $N(Y)$ such that for every

$N \in \Omega'$, (Y,N) is unique predual. We refer to [25] for the proofs.

VII. Automatic ω^*–continuity Applications.

The terminology "automatic continuity" is used when the continuity of a map follows from a weaker condition, such as measurability, and some kind of algebraic regularity – such as linearity; this is of course an important direction of research. We will consider in this section some results of automatic ω^*–continuity for linear map between dual spaces. For establishing such results, we have to overcome an obvious obstruction: we can't find "big" subsets of X^* which are translation invariant and ω^*–Baire topological spaces. Our techniques, however, will lead to some positive results. We start with the simple

Proposition VII.1: *Let Y be a Banach space such that there exists a unique projection* $\pi\colon Y^{***} \to Y^*$ *with* $\|\pi\| = 1$ *and* $\mathrm{Ker}(\pi)$ ω^**–closed. Then every bijective isometry of* Y^* *is* $(\omega^*–\omega^*)$*–continuous.*

Proof: Let J be such a bijective isometry. J is $(\omega^*–\omega^*)$–continuous if and only if $J^*(Y) \subseteq Y$, if and only if $J^{**}(Y^\perp) \subseteq Y^\perp$. If π is the projection of kernel $Y^\perp$, then $\pi_0 = J\pi(J^{**})^{-1}$ is also a projection with $\|\pi_0\| = 1$ and $\mathrm{Ker}(\pi_0)$ ω^*–closed; hence $\pi_0 = \pi$ and $\mathrm{Ker}(\pi_0) = J^{**}(Y^\perp) = \mathrm{Ker}(\pi) = Y^\perp$.

■

Corollary VII.2: *Let Y be a Banach space such that* **one** *of the following conditions is satisfied:*

(a) 0_{Y^*} *is a* ω^**–closed vector subspace of* Y^{***}.

(b) Y *satisfies the condition* $(*)$ *of* V.1.

Then every bijective isometry of Y^* *is* $(\pi^*-\omega^*)$ *continuous.*

The proof is immediate since what we showed under each of these conditions is that Y satisfies the assumption of VII.1. Before stating the next result, let us recall that a bounded operator $T \in L(X)$ on a complex Banach space X is said to be *hermitian* if for every $\varphi \in L(X)^*$ such that $\|\varphi\| = 1 = \varphi(Id_X)$, one has $\varphi(T) \in \mathbb{R}$. With this notation, one has:

Corollary VII.3: *Under the assumptions of* VII.2 *every hermitian operator on* Y^* *is* $(\omega^*-\omega^*)$*–continuous.*

Proof: If $T \in L(Y^*)$ is hermitian then for every $t \in \mathbb{R}$, the operator $J_t = \exp(itT)$ is a bijective isometry ([5], p. 46) hence it is $(\omega^*-\omega^*)$–continuous. But we have

$$T = \lim_{t \to o} t^{-1}[\exp(itT) - Id_{Y^*}]$$

and the limit is uniform; hence T is a uniform limit of $(\omega^*-\omega^*)$–continuous operators, and therefore it is itself $(\omega^*-\omega^*)$–continuous.

∎

The assumption of being an isometric or a hermitian operator is of course quite strong. We will now consider more subtle results, which rely on a much weaker topological assumption.

Theorem VII.4: *Let* Y *and* Z *be Banach spaces, and let* $T: Y^* \longrightarrow Z^*$ *be a bounded operator such that* Ker(T) *and* $T(Y_1^*)$ *are* ω^**–closed. Then for every* ω*–Cauchy sequence* (y_n) *in* Y^*, *one has* $T(\omega^*\text{–lim}(y_n)) = \omega^*\text{–lim}\ T(y_n)$.

Proof: If we let $V = \text{Ker}(T) = W^{\perp}$, the operator T factors through the canonical quotient map Q from Y^* onto W^*; we let $T = SQ$, and $y_n' = Q(y_n)$. What we have to show is

$$S(\omega^*\text{–lim}(y_n')) = \omega^*\text{–lim}(S(y_n'))$$

and S is now an injective operator from W^* into Z^*. Without loss of generality, we may assume that $\omega^*\text{–lim}(y_n') = 0$.

We denote by t the limit of (y_n') in (W^{***},ω^*); t belongs to $W^{\perp}$ and is of the first Baire class on (W^{**},ω^*), hence by I.4. one has

$$\cap\, \mathcal{B} = \{0\}$$

where $\mathcal{B}$ denotes the following family of balls in W^*

$$\mathcal{B} = \{B(y, \|t-y\|) \mid y \in W^*\}.$$

Since S is injective, we have

$$\cap\{S(B) \mid B \in \mathcal{B}\} = \{0\}$$

We consider now the following families of convex subsets of Z^*:

$$\mathcal{B}' = \{S(B) + \varepsilon Z_1^* \mid B \in \mathcal{B},\ \varepsilon > 0\}$$
$$\mathcal{B}'' = \{B(y', \|S^{**}(t) - y'\|) \mid y' \in Z^*\}$$

Since $S(W_1^*)$ is ω^*–closed by assumption, $\mathcal{B}'$ consists of ω^*–compact sets; $\mathcal{B}''$ is a set of balls in Z^* which are of course ω^*–compact. If we denote by $\tilde{C}$ the ω^*–closure in the third dual of a subset of a dual, we clearly have

$$\forall B \in \mathcal{B},\quad \widetilde{S(B)} = S^{**}(\tilde{B})$$

and since $t \in \tilde{B}$ for every $B \in \mathcal{B}$, we have that $S^{**}(t) \in \widetilde{S(B)}$ for every $B \in \mathcal{B}$; on the other hand, it is obvious that $S^{**}(t) \in \tilde{B}''$ for every $B'' \in \mathcal{B}''$. Thus we have

$$S^{**}(t) \in \cap\{\tilde{C} \mid C \in \mathcal{B}'' \cup \{S(B) \mid B \in \mathcal{B}\}\}$$

In that situation, it is easy to deduce from the local reflexivity principle [41] that every finite subfamily of $\mathcal{B}' \cup \mathcal{B}''$ has a non–empty intersection; but $(\mathcal{B}' \cup \mathcal{B}'')$ consists of ω^*–compact sets, hence this implies that

$$\cap(\mathcal{B}' \cup \mathcal{B}'') \neq \emptyset$$

but clearly

$$\cap\mathcal{B}' = \cap\{S(B) \mid B \in \mathcal{B}\} = \{0\}$$

hence we have $0 \in \mathcal{B}''$; but

$$S^{**}(t) = \lim_{n\to\infty} S(y_n')$$

in (Z^{***},ω^*), hence $S^{**}(t)$ is of the first Baire class on (Z^{**},ω^*), and by I.4 $\cap\mathscr{B}''$ contains at most one point; that is,

$$\cap\mathscr{B}'' = \{0\}$$

Now since the projection π from Z^{***} onto Z^* of kernel $Z^{\perp}$ has norm one, $\pi(S^{**}(t)) \in \cap\mathscr{B}''$, hence $\pi(S^{**}(t)) = 0$; and since

$$\pi(S^{**}(t)) = \omega^*\text{–}\lim S(y_n')$$

the conclusion follows.

■

Remark VII.5: For sake of shortness, we limited ourselves to the case of ω–Cauchy sequences, where I.4 applies. The same argument shows that if Y^* and Z^* satisfy the assumption of I.5 any operator $T: Y^* \longrightarrow Z^*$ such that Ker(T) and $T(Y_1^*)$ are ω^*–closed is $(\omega^*\text{–}\omega^*)$–continuous. This is the case for instance if Y and Z have the Radon–Nikodym property.

From this result follows the

Corollary VII.6: *Let* Y *and* Z *be Banach spaces, and let* A *be a bounded subset of* Y^* *which does not contain a sequence equivalent to* $\ell^1(\mathbb{N})$. *If* $T: Y^* \longrightarrow Z^*$ *is such that* Ker(T) *and* $T(Y_1^*)$ *are* ω^**–closed, then the restriction of* T *to* $\overline{A}^*$ *is* $(\omega^*\text{–}\omega^*)$*–continuous from* $\overline{A}^*$ *to* $T(\overline{A}^*) = \overline{T(A)}^*$.

Proof: We prove the result under the assumption Y separable; Haydon's result [35] permits to use a similar argument in the general case.

If (a_n) is a sequence in A and $a = \omega^*\text{–lim}(a_n) \in \overline{A}^*$, we can assume by Rosenthal's theorem [44] that (a_n) is ω–Cauchy; now it follows from VIII.4 that $T(a) = \omega^*\text{–lim}\, T(a_n)$.

This implies that the restriction of T to $\overline{A}^*$ is $(\omega^*-\omega^*)$–continuous; indeed if not, there is $a \in \overline{A}^*$, a sequence (a_k) in $\overline{A}^*$ and an open neighborhood V of a such that for every $k \geq 1$, $T(a_k) \notin \overline{V}^*$; for every k, we may write $a_k = \lim_n a'_{n,k}$ with $a'_{n,k} \in A$; and since

$$\omega^*\text{–}\lim_n T(a'_{n,k}) = T(a_k) \notin \overline{V}^*$$

we may assume that $T(a'_{n,k}) \notin \overline{V}^*$ for every n,k; but there is a sequence (n_k) such that

$$a = \omega^*\text{–}\lim_k (T(a'_{n_k,k})).$$

This contradicts the fact that $T(a'_{n_k,k}) \notin V$ for every k.

Hence T is $(\omega^*-\omega^*)$–continuous from $\overline{A}^*$ into $T(\overline{A}^*)$; and $T(\overline{A}^*) = \overline{T(A)}^*$ follows by ω^*–compactness of $\overline{A}^*$.

■

Remark VII.7: VII.4 and VII.6 apply in particular when T is an isometric bijection from Y^* into itself, since then Ker T = {0} and $T(Y_1^*) = Y_1^*$ are obviously ω^*–compact. In this situation, VII.6 can be shown by using the ball topology b_{Y^*} of Y^*, since by [20] b_{Y^*} coincide with ω^* on $\overline{A}^*$, and an isometric bijection is obviously a b_{Y^*}–homeomorphism.

We will see now that in the class of spaces which satisfy (∗) of V.1 an interesting characterization of conjugate operators is available; if moreover (∗∗) is true, we can also use a measurability assumption.

Theorem VII.8: *Let* Y *be a Banach space which satisfies the assumption* (∗) *of* V.1 *and* Z *be an arbitrary Banach space. For any bounded operator* $T:Y^* \longrightarrow Z^*$, *the following are equivalent:*

(i) *there exists* $T_0: Z \longrightarrow Y$ *such that* $T_0^* = T$ (*that is,* T *is* $(\omega^*-\omega^*)$*–continuous*).

(ii) $\mathrm{Ker}(T)$ *and* $T(Y_1^*)$ *are* ω^**–closed.*

If moreover (∗∗) *is satisfied, then* (i) *and* (ii) *are also equivalent to:*

(iii) T *is* $(\omega^*-\omega^*)$*–strongly Baire measurable.*

(iv) T *is* $(\omega^*-\omega^*)$*–Borel.*

For the definition of the strong Baire measurability, see the proof of V.3.(ii).

Proof:

(i) ⇒ (ii) is always true by compactness of (Y_1^*,ω^*).

(ii) ⇒ (i): By VII.4 we have

$$T(\omega^*\text{–}\lim(y_n)) = \omega^*\text{–}\lim(T(y_n))$$

for every ω–Cauchy sequence (y_n) in Y^*. We have to show that for every $x \in Z$, $T^*(x)$

belongs to Y. For any ω–Cauchy sequence (y_n) in Y^*, we have

$$\begin{aligned}\lim \langle T^*(x), y_n\rangle &= \lim \langle x, T(y_n)\rangle \\ &= \langle x, \omega^*\text{–}\lim T(y_n)\rangle \\ &= \langle x, T(\omega^*\text{–}\lim(y_n))\rangle \\ &= \langle T^*(x), \omega^*\text{–}\lim(y_n)\rangle\end{aligned}$$

and now property $(*)$ of V.1 shows that $T^*(x) \in Y$. Note that the operator $T_0 = Z \longrightarrow Y$ such that $T_0^* = T$ is simply the restriction of T^* to Z.

(i) $\Longrightarrow$ (iv) is obvious

(iv) $\Longrightarrow$ (iii) is classical and easy.

It remains to show that under the assumption $(**)$ of V.3 the implication (iii) $\Longrightarrow$ (i) is true. Pick an $x \in Z$; if T is $(\omega^*\text{–}\omega^*)$ strongly Baire measurable, then $T^*(x) = x \circ T$ is strongly Baire measurable on (Y^*, ω^*); and now V.3(ii) shows that $T^*(x) \in Y$.

∎

The above result applies to the numerous examples V.4. Let us mention an important special case in the following

Corollary VII.9: *Let* W *be a von Neumann algebra — or more generally a* JBW^**–triple. Let* Z *be an arbitrary Banach space. for every bounded operator* $T: W \longrightarrow Z^*$ *the following are equivalent:*

(i) *There exists* $T_0: Z \longrightarrow W_*$ *such that* $T_0^* = T$.

(ii) $T(W_1)$ *and* $Ker(T)$ *are* ω^**–closed.*

(iii) T *is* $(\omega^*\text{–}\omega^*)$*–strongly Baire measurable.*

(iv) T *is* $(\omega^*\text{–}\omega^*)$*–Borel.*

Remark VII.10: The existence of an inaccessible cardinal allows to construct by forcing Levy's model M of ZFC ([40], [48]) in which every map which can be defined with ordinals and real numbers is strongly Baire measurable; this fact, and VII.9, implies that if W is a von Neumann algebra or a JBW^*–triple, every bounded linear operator from W into itself "which you can write" will be $(\omega^*-\omega^*)$–continuous; that is, ultra–weakly continuous in the terminology of von Neumann algebras.

Up to now, most of the applications of the results and techniques we describe in this survey are actually consequences of results of automatic ω^*–continuity. Let us outline some of these applications.

Applications VII.11: 1) A remarkable theorem of Kaup [37] asserts that the symmetric Cartan domains of holomorphy are the unit balls of the JB^*–triples. Every dual JB^*–triple is actually a JBW^*–triple [3] and Horn [36] has shown (by using Theorem V.3) that JBW^*–triples have a unique predual. This shows that there is at most one topology of convex compact on the symmetric Cartan domains. It is interesting to compare the result of [3] on ω^*–continuity of the product on a dual JB^*–triple with the above VII.8, since a predual (even isomorphic) of JB^*–triple has the property (∗∗).

2) Corollary VII.3 is used in [30] for showing that if A is a $\mathbb{C}^*$–algebra with unit $\mathbf{1}$, then any bilinear product $(x*y)$ on A^{**} which satisfies $||x*y|| \leq ||x||\,||y||$, $\mathbf{1}*x = x*\mathbf{1} = x$ and which extends the product of A is actually the Arens product of A^{**}. This permits to show that the "operators of local reflexivity" from the finite–dimensional subspaces of A^{**} into A are "multiplicative at infinity" [30].

3) If X^* is a Banach lattice, then X is in general not a Banach lattice; an extreme example of this situation is provided by the Gurarii space [34]. However, if every bijective isomtery of X^* is $(\omega^*-\omega^*)$–continuous, X is indeed a Banach lattice: what we have to show is $x \in X \Longrightarrow |x| \in X$, where $|x|$ is taken in X^{**}; but it is easily seen that

there exists an isometric bijection J of X^* (a band symmetry) such that $|x| = J^*(x)$; if $J = J_0^*$, $|x| = J_0^{**}(x) = J_0(x) \in X$. This result applies for instance when X^{**} is w.s.c. or when X has the R.N.P. [31].

Let us mention that it is unknown whether a dual space which is a Banach lattice has at least one predual which is a Banach lattice.

4) The proposition VII.1 can be iterated: for instance if Y^{**} does not contain $\ell^1(\mathbb{N})$, every bijective J isometry of Y^{**} can be written $J = J_0^{**}$, where J_0 is a bijective isomtery of Y. This fact (and much more!) has been used by A. Sersouri in his comprehensive study on the spaces which are hyperplanes in their bidual and isometric to it [46]. Note also that if Y does not contain ℓ_n^1's uniformly, then every bijective isometry J of $Y^{(n)}$ may be written $J = J_0^{(n)}$.

5) If X and Y are reflexive spaces, then $L(X)$ and $L(Y)$ satisfy the assumptions of VII.1 [22] hence every isometric isomorphism between $L(X)$ and $L(Y)$ is $(\omega^*-\omega^*)$–continuous.

Let us conclude this section by observing that the main gist of the above results is that some kind of smoothness on a dual space permits to inverse the operation "transposition"; applying this to spaces gives (unique) preduals; applying this to operators gives automatic ω^*–continuity. An early result [11] along these lines: every M–projection in a dual space is ω^*–continuous, fits into this frame; it can actually be shown by using the techniques of the present work.

VIII. More non–trivial conditions for the existence or the characterization of the predual.

We already met earlier in this paper (§II and III) several properties which are equivalent to the existence of an isometric predual in certain classes of spaces (see II.1 and

III.2). By using different methods, we will see now that this list may be enlarged.

We start with the simple

Proposition VIII.1: *Let* X *be a Banach space. For any closed subspace* Y *of* X, *we denote by* $\mathscr{C}(Y)$ *the set of convex bounded* $\sigma(X,Y)$*–closed subsets of* X. *The following are equivalent:*

(i) Y *is an isometric predual of* X.

(ii) $\mathscr{C}(Y)$ *contains the balls of* X *and has the* $I.P_{f,\infty}$.

Recall that $\sigma(X,Y)$ denotes the topology of pointwise convergence on Y and that $I.P_{f,\infty}$ means that every family with empty intersection contains a finite subfamily with empty intersection.

Proof:

(i) ⟹ (ii) is clear by ω^*–compactness.

(ii) ⟹ (i): If $\mathscr{C}(Y)$ contains the balls of X then by I.1 Y is a norming subspace of X^*. Let Q be the canonical quotient map from X^{**} onto Y^*; since Y is norming, the restriction of Q to X is an isomtery.

Now Y is a predual of X if and only if $Q(X_1)$ is ω^*–closed in Y^*, if and only if $Q(X_1) = Y_1^*$. If not, let $y \in Y_1^* \backslash Q(X_1)$.

For any ω^*–closed convex neighborhood V of y in (Y_1^*, ω^*), the set $\varphi(V) = Q^{-1}(V \cap Q(X_1))$ is a $\sigma(X,Y)$–closed convex bounded subset of X; the family $\{\varphi(V)\}$ clearly has the finite intersection property since the neighborhoods are stable under finite intersection; but $y \notin Q(X_1)$ implies that $\cap \{\varphi(v)\}$ is empty, and this contradicts (ii).

■

In view of this result, the natural approach for trying to obtain non–trivial results is to weaken the condition (ii). This was already done in II.1 and II.2 where we simply had to use the balls instead of the whole set $\mathscr{C}(Y)$. In the next results, we will consider a different subset of $\mathscr{C}(Y)$, namely the $\sigma(X,Y)$–closed slices; that is, the sets of the form

$$S_{y,\epsilon} = \{x \in X_1 \mid y(x) \geq \|y\| - \epsilon\} \quad (y \in Y, \epsilon > 0)$$

Not surprisingly, James' techniques will play a crucial role in what follows; what is more surprising is that we will be able to dispense with the a priori assumption that Y is norming. We will say that Y is α–norming (for some $\alpha > 0$) if

$$\forall x \in X, \ \alpha\|x\| \leq \sup\{y(x)) \mid y \in Y_1\}$$

"1–norming" is simply "norming" as defined in I.1. With this terminology, the following holds [32]:

Theorem VIII.2: *Let* X *be a* ω–$\mathscr{K}$*–analytic Banach space – e.g.* X *separable. Let* Y *be an* α*–norming subspace of* X^*. *Then the following are equivalent:*

(i) Y *is an isometric predual of* X.

(ii) *Every* $y \in Y$ *attains its supremum on* X_1.

Let us observe, for making the link with VIII.1 that y attains its supremum on X_1 if and only if the slices $S_{y,\epsilon}$ $(\varepsilon > 0)$ have the $I.P_{f,\infty}$.

Proof:

(i) $\Rightarrow$ (ii): If Y is an isometric predual, then X_1 is $\sigma(X,Y)$–compact, and (ii) follows from the fact that every continuous function on a compact set attains its supremum.

(ii) $\Rightarrow$ (i): We consider as before $K = Q(X_1)$, where Q is the canonical quotient map from X^{**} onto Y^*. Since Y is α–norming, Q induces an isomorphism between X and $Q(X)$, hence K is norm–closed in Y^*. We need to show that K is ω^*–closed in (Y^*,ω^*); if not, let $y \in \overline{K}^* \setminus K$. By Hahn–Banach, there is $z \in Y_1^{**}$ such that

$$(1) \qquad z(y) = \alpha > \beta = \sup_K z$$

Since X is ω–$\mathcal{K}$–analytic, K is ω–$\mathcal{K}$–analytic as well; the restriction of z to K belongs to $\mathscr{C}((K,\omega))$ and belongs to the ω^*–closure of the set

$$C = \{x \in Y_1 \mid x(y) \geq (\alpha+\beta)/2\}$$

A fortiori, z belongs to the pointwise closure $\overline{C}$ on K of the restriction of C to K. The set $\overline{C}$ is a pointwise compact subset of $\mathscr{C}((K,\omega))$. Since (K,ω) is $\mathcal{K}$–analytic, there exists [53] a sequence (x_n) in C such that

$$(2) \qquad z(k) = \lim_{n\to\infty} x_n(k) \qquad \forall\, k \in K$$

By the assumption (ii), the set K satisfies:

$$\forall\, x \in Y,\ \exists\, k \in K \text{ s.t. } \sup_K (x) = x(k) = \sup_{\overline{K}^*} (x)$$

Now Simons' inequality ([47], Lemma 2) reads

(3) $$\inf\left[\left\{\sup_{\overline{K}^*}(g) \;\middle|\; g \in \mathrm{conv}(x_n)\right\}\right] \le \sup_{k\in K}\left[\overline{\lim}\,(x_n(k))\right]$$

By (1) and (2), we have

$$\overline{\lim}\,(x_n(k)) \le \beta \quad \forall\, k \in K$$

on the other hand, every $g \in \mathrm{conv}(x_n)$ belongs to C, hence satisfies $g(y) \ge (\alpha+\beta)/2$, and hence

$$\sup_{\overline{K}^*}(g) \ge g(y) \ge (\alpha+\beta)/2$$

but this contradicts (3), and this contradiction concludes the proof.

■

Let us mention that the above argument works also [32] if we assume instead that Y^{**} is the ω^*–sequential closure of Y (and no assumption on X). However, some assumption is needed on X or Y, as shown by the

Example VIII.3: The space $X = \ell^1(2^{\aleph_0})$ is considered as the space of discrete measure on $\Delta = \{0,1\}^{\mathbb{N}}$, and $\mathscr{C}(\Delta)$ is seen as a norming subspace Y of X^*; clearly, Y satisfies (ii) but it is not even an isomorphic predual of X.

Let us now gather a few consequences of VIII.2.

Corollary VIII.4: *Let* X *be a separable Banach space such that* 0_X *is a vector* *subspace of* X^{**}. *We denote by* N *the smallest norming subspace of* X^*. *The following are equivalent:*

(i) X *is isometric to a dual space.*

(ii) *Every* $y \in N$ *attains its supremum on* X_1.

Moreover N *is the only* α*–norming subspace of* X^* *which may consist of norm–attaining linear functionals.*

Proof: By ([20], Theorem V.6) (see I.3) 0_X is also ω^*–closed under these assumptions, and then $N = 0_X^{\top}$ is the smallest norming subspace of X^* by I.2 and by II.1. N is the only possible predual of X. The result now follows from VIII.2.

■

By II.3 the above corollary applies to the separable spaces X which does not contain $\ell^1(\mathbb{N})$.

Remark VIII.5: The property

(P): there is a unique norming subspace of Y^{**} which consists of norm–attaining linear functionals

is clearly stronger than "Y unique predual" and is frequently satisfied. It is however a strictly stronger property: indeed it has been shown to me by A. Louveau that there exists – under the continuum hypothesis – a dense subset L of the spectrum Ω of L^∞ such that for every countable disjoint subsets D_1 and D_2 of L one has $\overline{D}_1 \cap \overline{D}_2 = \emptyset$. Now let Z be the space of discrete measures supported by L. It is clear that every $\mu \in Z$ satisfies $\text{supp}(\mu^+) \cap \text{supp}(\mu^-) = \emptyset$ and thus any $\mu \in Z$ attains its norm. Z is norming since $\overline{L} = \Omega$; but $L^1 \cap Z = \{0\}$.

Our next application concerns the isomorphic preduals of $\ell^1(\mathbb{N})$. Let us denote by $\ell^\infty_a(\mathbb{N})$ the set of $y \in \ell^\infty(\mathbb{N})$ which attain their supremum on the unit ball of ℓ^1 equipped with its natural norm. Then one has:

Corollary VIII.6: *A Banach space* X *is isomorphic to an isometric predual of* $\ell^1(\mathbb{N})$ *if and only if* X *is isomorphic to an* α*–norming subspace of* $\ell^\infty(\mathbb{N})$ *contained in* $\ell^\infty_a(\mathbb{N})$.

Proof: If X is isomorphic to an α–norming subspace contained in $\ell^\infty_a(\mathbb{N})$ then by VIII.2 X is an isometric predual of $\ell^1(\mathbb{N})$. The converse implication is trivial.

■

The last result of this section connects the presence of "some kind of compactness" on X_1 with the existence of an isometric predual; more precisely, of a weaker topology of convex compact on X_1.

Corollary VIII.7: *Let* X *be a separable Banach space. If there exists a topology* τ *on the unit ball* X_1 *of* X *such that:*

(i) $Y = \{y \in X^*) \mid y$ *is continuous on* $(X_1,\tau)\}$ *is an* α*–norming subspace of* X^*.

(ii) *Every sequence* (x_n) *in* X_1 *has a subsequence* (x'_n) *which* τ*–converges in Cesaro mean to some* $h \in X_1$.

Then Y *is an isometric predual of* X.

Proof: By VIII.2 it suffices to show that every $y \in Y$ attains its norm. For any $n \geq 1$, pick $x_n \in X_1$ such that $y(x_n) > \|y\| - 1/n$; it is clear that the element $h \in X_1$ we obtain through condition (ii) satisfies $y(h) = \|y\|$.

■

Example VIII.8: If X is a subspace of L^1 and $\tau = L^0$ is the topology of convergence in measure then Komlos' theorem [39] implies (ii), hence VIII.7 applies. A similar situation is met, through a very different approach, in [27], with the concept of "H–spaces". If $X = H^1(D)$ then the space Y is the natural predual $\mathscr{C}(\mathbf{T})/A_0(D)$ of $H^1(D)$.

Let us conclude this section with the

Remarks VIII.9: 1) With the notation of VIII.1, if every $C \in \mathscr{C}(Y)$ is an intersection of balls and Y is norming, then by VIII.1 Y is an isometric predual of X if and only if the balls of X have the I.P$_{f,\infty}$. This condition on $\mathscr{C}(Y)$ is satisfied if Y is "quite convex", e.g. if the norm of Y is l.u.c.; let us observe that "Y has R.N.P." is not sufficient, as shown by the example $X = \ell^\infty(\mathbb{N})$, $Y = \ell^1(\mathbb{N})$.

Note that if Y is a predual of X and every $C \in \mathscr{C}(Y)$ is an intersection of balls, then clearly (X_1, b_x) is Hausdorff and thus III.2 applies which shows that Y is unique predual — and much more.

2) An alternative approach to VIII.2 is described in [1], where it is shown that the validity of VIII.2 relies on the completeness of X_1 for the Mackey uniform structure $\mu(X,Y)$ of uniform convergence on the convex $\sigma(Y,X)$–compact subsets of Y. The proofs are also using crucially James' techniques.

3) The ball topology permits to show ([20], Corollary VI.4) that if X is separable and if for every $z \in X^{**}$, $P(z) = \{x \in X \mid \|x'-x\| \leq \|x'-z\| \; \forall \; x' \in X\}$ contains *exactly one* point, then X is isometric to a dual space — and (X_1, b_x) is Hausdorff, hence III.2 applies.

IX. Spaces which are not unique preduals, and other counterexamples.

In this section we gather several simple facts which mark the limits of the positive results we have been showing so far.

(1) The simplest examples of spaces which are not unique preduals are the spaces $c_0(I)$, where I is any infinite set. For showing this, it suffices to notice that $\mathscr{C}(\hat{I})$, the space of continuous functions on the Alexandrov compactification of I discrete, is not isometric to $c_0(I)$ but $\mathscr{C}(\hat{I})^* = c_0(I)^* = \ell^1(I)$. Actually, the class of Asplund–Lindenstrauss spaces – that is, of isometric preduals of $\ell^1(I)$ – is considerably richer than the class of spaces $\mathscr{C}_0(\Gamma)$, where Γ is locally compact scattered (see [4]).

(2) It is easy to deduce from (1) that if K is an infinite compact set, then $\mathscr{C}(K)$ is not unique predual. Indeed $\mathscr{C}(K)^* = \mathscr{M}_d(K) \oplus_1 \ell^1(|K|)$, $\mathscr{C}(K)^*$ is isometric to Y^* where $Y = \mathscr{C}(K) \oplus_\infty c_0(|K|)$. But Y is not isometric to $\mathscr{C}(K)$ since 0 belongs to the ω^*–closure of $\mathrm{Ext}(Y_1^*)$.

If we pick $K = \Delta = \{0,1\}^{\mathbb{N}}$, we have an example of two spaces, $\mathscr{C}(\Delta)$ and $\mathscr{C}(\Delta) \oplus_\infty c_0(|\Delta|)$ which have isometric duals although one of them is separable and the other is not. Another example of this situation is provided by $\mathscr{C}(\Delta)$ and $\mathscr{C}(I)$ where I is the "two arrows" space, that is, the space $[0,1] \times \{0,1\}$ equipped with the lexicographical order.

(3) If K is an infinite compact set, there is a bijective isometry of $\mathscr{C}(K^*)$ which is not $(\omega^*–\omega^*)$–continuous. Indeed one has the

Claim: In any infinite compact set K, there is a sequence (k_n) which has a cluster point k which is distinct of every (k_n).

Proof of the claim: If there exists an inifinite closed subset F of K and $k \in F$

such that for every neighborhood V of k, $(F\backslash V)$ is finite, we take D any countable subset of $F\backslash\{k\}$ and we write $D = \{k_n \mid n \geq 1\}$; clearly (k_n) converges to k.

If not, we pick $k_1 \in K$ and an open neighborhood $V(k_1)$ of k_1 such that $K\backslash V(k_1)$ is infinite; we let $F_1 = K\backslash V(k_1)$ and we pick $k_2 \in F_1$ and $V(k_2)$ such that $F_1\backslash V(k_2) = F_2$ is infinite. We go on picking $k_{n+1} \in F_n$ and $V(k_{n+1})$ such that $F_{n+1} = F_n\backslash V(k_{n+1})$ is infinite. It is clear now that every cluster point k of (k_n) is distinct of every (k_n). This proves the claim.

Once we have such a sequence (k_n) and $k \in \overline{(k_n)}\backslash\{k_n \mid n \geq 1\}$, we let $D = \{k\} \cup \{k_n \mid n \geq 1\}$ and we define $\varphi:: K \longrightarrow K$ by:

$$\begin{cases} \varphi(x) = 0 & \text{if } n \in K\backslash D \\ \varphi(k) = k_1 & \\ \varphi(k_n) = k_{n+1} & \forall\, n \geq 1 \end{cases}$$

and we define $T: \mathscr{C}(K)^* \longrightarrow \mathscr{C}(K)^*$ by $T(\mu) = \varphi(\mu)$. φ is a Borel–isomorphism of K hence T is a bijective isometry of $\mathscr{C}(K)^*$; but $\{\varepsilon_k\}$ is a ω^*–cluster point of (ε_{k_n}) and $T(\varepsilon_k) = \varepsilon_{k_1}$ is not a ω^*–cluster point of $T(\varepsilon_{k_n})$, hence T is not $(\omega^*\text{–}\omega^*)$ continuous.

(4) If Y is an M–ideal in its bidual and Y is not reflexive, then Y is not strongly unique predual [27]; the idea of the proof is to use the fact that since $Y^{***} = Y^* \oplus_1 Y^\perp$, there is "a lot of room" for moving $Y^\perp$ into another ω^*–closed subspace $Z^\perp$ of Y^{***} such that $Y^{***} = Y^* \oplus Z^\perp$ and the projection of kernel $Z^\perp$ has norm one.

In the proof of [27] we construct a predual Z such that $Y/Y\cap Z$ is one–dimensional; along these lines, it is shown in [26] that if two preduals Y and Z differ only on a finite number of dimensions (that is, $(Y/Y\cap Z)$ is finite–dimensional), then $(Y\cap Z)$ contains an isomorphic copy of $c_0(\mathbb{N})$.

(5) If X is a w.s.c. Banach lattice, then X is norm–one complemented in X^{**} by a band projection. This provides examples, such as $X = L^1$, of spaces which satisfy the conditions (iii)–(iv) of II.1 but are not isomorphic to dual spaces. Note that if X is a separable w.s.c. Banach lattice then X is isometric to a dual Banach lattice – for the norm and the order structure – if and only if X has the R.N.P. [54]. The techniques of the present paper and of [54] are quite different. Note that the predual X_* constructed in [54] is not always order–continuous [55]; it is order–continuous if and only if X has a lattice $\sigma(X,X_*)$–l.s.c. equivalent norm which is Frechet–smooth on a dense set [31].

(6) If X is separable and norm–one complemented in X^{**} but X is not isometric to a dual space then by ([20], Theorem V.6) – see I.3 – 0_X is not a vector space; in other words, there are two norming hyperplanes of X^* whose intersection is not norming. This applies for instance to separable w.s.c. Banach lattices X which fail the R.N.P.

(7) The above remark (3) has a non–commutative generalization: If a $\mathbb{C}^*$–algebra A is not an ideal of A^{**}, there is a hermitian operator H on A^* which is not $(\omega^*\text{–}\omega^*)$–continuous [42]; then $J = \exp(iH)$ is a bijective isometry which is not $(\omega^*\text{–}\omega^*)$–continuous.

On the other hand, it is easily seen that every bijective isometry of Y^* is $(\omega^*\text{–}\omega^*)$–continuous when Y is an M–ideal of Y^{**}.

X. Open Problems.

We list in this final section some of the natural problems which remain open. Some of them should not be desperately hard. Problems (1) and (2) are the most interesting; I believe that both of them are difficult.

(1) Let Y be a Banach space which contains no isomorphic copy of $c_0(\mathbb{N})$. Is Y unique predual?

A positive answer would improve all the isomorphic results of the present article, and would essentially end up the theory. It seems unlikely, however, that such a general theorem could be true; we refer the reader who is not afraid of technicalities to ([26], §V) for a tentative counterexample. Note that it is also unknown whether Y is unique predual under the stronger assumption that it does not contain ℓ_n^∞'s uniformly. Let us also mention that a separable space Y which contains $c_0(\mathbb{N})$ is not unique predual for every norm, since by Sobczyk's theorem we can write $Y = Z \oplus c_0(\mathbb{N})$. If Z is equipped with a Gateaux–smooth norm $\|\ \|$, and Y with the ℓ^1–sum of this norm and the natural norm of $c_0(\mathbb{N})$, $Y^* = Z^* \oplus_\infty \ell^1(\mathbb{N})$; but if $V = (Z, \|\ \|) \oplus_1 \mathscr{C}(\bar{\mathbb{N}})$, V^* is isometric to Y^* and it is easily seen – by using the fact that there is a single ℓ^1–projection in V and in Y – that Y and V are not isometric.

In other words, a separable space which is unique predual for every equivalent norm cannot contain $c_0(\mathbb{N})$, and (1) asks the question of the converse.

(2) Is there a space which is unique predual but not strongly unique predual?

This question is widely open. I see no reason for the two notions to coincide and I have no idea about what a counterexample should look like. In other words, is it possible for a dual space Y^* to have all its preduals isometric but occupying different positions in Y^{**}?

(3) Is it true that a Banach space which has two non–isometric preduals has infinitely many non–isometric preduals?

Within the language of convex sets, this question reads: could we partition the class of convex sets into the three families: $C \in (C_0) - (C_1) - (C_\infty)$ – if there exist zero–exactly one–infinitely many – topologies of locally convex compact on C?

Let us mention that within the class of quasi–Banach spaces, (3) has a positive answer [56]; moreover there is a good characterization of quasi–Banach spaces which are unique preduals in their class [56].

(4) Is it true that: Y unique predual implies Y^{**} unique predual? Conjecture: No. A candidate counterexample is $Y = (\Sigma \oplus \ell_n^\infty)_1$.

(5) Is it true that: Y unique predual implies $L^1(Y)$ unique predual? See [8] for several positive results along these lines.

(6) If Z is a subspace of Y and both Z and Y/Z are unique preduals, is it true that Y is unique predual? What if Z and Y/Z have the property (**) of V.3 – that is, the property (X) of [26]?

(7) Is the property (**) of V.3 stable under G_δ–embeddings? See [19] where this notion is introduced and studied. What can be said of the spaces which G_δ–embed into L^1? Note that (**) is stable under "Tauberian" embeddings – that is, operators T such that T^{**} is injective.

(8) If X is a Banach lattice and is isometric to a dual Banach space, does there exist at least one predual of X which is a Banach lattice? Which is predual of X for the norm and for the order structure? Note that the order intervals of X are $\sigma(X,Y)$–compact for every isometric predual Y of X [31].

(9) We denote by $\mathcal{N}$ the set of norming subspaces of the dual X^* of X. Does there exist a Banach space X such that $(\cap\mathcal{N})$ separates X without norming X?

(10) If there is a norm–one projection from X^{**} onto X then clearly the balls of X have the $I.P_{f,\infty}$. The converse is open. The $I.P._{f,\infty}$ for the balls is equivalent to:

$$\forall z \in X^{**},\ P(z) = \{x \in X \mid \|x'-x\| \le \|z-x'\|\ \forall x' \in X\} \neq \emptyset$$

Note that by [20], if X is separable and X enjoys "some kind of smoothness" (see §II and III) which implies that $P(z)$ has exactly one element for every $z \in X^{**}$, then the projection which send z to the unique element of $P(z)$ is linear and has norm one – and even a ω^*–closed kernel ([20], Theorem II.8 and II.9).

The problem one meets with the general question is to select $\pi(z) \in P(z)$ in a linear way.

(11) The space $\ell^1(\mathbb{N})$ has a huge supply of isomorphic preduals and it seems difficult to characterize which equivalent norms on ℓ^1 are dual norms. On the other hand, when a space does not contain $\ell^1(\mathbb{N})$, II.1 and III.2 provide us with some characterizations. It would be interesting to know whether the roots of these results lie into the topological complexity of the set of dual norms. If X is a separable Banach space and if we identify a norm with its unit ball, we may put on the set $N(X)$ of equivalent norms on X the Borel structure Σ induced by the Effros–Borel structure on the closed subsets of the Polish space X (see [9]); and then $(N(X),\Sigma)$ becomes a Lusin measurable space. Now the question reads: is the set of dual norms on $\ell^1(\mathbb{N})$ Lusin, or Souslin, in $(N(\ell^1),\Sigma)$? (Conjecture: No). What if X does not contain $\ell^1(\mathbb{N})$?

(12) If W is a von Neumann algebra, is $\sigma(W,W_*)$ the unique compact topology on W_1 such that for every $\lambda \in [0,1]$, the map $g_\lambda(x,y) = \lambda x + (1-\lambda)y$ is separately continuous from $W_1 \times W_1$ to W_1? The answer is yes if W is commutative and W_* is separable ([20], §VII) and it is natural to believe that this extends to the general case.

(13) By V.4(5) the space $H^\infty(D)$ has a unique predual. Is it also true for $H^\infty(D^2)$? Or more generally for $H^\infty(R)$ where R is a Riemann surface?

(14) Let U be the space of functions on T which are uniform limit of their Fourier expansion, equipped with its natural norm. Is U^* unique predual of U^{**}?

(15) If X is "almost uniformly smooth" (see [17]) and if for every bounded set C, the Chebyshev radii of C in X and X^{**} are the same, then X is a dual space [17]. Is it true under the assumption that X is Asplund, or merely that X does not contain $\ell^1(\mathbb{N})$?

REFERENCES

[1] M. ABD–ES–SALAM: On Banach spaces containing ℓ^1 and linear Kelley spaces, Ph.D., Columbia, Missouri (1986).

[2] T.ANDO: On the predual of H^∞, Commentationes Matehmaticae Special, I, Warsaw (1978).

[3] T. BARTON, R.M. TIMONEY: Weak* continuity of Jordan triple products and its applications, Math. Scand. 59, 2 (1986), 177–191.

[4] Y. BENYAMINI, J. LINDENSTRAUSS: A predual of ℓ^1 which is not isomorphic to a $\mathscr{C}(K)$ space, Israel J. of Maths, 13 (1972), 246–259.

[5] F.F. BONSALL, J. DUNCAN: Complete normed algebras, Springer–Verlag, Berlin (1973).

[6] J. BOURGAIN, D.M. FREMLIN, M. TALAGRAND: Pointwise compact sets of Baire–measurable functions, Amer. J. of Maths, 100, 4 (1978), 845–886.

[7] L. BROWN, T. ITO: Some non–quasireflexive spaces having unique isomorphic preduals, Israel J. Maths 20 (1975), 321–325.

[8] M. CAMBERN, P. GREIM: Uniqueness of preduals for spaces of continuous vector functions, Canad. Math. Bulletin, to appear.

[9] J.P.R. CHRISTENSEN: Topology and Borel structure, North Holland Math. Studies (1974).

[10] H.H. CORSON, J. LINDENSTRAUSS: On weakly compact subsets of Banach spaces, Proceedings of the A.M.S. 17 (1966), 407–412.

[11] F. CUNNNINGHAM, E.G. EFFROS, N.M. ROY: M–structure in dual Banach spaces, Israel J. Maths. 14, 3 (1973), 304–308.

[12] R. DEVILLE: Thèse de 3ème Cycle, Université Paris VI (1984).

[13] J. DIESTEL: Geometry of Banach spaces, Selected topics, Lecture Notes 485, Springer–Verlag (1975).

[14] J. DIXMIER: Sur certains espaces considerérés par M.H. Stone, Summa Brasil. Math. 2 (1951), 151–182.

[15] J. DIXMIER: Les algèbres d'opérateurs dans l'espace Hilbertien, Cahiers Scientifiques, Gauthier–Villars (1969).

[16] G.A. EDGAR: An ordering of Banach spaces, Pacific J. Maths, 108, 1 (1983), 83–98.

[17] C. FINET: Une class d'espaces de Banach à predual unique, Quarterly J. Maths, 35 (1984), 403–414.

[18] N. GHOUSSOUB, G. GODEFROY, B. MAUREY, W. SCHACHERMAYER: Some topolological and geometrical structures in Banach spaces, Memoirs of the A.M.S. 378 (1987).

[19] N. GHOUSSOUB, B. MAUREY: H_δ–embeddings in Hilbert space and optimization on G_δ–sets, Memoirs of the A.M.S. 349 (1986).

[20] G. GODEFROY, N. KALTON: The ball topology and its applications, to appear.

[21] G. GODEFROY: Points de Namioka, espaces normants, applications à la thèorie isomètrique de la dualité, Israel J. Maths, 38, 3 (1981), 209–220.

[22] G. GODEFROY, P.D. SAPHAR: Duality in spaces of operators and smooth norms on Banach spaces, to appear in Illinois J. Maths (1988).

[23] G. GODEFROY, B. MAUREY: Normes lisses et normes anguleuses sur les espaces de Banach séparables, unpublished preprint.

[24] G. GODEFROY: Metric characterization of first Baire class functions and everywhere octahedral norms, to appear.

[25] G. GODEFROY: Quelques remarques sur l'unicité des préduaux, Quarterly J. Maths, 2, 35 (1984), 147–152.

[26] G. GODEFROY, M. TALAGRAND: Nouvelles classes d'espaces de Banach à predual unique, Séminaire d'Ana. Fonctionnelle de l'Ecole Polytechnique (1980/81).

[27] G. GODEFROY: Sous–espaces bien disposés de L^1 Applications, Transactions of the A.M.S. 286, 1 (1984), 227–249.

[28] G. GODEFROY, P. SAAB: Quelques espaces de Banach ayant les propriétés (V) on (V^*) de A. Pelczynski, Note aux C.R.A.S. Paris, 303, 1, 11 (1986), 503–506.

[29] G. GODEFROY: Parties admissibles d'un espace de Banach. Applications. Ann. Sci. de l'Ec. Norm. Sup. 16, 4 (1983), 109–122.

[30] G. GODEFROY, B. IOCHUM: Arens–regularity of Banach algebras and the geometry of Banach spaces, J. of Funct. Ana. 79 (1988), to appear.

[31] G. GODEFROY: Symétries isométriques. Applications à la dualité des espaces réticules, Israel J. Maths, 44, 1 (1983), 61–74.

[32] G. GODEFROY: Boundaries of a convex set and interpolation sets, Math. Ann, 277 (1987), 173–184.

[34] V.I. GURARII: Space of universal disposition, isotopic spaces and the Mazur problem on rotations of Banach spaces, Sibirskii Mat. Zhurnal, 7 (1966), 1002–1013.

[35] R. HAYDON: Some more characterizations of Banach spaces containing $\ell^1(\mathbb{N})$, Math. Proc. Cambridge Phil. Soc. 80 (1976), 269–276.

[36] G. HORN: Characterization of the predual and ideal structure of a JBW^*–triple, Math. Scand. 61, 1 (1987), 117–133.

[37] W. KAUP: A Riemann mapping theorem for bounded symmetric domains in Banach spaces, Math. Z. 183 (1983), 503–529.

[38] KNAUST, T. ODELL: The space JH has the property (u), to appear.

[39] J. KOMLOS: A generalization of a problem of Steinhaus, Acta. Math. Acad. Sci. Hungar. 18 (1967), 217–229.

[40] A. LEVY: Definability in axiomatic set theory I, in Logic, Methodology and Philosophy of Science, Y. Bar–Hillel ed., North Holland (1965), 127–151.

[41] J. LINDENSTRAUSS, H.P. ROSENTHAL: The $\mathscr{L}^P$–spaces, Israel J. of Maths, 7 (1969), 325–349.

[42] K. MATTILA: A class of hypernormal operators and ω^*–continuity of hermitian operators, Arkw für mathematik, 25, 2 (1987), 265–274.

[43] E. ODELL, H.P. ROSENTHAL: A double–dual characterization of separable Banach spaces containing ℓ^1, Israel J. of Maths, 20 (1975), 375–384.

[44] H.P. ROSENTHAL: A characterization of Banach spaces containing ℓ^1, Proc. Nat. Acad. Sci. U.S.A. 71 (1974), 2411–2413.

[45] H.H. SCHAEFER: Banach lattices and positive operators, Springer–Verlag, 215 (1974).

[46] A. SERSOURI: Structure theorems for some quasi–reflexive Banach spaces, Transactions of the A.M.S., to appear.

[47] S. SIMONS: A convergence theorem with boundary, Pacific J. Maths, 40 (1972), 703–708.

[48] R.M. SOLOVAY: A model of set theory in which every set of reals is Lebesgue measurable, Annals of Math 2, 92 (1970), 1–56.

[49] C. STEGALL: The Radon–Nikodym property in conjugate Banach spaces, Transactions of the A.M.S. 206 (1975), 213–223.

[50] M. TAKESAKI: On the conjugate space of an operator algebra, Tohoku Math. J. 10 (1958).

[51] M. TALAGRAND: Propriété de Baire forte et fonctions continues, Fundamenta Math.

[52] M. TALAGRAND: A new type of affine Borel functions, Math. Scand. 54, 2 (1984), 183–188.

[53] M. TALAGRAND: Espaces de Banach faiblement $\mathscr{K}$–analytiques, Ann. Maths 110 (1979), 407–438.

[54] M. TALAGRAND: La structure des Banach réticulés ayant la propriété de Radon–Nikodym, Israel J. Maths, 44, 3 (1983), 213–220.

[55] M. TALAGRAND: Dual Banach lattices and Banach lattices with the Radon–Nikodym property, Israel J. Maths, 38 (1981), 46–50.

[56] E. WERNER: Quasi–Banach spaces which are unique preduals, Math. Annalen, to appear.

[57] E. WERNER: A new proof of the theorem of Odell–Rosenthal, preprint.

Contemporary Mathematics
Volume **85**, 1989

The ball topology and its applications

G. GODEFROY AND N. J. KALTON[1]

Abstract. *We study the coarsest topology on a Banach space X such that every norm-closed ball is closed. We show under certain hypotheses, in particular if X does not contain ℓ_1, that on the closed unit ball of X this topology coincides with a vector topology. This has applications to the study of uniqueness of compact "consistent" topologies on the unit ball and to the uniqueness of preduals. Numerous other applications are given, including an extension of a result of Corson and Lindenstrauss: every weakly compact set can be expressed as an intersection of finite unions of closed balls.*

1. Introduction. This paper is motivated by two general problems. The first is to determine conditions on a Banach space X so that it is isometrically a dual space in a unique way. We say that X has a unique predual (UPD) if there is precisely one closed subspace $E \subset X^*$ so that X can be identified with E^* (or equivalently the closed unit ball X_1 of X is $\sigma(X, E)-$compact). For results on this general problem, see, for example, [4], [5], [6] and [11]. One can rephrase the condition that X has a unique predual; X has (UPD) if there is exactly one compact Hausdorff topology on X_1 induced by a locally convex linear topology on X.

The second problem is to determine conditions on X so that for any Hausdorff linear topology τ on X with X_1 $\tau-$compact we have that $\tau \mid X_1$ agrees with a locally convex

[1] The second author was supported by N.S.F. grants DMS-8301099 and DMS-8601401

AMS Classifications 46B05, 46B10, 46B20

Key words: Banach space, norm, isometric, norming subspace, unique predual.

linear topology on X. This problem was originally studied in [13] and was motivated by the question of the validity of the Krein-Milman Theorem in non-locally convex F-spaces. It was shown by J. W. Roberts [27] that there exists an absolutely convex compact convex set with no extreme points. Thus there exists a Banach space X which is not a dual space but which admits a linear topology τ so that (X_1, τ) is compact. However it is shown in [13] that if X is reflexive or has a separable dual then any linear topology τ so that X_1 is $\tau-$compact agrees with a locally convex topology on X_1. Recently these results have been extended by Sersouri [30].

On comparing the results known for these two problems, one can deduce that for some Banach spaces X (e.g. if X has a separable dual) any two Hausdorff linear topologies which make X_1 compact must agree on X_1.

Let us say that a topology τ on X is *consistent* if the maps

$$g_\lambda(x, y) = \lambda x + (1 - \lambda)y$$

are separately continuous on $X_1 \times X_1$ for $0 \leq \lambda \leq 1$. Motivated by the preceding discussion we seek conditions on X so that X_1 admits exactly one compact Hausdorff consistent topology. In this case we say that X has the *compact uniqueness property (CUP)*. In such spaces the geometry of the unit ball predetermines the compact Hausdorff topology on it. We show in this paper that many of the known conditions for X to have (UPD) in fact force the stronger (CUP); at the same time our methods enable us to produce new conditions implying (UPD) and (CUP).

In order to study this question we introduce the *ball topology* on X denoted by b_X. b_X is the coarsest topology so that every closed ball $B(x, \rho) = \{u : \|u - x\| \leq \rho\}$ is closed in b_X. Thus a point $x_0 \in X$ has a base of neighborhoods of the form

$$V = X \setminus \cup_{i=1}^n B(x_i, \rho_i)$$

where $x_1, \ldots, x_n \in X$, $n \in \mathbf{N}$ and $\|x_0 - x_i\| > \rho_i$. The ball topology seems to have some intrinsic interest and we study it in some detail. Let us note that it was first employed by Corson and Lindenstrauss [2] (cf. Theorem 8.2 below).

Let us make some initial remarks. First we note that b_X is a Hausdorff topology but that it is a T_1-topology. We also note that:

(1) For fixed $y \in X$ the map $x \to x + y$ is b_X-continuous.

(2) For fixed $\lambda > 0$ the map $x \to \lambda x$ is b_X-continuous.

(3) The map $x \to -x$ is b_X-continuous.

It will be convenient to call a topology on a linear space X a *prelinear* topology if it satisfies (1) and (2) above. Thus b_X is a prelinear topology.

We also make some further observations which we use later without comment:

(4) For any $y \in X$ the map $x \to \|x - y\|$ is b_X-lower-semi-continuous (b_X-l.s.c.).

(5) If X is a subspace of E then $b_E \mid X$ is finer than b_X.

(6) If X is separable then b_X satisfies the second axiom of countability ([18] p. 48). In particular every $x \in X$ has a countable base of neighborhoods.

Note that, of course, b_X depends on the norm on X; however, throughout most of the paper, we will suppress mention of the given norm.

The key idea of the paper is to study the restriction of b_X to certain subsets of X, especially the unit ball X_1. In certain circumstances, we are able to show that the restriction of b_X is a "reasonable" topology (e.g. Hausdorff or regular) and from such results we obtain our main theorems. Such an approach was first used in [2].

Let us now summmarize the results of the paper. In Section 2, we discuss some preliminary results on the topology b_X restricted to X_1. We give conditions so that (X_1, b_X) is Hausdorff or compact and discuss the continuity of linear functionals restricted to (X_1, b_X). In Section 3, we prove a technical result characterizing absolutely convex sets C so that (C, b_X) is a regular topological space; we show that this can happen if and only if there is a Hausdorff locally convex linear topology on the span of C whose restriction to C agrees with b_X. We also show that (C, b_X) is regular when C is a Rosenthal set (i.e. contains no basic sequence equivalent to the unit vector basis of ℓ_1).

Our first application is in Section 4. An operator $T : X \to Y$ where Y is an arbitrary topological vector space is called quasi-convex [13] if the continuous affine functionals on $\overline{T(X_1)}$ separate the points of $\overline{T(X_1)}$. We improve considerably on results in [13] by showing that if X does contain ℓ_1 then every operator on X is quasi-convex. Combined with a recent result of Sersouri [30] this characterizes spaces not containing ℓ_1.

In Section 5, we study spaces X for which b_X is *locally linear* i.e. (X_1, b_X) is regular. Any Banach space which does not contain ℓ_1 has this property; more generally if X does not contain ℓ_1 and $X \subset Y \subset X^{**}$ then b_Y is locally linear. An important property of such spaces is that the intersection of all norming subspaces of X^* is itself a norming subspace (the minimal norming subspace). In Section 6, we apply our results to the properties (UPD) and (CUP). This leads to some new results on the property (UPD).

Our methods fail for $C(K)$−spaces; in Section 7, we discuss some partial results on (CUP) for these spaces.

In Sections 8-9 we discuss some other aspects of the ball topology. For example, we extend a result of Corson and Lindenstrauss [2] by showing that a weakly compact set is always "ball-generated" (i.e. can be expressed as an intersection of finite unions of balls). In Section 10 we gather some miscellaneous remarks and open problems.

NOTATION: As above, the closed unit ball of a Banach space X is denoted by X_1. The closed ball center x and radius ρ is denoted by $B(x, \rho)$. All Banach spaces are assumed real. If X and Y are in duality then $\sigma(X, Y)$ denotes the topology on X induced by pointwise convergence on Y. The dual of X is X^*; the weak topology $\sigma(X, X^*)$ is denoted by w and the weak*-topology $\sigma(X^*, X)$ is denoted by w^*. The cardinality of a set A is denoted by $|A|$.

ACKNOWLEDGEMENT: This work was done while the first author was visiting the University of Missouri-Columbia in 1985/6 and 1987/8. It is his pleasure to thank the Department of Mathematics for its wonderful hospitality.

2. Some general results. Ball continuous linear forms. Our first lemma concerns acceleration of convergence in certain prelinear topologies. It will paly a central role in our study of the ball topology.

LEMMA 2.1. *Let X be a Banach space and let α be any prelinear topology on X. Suppose 0 has a countable base of neighborhoods in the space (X_1, α). Suppose (x_n) is any bounded sequence in X and x is an α-cluster point of (x_n). Then there is a subsequence (u_m) of (x_n) so that if $v_n \in co\{u_i : i \geq n\}$ then we can find a sequence $w_n \in co(\{u_i : i \geq n\} \cup \{x\})$ with $\|w_n - v_n\| \to 0$ and $w_n \to x$ for the topology α.*

PROOF: Let us first note that the continuity properties of addition and scalar multiplication imply that it suffices to prove the lemma when $\|x_n\| \leq 1$ for all n and $x = 0$. Let $(U_i)_{i\geq 1}$ be a base of open α-neighborhoods of 0 in X_1, with $U_1 = X_1$.

Let $\mathbf{N}_0 = \mathbf{N} \cup \{0\}$ and let Λ be the space of all finitely non-zero sequences $\lambda = (\lambda_n)$ in $\mathbf{N}_0$ such that $\sum \lambda_n 2^{-n} \leq 1$. Then set Λ_k to be the subset of $\lambda \in \Lambda$ such that $\lambda_j = 0$ if $j \leq k$ and let Λ_k^n be the set of $\lambda \in \Lambda_k$ such that $\lambda_j = 0$ for $j \geq n+1$.

We shall construct (u_n) by induction so that if

$$w_n = \sum_{k=0}^{\infty} \lambda_k^{(n)} 2^{-k} u_k$$

where $\lambda^{(n)} = (\lambda_k^{(n)}) \in \Lambda_n$ then $w_n \to 0$ (α). This will establish the lemma. Indeed if $v_n \in \mathrm{co}\{u_i : i > n\}$ let us write $v_n = \sum_{i>n} \mu_i u_i$ where $\mu_i \geq 0$, $\sum \mu_i = 1$, and the μ_i are eventually zero. Set $\lambda_i^{(n)} = 0$ for $i \leq n$ and for $i > n$ set $\lambda_i^{(n)} = \sup(k \in \mathbf{N}_0 : k.2^{-i} \leq \mu_i)$. Now $\lambda^{(n)} \in \Lambda_n$ and if we define w_n as above then $w_n \in \mathrm{co}(\{u_i : i > n\} \cup \{0\})$ and

$$\|v_n - w_n\| \leq (\sum_{i>n} 2^{-i}) \sup_i \|u_i\|$$

so that the sequence w_n satisfies the conclusions of the theorem.

It suffices to construct (u_n) so that if $1 \leq m \leq n$ and $\lambda \in \Lambda_m^n$ then $\sum \lambda_k 2^{-k} u_k \in U_m$.

Pick $u_1 = x_1$; the conditions are then satisfied since Λ_n^n reduces to the zero sequence for every n and hence for $n = 1$. Next suppose $(u_1, \ldots, u_n)$ have been selected to satisfy the inductive hypothesis. If $\lambda \in \Lambda$ then in the topology α,

$$\lim_{p \to \infty} (\sum_{k=1}^{n} \lambda_k 2^{-k} u_k + \lambda_{n+1} 2^{-(n+1)} x_p) = \sum_{k=1}^{n} \lambda_k 2^{-k} u_k.$$

As the set of points $\{\sum_{k=1}^{n} \lambda_k 2^{-k} u_k : \lambda \in \Lambda\}$ is finite and the range of values of λ_{n+1} is finite we can choose p large enough so that $p > q$ where $u_n = x_q$ and

$$\sum_{k=1}^{n} \lambda_k 2^{-k} u_k + \lambda_{n+1} 2^{-(n+1)} x_p \in U_m$$

whenever $1 \leq m \leq n+1$ and $\lambda \in \Lambda_m^{n+1}$. If we set $u_{n+1} = x_p$ this completes the induction and establishes the lemma.

We shall call a sequence (u_n) which satisfies the conclusion of the Lemma above *convex clustering* at x for the topology α. If we take $\alpha = b_X$ then the conclusion may be strengthened by noting that if $\|w_n - v_n\| \to 0$ and $w_n \to x$ in b_X then $v_n \to x$ in b_X. In fact $v_n \to x$ for b_X is equivalent to the statement that for any $y \in X$ we have $\liminf \|v_n - y\| \geq \|x - y\|$.

LEMMA 2.2. *Let α be a prelinear topology on a Banach space X such that the unit ball X_1 is α-closed. Suppose (u_n) is a sequence which is convex clustering at x for α. Then for every weak*-cluster point x^{**} of (u_n) and every $t \in X$, we have $\|x^{**} - t\| \geq \|x - t\|$.*

PROOF: Clearly every norm-closed ball is also α-closed and the norm is α-lower-semi-continuous. Thus for every $t \in X$,

$$\lim_{n \to \infty} d(t, \mathrm{co}\{u_k : k \geq n\}) \geq \|x - t\|.$$

Indeed, if not, we can find $t \in X$, $\epsilon > 0$ and a sequence $v_k \in \mathrm{co}\{u_i : i \geq k\}$ such that $\|v_k - t\| \leq \|x - t\| - \epsilon$ for every k. Then, if w_k is given as in Lemma 2.1 we have that for large enough k, $\|v_k - w_k\| < \epsilon/2$. Thus $\|w_k - t\| \leq \|x - t\| - \epsilon/2$. This is impossible, since $w_k \to x$ for α and the norm is lower-semi-continuous.

Now let x^{**} be any cluster point of (u_n) in (X^{**}, w^*). For every $n \geq 1$, $\delta > 0$ there exists $y \in \mathrm{co}\{u_k : k \geq n\}$ with $\|y - t\| \leq (1+\delta)\|x^{**} - t\|$. Indeed, if not, by the Hahn-Banach separation theorem, there exists $f \in X^*$ with $\|f\| = 1$ so that for every $k \geq n$ $f(u_k - t) \geq (1+\delta)\|x^{**} - t\|$. But this implies that $f(x^{**} - t) \geq (1+\delta)\|x^{**} - t\|$ which is a contradiction.

We conclude that $\|x^{**} - t\| \geq d(t, \mathrm{co}\{u_k : k \geq n\})$ for all n and the lemma follows.

Note that Lemmas 2.1 and 2.2 apply to $\alpha = b_X$ when X is separable. The following lemma is proved for completeness (see [4]).

LEMMA 2.3. *Let X be a Banach space and suppose $x^{**} \in X^{**}$. The following conditions are equivalent:*

(1) $\|x^{**} - x\| \geq \|x\| \qquad \forall x \in X.$

(2) Ker $x^{**} \cap X_1^*$ is w^*-dense in X_1^*.

PROOF: For any $x \in X$ and any closed subspace $Y \subset X^*$ one has $\|x|_Y\| = d(x, Y^\perp)$ since Y^* is isometric to $X^{**}/Y^\perp$. If we apply this remark to $Y = \ker x^{**}$ we obtain that (1) is equivalent to $\|x|_{\ker(x^{**})}\| = \|x\|$ for all $x \in X$. The Hahn-Banach Theorem shows that this is equivalent to (2).

Let us recall that a closed subspace Y of X^* is said to be *norming* if

$$\|x\| = \sup\{y(x) : y \in Y, \|y\| \leq 1\} \qquad \forall x \in X.$$

Equivalently, Y is norming if $Y \cap X_1^*$ is w^*-dense in X_1^*.

Let us denote by N_X the intersection of all norming subspaces in X^*. N_X is a closed subspace of X^* which can reduce to $\{0\}$ (e.g. if $X = \ell_1$). If N_X is itself norming then X^* has a *minimal norming subspace.*

THEOREM 2.4. *Let X be a Banach space and suppose $f \in X^*$. Then:*

(1) If f is continuous on (X_1, b_X) then $f \in N_X$.

(2) If $f \in N_X$ and X is separable then f is continuous on (X_1, b_X).

PROOF: (1) Suppose $f \in X^*$ is continuous on (X_1, b_X). If Y is a (closed) norming subspace of X^* then the closed balls of X are also $\sigma(X,Y)$−closed and so $\sigma(X,Y)$ is finer than b_X. Therefore f is continuous on $(X_1, \sigma(X,Y))$. Then $\ker(f) \cap X_1$ is $\sigma(X,Y)$−closed; if u is such that $f(u) = 1$ it quickly follows from the Hahn-Banach theorem that for any $\epsilon > 0$ there exists $g \in Y$ with $g(u) = 1$ and $\|g|_{\ker(f)}\| \leq \epsilon$. Now by the Hahn-Banach theorem there exists $h \in X^*$ with $\|h\| \leq \epsilon$ and $g = h$ on $\ker(f)$. Hence $g - h = \alpha f$ for some α. Further $|\alpha - 1| \leq \epsilon\|u\|$ so that $\|f - g\| \leq \epsilon(\|u\|\|f\| + 1)$. Since Y is closed, $f \in Y$.

(2) If X is separable then (X_1, b_X) satisfies the second axiom of countability. Thus if f is not continuous on (X_1, b_X) there is a sequence (x_n) in X_1 so that $x_n \to 0$ (b_X) and $\lim f(x_n) = \lambda \neq 0$. By Lemma 2.1 we may suppose (x_n) convex clustering at 0 and by Lemma 2.2 we then have, for any w^*-cluster point x^{**}, $\|x^{**} - t\| \geq \|t\|$ for every $t \in X$. By Lemma 2.3, $\ker(x^{**}) = Y$ is a norming subspace of X^*; however, $x^{**}(f) = \lambda$ so that $f \notin N_X$.

In order to classify more precisely those $f \in X^*$ which are b_X−continuous on X_1 we require a rather more technical result.

PROPOSITION 2.5. *Let X be a Banach space and suppose $f \in X^*$. The following conditions are equivalent:*

(1) f is b_X-continuous on X_1.

(2) There is a separable closed subspace E of X so that for every separable closed subspace F of X with $F \supset E$, $f|_F$ is b_F−continuous on F_1.

PROOF: (1) $\Rightarrow$ (2) If f is b_X-continuous on X_1, then for every $n \in \mathbf{N}$, there is a finite family of balls $\mathcal{F}_n = \{B_{n,1}, \ldots, B_{n,k(n)}\}$ so that $0 \in V_n = X \setminus \cup_j B_{n,j}$ and $|f(x)| < 1/n$ whenever $x \in V_n \cap X_1$. Let C_n be the set of all centers of the balls of $\mathcal{F}_n$. It is clear that we may take E to be the closed linear span of $\cup C_n$.

(2) $\Rightarrow$ (1). Assume f is not b_X –continuous on X_1 and that $E \subset X$ is a separable closed subspace. Fix $\epsilon > 0$ so that $X_1 \cap f^{-1}(-\epsilon, \epsilon) = A$ is not a b_X –neighborhood of 0 in X_1.

We inductively define a sequence $x_n \in X_1$. Let $E_0 = E$ and $E_n = \operatorname{span}(E; x_1, \ldots, x_n)$ $(n \geq 1)$; let $E_n^{(0)}$ be a dense countable subset of E_n. For $k \geq 0$ we let $(V_n^k)_{n \geq 1}$ be an ordering of sets of the form $X \setminus \cup_{j=1}^{l} B(u_j, \nu_j)$ where $u_j \in E_k^{(0)}$ $(1 \leq j \leq l)$ and $\nu_j \in \mathbf{Q}$ satisfy $\nu_j < \|u_j\|$.

We now describe the inductive construction. Pick $x_1 \in (V_1^0 \cap X_1) \setminus A$. If $x_1, \ldots, x_{n-1}$ have been determined choose $x_n \in (X_1 \cap V_n^0 \cap V_{n-1}^1 \cap \ldots \cap V_1^{n-1}) \setminus A$.

Let F be the closed linear span of E and $(x_n : n \geq 1)$. The sets $X_1 \cap V_n^0 \cap \ldots \cap V_1^{n-1}$ $(n \geq 1)$ form a base of b_F- neighborhoods of 0. Thus $F_1 \cap f^{-1}(-\epsilon, \epsilon)$ is not a b_F –neighborhood of 0 relative to F_1, i.e. f is not b_F-continuous on F_1.

With 2.4 and 2.5 we have now a complete characterization of the elements of X^* which are b_X-continuous on X_1. Let us give some more concrete conditions. We denote by $\mathcal{C}(w^*, w)$ (resp. $\mathcal{C}(w^*, |\cdot|)$) the set of points of continuity of the identity Id:$(X_1^*, w^*) \to (X_1^*, w)$ (resp. Id:$(X_1^*, w^*) \to (X_1^*, \|\cdot\|)$).

THEOREM 2.6. *Let X be a Banach space and suppose $f \in X^*$. If $f \in \mathcal{C}(w^*, w)$ and if f has a countable base of neighborhoods in (X_1^*, w^*), then f is continuous on (X_1, b_X). This is true in particular if:*

(1) $f(x) = \|f\| = \|x\| = 1$ where $x \in X$ is a point of Gateaux-smoothness of the norm of X^{**}.

(2) f belongs to $\mathcal{C}(w^*, |\cdot|)$.

PROOF: Let f be in $C(w^*, w)$ with a countable basis of w^*-neighborhoods $(V_n)_{n\geq 1}$. We may assume each V_n is defined by a finite number of elements $(x_j^n)_{j\in I_n}$ of X. Let E be the separable closed subspace of X generated by $\{x_j^n : j \in I_n, n \geq 1\}$. It is clear that if F is any subspace of X containing E then $f|_F$ is still a point of continuity of $\mathrm{Id}:(F_1^*, w^*) \to (F_1^*, w)$. Such a point belongs to all norming subspaces Y of F^* since the weak*-closure of Y_1 is F_1^*; hence, if F is also separable then by Theorem 2.4, $f|_F$ is continuous on (F_1, b_F). Proposition 2.5 now shows that f is continuous on (X_1, b_X).

If (1) is satisfied then f is *-exposed in E_1^{***} by x and thus $E_1^{***} \cap (f + E^\perp) = \{f\}$. An easy compactness argument ([4], p.211) shows that this implies $f \in C(w^*, w)$; moreover, the family $V_n = \{t \in X_1^* : t(x) > 1 - \frac{1}{n}\}$ is a base of neighborhoods of f in (X_1^*, w^*). Finally (2) is clear.

REMARKS 2.7: (1) If X is separable then by 2.6 and the weak*-metrizability of X_1^* every $f \in C(w^*, w)$ is b_X-continuous on X_1.

(2) If $f(x) = \|f\| = \|x\| = 1$ and x is a point of Frechet-smoothness of the norm of X, then f is b_X-continuous on X_1. This is a special case of 2.6(1) or of 2.6(2); but a direct proof is available. Indeed, for every $y \in X_1$,

$$\|x + \tau y\| - 1 = \tau f(y) + o(|\tau|).$$

Hence, f is a uniform limit on X_1 of a sequence of b_X-l.s.c. functions, namely $g_n(x) = n(\|x + n^{-1}y\| - 1)$. Thus f is l.s.c. on (X_1, b_X). A similar argument shows $(-f)$ is also l.s.c. and so f is actually continuous.

Next we shall investigate the circumstances under which (X_1, b_X) is either Hausdorff or compact. Let us recall that X has the *finite-infinite intersection property* $(IP_{f,\infty})$ (see [21]) if for every collection $(B_\alpha : \alpha \in I)$ of closed balls in X such that $\cap B_\alpha = \emptyset$ there is a finite subset $\mathcal{F} \subset I$ so that $\cap\{B_\alpha : \alpha \in \mathcal{F}\} = \emptyset$. We also introduce the notation $D(x^{**}) = \{x \in X : \|x - t\| \leq \|x^{**} - t\| \ \forall t \in X\}$ for $x^{**} \in X^{**}$.

THEOREM 2.8. *Let X be a Banach space. The following conditions on X are equivalent:*

(1) X_1 is b_X −compact.

(2) X has $IP_{f,\infty}$.

(3) For every $x^{**} \in X^{**}$, $D(x^{**}) \neq \emptyset$.

PROOF: The equivalence of (1) and (2) is simply Alexander's subbase theorem (Kelley [18], p. 139).

(3) ⇒ (2). Let $(B_\alpha)_{\alpha \in I}$ be a collection of closed balls so that for every finite $\mathcal{F} \subset I$, $\cap_{\mathcal{F}} B_\alpha \neq \emptyset$. By weak*-compactness there exists x^{**} is the intersection of their weak*-closures. Then if $x \in D(x^{**})$ we have $x \in \cap B_\alpha$.

(2) ⇒ (3). If $x^{**} \in X^{**}$ let $\mathcal{B}$ be the collection of balls $B(x,\rho)$ where $\rho > \|x^{**} - x\|$. By the Principle of Local Reflexivity ([22] p.33) $\mathcal{B}$ has the finite-intersection property and so $D(x^{**}) = \cap \mathcal{B} \neq \emptyset$.

THEOREM 2.9. *Let X be a Banach space. Consider the following three statements.*

(1) (X_1, b_X) is Hausdorff.

(2) For every $x \neq y \in X_1$ there is a finite covering of X_1 by closed balls $B_1, \ldots, B_n$ so that no B_i contains both x and y.

(3) $|D(x^{**})| \leq 1$ for every $x^{**} \in X^{**}$.

Then (1) and (2) are equivalent and imply (3). If X is separable then (1), (2) and (3) are equivalent.

PROOF: The equivalence of (1) and (2) is trivial.

(2) ⇒ (3). Suppose $\|x^{**}\| \leq 1$ and $x, y \in D(x^{**})$ with $x \neq y$. Then for any covering of X_1 by closed balls $B_1, \ldots, B_n$, x^{**} is in the weak*-closure of some B_i. But then $D(x^{**}) \subset B_i$ and so $\{x, y\} \subset B_i$.

(3) ⇒ (1) when X is separable. Since each point has a countable base of neighborhoods, if (X_1, b_X) is not Hausdorff there is a sequence $x_n \in X_1$ with two distinct limits

x, y. Applying Lemma 2.1 twice we may suppose x_n is convex clustering at both x and y. If x^{**} is any weak*-cluster point of the sequence x_n then $\{x, y\} \subset D(x^{**})$ by Lemma 2.2. This produces the desired contradiction and concludes the proof.

We close this section with a simple lemma which will be used frequently.

LEMMA 2.10. *Let X be a Banach space and let C be an absolutely convex subset of X. Then the b_X −closure of C is also absolutely convex.*

PROOF: Assume $x, y \in \overline{C}$ and that $x_\alpha \to x$, $y_\beta \to y$ in b_X where $x_\alpha \in C$, $y_\beta \in C$. Then for fixed β and $0 < \lambda < 1$, $\lambda x_\alpha + (1-\lambda)y_\beta$ converges to $\lambda x + (1-\lambda)y_\beta$ which is therefore in $\overline{C}$. Now letting β vary we similarly obtain that $\lambda x + (1-\lambda)y \in \overline{C}$. As $\overline{C}$ is clearly symmetric this concludes the proof.

3. Localization of the ball topology to an absolutely convex set. In the next theorem we shall suppose that X is a Banach space and that C is a bounded absolutely convex subset of X. The linear span of C is denoted by Y. Y is a normed space if we equip it with the guage functional of C as a norm. With this notation we have the following fundamental result:

THEOREM 3.1. *The following assertions are equivalent:*

(1) There is a Hausdorff locally convex linear topology ν on Y so that $\nu \mid C = b_X \mid C$.

(2) 0 has a base of closed neighborhoods in (C, b_X).

(3) (C, b_X) is regular.

(4) There is a closed subspace M of Y^* which separates Y and so that on C we have $\sigma(Y, M) = b_X$.

(5) If V is neighborhood of 0 in (C, b_X) there is a further neighborhood W of 0 so that $(W + 2W) \cap C \subset V$.

(6) 0 has a base of absolutely convex neighborhoods in (C, b_X).

PROOF: We first establish the equivalence of (2),(5) and (6).

(2) $\Rightarrow$ (6). Let V be a closed neighborhood of 0 in (C, b_X). Then there is a closed neighborhood V_1 of 0 in $(2C, b_X)$ so that $V_1 \cap C \subset V$. Let W be an absolutely convex weak neighborhood of 0 in $2C$ with $W \subset V_1$. By Lemma 2.10, the closure of W in the topology b_X, which we denote W_b, is absolutely convex. Further $W_b \subset V_1$ and thus $W_b \cap C \subset V_1 \cap C \subset V$. It therefore suffices to show that $W_b \cap C$ is a neighborhood of 0 in (C, b_X).

Assume it is not. Then we construct by induction a sequence $(x_n)_{n \geq 1}$ in $C \setminus W_b$ so that $x_i - x_j \notin W_b$ for $i \neq j$. Indeed pick first any $x_1 \in C \setminus W_b$. If $x_1, \ldots, x_n$ have been chosen there is a b_X-neighborhood U of 0 so that $(x_k + U) \cap W_b = \emptyset$ for $k = 1, 2, \ldots, n$. Assuming W_b is not a neighborhood of 0 in (C, b_X) we have $U \cap C \not\subset W_b$ and we may pick x_{n+1} so that $-x_{n+1} \in (U \cap C) \setminus W_b$. Thus $x_i - x_{n+1} \notin W_b$. This completes the inductive construction of (x_n).

Now C is totally bounded for the weak topology and hence 0 is a closure point of $\{x_i - x_j : i \neq j\}$ for the weak topology. Since W is a weak neighborhood of 0 in $2C$ there exist $i \neq j$ so that $x_i - x_j \in W$. This contradiction completes the proof of this implication.

(6) $\Rightarrow$ (5). Let V be an absolutely convex neighborhood of 0 in (C, b_X). Then there is an absolutely convex neighborhood U of 0 in $(3C, b_X)$ with $U \cap C \subset V$. If $W = \frac{1}{3}U$ then W satisfies (5).

(5) $\Rightarrow$ (2). Let V be a neighborhood of 0 in (C, b_X). Let W be a neighborhood such that $(W + 2W) \cap C \subset V$. Let W_b be the closure of W in b_X. and suppose $x \in W_b \cap C$. Then since $C \subset x - 2C$ we have $x \in W_b \cap (x - 2C)$. However $x - 2W$ is a neighborhood of x in $x - 2C$ and $W \subset C \subset x - 2C$ and so $(x - 2W) \cap W \neq \emptyset$ or $x \in W + 2W$. Thus $W_b \cap C \subset V$ and so 0 has a base of closed neighborhoods in (C, b_X).

(5) + (6) $\Rightarrow$ (1). Let $\mathcal{V}$ be a base of absolutely convex neighborhoods of 0 in (C, b_X). Let $\mathcal{A}$ be the family of sets A of the form

$$A = \bigcup_{N=0}^{\infty} \sum_{n=0}^{N} 2^n V_n$$

where $V_n \in \mathcal{V}$, $n = 0, 1, 2, \ldots$. It is easily checked that $\mathcal{A}$ is a base of a locally convex topology ν on Y. Note that each V_n is absorbent for Y since the ball topology is coarser than the weak topology of X. We observe that on C, b_X is finer than ν. To complete the proof we show that b_X coincides on C with ν. This will in turn show that ν is Hausdorff since b_X is T_1.

Let $V \in \mathcal{V}$; we will construct $A \in \mathcal{A}$ with $A \cap C \subset V$. Pick first any $W_0 \in \mathcal{V}$ with $(W_0 + 2W_0) \cap C \subset V$. Then inductively pick $(W_n)_{n \geq 1}$ with $W_n \in \mathcal{V}$ and $(W_n + 2W_n) \cap C \subset W_{n-1}$. Let $A = \cup \sum_{n=0}^{N} 2^n W_n$. If $x \in A \cap C$ there exists N so that $x = \sum_{n=0}^{N} 2^n w_n$ with $w_n \in W_n$ for $0 \leq n \leq N$.

For $0 \leq k \leq N$, set

$$v_k = \sum_{n=k}^{N} 2^{n-k} w_n$$

Then for $1 \leq k \leq N-1$ we have $v_k = w_k + 2v_{k+1}$ while $v_N = w_N$. Thus $v_N \in W_N \subset W_{N-1}$ and by an easy induction we obtain $v_k \in W_{k-1}$ for $1 \leq k \leq N$. Thus $v_1 \in W_0$ and $x = w_0 + 2v_1 \in (W_0 + 2W_0) \cap C \subset V$. Hence $A \cap C \subset V$ as required.

(1) $\Rightarrow$ (4). Clearly b_X is coarser than the weak topology on X. Hence if x_n is any sequence in C, 0 is a closure point of the set $\{\frac{1}{2}(x_i - x_j) : i \neq j\}$. This shows that C is precompact for ν and so ν agrees on C with its weak topology and hence with the topology $\sigma(Y, M)$ where M is the collection of all linear functionals on Y whose restrictions to C are b_X-continuous. It is straightforward then to show that M is a closed linear subspace of Y^*. M must also separate the points of Y since ν is Hausdorff.

The remaining implications (4) $\Rightarrow$ (3) $\Rightarrow$ (2) are obvious.

DEFINITION 3.2: Let X be a Banach space. An absolutely convex subset C of X is called a *Rosenthal* set if it is bounded and contains no basic sequence equivalent to the unit vector basis of ℓ_1.

Note that by the Rosenthal theorem [29], C is a Rosenthal set if and only if every sequnece in C has a weakly Cauchy subsequence.

THEOREM 3.3. *Let X be a Banach space and let C be an absolutely convex Rosenthal subset of X. Then (C, b_X) is regular and hence satisfies (1)-(6) of Theorem 3.1.*

PROOF: The proof will be carried out in two steps.

STEP 1. *Suppose Y is a separable Banach space not containing ℓ_1, and that X is a separable Banach space. Let $j : Y \to X$ be a bounded linear injection. Then there is a closed separating subspace M of Y^* so that j is a homeomorphism between $(Y_1, \sigma(Y, M))$ and $(j(Y_1), b_X)$.*

PROOF OF STEP 1: We define a subset Z of Y^{**} by

$$Z = \{y^{**} \in Y^{**} : \ker(j^{**}(y^{**})) \cap X_1^* \text{ is } w^* - \text{dense in } X_1^*\}.$$

Since ℓ_1 does not embed into Y, $j^{**}y^{**}$ is of first Baire class on (X_1^*, w^*) and thus $\ker(j^{**}y^{**}) \cap X_1^*$ is a $w^* - G_\delta$ subset of X_1^*. A Baire category argument now shows that Z is a vector subspace of Y^{**} and that Z is weak* sequentially closed (see [4], p.214). Now Y_1^{**} is a pointwise compact set of first Baire class functions on (Y_1^*, w^*) and so by results in [1], $Z \cap Y_1^{**}$ is w^*-closed; thus, by the Banach-Dieudonné theorem, Z is weak*-closed. Let M be the annihilator of Z in Y^*. Note that $M^\perp = Z$ and that M separates Y since $M^\perp \cap Y = Z \cap Y = \{0\}$.

Now suppose $f \in M$. we show that $f \circ j^{-1}$ is continuous on $(j(Y_1), b_X)$. Since X is separable, it is enough to show that if $y_n \in Y_1$, $y \in Y_1$, and $j(y_n) \to j(y)$ for b_X then $f(y_n) \to f(y)$. By Lemma 2.1 we can assume that $(j(y_n))$ is convex clustering at $j(y)$ and by [24] and [29] that y_n converges weak* to y^{**} in Y^{**}. By Lemma 2.2 we have

$$\|j^{**}(y^{**}) - x\| \geq \|j(y) - x\| \qquad \forall x \in X.$$

Now by Lemma 2.3, $\ker(j^{**}(y^{**}-y))\cap X_1^*$ is weak*-dense in X_1^*. Thus $y^{**}-y\in Z$ and so $f(y^{**}-y)=0$. Thus $f(y_n)\to f(y)$ and $f\circ j^{-1}$ is b_X-continuous on $j(Y_1)$. Thus the map $j^{-1}:(j(Y_1),b_X)\to(Y_1,\sigma(Y,M))$ is continuous.

Conversely suppose y_α is a net in Y_1 converging to some $y\in Y_1$ for $\sigma(Y,M)$. We show that $j(y_\alpha)\to j(y)$ in b_X. It suffices to consider the case when y_α converges to some y^{**} in Y^{**} for the weak* topology. Then $f(y^{**}-y)=0$ for every $f\in M$ and so $y^{**}-y\in M^\perp=Z$. Therefore $\ker(j^{**}(y^{**}-y))\cap X_1^*$ is w^*-dense in X_1^* and by Lemma 2.3 for every $x\in X$ we have

$$\|j^{**}(y^{**})-x\|\geq\|j(y)-x\|.$$

Thus

$$\|j(y)-x\|\leq\liminf\|j(y_\alpha)-x\|$$

so that $j(y_\alpha)$ converges to $j(y)$ in b_X as required.

STEP 2: COMPLETION OF THE PROOF OF THEOREM 3.3.

By the factorization technique of [3] (cf. [26]) there is a Banach space Y, not containing ℓ_1, and a bounded linear injection $j:Y\to X$ with $j(Y_1)\supset C$.

Let us assume C is not regular in the topology b_X. Then by Theorem 3.1, there is a b_X-neighborhood of 0 relative to C, say V, such that for every b_X-neighborhood W of 0 relative to C we have $(W+2W)\cap C\not\subset V$. We may suppose that

$$V=C\bigcap(X\setminus\bigcup_{i=1}^n B(u_i,\nu_i)).$$

Let $E_0=\operatorname{span}\{u_1,\dots,u_n\}\subset X$ and let $F_0=\{0\}\subset Y$.

We now construct, by induction two increasing sequences of finite-dimensional subspaces $(E_n)_{n\geq0}$ in X and $(F_n)_{n\geq0}$ in Y. Let $E_n^{(0)}$ denote a dense countable subset of E_n (when E_n has been determined). Let $(V_{n,k})_{k\geq1}$ be an ordering of all sets of the form $X\setminus\cup_{l=1}^m B(x_l,r_l)$ where $x_1,\dots,x_m\in E_n^{(0)}$, $\|x_l\|>r_l$ and $r_l\in\mathbf{Q}_+$.

Now suppose $E_0, \ldots, E_N$, $F_0, \ldots, F_N$ have been determined. Set

$$W_N = C \cap V_{0,N+1} \cap V_{1,N} \cap \ldots \cap V_{N,1}.$$

There exist $y_{N+1}, z_{N+1} \in Y_1$, so that $j(y_{N+1})$, $j(z_{N+1}) \in W_N$ and $j(y_{N+1} + 2z_{N+1}) \in C \setminus V$. We then set $F_{N+1} = \text{span}\{F_N, y_{N+1}, z_{N+1}\}$ and $E_{N+1} = \text{span}\{E_N, j(y_{N+1}), j(z_{N+1})\}$.

Let E be the closed linear span of $\cup E_n$ and let F be the closed linear span of $\cup F_n$. Note that j maps F into E and both are separable. Now we can apply Step 1. On $j(F) \cap C$ b_E agrees with a linear topology. Now $V \cap j(F)$ is a b_E-neighborhood of 0 in $C \cap j(F)$. Now $(W_N : N \geq 1)$ is a base of b_E-neighborhoods of 0 in C. Hence there exists N so that

$$(W_N + 2W_N) \cap C \cap j(F) \subset V \cap j(F).$$

However $j(y_{N+1})$, $j(z_{N+1}) \in W_N$ and $j(y_{N+1}) + 2j(z_{N+1}) \in C \setminus V$. Clearly $j(y_{N+1}) + 2j(z_{N+1}) \in C \cap j(F)$ which yields a contradiction.

We now state a useful extension lemma.

LEMMA 3.4. *Let C be a bounded absolutely convex subset of a Banach space, and let f be b_X-continuous affine function on C. Then there is a b_X-continuous affine function $\overline{f}$ defined on the b_X-closure $\overline{C}$ of C, so that $\overline{f}(x) = f(x)$ for $x \in V$.*

PROOF: Note that by Lemma 2.10, $\overline{C}$ is absolutely convex.

Now if $x \in \overline{C}$ define $\overline{f}(x) = \liminf_{y \to x} f(y)$. Then $\overline{f}$ is l.s.c on $(\overline{C}, b_X)$ and $\overline{f} \mid C = f$.

We claim that $\overline{f}$ is convex. Suppose first that $x \in C$, $y \in \overline{C}$ and $0 \leq \lambda \leq 1$. Suppose $y_\alpha \in C$ is a net chosen so that $y_\alpha \to y$ (b_X) and $f(y_\alpha) \to \overline{f}(y)$. Then $\lambda x + (1-\lambda)y_\alpha \to \lambda x + (1-\lambda)y$ for b_X and so

$$\begin{aligned}\overline{f}(\lambda x + (1-\lambda)y) &\leq \liminf f(\lambda x + (1-\lambda)y_\alpha) \\ &= \lambda f(x) + (1-\lambda)\liminf f(y_\alpha) \\ &= \lambda \overline{f}(x) + (1-\lambda)\overline{f}(y_\alpha).\end{aligned}$$

Next suppose $x, y \in \overline{C}$. Choose a net $x_\alpha \in C$ so that $x_\alpha \to x$ and $f(x_\alpha) \to \overline{f}(x)$. Repeating the above reasoning we have

$$\begin{aligned}\overline{f}(\lambda x + (1-\lambda)y) &\leq \liminf \overline{f}(\lambda x_\alpha + (1-\lambda)y) \\ &\leq \liminf(\lambda f(x_\alpha) + (1-\lambda)\overline{f}(y)) \\ &= \lambda \overline{f}(x) + (1-\lambda)\overline{f}(y).\end{aligned}$$

Since $x \to -x$ is an affine homeomorphism of $(\overline{C}, b_X)$ the function $\overline{f}(-x)$ is also convex and l.s.c.

Now if $x \in C$, $\overline{f}(x) + \overline{f}(-x) = 2f(0)$. Hence by lower-semi-continuity we have $\overline{f}(x) + \overline{f}(-x) \leq 2f(0)$ for every $x \in \overline{C}$. By convexity we have therefore $\overline{f}(x) + \overline{f}(-x) = 2f(0)$ and thus $\overline{f}(x) = 2f(0) - \overline{f}(-x)$ is both concave and u.s.c. Hence $\overline{f}$ is affine and continuous.

THEOREM 3.5. *Let X be a Banach space and let C be an absolutely convex Rosenthal subset of X. Then the b_X-continuous affine functionals on $\overline{C}$ separate the points of $\overline{C}$. In particular, $\overline{C}$ is Hausdorff for the topology b_X.*

PROOF: Suppose $x, y \in \overline{C}$ with $x \neq y$. Let A be the absolutely convex hull of $C \cup \{x, y\}$. Then A is a Rosenthal set and so by 3.3 and 3.1 there is an affine functional f on A which is continuous for b_X and such that $f(x) \neq f(y)$. f then extends to $\overline{C}$ by 3.4.

4. An application: quasi-convex operators. Let X be a Banach space and suppose E is a topological vector space (not assumed to be locally convex). In [13] an operator $T : X \to E$ is called *quasi-convex* if the set $C = \overline{T(X_1)}$ has the property that the continuous affine functions on C separate the points of C.

It is shown in [13] that if X is reflexive then every operator is quasi-convex, while if X^* is separable every compact operator is quasi-convex. An example of Roberts [27] shows that there exists a compact absolutely convex subset of L_p, $(p < 1)$ with no extreme points and hence there is a Banach space B and a compact operator $T : B \to L_p$ which is not quasi-convex.

Recently A. Sersouri [30] has improved the results of [13] by showing that if X is separable and does not contain ℓ_1 then every compact operator on X is quasi-convex. He also shows that if X is a Banach space containing ℓ_1 then there is a non-quasi-convex operator on X. However from [17] it can be shown that every compact operator on a $C(K)$−space is quasi-convex. We remark that there is a compact semi-embedding of ℓ_1 into L_p $(p < 1)$ which is not quasi-convex [15].

Our main result combined with the work of Sersouri shows that a Banach space X has the property that every operator on X is quasi-convex if and only if X contains no copy of ℓ_1.

THEOREM 4.1. *Let X be a Banach space not containing ℓ_1 and let E be a Hausdorff topological vector space. Then any continuous operator $T : X \to E$ is quasi-convex.*

PROOF: We may assume that E is complete. Let C be the closure of $T(X_1)$ and let Z be the linear span of C equipped with the guage functional of C. Then Z is a Banach space and the ball topology b_Z is coarser than the topology induced on Z by the topology of E.

Since X does not contain ℓ_1, $T(X_1)$ is a Rosenthal set in Z. The b_Z−closure of $T(X_1)$ includes the E−closure, i.e. C. By Theorem 3.5, the b_Z−continuous affine functionals on C separate the points of C. The result now follows.

5. Banach spaces for which the ball topology is locally linear. We shall say that the ball topology b_X is locally linear if the unit ball X_1 is regular for the b_X-topology and hence satisfies the equivalent conditions of Theorem 3.1.

PROPOSITION 5.1. *The ball topology is locally linear if and only if the set of $f \in X^*$ which are b_X−continuous on X_1 is a norming subspace of X^*.*

PROOF: Let $M \subset X^*$ be the set of linear functionals which are b_X−continuous on X_1.

Then M is a closed linear subspace of X^*. If b_X is locally linear then $\sigma(X,M)$ agrees with b_X on X_1. If $\lambda \geq 1$, $\lambda^{-1}X_1$ is $\sigma(X,M)$-closed in X_1 and hence X_1 is $\sigma(X,M)-$closed in λX_1; thus X_1 is $\sigma(X,M)-$closed and so M is norming. Conversely if M is norming then $\sigma(X,M)$ is finer than b_X so that $\sigma(X,M) = b_X$ on X_1.

THEOREM 5.2. *If (X, b_X) is locally linear then X^* contains a minimal norming subspace.*

If X is separable then (X, b_X) is locally linear if and only if X^* contains a minimal norming subspace.

PROOF: Let N_X denote the intersection of all norming subspaces in X^* and let M denote the subspace of all linear functionals which are b_X −continuous on X_1. If b_X is locally linear then $N_X \subset M$ by Proposition 5.1, but by Theorem 2.4 we have $N_X \supset M$. Hence $N_X = M$ is norming.

If X is separable we have $N_X = M$ in general by Theorem 2.4 and so the converse also follows.

Our next theorem lists some easy examples of Banach spaces for which b_X is a locally linear topology.

THEOREM 5.3. *If X is a Banach space satisfying any one of the following conditions then b_X is locally linear and hence X^* has a minimal norming subspace.*

(1) The norm on X is Frechet-smooth on a dense set.

(2) $X = \mathcal{L}(E)$ where E is a reflexive Banach space.

(3) X is the dual of a Banach space with RNP.

(4) X contains no subspace isomorphic to ℓ_1.

PROOF: (1) This follows from Theorem 2.6 (and Remark 2.7) combined with Proposition 5.1.

(2) This follows from 2.6 and [10], combined with Proposition 5.1.

(3) Let $X = E^*$ where E has RNP. If $f \in E \subset X^*$ is a strongly exposed point of E_1 then f is continuous on (X_1, b_X) by Theorem 2.6. It follows that every $f \in E$ is continuous on (X_1, b_X) [25] and so b_X is locally linear by Proposition 5.1.

(4) Theorem 3.3.

We remark that it was shown in [4] that if X is a separable Banach space not containing ℓ_1 then X^* has a minimal norming subspace. Thus (4) of the above theorem removes the separability assumption. We shall, however, extend (4) to prove a much more general result.

THEOREM 5.4. *Let E be a Banach space such that E^* contains a norming subspace Y which contains no subspace isomorphic to ℓ_1. Let X be any closed subspace of E^* with $X \supset Y$. Then b_X is locally linear, and X^* has a minimal norming subspace.*

PROOF: Let $M \subset X^*$ be the subspace of functionals continuous on (X_1, b_X). We show that M is norming. Indeed suppose $x \in X, \|x\| = 1$ and $\epsilon > 0$. Let $Z = \operatorname{span}\{Y, x\}$. Then b_Z is locally linear by Theorem 5.3(4). Hence there exists $\phi \in Z_1^*$ with $\phi(x) > 1 - \epsilon$ and so that ϕ is b_Z-continuous on Z_1; in particular ϕ is continuous for the restriction of b_X to Z_1.

Now b_X is coarser than the topology $\sigma(E^*, E)$ on X and Y_1 is $\sigma(E^*, E)$-dense in E_1^*. Hence the b_X-closure of Z_1 is X_1. By Lemma 3.4, ϕ extends to a b_X-continuous affine function on X_1; since $\phi(0) = 0$ this implies that there is a linear functional f on X with $f = \phi$ on Z and $f \mid X_1$ is b_X-continuous. Clearly if $u \in X_1$ then $|f(u)| \leq \sup_{z \in Z_1} |f(z)| \leq 1$ so that $f \in X_1^*$. Hence $f \in M$, $\|f\| \leq 1$ and $f(x) \geq 1 - \epsilon$. Thus M is norming and the proof is complete.

We emphasize a special case (see Theorem 9.3 later):

COROLLARY 5.5. *Let Y be a Banach space not containing ℓ_1 and let X be a subspace of Y^{**} with $X \supset Y$. Then b_X is locally linear and X^* contains a minimal norming subspace.*

PROOF: Take $E = Y^*$.

Let us recall that a Banach space X has a *unique predual* (X has (UPD)) if there is exactly one norming subspace $M \subset X^*$ so that X_1 is $\sigma(X, M)$–compact. If X^* has a minimal norming subspace and X is a dual space then it is easy to see that X has a unique predual. Thus the results of this section can be employed to show that a wide range of spaces have (UPD). However we shall examine the more general question of the existence of compact topologies on X_1 consistent with the convex structure in Section 6.

Let us conclude with an application to the set $C_X = \{x^{**} \in X^{**} : \|x^{**} - x\| \geq \|x\| \ \forall x \in X\}$. By Lemma 2.3, C_X is the set of $x^{**} \in X^{**}$ so that $\ker x^{**}$ is norming. It is shown in [4] that C_X is a weak*-closed subspace of X^{**} whenever X is separable and does not contain ℓ_1.

THEOREM 5.6. *Let X be a Banach space. Consider the following statements:*

(1) b_X is locally linear.

(2) C_X is a weak-closed linear subspace of X^{**}.*

(3) C_X is a linear subspace of X^{**}.

In general (1) $\Rightarrow$ (2) $\Rightarrow$ (3). *If X is separable then (1), (2) and (3) are equivalent.*

PROOF: As usual let N_X be the intersection of all norming subspaces of X.

(1) $\Rightarrow$ (2). Clearly $C_X \subset N_X^\perp$. Suppose $x^{**} \in N_X^\perp$ and $x \in X$. Then there is a net $x_\alpha \in X$ so that $\|x_\alpha - x\| \leq \|x^{**} - x\|$ and $x_\alpha \to x^{**}$ (w^*). Clearly, $x_\alpha \to 0$ for the topology $\sigma(X, N_X)$ and hence in b_X. Thus $\|x\| \leq \liminf \|x_\alpha - x\| \leq \|x^{**} - x\|$ and so $x^{**} \in C_X$. Hence $C_X = N_X^\perp$ is a weak*-closed linear subspace.

(2) $\Rightarrow$ (3). Obvious.

(3) $\Rightarrow$ (1) for X separable. If b_X is not locally linear then by Theorem 3.1 there exist sequences $(x_n), (y_n)$ so that $\|x_n\| \leq 1$, $\|y_n\| \leq 1$, $\|x_n + 2y_n\| \leq 1$ and $x_n \to 0$ (b_X), $y_n \to 0$ (b_X) but $x_n + 2y_n \notin V$ for some fixed b_X- neighborhood, V, of 0.

By passing to subsequences we may suppose that both (x_n) and (y_n) are convex clustering in the ball topology at 0 (Lemma 2.1). It then follows that for any weak*-cluster points x^{**}, y^{**} of $(x_n), (y_n)$, respectively we have $x^{**}, y^{**} \in C_X$ by Lemma 2.2. Hence $x^{**} + 2y^{**} \in C_X$. Now fix $x \in X$. Then

$$\|x\| \leq \|x^{**} + 2y^{**} - x\| \leq \liminf \|x_n + 2y_n - x\|$$

so that $(x_n + 2y_n)$ converges to 0 in the ball topology contrary to assumption.

REMARKS 5.7: In particular is a weak*-closed linear subspace of X^{**} if X contains no copy of ℓ_1, removing the separability assumption in [4].

6. Compact topologies on the unit ball of X. Let C be a convex subset of a linear space E. For $0 \leq \lambda \leq 1$ define the map $g_\lambda : C \times C \to C$ by $g_\lambda(x, y) = \lambda x + (1 - \lambda)y$. Let τ be a topology on C. Then we shall say that τ is *consistent* if the map g_λ is separately continuous for $0 \leq \lambda \leq 1$ and *strongly consistent* if g_λ is jointly continous for $0 \leq \lambda \leq 1$.

We observe that compact Hausdorff convex sets with a strongly consistent topology have been studied in the literature. Let us mention the theorem of Lawson [20] and Roberts [28] that a compact Hausdorff convex set with a strongly consistent topology such that each point has a base of convex neighborhoods is affinely homeomorphic to a compact convex subset of a locally convex linear topological space.

As we have noted, a Banach space X has a unique predual (UPD) if there is exactly one compact Hausdorff topology on X_1 induced by a locally convex linear topology on X. In view of the Lawson-Roberts theorem one can rephrase this: X has (UPD) if X_1 admits exactly one compact Hausdorff strongly consistent topology so that each point has a base of convex neighborhoods.

We extend these ideas naturally by defining X to have the *compact uniqueness property (CUP)* if there is exactly one compact Hausdorff consistent topology on X_1. If X is a dual space then (CUP) implies (UPD), but in general there is no obvious relationship between these properties.

LEMMA 6.1. *Let X be a Banach space and suppose τ is a compact Hausdorff consistent topology on X_1. Define a topology τ^* on X by $U \in \tau^*$ if and only if $\lambda U \cap X_1 \in \tau$ for every $\lambda > 0$. Then τ^* is a prelinear topology on X which is finer than b_X and $\tau^* \mid X_1 = \tau$. (We do not assert that τ^* is Hausdorff.)*

PROOF: First note that if $0 < \lambda < 1$ then $h_\lambda(x) = \lambda x$ is a homeomorphism of X_1 onto λX_1 in τ. Thus the sets λX_1 are τ−closed for $0 < \lambda < 1$.

Now we show that $\tau^* \mid X_1 = \tau$. Clearly τ^* is coarser than τ on X_1. Conversely suppose $V = V_0 \subset X_1$ is τ−open. For any τ−open set W the set $\frac{1}{2}W$ is τ−open in $\frac{1}{2}X_1$ and so there is a τ−open set W' with $W' \cap \frac{1}{2}X_1 = \frac{1}{2}W$ or $2W' \cap X_1 = W$. Hence, by induction we can find a sequence $(V_n)_{n\geq 0}$ of τ−open sets in X_1 so that $V_n = 2V_{n+1} \cap X_1$ for $n \geq 0$. By induction we have $2^n V_n \cap 2^k X_1 = 2^k V_k$ for $n \geq k$. Thus the sequence $2^n V_n$ is increasing. Let $U = \cup 2^n V_n$. Then $U \cap 2^k X_1 = 2^k V_k$ for $k \geq 0$, and, in particular $U \cap X_1 = V$.

Now suppose $\lambda > 0$. Pick k so that $\lambda > 2^{-k}$ and write $\mu = \lambda^{-1}2^k$. Then $\lambda U \cap X_1 = \mu^{-1}(2^{-k}U \cap \mu X_1)$. Now $2^{-k}U \cap X_1 = V_k$ so that $\lambda U \cap X_1 = \mu^{-1}(V_k \cap \mu X_1) = h_\mu^{-1}(V_k \cap \mu X_1)$ is τ−open. Thus U is τ^*−open. We thus have proved that τ and τ^* agree on X_1.

To show τ^* is prelinear note that the map $x \to \lambda x$ is obviously τ^*-continuous for $\lambda > 0$. Also τ^* is the finest topology agreeing with itself on the sets mX_1 for $m \in \mathbf{N}$. If $x \in X_1$ the map $y \to \frac{1}{2}(x+y)$ is continuous on each mX_1 and hence is τ^*-continuous. It follows easily that τ^* is prelinear.

Finally since λX_1 is τ-closed for $0 < \lambda \leq 1$, the set X_1 is τ^*-closed and hence every ball is τ^*-closed so that τ^* is finer than the ball topology b_X.

The following lemma is essentially implied by the work of Waelbroeck [34] (cf. also Turpin [33]). However, as our formulation is rather different, we include a proof for completeness.

LEMMA 6.2. *τ^* is a linear topology if and only if τ is strongly consistent.*

PROOF: One direction is clear. For the other, let us suppose that τ is strongly consistent. We will show that the map $(\lambda, x) \to \lambda x$ is jointly continuous on $[-1,1] \times (X_1, \tau)$. For, if not, there is a net $x_\alpha \in X_1$ and $\lambda_\alpha \in [-1,1]$ such that $x_\alpha \to x$ and $\lambda_\alpha \to \lambda$ but $\lambda_\alpha x_\alpha \to y$ in τ where $y \neq \lambda x$. But then $\frac{1}{2}(\lambda_\alpha x_\alpha) \to \frac{1}{2}y$. Hence $\frac{1}{2}((\lambda_\alpha - \lambda)x_\alpha + \lambda x_\alpha) \to \frac{1}{2}y$. Now $\frac{1}{2}(\lambda x_\alpha + (-\lambda)x_\alpha) \to 0$ and $|\lambda| x_\alpha \to |\lambda| x$ by the consistency of τ; hence, the strong consistency of τ implies that the only possible limit point of λx_α is λx and hence by compactness $\lambda x_\alpha to \lambda x$. For some β we have $|\lambda_\alpha - \lambda| \leq 1$ for $\alpha \geq \beta$. Let u be any cluster point of the net $\{(\lambda_\alpha - \lambda)x_\alpha : \alpha \geq \beta\}$. Then by the strong consistency of τ we have $\frac{1}{2}(u + \lambda x) = \frac{1}{2}y$. However for any given $\epsilon > 0$ we have $(\lambda_\alpha - \lambda)x_\alpha \in \epsilon X_1$ eventually. Hence $u \in \epsilon X_1$ for every $\epsilon > 0$ and hence $u = 0$ leading to a contradiction and proving joint continuity.

From this follows that 0 has a base of balanced absorbent sets in τ; further, by joint continuity of the map $(x,y) \to \frac{1}{2}(x+y)$ given any $\tau-$neighborhood V we can a further $\tau-$neighborhood W with $W + W \subset 2V$.

Consider sets of the form $A = \cup_{n=0}^{\infty} \sum_{k=0}^{n} 4^k V_k$, where each V_k is a $\tau-$ neighborhood of 0 in X_1. Such sets A form a base at 0 for a linear topology ν on X.

Suppose each V_k is $\tau-$open and let $A_n = \sum_{k=0}^{n} 4^k V_k$. then

$$A_n \cap 2^{2n-1} X_1 = \bigcup_{x \in A_{n-1}} (x + 4^n V_n) \cap 2^{2n-1} X_1$$

for $n \geq 1$.

If $V_n = W_n \cap X_1$ where W_n is τ^*-open then

$$(x + 4^n W_n) \cap 2^{2n-1} X_1 = (x + (4^n W_n \cap 4^n X_1)) \cap 2^{2n-1} X_1$$

for every $x \in A_{n-1}$ since then $\|x\| < \frac{1}{3}4^n$. Hence $A_n \cap 2^{2n-1} X_1$ is relatively τ^*-open in $2^{2n-1} X_1$. Thus $A_n \cap \lambda X_1$ is relatively τ^*-open in λX_1 for $0 \leq \lambda \leq 2^{2n-1}$. Hence $A \cap \lambda X_1$

is relatively τ^* −open in λX_1 for every $\lambda \geq 0$. Thus A is τ^* −open and this implies that ν is coarser than τ^*.

Conversely suppose U is a τ^* −open neighborhood of 0. We construct a sequence V_k, $k \geq 0$ of τ−open neighborhoods of 0 by induction so that $\overline{A_n} \subset U$ where $A_n = \sum_{k=0}^{n} 4^k V_k$ and the closure in taken in τ^*.

For $n = 0$ this is possible since a compact Hausdorff space is regular. For $n > 0$ we utilize the remark that addition is jointly continuous for τ^* on norm bounded sets and so if $C_1, \ldots C_m$ are norm bounded and τ^*-compact then $C_1 + \cdots + C_m$ is τ^*-compact. Suppose then (V_k) have been constructed for $k \leq n$ with $\overline{A_n} \subset U$. If the next inductive step cannot be carried out then for every neighborhood W of 0 in (X_1, τ) we would have $\overline{A_n} + 4^{n+1}\overline{W} \not\subset U$. Then by the compactness of $4^{n+2}X_1$ we would have the existence of a point $x \in \overline{A_n} + 4^{n+1}\overline{W}$ for all such W. A simple compactness argument (since addition is jointly continuous for τ^* on norm-bounded sets) now shows that $x \in \overline{A_n}$, a contradiction. This show the induction can be carried out and leads to the conclusion that U is a ν−neighborhood of 0. Since τ^* is prelinear this rapidly shows that $\tau^* = \nu$.

REMARK: If 0 has a base of convex neighborhoods in τ then τ^* is locally convex; this proves a special case (for absolutely convex sets) of the Lawson-Roberts theorem quoted above. In general (see Proposition 7.16 of [16]) 0 has a base of p−convex neighborhoods for any $p < 1$ and so τ^* is locally p−convex for every $p < 1$.

THEOREM 6.3. *Let X be a Banach space so that (X_1, b_X) is Hausdorff. Then the following conditions on X are equivalent:*

(1) X is a dual space.

(2) X has (UPD).

(3) X has (CUP).

(4) X has $(IP_{f,\infty})$.

(5) There is a norm-one projection of X^{**} onto X.

PROOF: The implications (2) $\Rightarrow$ (1) $\Rightarrow$ (5) $\Rightarrow$ (4) are trivial.

(4) $\Rightarrow$ (3). By (4) X_1 is b_X −compact (Theorem 2.8). Now if τ is any consistent compact Hausdorff topology on X_1 then $\tau^* \geq b_X$ and hence on X_1 $\tau = \tau^* = b_X$.

(3) $\Rightarrow$ (4). If τ is any consistent topology for which X_1 is compact then in τ^* every ball is compact and the finite-intersection property implies $(IP_{f,\infty})$.

(4) $\Rightarrow$ (2). Again by Theorem 2.8, X_1 is compact for the ball topology and therefore regular. Thus b_X is locally linear and Theorem 3.1 implies the existence of a locally convex vector topology ρ on X so that $\rho \mid X_1 = b_X$. Thus X is a dual space and since (3) holds we have (2).

Let us consider some special cases of Theorem 6.3.

COROLLARY 6.4. *Let X be a separable Banach space. If $D(x^{**}$ has exactly one member for every $x^{**} \in X^{**}$ then X is a dual space and has both (CUP) and (UPD).*

PROOF: By Theorems 2.8 and 2.9.

We also note that if b_X is a locally linear topology then (X_1, b_X) is Hausdorff and so the results of Section 4 apply.

COROLLARY 6.5. *A dual space has (CUP) if either (a) the norm on X is Frechet-smooth on a dense set, (b) $X = \mathcal{L}(E)$ where E is reflexive or (c) X is the dual of a space with (RNP).*

COROLLARY 6.6. *Let E be a Banach space not containing ℓ_1 and let X be a closed subspace of E^{**} containg E. Then the following are equivalent:*

(1) X is a dual space.

(2) X has (CUP).

(3) X has $(IP_{f,\infty})$.

(4) X has (UPD).

(5) There is a norm-one projection of X^{**} onto X.

In particular these equivalences apply to the case $X = E$.

REMARKS: (1) Let X be a separable Banach space not containing ℓ_1. In [5] it is shown that X is a dual space if and only if X has $(IP_{f,\infty})$; in [6] it is shown that X^{**} has (UPD). A. Sersouri [30] has shown that there is at most one strongly consistent compact topology on X_1. The above Corollary improves all these results; note that no assumptions of separability are now required.

(2) It is not known if $(IP_{f,\infty})$ already implies the existence of a norm-one projection of X^{**} onto X in general.

7. C(K)-spaces. If K is compact Hausdorff space, the ball topology on $C(K)$ is Hausdorff if and only if K has a dense set of isolated points (e.g. $K = \beta\mathbf{N}$, $C(K) = \ell_\infty$). However a classical result of Grothendieck [11] asserts that $C(K)$ has (UPD) if and only if $C(K)$ is a dual space (if and only if K is hyperstonean). In this section we give some weaker analogues of the results of Section 6 for $C(K)$-spaces. The first result is a direct consequence of results in [17].

THEOREM 7.1. *Let K be comapct Hausdorff. Then any strongly consistent compact Hausdorff topology on the closed unit ball B of $C(K)$ is locally convex. Hence if B admits a strongly consistent compact Hausdorff topology, it is unique.*

PROOF: By Lemma 6.2, we may suppose that $C(K)$ admits a Hausdorff vector topology τ^* such that B is τ^*-compact. It suffices to show that the τ^*-continuous affine functionals on B separate the points of B. Suppose $f_1, f_2 \in B$ with $f_1 \neq f_2$. Since τ^* is locally p-convex for any $p < 1$ (Lemma 6.2 and the following remarks) there is a quasi-Banach space X and a linear operator $S : C(K) \to X$ so that S is τ^*-continuous and $Sf_1 \neq Sf_2$. The proof is completed by showing that the operator S is quasi-convex in the sense of Section 4. For then there exists a continuous affine functional h on $S(B)$ with $h(Sf_1) \neq h(Sf_2)$ and $h \circ S$ is τ^*-continuous on B.

First we note that S is compact on $C(K)$ and $S(B)$ is closed. By a representation theorem due to Thomas [32] (cf. also [14]) there is a regular X-valued vector measure μ defined on the Borel sets $\mathcal{B}$ of K so that $S\phi = \int \phi\, d\mu$ for all $\phi \in C(K)$. In particular the set $\overline{\text{co}}\mu(\mathcal{B})$ coincides with the closed set $S\{\phi : 0 \leq \phi \leq 1\}$ and is compact. Now by [17] Theorem 5.1 μ has a control measure and by [13] Theorem 4.1 the set $\overline{\text{co}}\mu(\mathcal{B})$ is quasi-convex so that the operator S is also quasi-convex.

The last part follows from the cited result of Grothendieck.

In view of this result it is appropriate to ask if dual $C(K)$-spaces have (CUP). However the argument given above does not seem to extend to topologies which are merely consistent. We can however prove a partial result, which depends on the following proposition. Note that the ball topology is not Hausdorff on the ball of $C(K)$ in general.

PROPOSITION 7.2. *Let K be a compact Hausdorff space and let Ω be a comeager subset of K. Let $(f_n)_{n\geq 1}$ be a bounded sequence in $C(K)$ which converges pointwise on Ω to some $f \in C(K)$. Then f is the unique limit of (f_n) for the ball topology.*

PROOF: It is easy to see that $f_n \to f$ in the ball topology. For uniqueness, first define, for any $k, N \geq 1$

$$\Omega_{N,k} = \{x \in K : |f_m(x) - f_n(x)| \leq \frac{1}{k} \quad \forall m, n \geq N\}.$$

For any k, $\Omega \subset \cup_{N\geq 1}\Omega_{N,k}$ and each $\Omega_{N,k}$ is closed. Thus by Baire's theorem the set $\Omega \cap (\cup_{N\geq 1}\text{int}\Omega_{N,k})$ is dense in Ω. Again, by Baire's theorem, the set

$$\Omega' = \bigcap_{k\geq 1}\left(\bigcup_{N\geq 1}\text{int}\Omega_{N,k}\right)\bigcap \Omega$$

is dense in Ω.

Suppose $x \in \Omega'$ and $\epsilon > 0$ are given. Then there is an open set V containing x, and $N \geq 1$ such that:

$$|f_n(x') - f_m(x')| < \frac{\epsilon}{3} \quad \forall x' \in V,\ \forall m, n \geq N.$$

We can further find an open neighborhood V' of x contained in V so that

$$|f_N(x') - f_N(x)| < \frac{\epsilon}{3} \qquad \forall x' \in V'.$$

Then we have

$$|f_m(x') - f(x)| < \epsilon \qquad \forall x' \in V', \ \forall m \geq N.$$

Now ϕ be any limit of the sequence (f_n). For any $x \in \Omega'$ and $\epsilon > 0$ we construct V' as above and then there is a function $h \in C(K)$ with supp $h \subset V'$ and $0 \leq h(x') \leq h(x) = 1$ fro $x' \in V'$. For any $k \geq 0$,

$$\|\phi + kh\| \leq \liminf \|f_n + kh\|.$$

If $x' \in V'$ then for $n \geq N$,

$$\begin{aligned} f_n(x') + kh(x') &\leq f(x) + \epsilon + kh(x) \\ &\leq f(x) + \epsilon + k. \end{aligned}$$

Choose any $k_0 > 2 \sup \|f_n\|$. For every n,

$$\sup_{V'} |f_n + k_0 h| > \sup \|f_n\| \geq \sup_{k \setminus V'} |f_n + k_0 h|.$$

Thus

$$\|\phi + k_0 h\| \leq \liminf \sup_{V'} |f_n + k_0 h| \leq f(x) + k_0 + \epsilon$$

and so

$$\phi(x) \leq f(x) + \epsilon.$$

As $\epsilon > 0$ is arbitrary and Ω' is dense in K we conclude that $\phi \leq f$; by a similar argument we have $\phi \geq f$.

THEOREM 7.3. *Let μ be a probability measure and let τ be a consistent compact metric topology on the ball of $L_\infty(\mu)$. Then τ coincides with the weak*-topology $\sigma(L_\infty, L_1)$ (and, of course, L_1 is separable).*

PROOF: Let K be the spectrum of $L_\infty(\mu)$. By Lemma 6.1 there is a prelinear topology τ^* on $C(K) = L_\infty(\mu)$ which induces τ on the closed unit ball B, and τ^* is finer than the ball topology.

We prove first that the L_1 –norm topology on the ball B is finer than τ. Indeed suppose $f_n \in B$ and $\|f_n - f\|_1 \to 0$. It suffices to show that for some subsequence g_n we have $g_n \to f$ fro τ. In fact we may pick g_n so that $g_n \to f$, μ – a.e. or equivalently on a comeager subset of K. Thus, by Proposition 7.2, f is the only limit point of any subsequence of g_n for the ball topology and thus $g_n \to f$ for the topology τ.

To complete the proof we suppose $f_n \to f$ (τ) and for some $\sigma(L_\infty, L_1)$-neighborhood V of f we have $f_n \notin V$ for all n. We may suppose that (f_n) is τ^* –convex clustering at f by Lemma 2.1. Since B is weakly compact in L_1 we may further suppose that f_n converges to some g for $\sigma(L_1, L_\infty)$ by Eberlein's theorem. Now there is a sequence $g_n \in \text{co}\{f_i : i \geq n\}$ so that $\|g_n - g\|_1 \to 0$. Hence there is a sequence $h_n \in \text{co}(\{f_i : i \geq n\} \cup \{f\})$ with $\|g_n - h_n\|_\infty \to 0$ and $h_n \to f$ in τ. However $\|h_n - g\|_1 \to 0$ and so we must have $f = g$. Now clearly $f_n \to g$ for the weak*-topology which contradicts the fact that $f_n \notin V$. This shows that the identity map from (B, τ) to $(B, \sigma(L_\infty, L_1))$ is continuous and completes the proof.

8. Ball-generated subsets of a Banach space. Let X be a Banach space and let A be a subset of X. We shall say that A is *ball-generated* if there is a family of sets $\{F_i : i \in I\}$ so that each F_i is a finite union of closed balls and $A = \cap F_i$.

Alternatively, A is ball-generated if and only if A is closed for the ball topology. However the definition above seems more suggestive. Our first result extends a result of

Corson and Lindenstrauss [2] who showed that a weakly compact subset of a separable reflexive space is ball-generated. Their proof is essentially to show that the unit ball is Hausdorff for the ball topology in such a space.

THEOREM 8.1. *Let X be a Banach space. Then any weakly compact subset of X is ball-generated.*

PROOF: Let W be weakly compact and let A be its closed absolutely convex hull, which is also weakly compact. Let C be the closure of A in the ball topology. By Theorem 3.5, the ball topology is Hausdorff on C. Since b_X is coarser than the weak topology, we conclude that W is compact for the ball topology and hence closed relative to C. Thus W is ball-generated.

THEOREM 8.2. *Let W be a subset of a Banach space which is ball-generated for every equivalent norm on X. Then W is weakly compact.*

PROOF: Suppose W is not weakly compact; since it is clearly bounded and norm-closed there exists a net w_α in W with a weak* limit $x^{**} \in X^{**} \setminus X$. Suppose $x_0 \in X \setminus W$. Let $M \subset X^*$ be the subspace of all x^* such that $x^{**}(x^*) = x^*(x_0)$. We define a new norm on X by

$$|||x||| = \sup\{|x^*(x)| : \ \|x^*\| \leq 1, \ x^* \in M\}.$$

Clearly $|||x||| \leq \|x\|$. Conversely if $x \in X$ there exists by the Hahn-Banach theorem $\phi \in X^{**}$ with $\|\phi\| = |||x|||$ and $\phi \mid M = x \mid M$. Thus $\phi = x + \lambda(x^{**} - x_0)$ for some real λ. Hence $d(x, E) \leq |||x|||$ where E is the span of $(x^{**} - x_0)$. Thus $|||.|||$ is an equivalent norm on X.

Now if $x^* \in M$ then $\lim_\alpha x^*(w_\alpha) = x^{**}(x^*) = x^*(x_0)$. Hence if $x \in X$, $|||x - x_0||| \leq \liminf |||x - w_\alpha|||$. Thus x_0 is a closure point of W in the ball topology induced by the new norm and W is not closed for this topology.

We may also ask when every weakly closed and bounded set is ball-generated.

THEOREM 8.3. *Let X be a Banach space. Consider the following statements.*

(1) Every norm-closed bounded convex set is ball-generated.

(2) Every weakly closed bounded set is ball-generated.

(3) X^* has no proper norming subspace.

Then (1) and (2) are equivalent and imply (3). If X is separable (1), (2) and (3) are equivalent.

PROOF: Clearly (2) implies (1). If (1) holds and $f \in X^*$ then for every closed interval $I \subset \mathbf{R}$, $f^{-1}(I) \cap X_1$ is b_X-closed and so f is continuous on (X_1, b_X). Thus by Theorem 2.4, $N_X = X^*$, i.e. X^* has no proper norming subspace and so (1) implies (3). Further b_X agrees with the weak topology on X_1 whence (1) also implies (2). If X is separable and (3) holds then Theorem 2.4 implies that each $f \in X^*$ is b_X-continuous on X_1, and so b_x agrees with the weak topology on X_1 whence (3) implies (2).

REMARKS 8.4: (1) Separability is probably not a necessary assumption for the full equivalence in Theorem 8.3. Any Banach space with the property that $\mathcal{C}(w^*, |\cdot|)$ separates X^{**} has the property that every weakly closed bounded set is ball-generated. This condition holds in most of the interesting examples.

(2) It is shown in [10] that for any Banach space X the condition that X^* contains no proper norming subspace is equivalent to

$$\bigcap_{x \in X} B(x, \|x^{**} - x\|) = \{x^{**}\}$$

for every $x^{**} \in X^{**}$. This property allows one to show, for example that if X^* has no proper norming subspace and X has (MAP) then X^* has (MAP) [10].

(3) If the weak topology agrees with b_X on X_1, then by the w^*-density of X_1 in X_1^{**} and by Lemma 3.4, $b_{X^{**}}$ agrees with the weak*-toplogy on X_1^{**}. In particular $b_{X^{**}}$ is

locally linear and the results of Section 6 apply to X^{**} (in particular X^{**} has (UPD) and (CUP)).

(4) Theorem 8.3 applies to spaces with the Mazur intersection property, for example to spaces with a Frechet-smooth norm (for recent results on the Mazur property see [35]). It applies also to Hahn-Banach smooth separable spaces, e.g. separable spaces which are M-ideals in their biduals, since $\mathcal{C}(w^*, w)$ is the unit sphere of the dual space (by [4]). Finally if X and Y are reflexive the space $\mathcal{K}(X,Y)$ of compact operators from X to Y equipped with operator norm satisfies 8.3 (e.g. by [10]).

We now characterize reflexive spaces.

THEOREM 8.5. *Let X be a Banach space. Then X is reflexive if and only if (1) every closed bounded convex set is ball-generated and (2) X has $(IP_{f,\infty})$.*

PROOF: If X is reflexive then (1) follows from Theorem 8.2 and (2) from weak compactness. Conversely if (1) holds then by the argument in 8.3, b_X agrees with the weak topology on X_1, while Theorem 2.8 implies that X_1 is b_X −compact.

Note that conditions (1) and (2) hold for any equivalent norm if they hold for just one norm. Also if X is non-reflexive and 1-complemented in X^{**} then 8.5 implies that X contains a closed convex set which is not ball-generated. This result (an improvement of [4] and [31]) means that these Banach spaces are somehow "rough" (see Section 10).

9. Characterizations of separable Banach spaces not containing ℓ_1. We already know (Theorem 3.5) that if X is a Banach space not containing ℓ_1 then the topology b_X is locally linear. We will show that, conversely, if X is separable and contains ℓ_1, there exists a norm on X such that X_1 is "very far" from being Hausdorff and *a fortiori* very far from being locally linear.

LEMMA 9.1. *Let X be a separable Banach space. The following are equivalent:*

(1) For every non-empty b_X-open subsets O_1 and O_2 of X_1, $O_1 \cap O_2 \neq \emptyset$.

(2) If $\{B_1, \ldots, B_n\}$ *is any finite covering of* X_1 *with closed balls there exists* j *so that* $X_1 \subset B_j$.

(3) For every finite-dimensional subspace F *of* X *and every* $\epsilon > 0$ *there exists* x *with* $\|x\| = 1$ *such that* $\|t + x\| \geq (1 - \epsilon)(\|t\| + \|x\|)$ *for every* $t \in F$.

(4) There exists $x^{**} \in X^{**} \setminus \{0\}$ *such that* $\|x^{**} + x\| = \|x^{**}\| + \|x\|$ *for every* $x \in X$.

PROOF: (1) $\Rightarrow$ (4). Let $(O_n)_{n \geq 1}$ be a base of the ball topology on X_1. By (1), we have $O_1 \cap O_2 \cap \ldots \cap O_n \neq \emptyset$ for every n. Hence we can pick $x_n \in O_1 \cap O_2 \cap \ldots \cap O_n$ for every n. It is clear that any $x \in X_1$ is a b_X-cluster point of any subsequence of (x_n). Thus by Lemma 2.1 and a diagonal argument, there is a sequence (t_n) in X_1 which is b_X-convex clustering to every $x \in X_1$. Let x^{**} be a w^*-cluster point of (t_n) in X_1^{**}. By Lemma 2.2 we have

$$\|t - x^{**}\| \geq \|t - x\| \quad \forall t \in X \quad \forall x \in X_1.$$

In particular, if $x = -t/\|t\|$, we have

$$\|t - x^{**}\| \geq \|t\| + 1.$$

Since $\|x^{**}\| \leq 1$, the result is proved.

(4) $\Rightarrow$ (3). This is clear by the Principle of Local Reflexivity.

(3) $\Rightarrow$ (2). Let $\{B(x_j, \rho_j)\}_{1 \leq j \leq n}$ be a covering of X_1 by closed balls. We must show that for some j, we have $\rho_j \geq \|x_j\| + 1$. Suppose the contrary; by (3) we can find x with $\|x\| = 1$ so that $\|x - x_j\| > \rho_j$ for every j. Then x belongs to no $B(x_j, \rho_j)$ which is a contradiction.

(2) $\Rightarrow$ (1). Let $O_1 = X_1 \setminus (B_1 \cup B_2 \cup \ldots \cup B_k)$ and $O_2 = X_1 \setminus (B_{k+1} \cup B_{k+2} \cup \ldots \cup B_n)$, where $B_1, \ldots, B_n$ are closed balls. If $O_1 \cap O_2 = \emptyset$ then $\{B_1, \ldots, B_n\}$ is a covering of X_1 and (2) implies that either $O_1 = \emptyset$ or $O_2 = \emptyset$. Since sets of this form are a basis for the topology on X_1, the result is proved.

It is now an immediate consequence of [9], Théorème III.1 and [8], Theorem 2.4, that we have

THEOREM 9.2. *Let X be a Banach space. Then X contains ℓ_1 if and only if there exists an equivalent norm on X which satisfies the equivalent conditions of 9.1.*

The above result concerns one given equivalent norm on X. There is no hope of having such a result for all equivalent norms on X. Indeed it is easily seen (with 2.4 and [22],p. 12) that on any separable Banach space X, there exists an equivalent norm such that b_X is locally linear. However, one has:

THEOREM 9.3. *Let X be a separable Banach space. The following are equivalent:*

(1) X does not contain ℓ_1.

(2) For every closed subspace Y of X^{**} with $X \subset Y \subset X^{**}$, Y_1 is b_Y −Hausdorff.

(3) For every closed subspace Y of X^{**} with $X \subset Y \subset X^{**}$, b_Y is locally linear.

PROOF: (1) $\Rightarrow$ (3) follows from Corollary 5.5.

(3) $\Rightarrow$ (2) is obvious.

(2) $\Rightarrow$ (1). If X is separable and contains ℓ_1 then by Maurey's theorem [23], there exists $x^{**} \in X^{**} \setminus \{0\}$ such that

$$\|x^{**} + x\| = \|x^{**} - x\| \quad \forall x \in X.$$

Let $Y = \text{span}(X \cup \{x^{**}\})$. We claim that Y_1 is not b_Y −Hausdorff. Indeed let $S : Y \to Y$ be the linear isometry such that $Sx = x$ for $x \in X$ and $S(x^{**}) = -x^{**}$. Then S is b_Y −continous since it is an isometric isomorphism. Let (x_α) be a net in X_1 which converges w^* and hence also in b_Y to x^{**}. Then Sx_α converges to $-x^{**}$ in b_Y and thus b_Y is not Hausdorff.

REMARKS 9.4:

(1) Let us mention as after 8.5, that the conditions (2) and (3) of 9.3 are satisfied for one norm if and only if they are satisfied for every equivalent norm.

(2) The above proof of (2) $\Rightarrow$ (1) together with Theorem 5.4 provides an alternative proof of one of the directions in Maurey's theorem [23] (actually, the easier one): if there exists $x^{**} \in X^{**} \setminus \{0\}$ such that $\|x^{**} - x\| = \|x^{**} + x\|$ for every $x \in X$, then X contains ℓ_1.

(3) By [3] there exists a lattice locally uniformly convex norm on c_0. Let X be the dual space of c_0 with this norm. Then b_X coincides with the w^*-topology. However X is a weakly sequentially complete lattice and so there is an isometric symmetry S on X^{**} such that $\ker(S - I) = X$ and $\ker(S + I) = c_0^{\perp}$. This shows that the only continuous linear forms on $(X_1^{**}, b_{X^{**}})$ are elements of c_0. Hence we have in particular an example of an absolutely convex set $C = (X_1, b_{X^{**}})$ so that the $b_{X^{**}}$- continuous linear forms on C separate C but not its closure X_1^{**}. This should be compared with Theorem 3.5.

10. Miscellaneous remarks. (1) We first observe that the ball topology may be used to introduce a comparison between norms on a given Banach space. If X is a Banach space and ν_1 and ν_2 are two equivalent norms on X then we say that ν_1 is *smoother* than ν_2 if the unit ball $B(\nu_2) = \{x : \nu_2(x) \leq 1\}$ is ball generated in (X, ν_1). equivalently ν_1 is smoother than ν_2 if and only if the associated ball topology b_{ν_1} is finer than b_{ν_2}.

If X contains no subspace isomorphic to ℓ_1 then ν_1 is smoother than ν_2 if and only if $N(\nu_1) \supset N(\nu_2)$ where $N(\nu)$ is the intersection of all norming subspaces of (X^*, ν^*). For separable spaces we can compare norms via the subsets $C(\nu)$ where

$$C(\nu) = \{x^{**} \in X^{**} : \nu^{**}(x^{**} - x) \geq \nu(x) \ \forall x \in X\}.$$

THEOREM 10.1. *If X is separable and ν_1 is smoother than ν_2 if and only if $C(\nu_1) \subset C(\nu_2)$.*

PROOF: Suppose ν_1 is smoother than ν_2 and that $x^{**} \in C(\nu_1)$. For $x \in X$ select a net $x_\alpha \in X$ so that $x_\alpha \to x^{**}$ (w^*) and $\nu_2(x_\alpha - x) \le \nu_2^{**}(x^{**} - x)$ for every α. Then for $y \in X$,

$$\liminf \nu_1(x_\alpha - y) \ge \nu_1^{**}(x^{**} - y) \ge \nu_1(y)$$

so that $x_\alpha \to 0$ in b_{ν_1}. Thus $x_\alpha \to 0$ in b_{ν_2} and so

$$\nu_2^{**}(x^{**} - x) \ge \liminf \nu_2(x_\alpha - x) \ge \nu_2(x)$$

so that $x^{**} \in C(\nu_2)$.

Conversely, suppose $C(\nu_1) \subset C(\nu_2)$. It suffices to show that b_{ν_1} is finer than b_{ν_2} on bounded sets. By Lemmas 2.1 and 2.2 it suffices to show that if (x_n) is a bounded sequence which is b_{ν_1}–convex clustering at x then $x_n \to x$ in b_{ν_2}. If x^{**} is any weak* cluster point of x_n we have $x^{**} - x \in C(\nu_1)$ by Lemma 2.1. Hence $x^{**} - x \in C(\nu_2)$. Now if $y \in X$

$$\liminf \nu_2(x_n - y) \ge \nu_2^{**}(x^{**} - y) \ge \nu_2(x - y)$$

so that $x_n \to x$ in b_{ν_2}.

Let us give an application to convolutions of norms. If ν_1 and ν_2 are two equivalent norms on X we define for $1 \le p < \infty$

$$\nu_1 *_p \nu_2(x) = \sup\{|x^*(x)| : \nu_1^*(x^*)^p + \nu_2(x^*)^p \le 1\}.$$

PROPOSITION 10.2. *If X is separable or does not contain ℓ_1, then for $1 \le p < \infty$, $\nu_1 *_p \nu_2$ is smoother than both ν_1 and ν_2.*

PROOF: If N is a norming subspace of $(X^*, (\nu_1 *_p \nu_2)^*)$ i.e. of $(X^*, ((\nu_1^*)^p + (\nu_2^*)^p)^{1/p})$ then it is easy to verify that N is norming in both (X^*, ν_1^*) and (X^*, ν_2^*). In fact if $x^* \in X^*$ there is a net $x_\alpha^* \in N$ with

$$\nu_1^*(x_\alpha^*)^p + \nu_2^*(x_\alpha^*)^p \le \nu_1^*(x^*)^p + \nu_2^*(x^*)^p$$

and $x_\alpha^* \to x^*$ (w^*). By passing to a subnet we may suppose that $\lim_\alpha \nu_j^*(x_\alpha^*)$ exists for $j = 1, 2$. Clearly we must have $\nu_j(x^*) = \lim_\alpha \nu_j^*(x_\alpha^*)$ for $j = 1, 2$ and hence N is norming in both spaces.

Now if X contains no copy of ℓ_1 the conclusion is immediate. If X is separable and $x^{**} \in C(\nu_1 *_p \nu_2)$ then ker x^{**} is norming in $(X^*, (\nu_1 *_p \nu_2)^*)$ and so $x^* \in C(\nu_j)$ for $j = 1, 2$ so that the result follows by the preceding theorem.

COROLLARY 10.3. *If X is separable or does not contain ℓ_1, and $\nu_1 * \nu_2$ is a dual norm then both ν_1 and ν_2 are dual norms with the same predual.*

If X does not contain ℓ_1 then it may be shown that ν is a dual norm if and only if ν is minimal for the ordering induced by smoothness, i.e. if ν_1 is less smooth than ν then ν and ν_1 induce the same ball topologies. In general this is false; if we equip ℓ_1 with the dual norm of an l.u.c norm on c_0 then this norm is strictly smoother than the usual norm on ℓ_1.

(2) Next we consider the question of when the $b_{X^{**}}$-topology restricts to the b_X-topology on X. This happens in a variety of situations, in particular when the b_X-topology agrees with the weak topology on X_1 (see Theorem 8.3). It may also be shown to be true for separable stable Banach spaces (see [19]). In fact $b_X = b_{X^{**}}|X$ if X is separable and satisfies a weak form of stability, i.e. for every pair (x_n), (y_n) of bounded sequences in X there exist $u_n \in \text{co}\{x_k : k \geq n\}$, $v_n \in \text{co}\{y_k : k \geq n\}$ and

$$(*) \qquad \lim_{n\to\infty} \lim_{k\to\infty} \|u_n + v_k\| = \lim_{k\to\infty} \lim_{n\to\infty} \|u_n + v_k\|.$$

However (*) can be shown to imply that X is weakly sequentially complete (by the same argument as [12]). Thus the example of c_0 shows that (*) does not characterize the property $b_X = b_{X^{**}}|X$, We do not know if any weakly sequentially complete Banach space has an equivalent norm satisfying (*).

(3) An *admissible* subset A of a Banach space X is one such that every x^{**} in the the weak*-closure of $A \subset X^{**}$ satisfies $|D(x^{**}| \leq 1$ ([7]). If X is separable then Theorem 2.9 can be adapted to show that b_X is Hausdorff on A. If τ is a prelinear topology on X so that X_1 is closed and A is compact then $\tau = b_X$ on A (see Section 6).

(4) There is an analogue of the characterization of weakly compact sets (8.2) for Rosenthal sets. If a set A is b_X –Hausdorff for every equivalent norm then A is a Rosenthal set (see [8], [9]).

(5) We conclude this work with some questions.

QUESTION A: Let C be an absolutely convex bounded subset of a Banach space X. If C is b_X –Hausdorff, is C b_X-regular?

A positive answer to Question A would imply, for separable X, a positive answer to the next question.

QUESTION B: Let X be a Banach space such that the intersection N_X of all norming subspaces of X^* separates X. Is N_X a norming subspace of X^*?

QUESTION C: Let C be a b_X –regular subset of X and let $\overline{C}$ be its b_X –closure. Is $\overline{C}$ b_X –regular? In particular what happens if C is a Rosenthal set? (See 3.5 and 9.4(3).)

QUESTION D: Let X be a Banach space with (UPD). Does X have (CUP) in general? What happens if X is a W*-algebra?

QUESTION E: Let X be an Asplund space. Does there exist an equivalent norm on X so that X^* has no proper norming subspace? Conversely if X^* has no proper norming subspace is X an Asplund space? For the separable case these questions have trivial positive answers.

QUESTION F: Suppose $(X^{**}, b_{X^{**}})$ is locally linear; does it follow that X cannot contain ℓ_1? Note that $(X^{**}, b_{X^{**}})$ is locally linear if and only if it is Hausdorff. (Compare 9.3.)

QUESTION G: Let $X = L_1[0,1]$. Is the usual norm minimal in the ordering introduced in this section?

References.

1. J. Bourgain, D. H. Fremlin and M. Talagrand, Pointwise compact sets of Baire measurable functions, Amer. J. Math. 100(1978) 845-886.

2. H. H. Corson and J. Lindenstrauss, On weakly compact subsets of Banach spaces, Proc. Amer. Math. Soc. 17(1966) 407-412.

3. W. B. Davis, T. Figiel, W. B. Johnson and A. Pelczynski, Factoring weakly compact operators, J. Functional Analysis 17(1974) 311-327.

4. G. Godefroy, Points de Namioka, espaces normantes, applications à la théorie isometrique de la dualité, Israel J. Math. 38(1981) 209-220.

5. G. Godefroy, Applications à la dualité d'une proprieté d'intersection de boules, Math. Zeit. 182(1983) 233-236.

6. G. Godefroy, Quelques remarques sur l'unicité des preduaux, Quart. J. Math. Oxford (2) 35(1984) 147-152.

7. G. Godefroy, Parties admissibles d'un espace de Banach, applications, Annales Sci. Ecole Norm. Sup. 16(1983) 109-122.

8. G. Godefroy, Metric characterizations of first Baire class functions and octahedral norms, to appear.

9. G. Godefroy and B. Maurey, Normes lisses et normes anguleuses sur les espaces de Banach separables, unpublished preprint.

10. G. Godefroy and P. D. Saphar, Duality in spaces of operators and smooth norms on Banach spaces, Illinois J. Math. to appear.

11. A. Grothendieck, Une caratérisation vectorielle metrique des espaces L^1, Canad. J. Math. 7(1955) 552-561.

12. S. Guerre and J. T. Lapreste, Quelques proprietes des espaces de Banach stables,

Israel J. Math. 39(1981) 247-254.

13. N. J. Kalton, Linear operators whose domain is locally convex, Proc. Edinburgh Math. Soc. 20(1977) 293-300.
14. N. J. Kalton, Isomorphisms between spaces of vector-valued continuous functions, Proc. Edinburgh Math. Soc. 26(1983) 29-48.
15. N. J. Kalton, unpublished.
16. N. J. Kalton, N. T. Peck and J. W. Roberts, *An F-space sampler,* London Math. Soc. Lecture Notes 89, Cambridge University Press, Cambridge 1985.
17. N. J. Kalton and J. W. Roberts, Uniformly exhaustive submeasures and nearly additive set functions, Trans. Amer. Math. Soc. 278(1983) 803-816.
18. J. L. Kelley, *General Topology,* van Nostrand, Princeton 1955.
19. J. L. Krivine and B. Maurey, Espaces de Banach stables, Israel J. Math. 39(1981) 273-295.
20. J. D. Lawson, Embeddings of compact convex sets and locally compact cones, Pacific J. Math. 66(1976) 443-453.
21. J. Lindenstrauss, *Extension of compact operators,* Mem. Amer. Math. Soc. 48, 1964.
22. J. Lindenstrauss and L. Tzafriri, *Classical Banach spaces I, Sequence spaces,* Springer Verlag, Berlin 1977.
23. B. Maurey, Types and ℓ_1-subspaces, Longhorn Notes, Texas Functional Analysis Seminar, Austin, Texas 1982/3.
24. E. Odell and H. P. Rosenthal, A double-dual characterization of separable Banach spaces containing ℓ_1, Israel J. Math. 20(1975) 375-384.
25. R. R. Phelps, Dentability and extreme points in Banach spaces, J. Functional Analysis 16(1974) 78-90.
26. L. H. Riddle, E. Saab and J. J. Uhl, Sets with the weak Radon-Nikodym Property in dual Banach spaces, Indiana Univ. Math. J. 32(1983) 527-541.

27. J. W. Roberts, Pathological compact convex sets in the spaces $L_p(0,1)$, $0 \leq p < 1$, The Altgeld Book, Illinois Functional Analysis Seminar, Urbana, Illinois 1976.

28. J. W. Roberts, The embedding of compact convex sets in locally convex spaces, Canad. J. Math. 30(1978) 449-454.

29. H. P. Rosenthal, A characterization of Banach spaces containing ℓ_1, Proc. Nat. Acad. Sci. U. S. A. 71(1974) 2411-2413.

30. A. Sersouri, Sur les operateurs quasi-convexes, to appear.

31. F. Sullivan, Dentability, smoothability and stronger properties in Banach spaces, Indiana Univ. Math. J. 26(1977) 545-553.

32. G. E. F. Thomas, On Radon maps with values in arbitrary topological vector spaces and their integral extensions, unpublished paper, Yale 1972.

33. P. Turpin, *Convexités dans les espaces vectoriels topologiques generaux,* Diss. Math. 131, Warsaw 1976.

34. L. Waelbroeck, *Topological vector spaces and algebras,* Springer Lecture Notes 230, Berlin 1971.

35. V. Zizler, Renormings concerning the Mazur intersection property of balls for weakly compact convex sets, Math. Ann. 276(1986) 61-66.

Equipe d'Analyse (Tour 46-0)
Université Paris VI
4 Place Jussieu
75252 Paris Cedex 05
France.

Department of Mathematics
University of Missouri-Columbia
Columbia
Missouri 65211
U.S.A.

Contemporary Mathematics
Volume **85**, 1989

ON THE CLOSEDNESS OF THE SUM OF CLOSED OPERATORS ON A UMD SPACE

Sylvie GUERRE
EQUIPE D'ANALYSE
UNIVERSITE PARIS VI
Tour 46 – 4ème Etage
4, Place Jussieu
75252 – PARIS CEDEX 05

ABSTRACT: We prove that if $(-A_k)_{k=1}^n$ are generators of analytic semi–groups on a UMD space X, which commute and have nice imaginary powers, then the sum $\sum_{k=1}^{n} A_k$ is closed. This gives the existence and uniqueness of a solution to the vector valued Cauchy problem:

$$\begin{cases} u' + \sum_{k=1}^{n-1} A_k u = f \text{ for all } f \in L^p(Y) \text{ when } Y \text{ is UMD.} \\ u(0) = 0 \end{cases}$$

This is an extension of the result of [DV] for two operators.

AMS (MOS) Subject Classification (1980), 47A05, 47A50, 47D10, 34G10.

Key words and phrases: Semi–groups of operators, vector valued Cauchy problem, unbounded operators, domains, resolvant sets, complex powers of positive operators, UMD–spaces, holomorphic function, residue, Γ–function.

INTRODUCTION: The subject of this paper was pointed out to me by J.B. Baillon. The starting point is a paper by G. Dore and A. Venni, [DV], where the closedness of the sum of two closed operators A and B on a UMD space X is proved under nice hypothesis on A and B. This abstract result solves the vector valued Cauchy problem:

$$\begin{cases} u' + Au = f: \text{ if } Y \text{ is UMD, then for all } f \in L^p(Y) \text{ this} \\ u(0) = 0 \end{cases}$$

differential equation has a unique solution belonging to $W^{1,p}(Y) \cap L^p(\mathscr{D}(A))$.

Unfortunately, the hypothesis on A and B which are needed to prove that $A + B$ is closed in [DV] do not pass from A and B to the sum $A + B$. So it is impossible to iterate Dore and Venni's theorem to study the closedness of the sum of more than two closed operators. Here, we are going to give a direct proof of the closedness of the sum of three operators on a UMD space. The generalization to the case of n operators is then easy to obtain but needs a lot of computations. As a corollary, we get also the existence and uniqueness of the solution of the vector valued Cauchy problem:

$$\begin{cases} u' + \sum_{k=1}^{n-1} A_k u \\ u(0) = 0 \end{cases}$$

for all $f \in L^p(Y)$ when Y is UMD.

This result is based on a representation of the function $\left[\sum_{k=1}^{n} a_k)\right]^{-1}$ for $a_k > 0$, $k = 1,\dots,n$, by complex integrals. This formula can pass to unbounded operators on a UMD space if these operators commute and have nice imaginary powers.

This representation is analogous to the representation of $(a+b)^{-1}$ given in [DV]. The hypothesis on $(A_k)_{k=1}^n$ called H_1, H_2, H_3, correspond to those of [DV].

THEOREM. *Let* $(A_k)_{k=1}^n$ *be* n *unbounded linear operators on a Banach space* X, *closed and with dense domains* $(\mathcal{D}(A_k))_{k=1}^n$. *Denote by* $\rho(A_k))_{k=1}^n$ *their resolvant sets. Suppose that:*

H_0: X *is a complex UMD Banach space*

H_1: $\mathbb{R}^- \cup \{0\} \subset \bigcap_{k=1}^n \rho(A_k)$ *and there exists* $M \geq 1$ *such that, for all* $\lambda \leq 0$:

$$\operatorname{Max}\left\{||(A_k-\lambda)^{-1}||, /k = 1,\ldots,n\right\} \leq \frac{M}{1+|\lambda|}$$

H_2: $\forall(\lambda_k)_{k=1}^n \in \prod_{k=1}^n \rho(A_k)$

$\{(A_k - \lambda_k)^{-1}/k = 1,\ldots,n\}$ *are mutually commuting*

H_3: $\forall\, s \in \mathbb{R}$, $\forall k = 1,\ldots,n$, A_k^{is} *belongs to* $\mathcal{L}(X)$,

the groups $s \longrightarrow A_k^{is}$ *are strongly continuous for all* $k = 1,\ldots,n$ *and there exist*

$$\begin{cases} K \geq 0 \\ 0 \leq \theta_{A_k} < \frac{\pi}{2}, \; k = 1,\ldots,n-1 \\ 0 \leq \theta_{A_n} \leq \frac{\pi}{2} \end{cases}$$

such that $||A_k^{is}|| \leq K\, e^{\theta_{A_k}|s|}$ *for* $k = 1,\ldots,n$.

Then $\sum_{k=1}^n A_k$ *is closed and* $\left[\sum_{k=1}^n A_k\right]^{-1} \in \mathcal{L}(X)$.

Remark: As it was noticed in [DV], H_1 gives the possibility to define A_k^λ for all $k \in \{1,\dots,n\}$ and $\lambda \in \mathbb{C}$. (cf. [T] and [DV] appendix). If in addition we suppose that H_3 is verified, then if $\mathcal{R}e\,\lambda \leq 0$, A_k^λ belong to $\mathcal{L}(X)$ for $k = 1,\dots,n$.

In order to make the presentation clearer, we are going to give the proof of this result only for three operators A, B, C. The generalization to n operators will follow easily with the formula given at the end of the proof (see "Extension").

We need three lemmas to prove this result. Lemma 1 and the application of the lemmas to prove the theorem are the main improvements that we gave to Dore and Venni's paper.

LEMMA 1. *Let* a,b,c *be three strictly positive real numbers. Then:*

$$(a+b+c)^{-1} = \frac{1}{(2\pi)^2}\int_{\alpha-i\infty}^{\alpha+i\infty}\int_{\beta-i\infty}^{\beta+i\infty} \Gamma(1-\lambda-\mu)\,\Gamma(\lambda)\,\Gamma(\mu)\,a^{-\lambda}b^{-\mu}c^{\lambda+\mu-1}\,d\mu\,d\lambda$$

where $\begin{cases} 0 < \alpha < 1 \\ 0 < \beta < 1 \\ 0 < \alpha+\beta < 1 \end{cases}$. *Moreover, this integral does not depend on* α *and* β.

Proof. The integral $I = \frac{1}{(2\pi)^2}\int_{\alpha-i\infty}^{\alpha+i\infty}\int_{\beta-i\infty}^{\beta+i\infty} \Gamma(1-\lambda-\mu)\,\Gamma(\lambda)\,\Gamma(\mu)\,a^{-\lambda}b^{-\mu}c^{\lambda+\mu-1}\,d\mu\,d\lambda$ is convergent because we know, cf [M], the estimations:

$$\forall (s,t) \in \mathbb{R}^2,\quad |\Gamma(s+it)| \leq e^{-\pi/2|t|}\,|t|^{s-\frac{1}{2}}\,(\sqrt{2\pi}+\epsilon).$$

On the other hand, I does not depend on α and β because the function to integrate is analytic in the domain $\begin{cases} 0 < \alpha < 1 \\ 0 < \beta < 1 \\ 0 < \alpha+\beta < 1 \end{cases}$ (cf. [E])

We recall the well known formula:

$$\forall z \in \mathbb{C}\backslash\mathbb{Z}^-,\ \Gamma(1+z) = z\Gamma(z) \quad \text{(cf. [E])}$$

and then we can write:

$$aI = \frac{1}{(2\pi)^2}\int_{\alpha-1-i\infty}^{\alpha-1+i\infty}\int_{\beta-i\infty}^{\beta+i\infty} \lambda\Gamma(-\lambda-\mu)\,\Gamma(\lambda)\,\Gamma(\mu)\,a^{-\lambda}b^{-\mu}c^{\lambda+\mu}\,d\mu\,d\lambda$$

$$bI = \frac{1}{(2\pi)^2}\int_{\alpha-i\infty}^{\alpha+i\infty}\int_{\beta-1-i\infty}^{\beta-1+i\infty} \mu\Gamma(-\lambda-\mu)\,\Gamma(\lambda)\,\Gamma(\mu)\,a^{-\lambda}b^{-\mu}c^{\lambda+\mu}d\mu\,d\lambda$$

$$cI = \frac{1}{(2\pi)^2}\int_{a-i\infty}^{a+i\infty}\int_{\beta-i\infty}^{\beta+i\infty} (-\lambda-\mu)\,\Gamma(-\lambda-\mu)\,\Gamma(\lambda)\,\Gamma(\mu)\,a^{-\lambda}b^{-\mu}c^{\lambda+\mu}\,d\mu\,d\lambda$$

Summing these three equalities, we get:

$$(a+b+c)I = -\frac{i}{2\pi}\int_{\beta-i\infty}^{\beta+i\infty} \mathcal{R}es_{\lambda=-\mu}\,[\lambda\Gamma(-\lambda-\mu)\,\Gamma(\lambda)\,\Gamma(\mu)\,a^{-\lambda}b^{-\mu}c^{\lambda+\mu}]d\mu$$

$$= -\frac{i}{2\pi}\int_{a-i\infty}^{a+i\infty} \mathcal{R}es_{\mu=-\lambda}\,[\mu\Gamma(-\lambda-\mu)\,\Gamma(\lambda)\,\Gamma(\lambda)\,\Phi(\mu)\,a^{-\lambda}b^{-\mu}c^{+\mu}]d\lambda$$

The residue of the Γ–function in $z = 0$ is 1 so we get:

$$(a+b+c)I = -\frac{i}{2\pi}\left[\int_{\beta-i\infty}^{\beta+i\infty}\mu\Gamma(-\mu)\,\Gamma(\mu)a^{\mu}b^{-\mu}d\mu + \int_{\alpha-i\infty}^{\alpha+i\infty}\lambda\Gamma(-\lambda)\,\Gamma(\lambda)a^{-\lambda}b^{\lambda}d\lambda\right]$$

Note that: $\lambda\Gamma(-\lambda)\,\Gamma(\lambda) = \Gamma(1-\lambda)\,\Gamma(\lambda) = \frac{\pi}{\sin\,\pi\lambda}$

Then, using the result of [DV], we get:

$$(a+b+c)I = (1+\tfrac{a}{b})^{-1} + (1+\tfrac{b}{a})^{-1} = 1.$$

LEMMA 2. *Let* A,B,C *satisfying* H_1,H_2,H_3. *Then:*

$$S = \frac{1}{(2\pi)^2}\int_{\alpha-i\infty}^{\alpha+i\infty}\int_{\beta-i\infty}^{\beta+i\infty} \Gamma(1-\lambda-\mu)\,\Gamma(\lambda)\,\Gamma(\mu)\,A^{-\lambda}B^{-\mu}C^{\lambda+\mu-1}d\mu\,d\lambda$$

belongs to $\mathscr{L}(X)$ *for all* $\begin{cases} 0 < \alpha < 1 \\ 0 < \beta < 1 \\ 0 < \alpha+\beta < 1 \end{cases}$

Moreover, S *does not depend on* α *and* β.

Proof: It is the analogous to Lemma 2.2 in [DV]. All operators appearing in this integral are bounded by H_1 and H_3. So it is enough to prove that the integral is absolutely convergent. This is an immediate consequence of H_3: actually, if we define $\lambda = \alpha + it$, $\mu = \beta + is$ we get the following estimates:

$$\|\Gamma(\lambda)A^{-\lambda}\| \le C_1 e^{(\theta_A-\pi/2)|t|}$$

$$\|\Gamma(\mu)B^{-\mu}\| \le C_2 e^{(\theta_B-\pi/2)|s|}$$

$$\|\Gamma(1-\lambda-\mu)C^{\lambda+\mu-1}\| \le C_3 e^{(\theta_C-\pi/2)|s+t|} \le C_3$$

where C_1, C_2, C_3 are polynomials in t and s.

The products of the right hand side functions is integrable on the domain and this implies the absolute convergence of this integral. S does not depend on α and β because the function $(\lambda,\mu) \longrightarrow \Gamma(1-\lambda-\mu)\,\Gamma(\lambda)\,\Gamma(\mu)\,A^{-\lambda}B^{-\mu}C^{\lambda+\mu-1}$ is analytic on

$$\begin{cases} 0 < \mathscr{R}e\,\lambda < 1 \\ 0 < \mathscr{R}e\,\mu < 1 \\ 0 < \mathscr{R}e\,\lambda + \mathscr{R}e\,\mu < 1 \end{cases}$$

see [E] and [T] or [DV] appendix.

In the sequel we define $\mathscr{D}(A+B+C) = \mathscr{D}(A) \cap \mathscr{D}(B) \cap \mathscr{D}(C)$.

LEMMA 3.

1) $\forall\, x \in \mathscr{D}(A+B+C)$, $S(A+B+C)x = x$

2) *If* $0 < \alpha < 1$, $0 < \beta < 1$, $0 < \alpha+\beta < 1$, *then*

   $\forall x \in \mathscr{D}(A^{1-\alpha}) \cap \mathscr{D}(B^{1-\beta})$, $Sx \in \mathscr{D}(A+B+C)$

   and $(A+B+C)Sx = x$.

3) $A+B+C$ *is closable and* $S = \overline{(A+B+C)}^{-1}$.

These results are extensions of Lemmas 2.4, 2.5 and 2.6 in [DV] and the proofs are similar. The only thing to note is that, under the hypothesis H_2, $\mathscr{D}(A) \cap \mathscr{D}(B) \cap \mathscr{D}(C)$ is dense in X by [DG] proposition 2.5.

Proof of the theorem.

Like in [DV], it remains to prove that for all $x \in X$, Sx belongs to $\mathscr{D}(A) \cap \mathscr{D}(B) \cap \mathscr{D}(C)$. Then this will show that $A+B+C = \overline{A+B+C}$ and thus that $A+B+C$ is closed and $S = (A+B+C)^{-1}$.

Let us prove for example that $Sx \in \mathscr{D}(C)$. By analyticity, we can write, for

$\epsilon_1,\epsilon_2 > 0$, $Sx = I^1x + I^2x + I^3x + I^4x$ with:

$$I^1x = \frac{-1}{(2\pi)^2}\int_{|t|>\epsilon_1}\int_{|s|>\epsilon_2}\Gamma(1-it-is)\Gamma(it)\Gamma(is)A^{-it}B^{-is}C^{it+is-1}x\,ds\,dt$$

$$I^2x = \frac{-1}{(2\pi)^2}\int_{|t|>\epsilon_1}\int_{-\pi/2}^{+\pi/2}\Gamma(1-it-\epsilon_2e^{i\varphi})\Gamma(it)\Gamma(\epsilon_2e^{i\varphi})A^{-it}B^{-\epsilon_2e^{i\varphi}}C^{it+\epsilon_2e^{i\varphi}-1}x\epsilon_2e^{i\varphi}d\varphi dt$$

$$I^3x = \frac{-1}{(2\pi)^2}\int_{-\pi/2}^{+\pi/2}\int_{|s|>\epsilon_2}\Gamma(1-\epsilon_1e^{i\theta}-is)\Gamma(\epsilon_1e^{i\theta})\Gamma(is)A^{-\epsilon_1e^{i\theta}}B^{-is}C^{\epsilon_1e^{i\theta}+is-1}x\epsilon_1e^{i\theta}dsd\theta$$

$$I^4x = \frac{-1}{(2\pi)^2}\int_{-\pi/2}^{+\pi/2}\int_{-\pi/2}^{+\pi/2}\Gamma(1-\epsilon_1e^{i\theta}-\epsilon_2e^{i\varphi})\Gamma(\epsilon_1e^{i\theta})\Gamma(\epsilon_2e^{i\varphi})A^{-\epsilon_1e^{i\theta}}B^{-\epsilon_2e^{i\varphi}}$$
$$C^{\epsilon_1e^{i\theta}+\epsilon_2e^{i\varphi}-1}x\epsilon_1e^{i\theta}\epsilon_2e^{i\varphi}d\theta\,d\varphi.$$

Using the fact that $z\Gamma(z) \longrightarrow 1$ when $z \longrightarrow 0$, and the hypothesis H_3, an easy computation shows that when $(\epsilon_1\epsilon_2) \longrightarrow (0,0)$, then $I^4x \longrightarrow \frac{1}{4}C^{-1}x$, which belongs to $\mathscr{D}(C)$.

We also get that when $\epsilon_1 \longrightarrow 0$, I^3x tends to:

$$I_3'x = \frac{1}{4\pi}\int_{|s|>\epsilon_2}\Gamma(1-is)\Gamma(is)B^{-is}C^{is-1}x\,ds$$

and when $\epsilon_2 \longrightarrow 0$, I^2x tends to:

$$I_2' x = \frac{1}{4\pi} \int_{|t|>\epsilon_1} \Gamma(1-it)\Gamma(it) A^{-it} C^{it-1} x \, dt.$$

The argument given in [DV] in the proof of theorem 2.1 shows that when $\epsilon_1 \longrightarrow 0$ $I'^2 x$ tends to a limit belonging to $\mathscr{D}(C)$, and when $\epsilon_2 \longrightarrow 0$, $I'^3 x$ tends to a limit belonging to $\mathscr{D}(C)$.

On the other hand, it is clear that $I^1 x$ belongs to $\mathscr{D}(C)$ and moreover, as C is closed, we get:

$$CI^1 x = \frac{-1}{(2\pi)^2} \int_{|t|>\epsilon_1} \int_{|s|>\epsilon_2} \Gamma(1-it-is)\Gamma(it)\Gamma(is) A^{-it} B^{-is} C^{it+is} ds \, dt$$

So to prove that Sx belongs to $\mathscr{D}(C)$, it is now enough to prove that $CI^1 x$ converges in X when $(\epsilon_1, \epsilon_2) \longrightarrow (0,0)$.

First of all, we can work with the integral

$$I^5 x = -\frac{1}{(2\pi)^2} \int_{\epsilon_1<|t|<r_1} \int_{\epsilon_2<|s|<r_2} \Gamma(1-it-is)\Gamma(it)\Gamma(is) A^{-it} B^{-is} C^{is+it} x \, ds \, dt$$

instead of $CI^1 x$ for any fixed numbers r_1 and r_2. Then, we know that $\Gamma(z) - \frac{1}{z}$ and $\Gamma(1+z) = z\Gamma(z)$ are analytic in 0 (cf. [E]). So an easy computation shows that we can write:

$$\Gamma(1-it-is)\Gamma(it)\Gamma(is) = \frac{-1}{st} + \frac{1}{2}\varphi(t) + \frac{1}{t}\varphi(s) + \psi(s,t)$$

where φ is analytic and bounded in a neighborhood of 0 and ψ is also analytic and bounded in a neighborhood of $(0,0)$. Again $I^5 x$ can be decomposed on the sum of four

double integrals, each of them being the product of two simple integrals. The integrals for which the function to integrate is analytic in 0 have obviously a limit. On the contrary, when a term in $\frac{1}{t}$ appears, we have to use the Hilbert transform. Let us give the details of the proof for the term in $\frac{1}{st}$:
we can write, by [DV] A 11:

$$\frac{1}{(2\pi)^2}\int_{\epsilon_1<|t|<r_1}\int_{\epsilon_2<s<r_2}\frac{1}{st}A^{-it}B^{-is}C^{it+is}x\,ds\,dt$$

$$=\frac{A^{-it'}B^{-is'}C^{it'+is'}}{(2\pi)^2}\int_{\epsilon_1<|t|<r_1}\int_{\epsilon_2<|t|<r_2}\frac{1}{st}A^{-i(t-t')}B^{-i(s-s')}C^{i(t-t')+i(s-s')}ds\,dt$$

$$=\frac{A^{-it'}C^{it'}}{(2\pi)^2}\int_{\epsilon_1<|t|<r_1}\frac{1}{t}A^{-i(t-t')}C^{i(t-t')}dt.\left[\frac{B^{-is'}C^{is'}}{2}\int_{\epsilon_2<|s|<r_2}\frac{1}{s}B^{-i(s-s')}C^{i(s-s')}x\,ds\right]$$

Since X is UMD (cf. B), we can choose s' such that the first simple integral converges when $\epsilon_2 \longrightarrow 0$. Then we can choose t' such that the second integral, applied to the limit of the first one, converges when $\epsilon_1 \longrightarrow 0$. We get the result by continuity of all operators which appear here, and this concludes the proof of the theorem for three operators.

EXTENSION.

The proof can easily be extended to the case of n operators $A_1,\ldots,A_n$: the formula in Lemma 1 is then, for n strictly positive real numbers $a_1,\ldots,a_n$:

$$(a_1+\dots+a_n)^{-1} = \frac{1}{(2\pi)^n}\int_{\alpha_1-i\infty}^{\alpha_1+i\infty}\dots\int_{\alpha_{n-1}-i\infty}^{\alpha_{n-1}+i\infty}\Gamma(1-\lambda_1+\cdots+\lambda_{n-1})\Gamma(\lambda_1)\dots\Gamma(\lambda_{n-1})$$

$$a_1^{-\lambda_1}\dots a_{n-1}^{-\lambda_{n-1}}a_n^{\lambda_1+\dots+\lambda_{n-1}-1}\,d\lambda_{n-1}\dots d\lambda_1$$

for $\begin{cases}0 < \alpha_j < 1 \\ \forall j = 1,\dots,n-1\end{cases}$ and $0 < \alpha_1 + \dots + \alpha_{n-1} < 1$.

And the hypothesis of the theorem imply that we get then a multiparameter analytic family $\{A_1^{\lambda_1}\dots A_n^{\lambda_n}/\lambda_1,\dots,\lambda_n \in \mathbb{C}^n,\ \mathcal{R}e\,\lambda_k \le 0\}$. See also [JM].

APPLICATION.

Let $1 < p +\infty$, $T > 0$ and Y be a UMD space. Suppose that $(A_k)_{k=1}^{n-1}$ are unbounded operators on Y, closed and with dense domains, such that:

1) $\forall\lambda \le 0,\ \|(A_k-\lambda)^{-1}\| \le \frac{M}{1+|\lambda|}$, $k = 1,\dots,n-1$
2) $\{A_k-\lambda_k)^{-1}/\lambda_k \in \rho(A_k)/k = 1,\dots,n-1\}$ are mutually commuting.
3) $s \longmapsto A_k^{is}$, $k = 1,\dots,n-1$ are strongly continuous groups in $\mathcal{L}(Y)$ such that $\|A_k^{is}\| \le Me^{\alpha_k|s|}$ with $0 < \alpha_k < \frac{\pi}{2}$, $k = 1,\dots,n-1$.

Then, if we define $\mathcal{A}_k$ on $L^p_{[0,T]}(\mathcal{D}(A_k))$, $k = 1,\dots,n-1$ by $(\mathcal{A}_k u)(t) = A_k(u(t))$, then we get the following result:

COROLLARY: *For all* $f \in L^p_{[0,T]}(Y)$, *the Cauchy problem*

$$\begin{cases} u'(T) + \sum_{k=1}^{n-1} A_k u(t) = f(t) & \textit{has a unique solution belonging to} \\ u(0) = 0 \end{cases}$$

$W^{1,p}_{[0,T]}(Y) \cap L^p_{[0,T]}\left(\bigcap_{k=1}^{n-1} \mathscr{D}(A_k)\right)$. *Moreover* u, $\sum_{k \in K} A_k u$ *for all* $K \subset \{1,\dots,n-1\}$ *and* u' *depend continuously on* f *in* $L^p(Y)$.

Proof. This is an immediate consequence of the preceding theorem and theorem 3.1 in [DV], where it is proved that if Y is UMD, the operator B on $L^p(Y)$ defined by $Bu = u'$ is such that: $\|B^{is}\| \le C(1+|s|^2)C^{\pi/2|s|}$. We apply the theorem with $A_1,\dots,A_{k-1},B$.

Remarks.

1) As it was remarked in [CL], the real p which appears in the corollary has no importance: if this L_p–regularity is true for one $p \in]1,+\infty[$, it is also true for all $p \in]1,+\infty[$.

2) In [L] another version of theorem 3.2 of [DV] (and thus of our corollary in certain cases) is proved when $Y = L_q$ under different assumptions on A (or on $(A_k)_{k=1}^{n-1}$).

3) This corollary has an interest when one knows each A_k, $k = 1,\dots,n-1$ separately and has no information on $\sum_{k=1}^{n-1} A_k$.

I am very grateful to J.B. Baillon for a lot of very fruitful conversations.

BIBLIOGRAPHY

[B] BURKHOLDER D.L.
A geometric condition that implies the existence of certain singular integrals of Banach space values functions. Confererence on harm. anal. in honor of Antoni Zygmund (Chicago 1981) p. 270–286 Belmont. Wadsworth 1983.

[CL] COULHON T. and LAMBERTON D.
Régularité L^p pour les équations d'évolution. Séminaire d'Analyse Fonctionnelle PARIS VI, VII, 1984–85.

[DG] DA PRATO G. and GRISVARD P.
Somme d'opérateurs linéaires et équations différentielles operationnelles. J. Math. pures et appl., 54, (1975) p. 305–387.

[DV] DORE G. and VENNI A.
On the closedness of the sum of two closed operators. Preprint.

[E] ERDELYI A.
The Harry Bateman project. Higher transcendental function (Vol. I) (p. 49, p. 256). McGraw Hill 1953.

[JM] JORGENSEN P.E.T. and MOORE R.T.
Operator commutation relations. Dordrecht (1984), D. Reidel Publishing Co.

[L] LAMBERTON D.
Equations d'évolution linéaires associées à des semi–groupes de contractions dans les espaces L^p. Journal of Funct. Analysis, vol. 72, n°2, (1987), p. 252–262.

[M] MELLIN H.
Abriss einer einheitlichen theorie der Gamma und der hypergeometrischen Funktionen. Math. Ann., 68, (1910) p. 305–337.

[T] TRIEBEL H.
Interpolation theory, function spaces, differential operators. Amsterdam, New York, Oxford. North Holland (1978).

Contemporary Mathematics
Volume **85**, 1989

Ultrapowers of rearrangement-invariant function spaces I[1]

RICHARD HAYDON AND PEI-KEE LIN[2]

Abstract. Let X be a rearrangement invariant function space of finite cotype. It is known that if $X = L_p$, then an ultrapower $\tilde{X}$ of X may be represented as a space $L_p(\tilde{\mu})$. In this article, we investigate the structure of $\tilde{X}$ for more general X, introducing the notion of elements "indicator type". We show that the band Z of $\tilde{X}$ generated by these elements may be represented as a generalized rearrangement-invariant function space on some product measure space $\Omega \times \Sigma$. This gives a partial description of the structure of $\tilde{X}$. However, the band Z is not necessarily the whole ultrapower, and the structure of the complementary band $Z^{\perp}$ is unclear to us.

1. Introduction: r.i.f.s. and g.r.i.f.s. It is the aim of this paper, optimistically numbered I, to give a partial description of the structure of ultrapowers of rearrangement-invariant function spaces. Some special cases are already well-understood. The best known of course is the result of Bretagnolle, Dacunha-Castelle and Krivine [1] that if X is a space of the type $L_p(\mu)$, then an ultrapower $\tilde{X}$ may be represented (in the sense of isometric lattice-isomorphism) as a space $L_p(\tilde{\mu})$. Dacunha-Castelle and Krivine [2] showed that ultrapowers of certain Orlicz spaces $L_\phi(\mu)$ allow representations of the form $L_\phi(\tilde{\mu}) \oplus L_p(\tilde{\nu})$. A fuller description of ultrapowers of Orlicz spaces is now known (see [4] for a proof): an ultrapower of any Orlicz space of nontrivial cotype may be represented as a Musielak-Orlicz space. The problem of how to represent ultrapowers of more general rearrangement-invariant function spaces, of Lorentz spaces in particular, has been posed for instance on page 83 of [5], a work

[1]Research partially supported by a grant from NSF to the Research Workshop of Banach Space Theory, University of Iowa, July 5-July 25, 1987.
[2]Research partially supported by NSF.
1980 *Mathematics subject classifications* (1985 *Revision*): 46B20,46B30,46E30
Keywords. ultrapower, rearrangement-invariant function space, band projection, indicator type
Part of the research that led to this paper took place while the authors attended the Research Workshop of Banach Space Theory, University of Iowa, July 5-July 25, 1987. They wish to thank the Mathematics Department for their hospitality.

to which the reader may be referred for an account of most of the theory of ultraproducts of Banach spaces as it existed in 1980.

We start by recalling the definition of an ultrapower of a Banach space. Let X be a Banach space, let I be a set and let $\mathcal{U}$ be a free ultrafilter on I. The *ultrapower* $X^I/\mathcal{U}$ of X over $\mathcal{U}$ is the quotient space of

$$\ell_\infty(I;X) = \{(x_i)_{i\in I} : x_i \in X \text{ for all } i \in I \text{ and } \|(x_i)\| = \sup_i \|x_i\| < \infty\}$$

by the closed subspace

$$\mathcal{N} = \{(x_i) \in \ell_\infty : \lim_{i\to\mathcal{U}} \|x_i\| = 0\}.$$

We shall denote the equivalence class which contains (x_i) by $(x_i)^\bullet$. In all that follows, we shall concern ourselves exclusively with the case $I = \mathbb{N}$, though we believe that all our results extend to the case of "countably incomplete" ultrafilters on arbitrary sets I (see [5] for definitions). For most applications, the crucial property of the ultrapower construction is that a Banach space W is *finitely representable* in X if and only if W embeds isometrically in some ultrapower of X. This fact will, however, be only marginally useful to us in the present paper, where our methods will mostly be based on a fairly careful analysis of representing sequences (x_n) for ultrapower elements.

Ultrapowers of Banach spaces tend to be very big objects, and representations of them as spaces of measurable functions necessarily involve non-σ-finite measure spaces. However, in such representation theorems, one can always arrange for the measure spaces to be *decomposable* and, since decomposable measure spaces are free of most of the pathologies normally associated with non-σ-finiteness, we shall from the start consider only measure spaces that are of this type. In doing this we follow the convention of [4], which may also be a useful reference for the following known facts about representations and ultrapowers of Banach lattices. If $(\Omega, \mathcal{F}, \mu)$ is a measure space (decomposable by what we have just been saying), a dense (order-) ideal of $L_0(\mu)$, equipped with a lattice-norm under which it is complete is called a *Köthe function space* on $(\Omega, \mathcal{F}, \mu)$. Any order-continuous Banach lattice allows a representation as a Köthe function space on some decomposable measure space. The Boolean algebra $\mathfrak{B}(X)$ of all *band projections* on a Köthe function space X may be identified with the *measure algebra* of the underlying measure space, an equivalence class $A^\bullet$ of measurable sets corresponding to the band projection $f \mapsto 1_A.f$. Although Köthe function spaces do not in general need to be order-continuous, it is known that an ultrapower $\tilde{X}$ of a Banach lattice is not representable as a Köthe function space unless it is order-continuous. Moreover, this is the case if and only if the original lattice X has finite cotype.

Let λ be Lebesgue measure on the real line. Recall (from [6] for example) that a Köthe function space X on $((0,\infty),\lambda)$ is said to be a *rearrangement-invariant function space* if the following conditions hold:

(i) If $f \in L_0(\lambda)$ and f^* is the decreasing rearrangement of f, then $f^* \in X \iff f \in X$ and, when this is the case, $\|f\| = \|f^*\|$;

(ii) The order-continuous dual X' is a norming subset of X^* and X is either minimal or maximal in X'';

(iii) $L_\infty(0,\infty) \cap L_1(0,\infty) \subseteq X \subseteq L_\infty(0,\infty) + L_1(0,\infty)$ and the inclusion maps are of norm one with respect to the natural norms in these spaces.

As pointed out in [6], condition (iii) is simply a normalization, equivalent to $\|1_{(0,1]}\| = 1$. Condition (ii) also need not concern us much since in this paper we shall be considering only spaces which are order-continuous (and hence both maximal and minimal).

If $(\Sigma, \mathcal{G}, \nu)$ is any measure space and f is a λ-measurable function we may define the decreasing rearrangement f^* to be the (unique, right-continuous) non-increasing function $(0,\infty) \to [0,\infty]$ which satisfies

$$\forall \alpha \in (0,\infty) \quad \lambda\{t \in (0,\infty) : f^*(t) \geq \alpha\} = \nu\{\sigma \in \Sigma : |f(\sigma)| \geq \alpha\}.$$

If X is a rearrangement-invariant function space then we may introduce a space $X(\Sigma, \mathcal{G}, \nu)$ consisting of all $f \in L_0(\nu)$ for which $f^* \in X$, equipped with the norm $\|f\|_{X(\Sigma,\mathcal{G},\nu)} =$

$\|f^*\|_X$.

The representation theorem that we are going to establish involves a generalization of the notion of rearrangement-invariant function space to the context of product measure spaces, where one considers a decreasing rearrangement in the "second coordinate". To be precise, let (Ω, μ) be a measure space, and Y be an order-continuous Köthe function space on $(\Omega \times (0,\infty), \mu \otimes \lambda)$. First of all, for $F \in L_0(\mu \otimes \lambda)$, we define a non-increasing rearrangement $F^* : \Omega \times (0,\infty) \to [0,\infty]$ by setting

$$F^*(\omega, t) = \left(F(\omega, \cdot)^*\right)(t).$$

We say that Y is a generalized rearrangement-invariant function space (g.r.i.f.s.) if, for $F \in L_0(\mu \otimes \lambda)$,

$$F \in Y \iff F^* \in Y \text{ with } \|F^*\| = \|F\| \text{ in this case.}$$

We now want to indicate how to define an analogue of $X(\Sigma, \mathcal{G}, \nu)$ for a g.r.i.f.s. Y on $\Omega \times (0,\infty)$. A little care is needed because of possible non-σ-finiteness. If $(\Omega, \mathcal{F}, \mu)$, $(\Sigma, \mathcal{G}, \nu)$ are measure spaces one can define a decomposable product measure space (the *c.l.d. product* in the terminology of [3]), one crucial property of which is that

$$(\mu \otimes \nu)(C) = \sup\{(\mu \otimes \nu)(C \cap (A \times B)) : \mu(A),\, \nu(B) < \infty\}.$$

This is the notion of product measure that we shall use. If $F \in L_0(\mu \otimes \nu)$ and F vanishes outside some set of the form $\Omega \times B$, with $\nu(B) < \infty$ then the formula

$$F^*(\omega, t) = \left(F(\omega, \cdot)^*\right)(t)$$

serves to define F^* a.e. To define F^* for a general F we set

$$F^* = \bigvee\{(1_{\Omega \times B} . F)^* : B \in \mathcal{G}, \nu(B) < \infty\},$$

where the sup is taken in the Dedekind complete lattice $L_0(\Omega \times (0,\infty))$. Having made this definition, it is clear that we shall set

$$Y(\Sigma, \mathcal{G}, \nu) = \{F \in L_0(\mu \otimes \nu) : F^* \in Y\}, \quad \|F\| = \|F^*\|.$$

If Y is a g.r.i.f.s. (or a r.i.f.s.) we write $\mathcal{D}_Y$ for the set of all *decreasing* elements of Y (that is, elements F with $F = F^*$). For any $\alpha > 0$, $D_\alpha : Y \to Y$ will denote the linear operator defined by

$$D_\alpha F(\omega, t) = F(\omega, t/\alpha).$$

It is also convenient sometimes to write ρ for the decreasing rearrangement function $F \mapsto F^*$. If F, G are elements of $\mathcal{D}_Y$ we define $F \oplus G$ to be $(F' + G')^*$, where F' and G' are disjoint elements of Y with $(F')^* = F$, $(G')^* = G$.

The main theorem of the paper identifies in an ultrapower of a r.i.f.s. of finite cotype a band that allows a natural representation as a g.r.i.f.s. on some product space $\Omega \times \Sigma$, and which, moreover, seems to be the maximal band that is so representable. We show, however, that this band is not necessarily the whole ultrapower. The structure of the complementary band is at present unclear, though we hope to return to this question in a later article, in which we will also give a more precise description of ultrapowers of X in the special case where X is a Lorentz space.

2. Ultrapowers of decreasing functions: atoms and indicators. For the rest of the paper we shall be dealing with a fixed rearrangement-invariant function space X on $(0,\infty)$ and a free ultrafilter $\mathcal{U}$ on $\mathbb{N}$; we assume that X has finite cotype, so that $\tilde{X} = X^I/\mathcal{U}$ is order-continuous. By results in Section I.f of [6], the assumption of finite cotype is equivalent to the assumption that X satisfies a *lower q-estimate* for some $q < \infty$, that is to say that there exists a constant C such that

$$C\,\|\sum_{j=1}^{n} x_n\|^q \geq \sum_{j=1}^{n} \|x_j\|^q$$

whenever $x_1, \ldots, x_n$ are disjoint elements of X. It is actually this property, rather than finite cotype that will be useful in our calculations. In particular we shall need the fact that, for the dilation operators D_α, one has $\|D_\alpha\| \to 0$ as $\alpha \downarrow 0$.

To simplify notation, we set $\mathcal{D} = \mathcal{D}_X$ and define

$$\tilde{\mathcal{D}} = \{(f_n)^\bullet : \forall n\ f_n \in \mathcal{D}\}.$$

On $\tilde{X}$ we may define dilation operators D_α by

$$D_\alpha((f_n)^\bullet) = (D_\alpha(f_n))^\bullet.$$

The operators $D_\alpha : \tilde{X} \to \tilde{X}$ evidently form a group of linear lattice isomorphisms. Evidently, if $f = (f_n)^\bullet \in \tilde{\mathcal{D}}$, we have $D_\alpha(f) \geq f$ whenever $\alpha \geq 1$. As a consequence of this, we may note that if f and g are disjoint elements of $\tilde{\mathcal{D}}$ then $D_\alpha f$ and $D_\beta g$ are disjoint for all $\alpha, \beta > 0$ (since if, say, $\alpha > \beta$ we have $D_\alpha(f) \wedge D_\beta(g) \leq D_\alpha(f) \wedge D_\alpha(g) = D_\alpha(f \wedge g)$). It is a bit less obvious that

$$\rho((f_n)^\bullet) = (f_n^*)^\bullet$$

well-defines a mapping $\tilde{X} \to \tilde{\mathcal{D}}$. In fact, this is the case, because of the important inequality

$$\|f^* - g^*\|_X \leq \|f - g\|_X,$$

established by Lorentz and Shimogaki [7]. We have $\rho \circ D_\alpha = D_\alpha \circ \rho$ on $\tilde{X}$ since this clearly true on X.

We now introduce some definitions and easily verified facts about g.r.i.f.s. which will help to motivate some of the things we do with ultrapowers. If $F \in \mathcal{D}_Y$ we can define the *support* of F as usual by $\operatorname{supp} F = \{(\omega, t) : F(\omega, t) \neq 0\}$; there is also a notion of "Ω-support", defined by Ω-$\operatorname{supp} F = \{\omega \in \Omega : \exists t\ F(\omega, t) \neq 0\}$. These two notions of support correspond to band projections S_F and P_F on Y given by

$$P_F(G) = \sup_{n \in \mathbb{N}} \{nF \wedge G\} \qquad S_F(G) = \sup_{n \in \mathbb{N}} \{n\, D_n\, F \wedge G\},$$

when $G \geq 0$. It is easy to see that if F, G are any two members of $\mathcal{D}_Y$ and α is any positive real number, then

(1) $S_F(G) \in \mathcal{D}_Y$;

(2) $S_{S_F(G)}(G) = S_F(G) = P_{S_F(G)}(G)$;

(3) $S_F \circ D_\alpha = D_\alpha \circ S_F$ and $S_{D_\alpha F} = S_F$;

(4) If $H \in Y_+$, then $S_F(H^*) = (S_F H)^*$;

(5) $G - S_F(G) \in \mathcal{D}_Y$;

(6) If $F \in \mathcal{D}_Y$ has the form $F(\omega, t) = \phi(\omega).1_{(0,\psi(\omega)]}(t)$ then F and $D_2(F) - F$ are disjoint;

(7) If $S = \bigvee\{S_F : F \text{ is as in } (6)\}$, then $S = I_Y$.

The functions of the type considered in (6) are particularly important to us, and will be called decreasing functions of *indicator type*.

For $f = (f_n)^\bullet \in \tilde{\mathcal{D}}$ we shall say that f is of *indicator type* if $D_2 f - f$ and f are disjoint. We shall write $\tilde{\mathcal{D}}_I$ for the set of all indicator type elements of $\tilde{\mathcal{D}}$. The following easy lemma shows that the number 2 plays no special role in our definition.

LEMMA 2.1. *If $(D_\alpha f - f) \wedge f = 0$ for some $\alpha > 1$ then $(D_\beta f - f) \wedge f = 0$ for all $\beta > 1$.*

PROOF: If $\beta < \alpha$ then $D_\beta(f) - f \leq D_\alpha(f) - f$ and so the assertion is trivial. If $\beta > \alpha$ we choose a natural number m such that $\alpha^m > \beta$. If $\gamma = \beta^{1/m}$ we then note that, for all $k \in \mathbb{N}$, $D_{\gamma^{k+1}}(f) - D_{\gamma^k}(f)$ is disjoint from $D_{\gamma^k}(f)$, and hence from f. Since

$$D_\beta(f) - f = \sum_{k=0}^{m-1} D_{\gamma^{k+1}}(f) - D_{\gamma^k}(f)$$

this gives the result. □

If $f \in \tilde{\mathcal{D}}$ we define band projections S_f and P_f by setting

$$S_f(g) = \sup_{n \in \mathbb{N}} \{n\, D_n\, f \wedge g\} \qquad P_f(g) = \sup_{n \in \mathbb{N}} \{g \wedge n\, f\},$$

when $g \geq 0$. We shall show S_f and P_f satisfy conditions equivalent to (1)-(5) above.

PROPOSITION 2.2. *Let f, g be in $\tilde{\mathcal{D}}$, and let α be a positive real. Then*

(1) $S_f(g) \in \tilde{\mathcal{D}}$ and $S_{S_f(g)} = S_f \circ S_g$;

(2) $S_{S_f(g)}(g) = S_f(g) = P_{S_f(g)}(g)$;

(3) $S_f \circ D_\alpha = D_\alpha \circ S_f$ and $S_{D_\alpha f} = S_f$;

(4) $S_f \circ \rho = \rho \circ S_f$;

(5) $g - S_f(g) \in \tilde{\mathcal{D}}$ and $S_{g - S_f(g)} = S_g - S_f \circ S_g$.

PROOF: The first four assertions are easy to establish. We shall therefore give a proof of (5) only. Let $(f_n)^\bullet$ (resp. $(g_n)^\bullet$, $(h_n)^\bullet$) be any representation of f (resp. g, $S_f(g)$), and for any $\alpha > 1$ let $A_{n,\alpha}$ denote the set $\{t : g_n(t) \geq D_\alpha h_n(t)\}$. Then

$$(g_n \wedge h_n)^\bullet = S_{S_f(g)}(g) = (g_n \wedge D_\alpha h_n)^\bullet \quad \text{and} \quad (h_n \cdot I_{A_{n,\alpha}}) \in \mathcal{N}.$$

Hence, if $\bar{h}_{n,\alpha} = g_n - (D_\alpha h_n \wedge g_n)$, then $g - h = (\bar{h}_{n,\alpha})^\bullet$.

Since X satisfies a lower q-estimate, for any $\epsilon > 0$ there exists a natural number $m \geq 2$ such that $\|D_{1/m}\| < \epsilon$. If $\{\hat{h}_{n,m}\}$ and $\{\tilde{h}_{n,m}\}$ are the sequences of functions defined by

$$\hat{h}_{n,m}(t) = \sup_{t' \geq mt} \{\bar{h}_{n,m}(t')\} \vee (g_n(t) - h_n(t)),$$
$$\tilde{h}_{n,m}(t) = \inf_{t' \leq t} \{\hat{h}_{n,m}(t')\} \in \mathcal{D}.$$

Note: if $t/m \leq t' < t$, then $\hat{h}_{n,m}(t') \geq g_n(t') - h_n(t') \geq g_n(t) - h_n(t/m) \geq g_n(t) - D_m h_n(t)$. On the other hand, if $t' \leq t/m$, then $\hat{h}_{n,m}(t') \geq \bar{h}_{n,m}(t) = g_n(t) - (D_m h_n(t) \wedge g_n(t))$. This implies that if $t' \leq t$, then for all t

$$(g_n(t) - D_m h_n(t)) \vee 0 \leq \hat{h}_{n,m}(t'), \qquad \text{and} \qquad (g_n(t) - D_m h_n(t)) \vee 0 \leq \tilde{h}_{n,m}(t).$$

Let $A'_{n,m}$ denote the set $\{t : mt \in A_{n,m}\} = A_{n,m}/m$. We claim that if $t \notin A'_{n,m}$, then $\tilde{h}_{n,m}(t) - (g_n(t) - (h_n(t) \wedge g_n(t))) \leq D_{1/m} g_n(t)$. Note: $\tilde{h}_{n,m} - (g_n - (h_n \wedge g_n)) \leq g_n - (g_n - (h_n \wedge g_n)) \leq h_n$. Hence, if the claim is true, then

$$0 \leq \Big(\tilde{h}_{n,m} - (g_n - (h_n \wedge g_n))\Big) \vee 0 \leq h_n \cdot 1_{A'_{n,m}} + D_{1/m} g_n,$$

and

$$\|(g-h)-(\tilde{h}_{n,m})^{\bullet}\| \le \|(h_n.1_{A'_{n,m}})^{\bullet}\| + \|(D_{1/m}\, g_n)^{\bullet}\| \le 2\|(D_{1/m}\, g_n)^{\bullet}\| \le 2\epsilon\|(g_n)^{\bullet}\|.$$

$$\text{(Note: } D_m\,(h_n \cdot 1_{A'_{n,m}}) \le g_n.)$$

This implies $g-h$ is in $\tilde{\mathcal{D}}$ (since ϵ is arbitrary and $\tilde{\mathcal{D}}$ is a closed set).

Now we prove our claim. If $t \notin A'_{n,m}$, then $g_n(mt) \le h_n(t)$.

Case 1. $\hat{h}_{n,m}(t) = (g_n(t)-h_n(t)) \vee 0$. Then $0 = \hat{h}_{n,m}(t) - ((g_n(t)-h_n(t)) \vee 0) \ge \tilde{h}_{n,m}(t) - ((g_n(t)-h_n(t)) \vee 0)$.

Case 2. $\hat{h}_{n,m}(t) = g_n(t') - h_n(t'/m)$ for some $t' \ge mt$. Then

$$\begin{aligned}
&\tilde{h}_{n,m}(t) - ((g_n(t)-h_n(t)) \vee 0)\\
\le&\hat{h}_{n,m}(t) - ((g_n(t)-h_n(t)) \vee 0)\\
\le&g_n(t') - h_n(t'/m) \le g_n(t') \le g_n(mt) = D_{1/m}\, g_n(t).
\end{aligned}$$

So we have proved our claim, and with it the first assertion of (5).

We now note that $S_{g-S_f(g)} \circ S_{S_f(g)} = 0$. So, for any $h \ge 0$,

$$\begin{aligned}
S_g(h) &= \sup_{n\in\mathbb{N}}\{n\,D_n g \wedge h\}\\
&= \sup_{n\in\mathbb{N}}\{n\,D_n(g - S_f(g) + S_f(g)) \wedge h\}\\
&= \sup_{n\in\mathbb{N}}\{(n\,D_n(g - S_f(g)) \wedge h) + (n\,D_n \circ S_f(g) \wedge h)\}\\
&= \sup_{n\in\mathbb{N}}\{(n\,D_n(g - S_f(g)) \wedge h)\} + \sup_{n\in\mathbb{N}}\{(n\,D_n \circ S_f(g) \wedge h)\}\\
&= S_{g-S_f(g)}h + S_{S_f(g)}h.
\end{aligned}$$

We have established the second assertion of (5). □

Recall from Section 1 that $\mathfrak{B}(\tilde{X})$ denotes the Boolean algebra of all band projections on $\tilde{X}$. Let $\mathfrak{S}_0$ denote the subset of $\mathfrak{B}(\tilde{X})$ consisting of all S_f with $f \in \tilde{\mathcal{D}}$, and let $\mathfrak{S}$ be the closed subalgebra of $\mathfrak{B}(\tilde{X})$ generated by $\mathfrak{S}_0$.

PROPOSITION 2.3. *With the above notation,*

(i) $\mathfrak{S}_0$ *is a subring of* $\mathfrak{B}(\tilde{X})$;

(ii) *if* A *is any subset of* $\mathfrak{S}_0$ *the band projection* $\bigwedge A$ *is in* $\mathfrak{S}_0$;

(iii) *if* A *is a countable subset of* $\mathfrak{S}_0$ *then* $\bigvee A$ *is in* $\mathfrak{S}_0$.

(iv) *every element of* $\mathfrak{S}$ *may be expressed in the form* $\bigvee A$ *for a suitable disjoint subset* A *of* $\mathfrak{S}_0$.

(v) *for all* $S \in \mathfrak{S}$ *and all* $\alpha > 0$ $D_\alpha \circ S = S \circ D_\alpha$.

PROOF: For $P, Q \in \mathfrak{B}(\tilde{X})$ we have $P \wedge Q = P \circ Q$ so 2.2.(1) shows that $\mathfrak{S}_0$ is closed under $\wedge$. It is clear that for $f, g \in \tilde{\mathcal{D}}$ we have $S_{f \vee g} = S_f \vee S_g$, so that $\mathfrak{S}_0$ is closed under $\vee$. Finally, 2.2.(5) shows that if S_f, S_g are in $\mathfrak{S}_0$ then so is $S_g - S_f \wedge S_g$. This completes the proof of (i). To prove (ii) we note that with the given hypotheses we may choose some $S_f \in A$ and set $g = \bigwedge\{S(f) : S \in A\}$; we then have $S_g = \bigwedge A$. For (iii), if f_n is a norm bounded sequence in $\tilde{\mathcal{D}}$ then we have

$$\bigvee_{n \in \mathbb{N}} S_{f_n} = S_{\sum_{n \in \mathbb{N}} 2^{-n} f_n}.$$

Finally (iv) follows from (i) and (ii), and (v) from (iv). □

Of most concern to us will be the set $\mathfrak{S}_I$ consisting of all S_f with $f \in \tilde{\mathcal{D}}_I$.

LEMMA 2.4.

(i) *If* $f \in \tilde{\mathcal{D}}_I$ *and* $S \in \mathfrak{S}$ *then* $Sf \in \tilde{\mathcal{D}}_I$;

(ii) *If* f *and* g *are disjoint elements of* $\tilde{\mathcal{D}}_I$, *then* $f + g \in \tilde{\mathcal{D}}_I$.

PROOF: Assertion (i) follows from 2.3.(v). To prove (ii) we note that $f \wedge [D_2(f+g) - (f+g)] \leq [f \wedge (D_2 f - f)] + [f \wedge D_2 g] = f \wedge (D_2 f - f) = 0$, since for $f, g \in \tilde{\mathcal{D}}$, $f \wedge g = 0$ implies $f \wedge D_2 g = 0$. □

PROPOSITION 2.5. *The set* $\mathfrak{S}_I = \{S_f : f \in \tilde{\mathcal{D}}_I\}$ *is an ideal of the Boolean algebra* $\mathfrak{S}$

PROOF: If $f \in \tilde{\mathcal{D}}_I$ and $S \in \mathfrak{S}$ we have $S_{Sf} = S \circ S_f$ and 2.4.(i) tells us that $Sf \in \tilde{\mathcal{D}}_I$. Thus to show that $\mathfrak{S}_I$ is an ideal it will be enough to show that it is closed under addition of disjoint members. Of course, this follows from 2.4.(ii). □

Although the fact is not strictly needed for what follows, let us first note that $\mathfrak{S}$ has atoms, that these atoms are all in $\mathfrak{S}_I$ and that they allow a straightforward characterization. Evidently, by 2.3.(iv), an atom of $\mathfrak{S}$ must be of the form S_f for some $f \in \tilde{\mathcal{D}}$, and S_f is an atom if and only if, for any $g \in \tilde{\mathcal{D}}$, we have $S_f \circ S_g = 0$ or $S_f = S_f \circ S_g = S_{S_f(g)}$.

PROPOSITION 2.6. *Suppose that S_f is an atom. Then there exists a sequence $g_n = a_n 1_{(0,\lambda_n]}$ such that $S_f = S_{(g_n)^\bullet}$. In particular, $S_f = S_g$ with $g \in \tilde{\mathcal{D}}_I$. Conversely, if $g_n = a_n 1_{(0,\lambda_n]}$ and $g = (g_n)^\bullet \neq 0$, then S_g is an atom.*

PROOF: Assume that $(f_n)^\bullet$ is a representation of f , $\|f\| = 1$ and S_f is an atom. Let $a_n = \inf\{s : \|f_n \wedge s\| \geq \frac{1}{4}\}$, $\lambda_n = \sup\{t : f_n(t) \geq a_n\}$, and $h_n = f_n \wedge a_n$. Then $\|h_n\| = \frac{1}{4}$. We claim that if $g_n = a_n 1_{(0,\lambda_n]}$, then $(g_n)^\bullet \neq 0$ (and so $S_{(g_n)^\bullet} \neq 0$). If this were not true, then

$$\begin{aligned}
1 = \|f\| = \|S_{(f_n)^\bullet} f\| &= \|S_{(h_n)^\bullet} f\| \qquad \text{since } (h_n)^\bullet \leq (f_n)^\bullet \text{ and } (h_n)^\bullet \neq 0 \\
&= \|P_{(h_n)^\bullet} f\| = \sup_{m \geq 0} \lim_{n \to \mathcal{U}} \|f_n \wedge m \cdot h_n\| \\
&\leq \sup_{m \geq 0} \lim_{n \to \mathcal{U}} \|m \cdot g_n + (h_n - g_n)\| \leq \|(h_n)^\bullet\| = \tfrac{1}{4}.
\end{aligned}$$

This is impossible. So $S_f = S_{(g_n)^\bullet}$.

Conversely, assume that $g_n = a_n 1_{(0,\lambda_n]}$ and $g = (g_n)^\bullet$. Suppose that $0 \neq h = (h_n)^\bullet \in \tilde{\mathcal{D}}$ such that $S_g h = h$ and $\|h\| = 1$. We claim that $S_g = S_h$ i.e. $S_h g = g$. Since $S_g h = h$, there exists $\alpha > 1$ such that $\|h - (\alpha \cdot D_\alpha g \wedge h)\| < \frac{1}{4}$. Let $b_n = \inf\{s : \|h_n \wedge s\| \geq \frac{1}{4}\}$ and let $\beta_n = \sup\{t : h(t) \geq b_n\}$. We claim that there is $\gamma > 0$ such that

(i) $0 < \lim_{\mathcal{U}} \beta_n / \lambda_n$,

(ii) $0 < \lim_{\mathcal{U}} b_n / a_n$.

If the claim were proved, then there exists $\gamma > 0$ such that $\gamma < \lim_{\mathcal{U}} \beta_n/\lambda_n$, and $\gamma < \lim_{\mathcal{U}} b_n/a_n$. This implies $g \leq \frac{1}{\gamma}.D_{1/\gamma}\, h$ and $S_g = S_h$.

Proof of (i). If $0 = \lim_{\mathcal{U}} \beta_n/\lambda_n$, then $\lim_{\mathcal{U}} \|a_n 1_{(0,\beta_n]}\| = 0$ and

$$\tfrac{3}{4} = 1 - \tfrac{1}{4} < \lim_{\mathcal{U}} \|\alpha \cdot D_\alpha(g_n) \wedge h_n\| \leq \lim_{\mathcal{U}} \|\alpha\, a_n\, 1_{(0,\beta_n]}\| + \lim_{\mathcal{U}} \|h_n \wedge b_n\| \leq \tfrac{1}{4}.$$

This is impossible. Hence, $0 < \lim_{\mathcal{U}} \beta_n/\lambda_n$.

Proof of (ii). If $0 = \lim_{\mathcal{U}} b_n/a_n$, then

$$0 < \lim_{\mathcal{U}} \|\alpha\, D_\alpha\, g_n \wedge h_n \wedge b_n\| \leq \lim_{\mathcal{U}} \|b_n \cdot 1_{(0,\alpha\lambda_n]}\| = 0.$$

We get a contradiction and we prove (ii). □

We shall shortly show that there exist nonzero elements of $\tilde{\mathcal{D}}_I$ which are disjoint from all atoms. In fact, the results achieve more than this, giving a precise characterization of indicator type elements that will be very useful later. The general idea is that in a linear combination of the type $\sum_{j \leq N} c_j . 1_{(0,\lambda_j]}$ the different terms behave as if they were disjoint, provided λ_{j+1}/λ_j is large for all j. We start with an easy calculation.

LEMMA 2.7. *Let $\{M_n\}$ and $\{\lambda_k^n\}$ be any sequences of positive real numbers such that*

(i) $\lim_{n \to \mathcal{U}} M_n = \infty$,

(ii) $\lambda_k^n \geq \lambda_{k-1}^n M_n$ *for all n and k.*

If $f_n = \sum_{k=1}^{K_n} a_k^n\, 1_{(\lambda_k^n, \lambda_{k+1}^n]}$ ($a_k^n > 0$ for all n and k) and if $f = (f_n)^\bullet \in \tilde{\mathcal{D}}$, then f is in $\tilde{\mathcal{D}}_I$.

PROOF: We may assume that $M_n > 2$ (so the intervals $(\lambda_k^n, 2\lambda_k^n]$ are pairwise disjoint). The inequality $D_2\, f_n - f_n \leq \sum_{k=1}^{K_n} a_k^n \cdot 1_{(\lambda_k^n, 2\cdot\lambda_k^n]}$, gives

$$\begin{aligned} &\|(D_2\, f_n - f_n) \wedge f_n\| \\ &\leq \Big\| \sum_{i=1}^{K_n - 1} a_{i+1}^n \cdot 1_{(\lambda_i^n, 2\cdot\lambda_i^n]} \Big\| \\ &\leq \Big\| \sum_{i=1}^{K_n - 1} a_{i+1}^n\, D_{1/(M_n - 1)} 1_{(\lambda_i^n, \lambda_{i+1}^n]} \Big\| \qquad \text{since } \lambda_i^n \leq \Big(\frac{1}{M_n - 1}\Big)(\lambda_{i+1}^n - \lambda_i^n) \\ &\leq \|D_{1/(M_n - 1)}(f_n)\|. \end{aligned}$$

Since the last term converges to 0, $D_2 f - f$ and f are disjoint and f is of indictor type. □

If the functions f_n considered in 2.7 are such that

$$\sup_{k \le K_n} \|a_k^n \cdot 1_{(0,\lambda_k^n]}\| \to 0 \text{ as } n \to \mathcal{U},$$

then f is disjoint from all atoms. This establishes the claim we made earlier. It will be of importance to us that every element of indicator type in $\tilde{\mathcal{D}}$ allows a representation of the type considered in 2.7. This will follow from a more general result which will be used later to say something about the band generated by indicator type elements.

PROPOSITION 2.8. *Let $f \in \tilde{\mathcal{D}}$ be an element such that $P_f \ne S_f$. Then there exists a non-zero $h \in \tilde{\mathcal{D}}_I$ with $h \le f$.*

PROOF: If $P_f(D_\gamma f) = D_\gamma f$ for some $\gamma > 1$, then for all $n \in \mathbb{N}$ $P_f(D_{\gamma^n} f) = D_{\gamma^n} f$. So the hypothesis tells us that if $\gamma > 1$ then $P_f(D_\gamma f) \ne D_\gamma f$, or, equivalently, that if $0 < \delta < 1$ then $P_{D_\delta f} f \ne f$. Set $g = P_{D_\delta f} f$ and choose representations $(f_n)^\bullet$, $(g_n)^\bullet$ with $g_n \le f_n$. We note that $f \wedge Rg = g$ for all $R > 1$. It follows there exists $\{a_n\}$ such that

(i) $a_n \to \infty$ as $n \to \mathcal{U}$,

(ii) for all n, $\|(f_n \wedge 2a_n g_n) - g_n\| < 1/a_n$.

For each n, we define the following two sets

$$A_n = \{t \in \mathbb{R} : \; 2g_n(t) > f_n(t)\}$$

$$A_n' = \{t \in \mathbb{R} : \; 2a_n g_n(t+) > f_n(t-)\}.$$

If $a_n > 1$, then

$$\max\left\{\|1_{A_n} \cdot (f_n - g_n)\|, \; \|1_{\mathbb{R}\setminus A_n} \cdot g_n\|\right\} \le \|(f_n \wedge 2\, a_n\, g_n) - g_n\| < 1/a_n, \tag{8}$$

$$\|1_{A_n'\setminus A_n} \cdot f_n\| \le 2\|1_{A_n'} \cdot (f_n - g_n)\| \le 2\|(f_n \wedge 2a_n\, g_n) - g_n\| < 2/a_n. \tag{9}$$

Note: A_n' is an open set for each n. If (α, β) is any component interval of A_n' such that $s \in (\alpha, \beta) \cap A_n \ne \emptyset$, then we have

$$f_n(\alpha-) \ge 2a_n g_n(\alpha+) \ge 2a_n g_n(s) > a_n f_n(s) \ge a_n f_n(\beta-).$$

Since $A_n \subseteq A'_n$, we may choose M_n and component intervals (α^n_j, β^n_j) $(1 \le j \le M_n)$ which satisfy the following conditions:

(i) $A_n \cap (\alpha^n_j, \beta^n_j) \neq \emptyset$,

(ii) $\beta^n_j \le \alpha^n_{j+1}$ for all n and $j \le M_n - 1$,

(iii) for all n,

$$\| 1_{A_n \setminus \bigcup_{j=1}^{M_n} (\alpha^n_j, \beta^n_j)} \cdot g_n \| < \frac{1}{a_n}. \tag{10}$$

Let

$$A''_n = \bigcup_{j=1}^{M_n} (\alpha^n_j, \beta^n_j),$$

$$h_n = \sum_{j=1}^{M_n} f_n(\beta^n_j) \cdot 1_{(0, \beta^n_j]}$$

$$k_n = 1_{\mathbb{R} \setminus A''_n} \cdot f_n = 1_{(0, \alpha^n_1]} \cdot f_n + \sum_{j=1}^{M_n - 1} 1_{[\beta^n_j, \alpha^n_{j+1}]} \cdot f_n + 1_{[\beta^n_{M_n}, \infty)} \cdot f_n.$$

Hence, if $a_n > 1$, then $h_n \le a_n f_n / (a_n - 1)$, $h_n + k_n \in \mathcal{D}$, and

$$\begin{aligned}
&\|k_n - (f_n - g_n)\| \\
&\le \|1_{\mathbb{R} \setminus A''_n} \cdot g_n\| + \|1_{A''_n} (f_n - g_n)\| \\
&\le \|1_{\mathbb{R} \setminus A_n} \cdot g_n\| + \|1_{A_n \setminus A''_n} \cdot g_n\| + \|1_{A'_n} \cdot (f_n - g_n)\| && \text{since } A'' \subseteq A' \\
&\le \frac{1}{a_n} + \frac{1}{a_n} + \|1_{A_n} \cdot (f_n - g_n)\| + \|1_{A'_n \setminus A_n} \cdot f_n\| + \|1_{\mathbb{R} \setminus A_n} \cdot g_n\| && \text{by (8) and (10)} \\
&\le \frac{2}{a_n} + \frac{1}{a_n} + \frac{2}{a_n} + \frac{1}{a_n} && \text{by (8) and (9)} \\
&= \frac{6}{a_n}.
\end{aligned}$$

So we have $k = (k_n)^\bullet = (f - g)$ and $(f - g) + (h_n)^\bullet \in \tilde{\mathcal{D}}$. Since $(f - g) \le (f - g) + (h_n)^\bullet$ and $D_\delta (f - g) \le D_\delta f$, we also have

(i) $D_\delta (f - g) \le (f - g) + (h_n)^\bullet$.

(ii) $0 \le D_\delta (f - g) \wedge (f - g) \le D_\delta f \wedge (f - g) = 0$.

This implies $D_\delta(f-g) \le (h_n)^\bullet = h$. Since by hypothesis $f-g \ne 0$, we also have $h \ne 0$.

Now, we claim that h is of indicator type.

$$\begin{aligned}
&\|(D_2 h_n - h_n) \wedge h_n\| \\
&\le \| \sum_{j=1}^{M_n-1} \sum_{\ell=j+1}^{M_n} f_n(\beta_\ell^n) I_{[\beta_j^n, 2\beta_j^n)} \| \\
&\le \| \sum_{j=1}^{M_n-1} \frac{1}{a_n-1} f_n(\beta_j^n) I_{[\beta_j^n, 2\beta_j^n)} \| \le \frac{1}{a_n-1} \|h_n\|.
\end{aligned}$$

Since $\lim_{n\to\mathcal{U}} a_n = \infty$, h is of indicator type and we prove our claim. □

PROPOSITION 2.9. *If f is in $\tilde{\mathcal{D}}_I$ then f allows a representation as $f = (f_n)^\bullet$ where*

$$f_n = \sum_{k=1}^{K_n} a_k^n \cdot 1_{(0,\lambda_k^n]},$$

and $M_n = \inf_{k<K_n} \lambda_{k+1}^n / \lambda_k^n \to \infty$ as $n \to \mathcal{U}$.

PROOF: It is not hard to see that the set of f which allow representations of the given kind is closed in $\tilde{X}$. We shall therefore just show how to approximate arbitrary elements of $\tilde{\mathcal{D}}_I$ with ones of this special sort. Given any indicator type $f \in \tilde{\mathcal{D}}$ and any $\delta \in (0,1)$, we may carry out the construction of 2.8 and obtain an indicator type element h, having a representation of the desired sort, and satisfying $D_\delta(f-g) \le h \le f$. Now because of the assumption that f is of indicator type, we have $g = P_{D_\delta(f)}(f) = D_\delta(f)$, and the elements $D_{\delta^k}(f - D_\delta(f))$ $(k \ge 1)$ are disjoint. Thus, since each one of them is $\le h$, so is the sum $\sum_{k=1}^{\infty} D_{\delta^k}(f - D_\delta(f))$. This, of course, is exactly $D_\delta(f)$ and we have proved that $D_\delta(f) \le h \le f$, which gives the required approximation when δ is close to 1. □

REMARK 2.10. *Let f and f_n be as in Proposition 2.9. Then*

$$\sum_{k=2}^{K_n} a_k^n \cdot 1_{(0,\lambda_{k-1}^n]} \le D_{1/M_n} \sum_{k=1}^{K_n} a_k^n \cdot 1_{(0,\lambda_k^n]}.$$

Hence, if $f'_n = \sum_{k=1}^{K_n} a_k^n \cdot 1_{(\lambda_{k-1}^n, \lambda_k^n]}$ $(\lambda_0^n = 0$ *for all* $n)$, *then* $f = (f'_n)^\bullet$ *is a representation of* f.

3. Representation as a g.r.i.f.s. In this section, we locate a sublattice of $\tilde{X}$ which may be represented as a g.r.i.f.s. on $\Omega \times (0, \infty)$ for a suitable measure space Ω. We define W to be the closed sublattice of $\tilde{X}$ generated by $\tilde{\mathcal{D}}_I$. Recall from Section 1 that the Boolean algebra $\mathfrak{B}(W)$ of all band projections on W is a measure algebra (because W is order-continuous). Let $\mathfrak{A}_I$ denote the closed subalgebra of $\mathfrak{B}(W)$ generated by the projections $S_f \quad (f \in \tilde{\mathcal{D}}_I)$. Since it is a closed subalgebra of a measure algebra, it is itself a measure algebra. Let us choose a measure space $(\Omega, \mathcal{F}, \mu)$ such that the measure algebra $\mathfrak{A}_\mu$ is isomorphic to $\mathfrak{A}_I$. Note that the ring of σ-finite elements of $\mathfrak{A}_\mu$ is isomorphic to the ring $\mathfrak{S}_I$ considered in the previous section. If $A \in \mathcal{F}$ and the element of $\mathfrak{A}_I$ which corresponds to $A^\bullet \in \mathfrak{A}_\mu$ is S then we shall write $1_A.f$ for $S(f)$; this action of indicator functions on W extends to give W the structure of a module over $L_\infty(\mu)$.

We are eventually going to show that W may be regarded as a g.r.i.f.s. over $(\Omega, \mathcal{F}, \mu)$. We start, however, with three lemmas that make us of the representation for indicator type decreasing functions given in 2.9.

LEMMA 3.1. *Let* f *be an element of* $\tilde{\mathcal{D}}_I$ *and let* $0 = a_0 < a_1 < \cdots < a_m$; $c_1, \ldots, c_m$ *be positive reals. Let* h *be the element of* $\tilde{X}$ *given by*

$$h = \sum_{j=1}^{m} c_j.(D_{a_j} - D_{a_{j-1}})(f).$$

Then $\rho(h)$ *is what it should be, namely,*

$$\rho(h) = \sum_{j=1}^{m} c_{\pi(j)}.(D_{b_j} - D_{b_{j-1}})(f),$$

where π *is a permutation such that* $(c_{\pi(j)})$ *is nonincreasing, and* $b_j = \sum_{i \le j}(a_{\pi(i)} - a_{\pi(i)-1})$.

PROOF: This is not too difficult, using the representation given by Proposition 2.9 and Remark 2.10. □

LEMMA 3.2. *Let f be in $\tilde{\mathcal{D}}_I$ and let R be the set of all linear combinations of elements of the form $(D_b - D_a) \circ S_g(f)$, with $g \in \tilde{\mathcal{D}}_I$ and $a < b$ positive reals. Then R is a norm-dense sublattice of $S_f[W]$.*

PROOF: We leave it to the reader to check that R (the notation stands for "rectangles" of course) is a linear sublattice. It is obvious that R is closed under dilation operators D_α. To establish density, it is enough to show that g is in the closure of R whenever $g \in \tilde{\mathcal{D}}$ and $g \leq f$.

By Proposition 2.9 and Remark 2.10, we can represent f as $(f_n)^\bullet$ and $(f'_n)^\bullet$ where

$$f_n = \sum_{k=1}^{K_n} c_k^n \cdot 1_{(\lambda_{k-1}^n, \lambda_k^n]},$$

$$f'_n = \sum_{k=1}^{K_n} c_k^n \cdot 1_{(0, \lambda_k^n]},$$

and $M_n = \inf_{k<K_n} \lambda_{k+1}/\lambda_k \to \infty$ as $n \to \mathcal{U}$. Since $g \in \tilde{\mathcal{D}}$ and $g \leq f$, then we may choose a representation of g as $(g_n)^\bullet$ with $f_n \geq g_n \in \mathcal{D}$ for each n. We now define decreasing functions $\phi_k^n : [0,1] \to [0,1]$ by setting

$$\phi_k^n(s) = \begin{cases} (c_k^n)^{-1} g_n(s\lambda_k^n) & \text{if } s\lambda_k^n > \lambda_{k-1}^n \\ (c_k^n)^{-1} g_n(\lambda_{k-1}^n) & \text{otherwise.} \end{cases}$$

Let $\mathcal{E}$ be the set of all decreasing functions from $[0,1]$ to itself. Then $\mathcal{E}$ is sequentially compact for the pointwise topology, and hence compact for the topology of convergence in measure. So, given $\eta > 0$, there exists a finite subset $\{\psi_1, \ldots, \psi_l\}$ such that, for all $\phi \in \mathcal{E}$ there exists i such that $d_0(\phi, \psi_i) < \eta$, that is to say, $|\phi(t) - \psi_i(t)| \leq \eta$ except on a set of measure at most η. We may, moreover, assume that each ψ_i is of the form

$$\psi_i = \sum_{d=1}^{H_i} w_d^i . 1_{(0, a_d^i]}$$

where $w_d^i \geq 0$, $\sum_d w_d^i \leq 1$ and $0 < a_1 < \ldots < a_{H_i}$. For each n, we partition the set $\{1, 2, \ldots, K_n\}$ into subsets U_i^n such that $d_0(\phi_k^n, \psi_i) < \eta$ whenever $k \in U_i^n$. We now define

$$f_n^i = \sum_{k \in U_i^n} c_k^n . 1_{(0, \lambda_k^n]}$$

and put $f^i = (f^i_n)^\bullet$. Clearly, $f^1, \ldots, f^l$ are disjoint elements of $\tilde{\mathcal{D}}_I$ with $\sum_{i=1}^{l} f^i = f$.

We also define

$$\begin{aligned} g^i_n(t) &= \sum_{k\in U^n_i} \psi_i(t/\lambda^n_k)\, c^n_k\, 1_{(0,\lambda^n_k]}(t) \\ &= \sum_{k\in U^n_i} \sum_{d=1}^{H_i} 1_{(0,a^i_d]}(t/\lambda^n_k)\, w^i_d\, c^n_k\, 1_{(0,\lambda^n_k]}(t) \\ &= \sum_{d=1}^{H_i} w^i_d \sum_{k\in U^n_i} c^n_k\, 1_{(0,a^i_d\lambda^n_k]}(t) \\ &= \sum_{d=1}^{H_i} w^i_d\, D_{a^i_d} f^i_n(t) \end{aligned}$$

so that $g^i = (g^i_n)^\bullet = \sum_{d=1}^{H_i} w^i_d\, D_{a^i_d}(f^i) \in R$.

To finish the proof we need to show that $\|g - \sum_{i=1}^{l} g^i\|$ is small. Let us write

$$h_n(t) = \sum_{k=1}^{K_n} c^n_k\, \phi^n_k(t/\lambda^n_k)\, 1_{(0,\lambda^n_k]}(t).$$

From the way we defined ϕ^n_k, we have

$$|h_n - g_n| \le \sum_{k=1}^{K_n} \Big(\sum_{j>k} c^n_j\Big) 1_{(\lambda^n_{k-1},\lambda^n_k]} + \sum_{k=2}^{K_n} c^n_k\, 1_{(0,\lambda^n_{k-1}]}$$

and so

$$\|h_n - g_n\| \le 2\,\|D_{1/(M_n-1)}(f'_n)\|.$$

Thus we actually have $g = (h_n)^\bullet$.

We now estimate $\|h_n - \sum_i g^i_n\|$. We note that the inequality $d_0(\phi^n_k, \phi_i) < \eta$, for $k \in U^n_i$, implies the inequality $(\phi^n_k - \psi_i)^* \le \eta 1_{[0,1]} + 1_{[0,\eta]}$. Thus we have

$$\begin{aligned} \|h_n - \sum_i g^i_n\| &= \Big\|\sum_{i=1}^{l} \sum_{k\in U^n_i} c_k\big(D_{\lambda^n_k}(\phi^n_k - \psi_i)1_{(0,\lambda^n_k]}\big)\Big\| \\ &\le \Big\|\sum_{i=1}^{l} \sum_{k\in U^n_i} c_k\big(D_{\lambda^n_k}(\phi^n_k - \psi_i)1_{(0,\lambda^n_k]}\big)^*\Big\| \\ &\le \Big\|\sum_{i=1}^{l} \sum_{k\in U^n_i} c_k\big(\eta 1_{(0,\lambda^n_k]} + 1_{(0,\eta\lambda^n_k]}\big)\Big\| \\ &\le \|\eta f'_n + D_\eta(f'_n)\| \le (\eta + \|D_\eta\|)\, \|f'_n\|. \end{aligned}$$

This achieves what we want, since $\|D_\eta\| \to 0$ as $\eta \downarrow 0$. □

LEMMA 3.3. *Let h be in $\tilde{X}$. Then $\rho(h) \in \tilde{\mathcal{D}}_I$ if and only if h has a representation as $(h_n)^\bullet$ with*

$$h_n = \sum_{j=1}^{K_n} c_j^n \, 1_{A_j^n} \,,$$

the A_j being disjoint measurable subsets of $(0,\infty)$ with $M_n = \inf_{j<K_n} \lambda(A_{j+1}^n)/\lambda(A_j^n) \to \infty$ as $n \to \mathcal{U}$. If h is of this type and $g \in \tilde{X}_+$ is such that $g \le Rh$ for some real R, then

$$\rho(h-g) = \rho(\rho(h) - \rho(g)).$$

PROOF: The first assertion follows easily from the fact that h must have a representation as $(h_n)^\bullet$, the h_n^* being like the f_n of Lemma 2.7.

To prove the second assertion, first we note

(i) we may assume that $h \in \tilde{\mathcal{D}}_I$ and $c_{j+1}^n < c_j^n$ for all $j < K_n$, or, equivalently that $A_j^n = 1_{(a_{j-1}^n, a_j^n]}$ where $a_j^n = \sum_{\ell=1}^{j} \lambda(A_\ell^n)$ and $a_0^n = 0$;

(ii) we may choose a representation of g as $(g_n)^\bullet$ with $0 \le g_n \le Rh_n$;

(iii) for each n and j, if $g_n 1_{A_j^n}$ and $g_n' 1_{A_j^n}$ have the same distribution, then

$$\left(c_j^n.1_{A_j^n} - g_n.1_{A_j^n}\right)^* = \left(c_j^n.1_{A_j^n} - g_n'.1_{A_j^n}\right)^*.$$

Let $B_j^n = \{t : g_n(t) > R\,c_{j+1}^n \text{ and } g_n(t) \le R\,c_j^n\}$, and let g_n' be a function such that $(g_n' \, 1_{A_j^n})^* = (g_n \, 1_{B_j^n})^* \cdot 1_{(0,\lambda(A_j^n))}$. We claim that

$$\left\|(h_n - g_n)^* - (h_n - g_n')^*\right\| \le 4\,R\,\left\|D_{1/(M_n-1)}\right\| \, \|h\|.$$

Suppose that the claim were proved. Since $(g_n)^*$ and g_n have the same distribution, we have

$$\begin{aligned}
&\left\|(h_n - g_n)^* - (h_n - (g_n)^*)^*\right\| \\
\le &\left\|(h_n - g_n)^* - (h_n - g_n')^*\right\| + \left\|(h_n - (g_n)^*)^* - (h_n - g_n')^*\right\| \\
\le &8\,R\,\left\|D_{1/(M_n-1)}\right\| \, \|h\|.
\end{aligned}$$

So $\rho(h-g) = \rho(\rho(h) - \rho(g))$.

Now, we prove our claim. Let $C_j^n = B_j^n \cap A_j^n$. Since $\lambda(B_j^n) \le \sum_{\ell=1}^{j} \lambda(A_j^n) = a_j^n$, there exist $E_j^n \subseteq C_j^n$ and $F_j^n \subseteq A_j^n$ such that

(iv) $\lambda(C_j^n \setminus E_j^n) \le a_{j-1}^n$;

(v) $g_n 1_{E_j^n}$ and $g_n' 1_{F_j^n}$ have the same distribution.

By (iii), we may assume that $F_j^n = E_j^n$ and $g_n 1_{E_j^n} = g_n' 1_{E_j^n}$. Hence, if $b_j^n = \sum_{\ell=0}^{j} a_\ell^n$ and $E = \bigcup_{j=1}^{K_n} E_j^n$, then

$$\begin{aligned}
&\|(h_n - g_n)^* - (h_n - g_n')^*\| \\
\le &\|(h_n - g_n) - (h_n - g_n')\| = \|g_n - g_n'\| \\
\le &\|g_n 1_{\mathbb{R}\setminus E}\| + \|g_n' 1_{\mathbb{R}\setminus E}\| \\
\le &\|R \sum_{j=2}^{K_n} c_{j-1}^n 1_{A_j^n}\| + R\|\sum_{\ell=1}^{K_n-1} c_{\ell+1}^n 1_{(b_{\ell-1}^n, b_\ell^n]}\| \\
&+ 2R\|\sum_{\ell=1}^{K_n-1} c_{\ell+1}^n 1_{(b_{\ell-1}^n, b_\ell^n]}\| \\
\le &4R\|\sum_{\ell=1}^{K_n-1} c_{\ell+1}^n 1_{(b_{\ell-1}^n, b_\ell^n]}\| \\
\le &4R\|D_{1/(M_n-1)}\|\,\|h\| \qquad \text{since} \quad a_j^n \le \frac{\lambda(A_{j+1}^n)}{M_n - 1}.
\end{aligned}$$

We have proved our claim. □

PROPOSITION 3.4. *Let f be in $\tilde{D}_I$ and let the band projection S_f correspond to the subset Ω_f of Ω. There exist a g.r.i.f.s. Y_f on $\Omega_f \times (0,\infty)$ and an isometric lattice isomorphism $\Phi_f : S_f[W] \to Y_f$ which satisfy*

(i) $\Phi_f(f) = 1_{\Omega_f \times (0,1]}$;

(ii) $\Phi_f(\rho(g)) = (\Phi_f(g))^* \quad (g \in S_f[W])$;

(iii) $\Phi_f \circ D_\alpha = D_\alpha \circ \Phi_f \quad (\alpha > 0)$;

(iv) $\Phi_f(\phi.g) = \phi.\Phi_f(g) \quad (g \in S_f[W],\ \phi \in L_\infty(\Omega_f))$.

Moreover, Y_f and Φ_f is uniquely determined by these conditions.

PROOF: We start by defining Φ_f on the sublattice R, taking

$$\Phi_f(1_A.(D_b - D_a)(f)) = 1_{A\times(a,b]},$$

and extending linearly. Recall that by $1_A.g$, where $g \in W$, we mean $S(g)$, where $S \in \mathfrak{A}_I$ corresponds to the element $A^\bullet$ of the measure algebra $\mathfrak{A}_\mu$. This having been said, it is clear that $\Phi_f(1_B.h) = 1_B.\Phi_f(h)$ when $h \in R$. Properties (i) and (iii) clearly hold, and (ii) holds by Lemma 3.1.

The density result 3.2 now allows us to extend Φ_f from R to $S_f[W]$ and obtain an isometric isomorphism from $S_f[W]$ onto some g.r.i.f.s. on $\Omega\times(0,\infty)$ with all the properties listed above. Moreover, properties (i), (iii) and (iv) clearly force Φ_f to have the given form on R and so we have the asserted uniqueness property. □

THEOREM 3.5. *Let X be a r.i.f.s. of finite cotype on $(0,\infty)$ and let W be the closed sublattice of the ultrapower $\tilde{X}$ generated by the set $\tilde{\mathcal{D}}_I$ of decreasing elements of indicator type. Then there exists a g.r.i.f.s. Y on $\Omega\times(0,\infty)$, for a suitable measure space $(\Omega,\mathcal{F},\mu)$, and an isometric lattice isomorphism $\Phi : W \to Y$, which satisfies:*

(i) $\Phi(\rho(g)) = (\Phi(g))^* \quad (g \in W)$;

(ii) $\Phi \circ D_\alpha = D_\alpha \circ \Phi \quad (\alpha > 0)$.

PROOF: We take $(\Omega,\mathcal{F},\mu)$ as in the introductory paragraph. We fix an arbitrary maximal disjoint family $(f^\alpha)_{\alpha\in\mathcal{A}}$ in $\tilde{\mathcal{D}}_I$ and let $(\Omega_\alpha)_{\alpha\in\mathcal{A}}$ be a decomposition of Ω into sets Ω_α corresponding to the band projections S_{f^α}. We apply Proposition 3.4 to each of the bands $S_{f^\alpha}[W]$ and obtain isomorphisms Φ_{f^α}. If we define $\Phi : W \to L_0(\Omega\times(0,\infty))$ by $\Phi(h) = \sum_\alpha \Phi_{f^\alpha}\circ S_{f^\alpha}(h)$ then Φ is an isomorphism of W onto some g.r.i.f.s. and satisfies conditions (i) and (ii). □

Having established that the sublattice W of $\tilde{X}$ can be regarded as a g.r.i.f.s. Y on

$\Omega\times(0,\infty)$, we are now ready to show that a much larger sublattice may be represented as $Y(\Sigma,\mathcal{G},\nu)$ for a suitable measure space $(\Sigma,\mathcal{G},\nu)$. Recall that $\rho:\tilde{X}\to\tilde{\mathcal{D}}$ was defined by $\rho((f_n)^\bullet)=(f_n^*)^\bullet$. We define Z to be $\rho^{-1}[W]$, that is to say $Z=\{h\in\tilde{X}:\rho(h)\in W\}$. Although W is not itself a band in $\tilde{X}$, we do have

$$\tilde{\mathcal{D}}\ni g\le f\in W\cap\tilde{\mathcal{D}}\implies g\in W$$

by the proof of Lemma 3.2. Thus Z is a band in $\tilde{X}$. The proof of the final representation theorem, a significant part of which involves finding some sort of "product measure" structure in Z, has much in common with methods developed in [4]. We hope that the reader will excuse the sketchy nature of parts of the proof given here.

THEOREM 3.6. *There exist a measure space $(\Sigma,\mathcal{G},\nu)$ and an isometric lattice-isomorphism $\Psi:Z\to Y(\Sigma,\mathcal{G},\nu)$ satisfying $(\Psi(h))^*=\Phi(\rho(h))$ $(h\in Z)$.*

PROOF: For each $f\in\tilde{\mathcal{D}}_I$ we note that $\{h\in Z:\rho(h)\in S_f[W]\}$ is a band in Z; let R_f be the corresponding band projection. Then these projections form a ring isomorphic to $\mathfrak{S}_I$ and the closed subalgebra of $\mathfrak{B}(Z)$ that they generate is isomorphic to $\mathfrak{A}_I$. We recall that $(\Omega,\mathcal{F},\mu)$ was chosen so that its measure algebra was isomorphic to $\mathfrak{A}_I$ and see therefore that we can endow Z with the structure of a module over $L_\infty(\mu)$, extending what we have already defined on W, and satisfying $\rho(\phi.h)=\phi.\rho(h)\quad(h\in Z,\ \phi\in L_\infty(\mu))$.

We now simplify notation by writing $\sigma=\Phi\circ\rho$ so that $\sigma:Z\to Y$ satisfies

(i) $|g|\le|h|\implies\sigma(g)\le\sigma(h)$;

(ii) $\sigma(\phi g)=|\phi|\sigma(g)\quad(\phi\in L_\infty(\mu))$;

(iii) $\sigma(g+h)=\sigma(g)\oplus\sigma(h)$ whenever $|g|\wedge|h|=0$;

(iv) $\sigma(h)\ge g\in\mathcal{D}_Y\implies$ there is k such that $0\le k\le|h|$ and $\sigma(k)=g$.

We shall say that an element h of Z_+ is an *indicator* and write $h\in\mathfrak{J}$ if $\sigma(h)$ is the indicator of some subset of $\Omega\times(0,\infty)$. Note that this is not the same notion as "function of indicator *type*" as used earlier on, though for $g\in\mathfrak{J}$ the element $\rho(g)$ of $\tilde{\mathcal{D}}$ *is* necessarily of indicator type. In fact, $\rho(g)$ must be of a rather special form since $\Phi(\rho(g))$ must be of the type $(\omega,t) \mapsto 1_{(0,\psi(\omega)]}$ rather than the more general $(\omega,t) \mapsto \phi(\omega).1_{(0,\psi(\omega)]}$.

We are going to show that $\mathfrak{J}$ is a Boolean ring. By Lemma 3.3, if $0 \leq h \in \tilde{\mathcal{D}}_I$ and $0 \leq g \in \operatorname{band} h$ then $\rho(h-g) = \rho(\rho(h)-\rho(g))$. In particular, in the case where $\rho(g) = \rho(h)$, we must have $g = h$. It follows from this that if $g, h \in \mathfrak{J}$ and $\operatorname{band} g = \operatorname{band} h$ then $g = h$. (For these assumptions imply that $\sigma(g)$ and $\sigma(h)$ are indicator functions generating the same band in Y; thus $\sigma(g) = \sigma(h)$, which as we have just seen gives us $g = h$.) We note further that if $g, h \in Z_+$, $g \wedge h = 0$ and $g + h \in \mathfrak{J}$ then both g and h are in $\mathfrak{J}$. (This follows from (iii) and an obvious property of $\oplus$.) We are now in a position to prove that if $g, h \in \mathfrak{J}$ then $g \wedge h \in \mathfrak{J}$: indeed, let $g' = P_{g\wedge h}(g)$ be the projection of g onto the band generated by h and let $h' = P_{g\wedge h}(h)$ be defined similarly. Then g is the sum of the disjoint elements g', $g - g'$ and so $g' \in \mathfrak{J}$; similarly $h' \in \mathfrak{J}$. Thus g', $h' \in \mathfrak{J}$ generate the same band and are thus equal. Since it is clear that $g \wedge h \leq g' \leq g$ and $h' \leq h$, we have $g' = h' = g \wedge h$.

Recall that in the proof of 3.5 we fixed a maximal disjoint family (f^α) in $\tilde{\mathcal{D}}_I$ whose images under Φ were taken to be indicators $1_{\Omega_\alpha \times (0,1]}$. Since the sets Ω_α are necessarily σ-finite, we lose no generality by assuming that the measure μ was so chosen that all the sets Ω_α have measure 1. Now if $g \in \mathfrak{J}$ is such that $\rho(g) = f^\alpha$, or equivalently $\sigma(g) = 1_{\Omega_\alpha \times (0,1]}$, we may equip the ideal $\mathfrak{J}_g = \{h \in \mathfrak{J} : h \leq g\}$ with a measure θ defined by $\theta(h) = \int_{\Omega_\alpha \times (0,1]} \sigma(h) \mathrm{d}(\mu \otimes \lambda)$. The measure algebra $\mathfrak{J}_g$ has a closed subalgebra $\mathfrak{A}_g = \{1_A.g : \mathcal{F} \ni A \subset \Omega_\alpha\}$ isomorphic to the measure algebra of $\mu|\Omega_\alpha$. This is a situation in which one can apply a theorem of Maharam [8], a version of which may be found as Proposition 2.6 of [4]. We deduce that the measure algebra $(\mathfrak{J}_g, \theta)$ may be identified with a the measure algebra of a product $(\Omega_\alpha \times \Sigma_g\, , \mu|\Omega_\alpha \otimes \nu_g)$ for a suitable probability space $(\Sigma_g\, , \nu_g)$, which may be taken to be $\{0,1\}^{\mathfrak{m}}$ equipped with its usual product measure, $\mathfrak{m}$ being an appropriate cardinal. The theorem requires a hypothesis of "homogeneity": one needs to prove that, for every nonzero $h \in \mathfrak{J}_g$,

$$\min\{\#H : H \subset \mathfrak{J}_h \text{ and } H \cup \{h \cap k : k \in \mathfrak{A}_g\} \text{ generates } \mathfrak{J}_h\} = \mathfrak{m}.$$

In fact, in our case, this is satisfied with $\mathfrak{m}$ equal to the continuum $\mathfrak{c}$. This follows from a construction like the one presented in 7.22 and 7.23 of [4]. The same construction enables one to show that, for each α the band $R_{f^\alpha}[Z]$ may be expressed as the direct sum of $\mathfrak{c}$ disjoint bands $P_{g^\alpha_\beta}[Z]$ generated by elements g^α_β such that $\rho(g^\alpha_\beta) = f^\alpha$ for all $\beta \in \mathfrak{c}$. For each α and β there is a measure space $(\Sigma^\alpha_\beta, \nu^\alpha_\beta)$ (identifiable with $\{0,1\}^{\mathfrak{c}}$) such that $\mathfrak{J}_{g^\alpha_\beta}$ may be regarded as the measure algebra of $\Omega_\alpha \times \Sigma^\alpha_\beta$. We obtain therefore an embedding of $\mathfrak{J}$ into the measure algebra of the disjoint union $\bigsqcup \Omega_\alpha \times \Sigma^\alpha_\beta$, a measure space identifiable with the product $\Omega \times \Sigma$ where Σ is the disjoint union of $\mathfrak{c}$ copies of $\{0,1\}^{\mathfrak{c}}$. With a bit more work one may verify that this embedding extends to give the desired isomorphism of Z onto the g.r.i.f.s. $Y(\Sigma)$. □

The reader may be wondering where the canonically embedded copy of the r.i.f.s. X fits in with the g.r.i.f.s. structure of $Z \subseteq \tilde{X}$. Of course, X is the closed sublattice generated by the dilations $D_\alpha e$ of the element $e = (1_{(0,1]})^\bullet$. That is to say, in the notation we have been using, $X = S_e[W]$. The element e is an atom in the sense of Section 2 and may be taken to correspond to a single point ω_e of Ω, having non-zero measure. The band $R_e[Z]$, which is mapped by Ψ to the set of all elements of $Y(\Sigma, \mathcal{G}, \nu)$ that are carried by $\{\omega_e\} \times \Sigma$, may be identified with $X(\Sigma, \mathcal{G}, \nu)$.

4. The degenerate band. The representation theorem for the band Z generated by indicators naturally leaves one asking the question: is Z the whole of $\tilde{X}$? We do not know the answer to this in general, but note that in some cases at least it is not. It follows from 2.8 that for an element $g \in \tilde{X}$, $\rho(g)$ is disjoint from all indicators if and only if $P_k = S_k$ for all $k \in \tilde{\mathcal{D}}$ with $k \leq \rho(g)$. In some sense, it was the distinction between these two sorts of band projections that enabled us to introduce product structure in Section 3. Consequently, the collapsing of the two classes of band projections leads us to refer to the elements of the band $Z^\perp$ complementary to Z as *degenerate*. The structure of the degenerate band is still fairly unclear to us, though there seems to be some connection with eigenvectors of the dilation operators D_α and with the Boyd indices p_X and q_X.

Recall, from [6] for instance, that the Boyd indices p_X and q_X are defined by

$$p_X = \lim_{s\uparrow\infty} \frac{\log s}{\log \|D_s\|} = \sup_{s>1} \frac{\log s}{\log \|D_s\|}$$
$$q_X = \lim_{s\downarrow 0} \frac{\log s}{\log \|D_s\|} = \inf_{0<s<1} \frac{\log s}{\log \|D_s\|}$$

and that our assumption of finite cotype for X implies that $p_X < \infty$. The following result sums up our current knowledge about $Z^\perp$.

THEOREM 4.1. *There exist $f, g \in \tilde{\mathcal{D}}$ such that $D_s f = s^{1/p_X} f$ and $D_s g = s^{1/q_X} g$ for all $s > 0$. In the special case where $X = L_p$ (and so $p_X = q_X$ of course), such an f is disjoint from all elements of indicator type.*

PROOF: To simplify notation, we set $p = p_X$. We note that $2^{1/p} = \lim_{n\to\infty} \|D_2^n\|^{1/n}$, so that $2^{1/p}$ is the spectral radius of D_2. Since D_2 is a positive operator its spectral radius is in the spectrum, and is necessarily an approximate eigenvalue. Hence, there exist g_n's with

$$\|g_n\| = 1 \quad \text{and} \quad \lim_{n\to\infty} \|D_2\, g_n - 2^{1/p}\, g_n\| = 0.$$

If $g = (g_n)^\bullet$ then we certainly have $D_2(g) = 2^{1/p} g$ and so $D_2(\rho(g)) = \rho(D_2(g)) = 2^{1/p}\rho(g)$. We now modify $\rho(g) = (g_n^*)^\bullet$ so as to obtain f with the desired property.

Define

$$f_n = \int_1^2 t^{-1-\frac{1}{p}}\, D_t\, g_n^* dt,$$

and $f = (f_n)^\bullet$. We claim that $D_s f = s^{1/p} f$ for all $s > 0$. If $s = 2^n$ for some $n \in \mathbb{Z}$, then

$$\begin{aligned}
&\lim_{n\to\infty} \|D_s\, f_n - s^{1/p}\, f_n\| \\
&= \lim_{n\to\infty} \|D_s \int_1^2 t^{-1-\frac{1}{p}}\, D_t\, g_n^* dt - \int_1^2 t^{-1-\frac{1}{p}}\, D_t\, s^{1/p}\, g_n^* dt\| \\
&\le \lim_{n\to\infty} \int_1^2 t^{-1-\frac{1}{p}}\, \|D_t\, (D_s\, g_n^* - s^{1/p}\, g_n^*)\| dt = 0.
\end{aligned}$$

So we may assume that $1 < s < 2$.

$$\begin{aligned}
&\lim_{n\to\infty} \|D_s f_n - s^{1/p} f_n\| \\
&= \lim_{n\to\infty} \left\| \int_1^2 t^{-1-\frac{1}{p}} D_{st}\, g_n^* dt - \int_1^2 t^{-1-\frac{1}{p}} s^{1/p} D_t\, g_n^* dt \right\| \\
&\leq \lim_{n\to\infty} \left\| \int_s^{2s} t^{-1-\frac{1}{p}} s^{1/p} D_t\, g_n^* dt - \int_1^2 t^{-1-\frac{1}{p}} s^{1/p} D_t\, g_n^* dt \right\| \\
&= \lim_{n\to\infty} \left\| \int_2^{2s} t^{-1-\frac{1}{p}} s^{1/p} D_t\, g_n^* dt - \int_1^s t^{-1-\frac{1}{p}} s^{1/p} D_t\, g_n^* dt \right\| \\
&\leq \lim_{n\to\infty} \int_1^s t^{-1-\frac{1}{p}} s^{1/p} \| 2^{-1/p} D_{2t}\, g_n^* - D_t\, g_n^* \| dt = 0
\end{aligned}$$

The construction for q_X is similar.

Suppose that $X = L_p$. Then $p_X = p$. Let f be any nonzero function in $\tilde{\mathcal{D}}$ such that $D_2(f) = 2^{1/p} f$. We claim that if $h \in \tilde{\mathcal{D}}$ and $h \leq f$, then h cannot be a function of indicator type. Note that $S_h f$, $f - S_h f$ are disjoint, and that

$$2^{1/p} S_h f + (2^{1/p} f - 2^{1/p} S_h f) = 2^{1/p} f = D_2\, f = S_h D_2(f) + (D_2\, f - S_h(D_2\, f)).$$

So $2^{1/p}\, S_h f = D_2\,(S_h f)$, and we may assume $S_h\, f = f$. Hence,

$$\begin{aligned}
f = S_h f &= \lim_{n\to\infty} (2^{n/p} D_{2^n}(h) \wedge f) \\
&= \lim_{n\to\infty} D_{2^n}(2^{n/p} h \wedge \frac{f}{2^{n/p}}) \\
&= \lim_{n\to\infty} 2^{-n/p} D_{2^n}(2^{2n/p} h \wedge f).
\end{aligned}$$

Since $\|D_s\, f\| = s^{1/p} \|f\|$ for any $f \in L^p$, we have

$$\begin{aligned}
0 \leq \|f - P_h f\| &= \lim_{n\to\infty} \|2^{-n/p} D_{2^n}(f - P_h f)\| \\
&\leq \lim_{n\to\infty} \|2^{-n/p} D_{2^n}(f - (2^{2n/p} h \wedge f))\| = 0.
\end{aligned}$$

This implies $f = P_h f$. So

$$\begin{aligned}
D_2 h &\geq P_h\, D_2 h \\
&= P_f\, D_2 h \qquad\qquad P_f = P_h \text{ since } P_h f = f \text{ and } h \leq f \\
&= S_f D_2 h = D_2 h.
\end{aligned}$$

h is not of indicator type (otherwise, $D_2 h = h$). We have proved our claim. □

Remarks 4.2. (i) If $f \in \tilde{X}$ is such that $D_s f = s^{1/p} f$ for all $s > 0$ then we may construct a sequence (f_m) of disjoint elements of $\tilde{X}$ such that $\rho(f_m) = \rho(f)$ for all m. We note that

$$\rho(\sum_{i=1}^{m} a_i f_i) = \bigoplus_{i=1}^{m} a_i f = \bigoplus_{i=1}^{m} D_{|a_i|^p} f = D_{\sum_i |a_i|^p} f = (\sum_i |a_i|^p)^{1/p} f.$$

Thus the f_i form an ℓ_p-basis. It follows from this that, for any m and any $\epsilon > 0$ we may find in X disjointly supported elements $g_1 \dots, g_m$, with $g_i^* = g_1^*$ for all i, which are $(1+\epsilon)$-equivalent to the ℓ_p^m-basis. This result, in the case of p_X and q_X can be found in [6]. The authors of that work hinted at a proof using Krivine's theorem. The reader may have noticed that our proof of 4.1 used some ideas that occur in recent proofs of the same theorem.

(ii) It may be of interest to record an explicit formula for a degenerate element of $\tilde{X}$ when $X = L_p$. Let

$$f_n(t) = \begin{cases} \frac{1}{\log n^{1/p}}, & \text{if } t \le 1, \\ \frac{1}{(t \log n)^{1/p}}, & \text{if } n \ge t \ge 1, \\ 0, & \text{if } t > n. \end{cases}$$

Then $f_n \in L^p (= X)$ and $\frac{D_2\big((f_n)^\bullet\big)}{2^{1/p}} = (f_n)^\bullet$.

(iii) One should not, however, suppose that f must be degenerate whenever f is an eigenvector of D_α. Indeed, if $q < p < r$ and $X = L_q + L_r$ then the function $f(t) = t^{1/p}$ is in X.

REFERENCES

1. J. Bretagnolle, D. Dacunha-Castelle and J. L. Krivine, *Lois stables et espaces L^p*, Ann. Scient. Ec. Norm. Sup. (1969), 437–480.

2. D. Dacunha-Castelle and J. L. Krivine, *Applications des ultraproduits à l'étude des espaces et des algèbres de Banach*, Studia Math. **41** (1972), 315–334.

3. D. H. Fremlin, *Decomposable measure spaces*, Z. Wahrscheinlichkeitstheorie verw. Gebiete **45** (1978), 159–167.

4. R. G. Haydon, M. F. Levy and Y. Raynaud, "Randomly Normed Spaces," (to appear).

5. S. Heinrich, *Ultraproducts in Banach space theory*, J. Reine Angew. Math. **313** (1980), 72–104.

6. J. Lindenstrauss and L. Tzafriri, "Classical Banach Spaces II: Function Spaces," Vol. 97, Springer-Verlag Berlin Heidelberg New York, 1979.

7. G. G. Lorentz and T. Shimogaki, *Interpolation theorems for operators in function spaces*, J. Funct. Anal. **2** (1968), 31–51.

8. D. Maharam, *Decompositions of measure algebras and spaces*, Trans. Amer. Math. Soc. **69** (1950), 142–160.

Brasenose College, Oxford, England OX1 4AJ

Department of Mathematics, Memphis State University, Memphis, TN 38152

Contemporary Mathematics
Volume **85**, 1989

KMP, RNP, and PCP FOR BANACH SPACES

R.C. James

Abstract. It is shown that if X is isomorphic with $Y \oplus X$ and Y fails RNP, then X fails KMP. If F is a countable norming set of continuous linear functionals on Z and τ is the weakest topology for which each f in F is continuous, then RNP and KMP are equivalent for C if C is a bounded convex τ–compact subset of Z; if C fails RNP, then C contains a basic 2–tree. For a space X contained in a Banach space Y with an unconditional basis of finite–dimensional subspaces, RNP $\Longleftrightarrow$ KMP $\Longrightarrow$ PCP; if no subspace of Y is isomorphic with c_0, then RNP, KMP, PCP, and CPCP are equivalent; if $X = Y$, then each of these properties is equivalent to X not having a subspace isomorphic with c_0.

1. INTRODUCTION

A bounded closed convex subset K of a Banach space X has the *Radon–Nikodým property* (RNP) if, for any finite–measure space (S,Σ,μ) and any μ–continuous measure

---

1980 Mathematics Classification (1985 Revision): Primary 46B20
Key words and phrases: Krein–Milman and Radon–Nikodým properties, point–of–continuity property, bushes, topology determined by norming functionals, unconditional basis.

$\lambda: \Sigma \longrightarrow X$ with $\lambda(E)/\mu(E) \in K$ for each $E \in \Sigma$, there is a Bochner–integrable function $f: S \longrightarrow X$ such that $\lambda(E) = \int_E f\, d\mu$ for each $E \in \Sigma$. For X to have RNP means that the unit ball has RNP; X has RNP if X is a separable dual (e.g., if X is ℓ_1 or if X is reflexive).

A bounded closed convex subset K of a Banach space X has the *Krein–Milman property* (KMP) if each closed convex subset of K is the closure of the convex span of its extreme points. For X to have KMP means that the unit ball has KMP; X has KMP if X has RNP [13, Theorem 2], but it is not known if the converse is true.

A bounded closed convex subset K of a Banach space X has the *point–of–continuity property* (PCP) if, for each nonempty closed subset C of K, there is a point x of C such that the weak and norm topologies (restricted to C) coincide at x; K has the *convex–point–of–continuity property* (CPCP) if this condition is satisfied for all nonempty closed convex subsets of K. For X to have PCP (CPCP) means that the unit ball has PCP (CPCP).

A *bush* (*2–tree*) in a Banach space is a bounded partially ordered subset B for which each member has finitely many successors, B has a first member b_{11}, each member of B can be joined to b_{11} by a linearly ordered chain of successive members of B, and:

(i) Each member of B has at least two (exactly two) successors and is a (weighted) average of its successors.

(ii) There is a positive *separation constant* δ such that $\|y-x\| > \delta$ if y is a successor of x.

The successors of b_{11} will be denoted by $\{b_{2i}: 1 \le i \le d(2)\}$ and said to be of *order 2*. Inductively, the successors $\{b_{ni}: 1 \le i \le d(n)\}$ of members of B of order $n-1$ will be said to be of *order n*. A difference $\Delta_{nj} = b_{n+1,j} - b_{ni}$ between b_{ni} and a successor $b_{n+1,j}$ is a *difference of order n*.

A *branch* of a bush is an infinite linearly ordered subset whose first member is b_{11}; a *segment* is a linearly ordered subset. The *wedge* W_{ni} is the closure of the convex span of all followers of b_{ni}. The *wedge–intersection* corresponding to a branch β is the intersection of all wedges along β.

An *asymptotic subbush* of a bush B is a subset B_0 of $\overline{\text{con}}(B)$ which is a bush and for which there is a one–to–one correspondence with a subset of B which preserves partial ordering and for which $\lim_{n\to\infty}\|x_n - y_n\| = 0$ if $\{x_n\}$ is the set of elements along a branch in B_0 and $\{y_n\}$ is the sequence of corresponding members of B. An *approximate bush* is a set B^a that satisfies all the hypotheses for a bush except that (i) is replaced by "Each member of B^a has at least two successors and there is a sequence of positive numbers $\{\delta_n\}$ for which $\Sigma_1^\infty \delta_n < \infty$ and each member of B^a of order n differs from a convex combination of its successors by less that δ_n." The closure of the convex span of an approximate bush contains a bush [8].

There is a natural one–to–one correspondence between the branches of a bush and the points of a Cantor set on the real line with measure 1 [11, pp. 60–61]. The *branch point* corresponding to a branch β is the point in the Cantor set that corresponds to β.

A Banach space X (or a bounded closed convex subset of X) has RNP if and only if it does not contain a bush (for an easy proof that is easily adapted to the case of bounded closed convex subsets, see [10, Theorem 7, pg. 354]). It follows easily from this that X has RNP if and only if each bounded closed nonempty subset K of X is *dentable* [10, Theorem 11, pg. 359]; i.e., for each $\epsilon > 0$ there is an x in K for which x does not belong to

$$\overline{\text{con}}\{x+\Delta\colon\ x+\Delta \in K \text{ and } \|\Delta\| > \epsilon\}.$$

It is an easy exercise to use dentability to show that RNP implies PCP.

A *finite–dimensional decomposition* (FDD) of X is a sequence $\{V_n\}$ of finite–dimensional subspaces for which $X = \overline{\text{lin}}\{V_i\colon i \geq 1\}$ and each V_n is complementary to $\overline{\text{lin}}\{V_i\colon i \neq n\}$; $\{V_n\}$ is *basic* if each $x \in X$ is uniquely representable as $\Sigma_1^\infty a_i v_i$, where $v_n \in V_n$ for each n; $\{V_n\}$ is an *unconditionally basic finite–dimensional decomposition* (UBFDD) if this representation of x converges unconditionally. A space with an UBFDD is contained in a space with an unconditional basis [12, pg. 51]. A *basic bush* is a bush $\{b_{ni}\}$ for which $\{S_{ni}\}$ is a basic FDD if $\{S_{ni}\}$ is ordered lexicographically and S_{ni} is $\text{lin}\{b_{n+1,j}-b_{ni}\colon (n+1,j) \succ (n,i)\}$ for each (n,i). It is trivially true that each wedge–intersection of a basic bush is empty.

The following lemmas will be useful later to replace bushes by bushes that are easier to work with. Here and later, we use the notation $N(X,\epsilon)$ for the ϵ–neighborhood of 0 in the space X.

LEMMA 1.1. Suppose B is a bush in a Banach space X, F is a countable subset of X^*, and δ is a separation constant of B. Then for any $\bar{\delta} < \delta$, any $\eta > 0$, and any sequence $\{\beta_n\}$ of positive numbers, there is an asymptotic subbush B' of B and a bush B'' in X such that B' and B'' are isomorphic as partially ordered sets, both B' and B'' have separation constants greater than $\frac{1}{2}\bar{\delta}$, and:

(a) For each $f \in F$ and $\epsilon > 0$, there is an m such that $|f(s)| < \epsilon$ if s is a segment in B' whose first member has order greater than m.

(b) $B'' \subset \overline{\text{con}}(B) + N(X,\eta)$ and $\|\Delta'-\Delta''\| < \beta_n$ of Δ' and Δ'' are corresponding differences of order n in B' and B'', respectively.

(c) For each $f \in F$, there is an m such that $f \equiv 0$ on the set of differences in B'' with orders greater than m.

Proof. Let δ' be any number for which $\bar{\delta} < \delta' < \delta$. Let $\{\mu_n\}$ and $\{\beta_n\}$ be sequences of positive numbers for which $\Sigma_1^\infty \mu_n < 1$, $\Sigma_1^\infty \beta_n < \eta$, and $\beta_n < \frac{1}{2}(\delta' - \bar{\delta})$ for each n. Without loss of generality, we assume that $\|f\| = 1$ if $f \in F$. Order the members of F, so that $F = \{f_i\colon i \geq 1\}$. For each positive integer r, let Y_r be the intersection of the null spaces of $\{f_i\colon 1 \leq i \leq r\}$ and let X_r be a complement of Y_r in X. Since X_r is finite–dimensional, X_r has RNP. For each (n,i), let $b_{ni} = x_{ni}^r + y_{ni}^r$, where $x_{ni}^r \in X_r$ and $y_{ni}^r \in Y_r$. There is a natural way to associate each b_{ni} with a subset C_{ni} of a Cantor set C_{11} and to define a measure μ so that $\mu(C_{11}) = 1$ and there is an X_r–valued measure λ such that

$$\lambda(C_{ni}) = x_{ni}^r \cdot \mu(C_{ni}) \text{ for each } (n,i).$$

By definition, a subset W of C_{11} with positive measure is *ϵ–pure with respect to λ and μ* if

$$\left\|\frac{\lambda(E)}{\mu(E)} - \frac{\lambda(F)}{\mu(F)}\right\| < \epsilon$$

for all subsets E and F with positive μ–measure [14, Definition 2]. Choose a positive number θ so that if

$$\left\|\frac{\lambda(E)}{\mu(E)} - \frac{\lambda(F)}{\mu(F)}\right\| < \frac{1}{4}\beta_r \tag{1}$$

and $A \subset E \subset F$ with $\mu(A) < \theta\mu(E)$, then (1) is satisfied if E is replaced by E–A, F is replaced by F–A, and $\frac{1}{4}$ is replaced by $\frac{1}{2}$. Since X_r has RNP, C_{11} is the union of a countable set of disjoint $\frac{1}{4}\beta_r$–pure subsets [10, pg. 355]. This implies that we can have

$$C_{11} = \cup_1^\infty C^k - A,$$

where each $C^k - A$ is $\frac{1}{4}\beta_r$–pure, $\{C^k\}$ is a set of disjoint members of $\{C_{ni}\}$, and $\mu(A) < \frac{1}{2}\theta\mu_r$. Use induction on n to define V as the union of all C_{ni} for which C_{ni} is contained in some C^k and $\mu(C_{ni} \cap A) \geq \theta\mu(C_{ni})$, but C_{ni} is not contained in a previously chosen C_{mj} for $m < n$. Since $\mu(A) < \frac{1}{2}\theta\mu_r$, we have

$$\theta\mu(V) < \frac{1}{2}\theta\mu_r \text{ and } \mu(V) < \frac{1}{2}\mu_r.$$

If a branch β contains a segment $s = x_{mj}^r - x_{ni}^r$ with $\|s\| > \frac{1}{2}\beta_r$, then

$$\left\|\frac{\lambda(C_{mj})}{\mu(C_{mj})} - \frac{\lambda(C_{ni})}{\mu(C_{ni})}\right\| = \|x_{mj}^r - x_{ni}^r\| > \frac{1}{2}\beta_r. \tag{2}$$

Thus if the branch point $p(\beta)$ of β is contained in $C^k - A$ and if β contains infinitely many such disjoint segments, then there is an n for which the C_{ni} that contains $p(\beta)$ contains some C_{mj} for which (2) is satisfied, $p(\beta) \in C_{mj}$, and $C_{ni} \subset C^k$ for some k. If $\mu(C_{mj} \cap A) < \theta\mu(C_{mj})$, then it follows from the definition of θ and (2) that $C_{ni} - A$ is not $\frac{1}{4}\beta_r$–pure. Thus $C \subset V$ and therefore $p(\beta) \in V$. The branches with branch points not in V have only finitely many disjoint segments s with $\|s\| > \frac{1}{2}\beta_r$, so there is a number $c(r)$ such that the measure of the set of such branch points whose branch has a segment beginning after $c(r)$ with $\|s\| > \frac{1}{2}\beta_r$ is less than $\frac{1}{2}\mu_r$. This and $\mu(V) < \frac{1}{2}\mu_r$ imply there is a set $P_r \subset C_{11}$ such that $\mu(P_r) < \mu_r$ and, if $s = s_r + \sigma_r$ is a segment on any branch whose branch point is not in P_r, if $s_r \in X_r$ and $\sigma_r \in Y_r$, and if the first difference in s has order greater than $c(r)$, then

$$||s_r|| \le \frac{1}{2}\beta_r. \tag{3}$$

We can use the method used to prove Theorem 1 of [11] to obtain an approximate bush B_0 that avoids all branches with branch points in $\cup_1^\infty P_r$; whose members are in B; and, if a member of B_0 has order n, then the order of the corresponding member of B is at least $c(n)$. Moreover, this can be done so that the "errors in averaging" are arbitrarily small, which we take to be small enough that the asymptotic subbush B' obtained from B_0 by "averaging back" has the properties that B' has separation constant greater than $\frac{1}{2}\delta'$ and, if b' is any member of B' with order n and b is the corresponding member of B, then $||b'-b|| < \frac{1}{4}\beta_n$. This implies that, if s' is a segment in B' whose first member has order greater than n in B', then the norm of the difference between s' and the corresponding segment in B is less that $\frac{1}{2}\beta_n$. This and (3) imply that if the first member of a segment s' in B' has order greater than n in B' and if $s' = s_n'+\sigma_n'$ with $s_n' \in M_n$ and $\sigma_n' \in S_n$, then

$$||s_n'|| < \beta_n. \tag{4}$$

To see that (a) is satisfied, let $f \in F$ and let ϵ be any positive number. Choose n so that $\beta_n < \epsilon$ and $f = f_k$ with $k \le n$. Then if s' is a segment in B' whose first member has order greater than n, and if $s' = s_n'+\sigma_n'$ with $s_n' \in X_n$ and $\sigma_n' \in Y_n$, then $|f(s')| < \epsilon$ follows from (4).

We now obtain B'' from B'. Let Δ' be a difference of order n in B'. Then the first member of the segment in B that corresponds to Δ' has order greater than $c(n)$ and it follows from (4) that $||\delta_n|| < \beta_n$ if $\Delta' = \delta_n+\Delta_n$, where $\delta_n \in X_n$ and $\Delta_n \in Y_n$. Let

$\Delta'' = \Delta_n$. Then $||\Delta' - \Delta''|| < \beta_n$. When this has been done for all differences in B', no difference has been changed by as much as $\beta_n < \frac{1}{2}(\delta' - \bar{\delta})$, so the resulting bush B'' has separation constant greater than $\frac{1}{2}\delta' - \beta_n > \frac{1}{2}\bar{\delta}$. Also, no segment in B' is changed by more than $\Sigma_1^\infty \beta_n < \eta$, so $B'' \subset \overline{\mathrm{con}}(B) + N(X,\eta)$. Finally, if $k \leq n$, then $f_k \equiv 0$ on differences with orders greater than n.

<u>LEMMA 1.2</u>. Suppose B is a bush in a Banach space X, $\{\Omega_n\}$ is an FDD of a Banach space Y that contains X, and δ is a separation constant of B. Then for any $\bar{\delta} < \delta$, any $\eta > 0$, and any sequence of positive numbers $\{\beta_n\}$, there is a bush B''' in X with separation constant greater than $\frac{1}{4}\bar{\delta}$ for which $B''' \subset \overline{\mathrm{con}}(B) + N(X,\eta)$ and there is an increasing sequence of positive integers $\{r(n)\}$ such that, if

$$\Phi_0 = \{0\} \text{ and } \Phi_n = \mathrm{lin}(\cup_{r(n)}^{r(n+1)-1}\Omega_i) \text{ if } n \geq 1, \tag{5}$$

and if Δ''' is a difference of order n for B''', then

$$\Delta''' \in \mathrm{lin}(\Phi_{n-1}, \Phi_n) + N(Y, \beta_n). \tag{6}$$

<u>Proof</u>. For each positive integer r, let $T(r) = X \cap \overline{\mathrm{lin}}(\cup_{i=r}^\infty \Omega_i)$. Then $T(r)$ has a finite-dimensional complement $S(r)$ in X and there is a finite subset $F(r)$ of X^* that is total on $S(r)$ and identically zero on $T(r)$. Let

$$F = \cup_{r=1}^\infty F(r).$$

For this F, any $\bar{\delta} < \delta$, any $\eta > 0$, and any sequence of positive numbers $\{\beta_n\}$, let B'' be a bush in X as described in Lemma 1.1, with separation constant $\sigma > \frac{1}{2}\bar{\delta}$. The members of

B''' will be members of B'', so we will have $B''' \subset \overline{\text{con}}(B) + N(X,\eta)$. Let $r(1)$ be 1 and choose $r(2)$ so that, if Φ_1 is as in (5) for $n = 1$, and if Δ'' is a difference of order 1 in B'', then

$$\Delta'' \in \Phi_1 + N(Y,\beta_1). \tag{7}$$

Let the first members of B'' and B''' and their successors be identical, which implies that differences of order 1 of B''' are the same as differences of order 1 of B''. Then it follows from (7) that (6) is satisfied if $n = 1$.

The proof will be completed inductively. Let $m > 1$ be a positive integer. Assume that $\{\Phi_i: 1 \le i \le m\}$, and positive integers $\{r(i): 1 \le i \le m+1\}$ and $\{d(i): 1 \le i \le m\}$, have been chosen, and that (5) is satisfied for $1 \le n \le m$. Also assume that $d(1) = 1$, and if Δ'' is a difference of order $d(m)$ or greater in B'', then

$$\Delta'' \in \overline{\text{lin}}(\cup_{i=r(m)}^{\infty} \Omega_i), \tag{8}$$

which is satisfied trivially if $m = 1$. Finally, assume that all elements of order less than or equal to $m+1$ have been defined for B''' and that all differences Δ''' of order m satisfy (6) with $n{=}m$, and differences of order m in B''' are segments in B'' whose first members are of order $d(m-1)$ or $d(m-1) + 1$ and whose last members are of order $d(m)$ or $d(m) + 1$.

Choose $d(m+1) > d(m)+1$ so that each member of $F[r(m+1)]$ is zero on all differences in B'' that are of order $d(m+1)$ or greater. This implies that

$$\Delta'' \in T[r(m+1)] = X \cap \overline{\text{lin}}(\cup_{r(m+1)}^{\infty} \Omega_i), \tag{9}$$

if Δ'' is a difference of order $d(m+1)$ or greater in B''. Then (8) is satisfied for $m+1$.

Now choose $r(m+2)$ so that, if Φ_{m+1} is defined by (5) and if s'' is a sum of successive differences of order not greater than $d(m+1)+1$ in B'', then

$$s'' \in \mathrm{lin}(\cup_1^{r(m+2)-1}\Omega_i) + N(Y,\beta_{m+1}) = \mathrm{lin}(\cup_1^{m+1}\Phi_i) + N(Y,\beta_{m+1}). \tag{10}$$

Define members of order $m+2$ in B''' by adding to a member $b'''_{m+1,i}$ of order $m+1$ in B''' a segment that begins at the corresponding member of B'' and is a sum of successive differences that ends with a difference of order $d(m+1)$, if the norm of such a segment is greater than $\frac{1}{2}\sigma$, but otherwise continues to include one of the next differences so that the norm of the segment is greater than $\frac{1}{2}\sigma > \frac{1}{4}\bar{\delta}$. Repetition of this step will insure that the separation constant of B''' will be greater than $\frac{1}{4}\bar{\delta}$. Do this in all possible ways to obtain column $m+2$ of B'''. Let s'' be such a segment; i.e., a difference Δ''' in B''' of order $m+1$. Then it follows from (8) that

$$\Delta''' \in \overline{\mathrm{lin}}(\cup_m^\infty \Phi_i).$$

This and (10) imply that $\Delta''' \in \mathrm{lin}(\Phi_m,\Phi_{m+1}) + N(Y,\beta_{m+1})$.

2. EMPTY WEDGE–INTERSECTIONS

It is known that any bush has an asymptotic subbush all of whose wedge–intersections are empty [11, Theorem 4, pg. 77]. This will be used to obtain some simple corollaries of the next theorem.

THEOREM 2.1. A Banach space X fails KMP if it has subspaces $\{X^1, X^2, \cdots\}$ such that X^m and $\overline{\text{lin}}\{X^n: n \neq m\}$ are complementary in $\overline{\text{lin}}\{X^n: n \geq 1\}$ for each m, there are bushes $\{B^m\}$ for which $B^m \subset X^m$ for each m, and:

(a) All wedge–intersections of B^1 are empty.

(b) For each $m \geq 1$, there is a continuous linear map of $\text{lin}(B^{m+1})$ onto $\text{lin}(B^m)$, which is also an order–preserving isomorphism between B^{m+1} and B^m as partially ordered sets.

Proof. Construct a bush B in X as follows. Let $\{\epsilon_i\}$ be positive numbers with $\Sigma\epsilon_i < \frac{1}{2}\delta$, where δ is a separation constant of the bush B^1 in X^1. Order the members $\{b^1_{ni}\}$ of B^1, other than the first member, lexicographically. Suppose b^1_{ni} is the pth member of B^1. If b^1_{mj} is either b^1_{ni} or a follower of b^1_{ni}, add to each difference $\Delta^1_{mh} = b^1_{m+1,h} - b^1_{mj}$, where $b^1_{m+1,h}$ is a successor of b^1_{mj}, the product of ϵ_p and the corresponding difference in B^p; that is, replace Δ^1_{mh} by

$$\Delta^1_{mh} + \epsilon_p\Delta^p_{mh} = (b^1_{m+1,h} - b^1_{mj}) + \epsilon_p(b^p_{m+1,h} - b^p_{mj}).$$

This defines a bush B for which there is a separation constant at least as great as $\delta - \Sigma^\infty_1\epsilon_i > \frac{1}{2}\delta$. Also, all wedge–intersections of B are empty. For if w is in a wedge–intersection, then $w = x+y$ with $x \in X^1$ and $y \in \overline{\text{lin}}\{X^n: n > 1\}$. But then x belongs to a wedge–intersection of B^1.

We will show that $\overline{\text{con}}(B)$ has no extreme points. Suppose w is an extreme point. Since w is not in any wedge–intersection, there is an n for which no wedge W_{ni} contains w. Then w is the limit as $r \to \infty$ of convex combinations of $\{x_{ni}(r): 1 \leq i \leq d(n)\}$, where $d(n)$ is the number of wedges that start at column n and $x_{ni}(r)$ belongs to the wedge W_{ni}

of B. This implies that

$$w = \lim_{r\to\infty} \sum_{i=1}^{d(n)} \lambda_i^n x_{ni}(r), \quad 0 \le \lambda_i^n < 1, \quad \Sigma\lambda_i^n = 1. \tag{11}$$

Choose β for which $\lambda_\beta^n \neq 0$. We will obtain a contradiction by showing that $w \in W_{n\beta}$. Choose p for which $b_{n\beta}$ is the pth member of B^1. Then

$$x_{n\beta}(r) = \xi_{n\beta}(r) + S_{n\beta}(r) + T_{n\beta}(r),$$

where $\xi_{n\beta}(r) \in X^p$, $S_{n\beta}(r)$ is a sum of terms $\eta_{m\beta}(r) \in X^m$ with $1 \le m < p$, and $T_{n\beta}(r)$ is a finite sum of terms $\zeta_{m\beta}(r) \in X^m$ with $m > p$ and no two terms belonging to the same X^m. Note that the number of terms in $T_{n\beta}(r)$ may increase with r. Since the limit in (11) exists, it follows from complementation that $\lim_{r\to\infty}\xi_{n\beta}(r)$ exists and $\lim_{r\to\infty}\zeta_{m\beta}(r)$ exists for each $m > p$. It then follows from (b) that $\lim_{r\to\infty}\eta_{m\beta}(r)$ exists if $1 \le m < p$, so that $\lim_{r\to\infty}S_{n\beta}(r)$ exists. Since $\zeta_{m\beta}(r)$ is the product of ϵ_m and part of a convex combination of differences of members of B^m, we have

$$||\zeta_{m\beta}(r)|| \le 2\epsilon_m.$$

Since $\Sigma_1^\infty \epsilon_i < \infty$, it follows that $\lim_{r\to\infty}T_{n\beta}(r)$ exists. Now we know that $\lim_{r\to\infty}x_{n\beta}(r)$ exists. Since w is an extreme point, this limit is w. Thus $w \in W_{n\beta}$.

COROLLARY 2.2. Suppose X has mutually isomorphic subspaces $\{X^1, X^2, \cdots\}$ such that X^m and $\overline{\text{lin}}\{X_n\colon n \neq m\}$ are complementary in $\overline{\text{lin}}\{X^n\colon n \ge 1\}$ and X^1 fails

RNP. Then X fails KMP.

Proof. Since X^1 contains a bush, it contains a bush B^1 all of whose wedge–intersections are empty. For each m, we can use the isomorphism between X^1 and X^m to find a bush B^m in X^m so that all the hypotheses of the theorem are satisfied.

COROLLARY 2.3. If X is isomorphic with $Y \oplus X$ and Y fails RNP, then X fails KMP.

Proof. If Y fails RNP, then Y contains a bush and this bush has an asymptotic subbush with empty wedge–intersections. Repeated applications of $X \cong Y \oplus X$ shows that the hypotheses of Corollary 2.2 are satisfied, so X fails KMP.

The next corollary follows by letting $Y = X$ in Corollary 2.3 and using the known fact that RNP $\Rightarrow$ KMP. The result stated in this corollary was proved first by W. Schachermayer, using a much longer argument [17].

COROLLARY 2.4. If X is isomorphic with $X \oplus X$, then X has KMP if and only if X has RNP.

3. COMPACTNESS CRITERIA

It was shown by Huff and Morris [9] that RNP and KMP are equivalent for dual spaces. Stegall has shown that a dual space fails RNP if and only if it contains a 2–tree [19, pg. 222], although there are Banach spaces that fails RNP and therefore contain bushes, but do not contain trees [3]. Stegall's theorem is true for some subsets of a dual other than the unit ball: For a w^*–compact convex set C in a dual Banach space, RNP and KMP are equivalent and each is equivalent to C not containing a 2–tree [5, Theorem 4.2.13, pg. 91].

The next theorem shows that C being in a dual is not needed for proving that C contains a 2–tree. Instead, one can replace w^*–compactness of C by compactness defined by using a countable norming set of functionals and obtain a 2–tree whose differences in a certain order are arbitrarily close to being a monotone basic sequence. It is know that if X has CPCP and C is a bounded closed convex subset without RNP, then C "almost" contains a bush whose differences span a space with a basic FDD that is almost monotone [15, Theorem 3.9]. If X fails CPCP, then the nearly monotone bush can be in C but still need not be a tree (Corollary 4.4).

The 2–tree obtained in the next theorem will be used in the proof of Theorem 3.2 to obtain a closed convex subset of C without extreme points.

THEOREM 3.1. Let F be a countable norming set of continuous linear functionals on a Banach space X and let τ be the weakest topology for which each f in F is continuous. Let C be a subset of X that is bounded, convex, τ–compact, and fails RNP. Then for any positive number $\Theta < 1$, there is a positive number r such that, if $r_- < r < r_+$, then C contains a basic 2–tree $T = \{t_{mj}\}$ which has basis constant greater than Θ and satisfies:

(i) $\|t_{mj} - t_{11}\| < r_+$ for all (m,j);

(ii) $\|\Delta\| > \frac{1}{2} r_-$ if Δ is any difference between successive members of T;

(iii) If x belongs to the linear span of the first n basis elements and y belongs to the linear space of the remainder, then there is an f in F for which $|f(x+y)| \geq \Theta\|x\|$ unless $\|y\| > 3\|x\|$.

Proof. Symbols such as $s_{nj\to pk}$ will be used to denote the segment from b_{nj} to b_{pk} in a bush $B = \{b_{ni}\}$. For an arbitrary bush B, we let

$$R(B) = \inf_{(m,j)}\sup\{\|s_{mj\to pk} - s_{mj\to q\kappa}\|: \ (p,k) \text{ and } (q,\kappa) \text{ follow } (m,j)\}.$$

If B has separation constant σ, then $R(B) \geq \frac{1}{2}\sigma$, since if the difference of two segments has norm less than $\frac{1}{2}\sigma$, then the norm of the difference becomes greater than $\frac{1}{2}\sigma$ if one segment is extended by adding one more difference. Let B be a bush in C. Let δ be a separation constant of B and $\bar{\delta}$ be a positive number less than δ. For an arbitrary sequence $\{\beta_n\}$ of positive numbers, let B' and B'' be Banach spaces with separation constants greater than $\frac{1}{2}\bar{\delta}$, as described in Lemma 1.1. Then $B' \subset \overline{\text{con}}(B) \subset C$, but B'' need not be in C. Let r be an accumulation point of $R(B')$ as $\Sigma_1^\infty \beta_n \to 0$. Let r_-, ρ_-, ρ_+, and r_+ be any numbers for which $r_- < \rho_- < r < \rho_+ < r_+$. Choose $\{\beta_n\}$ so that, if B' and B'' are determined by $\{\beta_n\}$ and Lemma 1.1, and if $r' = R(B')$ and $r'' = R(B'')$, then

$$r_- < \rho_- < r' < \rho_+ < r_+, \quad r_- < \rho_- < r'' < \rho_+ < r_+,$$

$\Sigma_1^\infty \beta_n < \frac{1}{2}(r_+ - \rho_+)$, $\Sigma_1^\infty \beta_n < \rho_- - r_-$, and $2^n \Sigma_n^\infty \beta_i < \eta_n$ for each n, where $\Sigma_1^\infty \eta_n$ will be restricted later.

We will introduce a partially ordered set $Q \subset B''$. Then the members of Q will be replaced by members of a partially ordered subset P of B', and the members of P will be replaced by τ–limits of averages to obtain the desired tree T. Choose (n,i) in B'' for which

$$\sup\{\|s''_{ni\to pk}-s''_{ni\to q\kappa}\|: (p,k) \text{ and } (q,\kappa) \text{ follow } (n,i) < \rho_+. \tag{12}$$

Then every difference of segments in B'' beginning at b''_{ni} has norm less than ρ_+ and, for any (m,j), there exist (p,k) and (q,κ) following (m,j) such that $\|s''_{mj\to pk}-s''_{mj\to q\kappa}\| > \rho_-$. We start defining Q by letting $q_{11} = b''_{ni}$. Rather than complicating the argument with many more symbols, we shall use $\doteq$ and "nearly" to indicate approximations good enough for what comes later. Without loss of generality, assume that $-f \in F$ and $\|f\| = 1$ for each f in F.

Choose (p,k) and (q,κ) following (n,i) and $f_{11} \in F$ so that $\|s''_{ni\to pk}-s''_{ni\to q\kappa}\| > \rho_-$ and is "nearly" the sup in (12), and also

$$m_{11} = f_{11}(s''_{ni\to pk}-s''_{ni\to q\kappa}) \doteq \|s''_{ni\to pk}-s''_{ni\to q\kappa}\|. \tag{13}$$

Choose $N_1 > \max(k,\kappa\}$ for which $f_{11} \equiv 0$ on differences of order N_1 or greater in B''. If

$$f_{11}[(s''_{ni\to pk}+s''_{pk\to N_1x}) - (s''_{ni\to q\kappa}+s''_{q\kappa\to N_1y})] < m_{11}$$

for all (x,y) for which $(N_1,x) \succ (p,k)$ and $(N_1,y) \succ (q,\kappa)$, then an average over x and y contradicts the equality in (13). Therefore, there exist α and β such that

$$f_{11}[(s''_{ni\to pk}+s_{pk\to N_1\alpha}) - (s''_{ni\to q\kappa}+s''_{q\kappa\to N_1\beta})] = f_{11}(s''_{ni\to N_1\alpha}-s''_{ni\to N_1\beta}) \ge m_{11}.$$

It now follows that f_{11} "nearly" norms the difference between any convex combination of extensions of $s''_{ni\to N_1\alpha}$ and any convex combination of extensions of $s''_{ni\to N_1\beta}$, that this norm is greater than ρ_{-}, and that f_{11} is identically zero on the linear span of all differences in B'' with orders greater than N_1. With (17), this will be used to show that $\|\Delta^{t}_{11}\| > \frac{1}{2}r_{-}$ and Δ^{t}_{11} is "nearly" orthogonal to $\mathrm{lin}\{\Delta^{t}_{21}, \Delta^{t}_{23}, \Delta^{t}_{31}, \cdots\}$. Let

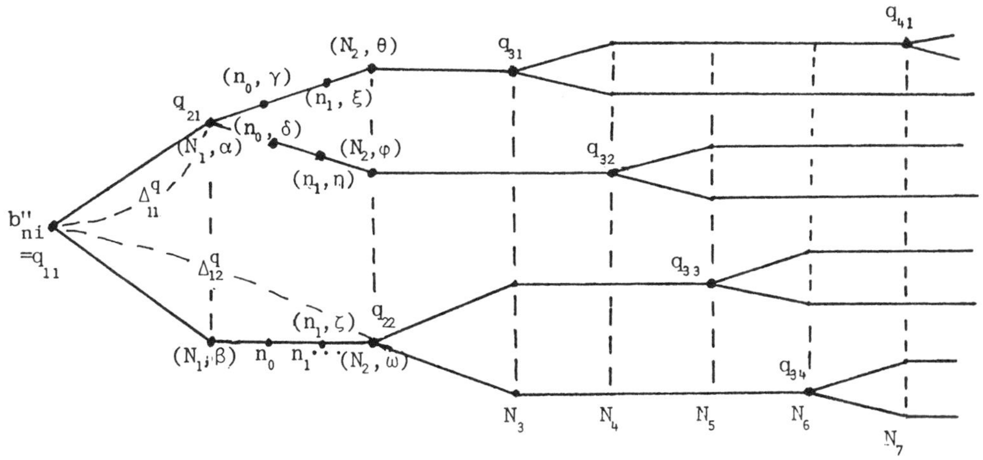

$$q_{21} = q_{11} + s''_{ni \to N_1\alpha} = q_{11} + \Delta^{q}_{11}.$$

Choose (p,k) and (q,κ) following (N_1,α) so that $\|s''_{N_1\alpha \to pk} - s''_{N_1\alpha \to q\kappa}\| > \rho_{-}$ and is "nearly" the supremum for all such (p,k) and (q,κ). Then find n_0, γ, δ, and f_{21} similarly to the method used to get N_1, α, β, and f_{11}, so that f_{21} "nearly" norms the difference between any convex combination of extensions of $s''_{N_1\alpha \to n_0\gamma}$ and any convex combination of extensions of $s''_{N_1\alpha \to n_0\delta}$, that this norm is greater than ρ_{-}, and that f_{21} is identically zero on the linear span of all differences in B'' with orders greater than n_0.

We now lay the foundation for using (17) and (18) to show that $\mathrm{lin}\{\Delta_{11}^t, \Delta_{21}^t\}$ is "nearly" orthogonal to $\mathrm{lin}\{\Delta_{23}^t, \Delta_{31}^t, \Delta_{33}^t, \Delta_{35}^t, \Delta_{37}^t, \Delta_{41}^t, \cdots\}$. Choose $\{(a_i, b_i)\colon\ 1 \le i \le A\}$ "nearly" dense in $\{(a,b)\colon\ |a| + |b| = 1\}$. For $i = 1$, choose (λ, j) following or equal to (n_0, γ), (μ, k) following or equal to (n_0, δ), (ν, r) following or equal to (N_1, β), and $g_1 \in F$ so that

$$M_1 = g_1\{a_1[\tfrac{1}{2}(s''_{ni\to N_1\alpha} - s''_{ni\to\nu r}) + \tfrac{1}{4}(s''_{N_1\alpha\to\lambda j} + s''_{N_1\alpha\to\mu k})] + b_1[\tfrac{1}{2}(s''_{N_1\alpha\to\lambda j} - s''_{N_1 a\to\mu k})]\} \tag{14}$$

$$\doteq \|a_1[\tfrac{1}{2}(s''_{ni\to N_1\alpha} - s''_{ni\to\nu r}) + \tfrac{1}{4}(s''_{N_1\alpha\to\lambda j} + s''_{N_1\alpha\to\mu k})] + b_1[\tfrac{1}{2}(s''_{N_1\alpha\to\lambda j} - s''_{N_1\alpha\to\mu k})]\|$$

is "nearly" the supremum for all such (λ, j), (μ, k), and (ν, r). Choose $n_1 > \max\{\lambda, \mu, \nu\}$ for which $g_1 \equiv 0$ on differences of order n_1 or greater in B''. If

$$g_1\{a_1[\tfrac{1}{2}(s''_{ni\to N_1\alpha} - s''_{ni\to n_1 z}) + \tfrac{1}{4}(s''_{N_1\alpha\to n_1 x} + s''_{N_1\alpha\to n_1 y}) + b_1[\tfrac{1}{2}(s''_{N_1\alpha\to n_1 x} - s''_{N_1\alpha\to n_1 y})]\} < M_1$$

for all $(n_1, x) \succ (\lambda, j)$, $(n_1, y) \succ (\mu, k)$, and $(n_1, z) \succ (\nu, r)$, then an average over x, y, and z contradicts the equality in (14). Therefore, there exist ξ, η, and ζ such that $(n_1, \xi) \succ (\lambda, j)$, $(n_1, \eta) \succ (\mu, k)$, $(n_1, \zeta) \succ (\nu, r)$, and

$$g_1\{a_1[\tfrac{1}{2}(s''_{ni\to N_1\alpha} - s''_{ni\to n_1\zeta}) + \tfrac{1}{4}(s''_{N_1\alpha\to n_1\xi} + s''_{N_1\alpha\to n_1\eta}) + b_1[\tfrac{1}{2}(s''_{N_1\alpha\to n_1\xi} - s''_{N_1\alpha\to n_1\eta})]\} \ge M_1. \tag{15}$$

Now treat (a_2, b_2) similarly, and determine n_2. Continue this step–by–step until $n_A = N_2$, θ, φ, and ω have been defined, as shown in the illustration. Let

$$q_{22} = q_{11} + s''_{ni \to N_2 \omega} = q_{11} + \Delta^q_{12}.$$

Any linear combination of Γ and Φ is "nearly" orthogonal to the linear span of all differences in B'' of order N_2 or greater, if

$$\Gamma = \frac{1}{2}(s''_{ni \to N_1 \alpha} - w) + \frac{1}{4}(u + v),$$

$$\Phi = \frac{1}{2}(u - v),$$

where u is a convex combination of segments beginning at (N_1, α) and passing through (N_2, θ), v is a convex combination of segments beginning at (N_1, α) and passing through (N_2, φ), and w is a convex combination of segments beginning at (n,i) and passing through (N_2, ω). To see this, recall that there is an i for which the linear combination of Γ and Φ is "nearly" a multiple of $a_i\Gamma + b_i\Phi$. It follows from convexity, and the norm in the ith case of (14) being "nearly" maximal, that $a_i\Gamma + b_i\Phi$ is "nearly" less than M_i. Also $g_i(a_i\Gamma + b_i\Phi) \geq M_i$ follows from the ith case of (15) and g_i being identically zero on differences of order n_i or greater in B''. The conclusion now follows, since we have $g_i(a_i\Gamma + b_i\Phi) \doteq \|a_i\Gamma + b_i\Phi\|$ and $g_i \equiv 0$ on differences of order N_2 or greater in B".

Continue inductively to define N_3 and q_{31}, N_4 and q_{32}, etc. Use four branches when defining N_3, five branches when defining N_4, etc. Then define P by replacing each q_{ni} by the corresponding member of B', which we designate by p_{ni}. Define a new set T as follows. Let S_1 be a sequence of positive integers for which

$$t_{11} = \tau\text{–}\lim_{\lambda \in S_1} [\text{average over } j \text{ of } \{p_{\lambda j}\}]$$

exists. Inductively, get sequences $\{S_m: m \geq 2\}$ such that, for each (m,j),

$$t_{mj} = \tau\text{-}\lim_{\lambda \in S_m} [\text{average over } k \text{ of } \{p_{\lambda k}: (\lambda,k) \succ (m,j)\}]$$

exists, and S_m is a subsequence of S_{m-1} for each m. Then $T \subset C$ and each t_{mj} is the average of its two successors in T. Let $\{\Delta^q_{mj}\}$, $\{\Delta^p_{mj}\}$, and $\{\Delta^t_{mj}\}$ denote corresponding differences in Q, P, and T. Then

$$t_{11}=p_{11}+\tau\text{-}\lim_{\lambda \in S_1}[\tfrac{1}{2}(\Delta^p_{11}+\Delta^p_{12})+\tfrac{1}{4}(\Delta^p_{21}+\Delta^p_{22}+\Delta^p_{23}+\Delta^p_{24})+\cdots+2^{-\lambda}(\Delta^p_{\lambda 1}+\cdots+\Delta^p_{\lambda,2^\lambda})];$$

$$t_{21}=p_{11}+\tau\text{-}\lim_{\lambda \in S_2}[\Delta^p_{11}+\tfrac{1}{2}(\Delta^p_{21}+\Delta^p_{22})+\cdots+2^{1-\lambda}(\Delta^p_{\lambda 1}+\cdots+\Delta^p_{\lambda,2^{\lambda-1}})];$$

$$\begin{aligned}\Delta^t_{11} = -\Delta^t_{12} = t_{21}-t_{11} = \tau\text{-}\lim_{\lambda \in S_2}[&\tfrac{1}{2}(\Delta^p_{11}-\Delta^p_{12})+\tfrac{1}{4}(\Delta^p_{21}+\Delta^p_{22}-\Delta^p_{23}-\Delta^p_{24})\\ &+\tfrac{1}{8}(\Delta^p_{31}+\Delta^p_{32}+\Delta^p_{33}+\Delta^p_{34}-\Delta^p_{35}-\Delta^p_{36}-\Delta^p_{37}-\Delta^p_{38})+\cdots\\ &+2^{-\lambda}(\Delta^p_{\lambda 1}+\cdots+\Delta^p_{\lambda,2^{\lambda-1}}-\cdots-\Delta^p_{\lambda,2^\lambda})].\end{aligned} \tag{17}$$

Similarly,

$$\begin{aligned}\Delta^t_{21} &= -\Delta^t_{22} = t_{31}-t_{21} = -(t_{32}-t_{21})\\ &= \tau\text{-}\lim_{\lambda \in S_3}[\tfrac{1}{2}(\Delta^p_{21}-\Delta^p_{22}) + \tfrac{1}{4}(\Delta^p_{31}+\Delta^p_{32}-\Delta^p_{33}-\Delta^p_{34}) + \cdots + 2^{-\lambda}(\quad)],\end{aligned} \tag{18}$$

$$\Delta^t_{23} = -\Delta^t_{24} = t_{33} - t_{22} = -(t_{34} - t_{22})$$

$$= \tau\text{–}\lim_{\lambda \in S_3}[\tfrac{1}{2}(\Delta^p_{23} - \Delta^p_{24}) + \tfrac{1}{4}(\Delta^p_{35} + \Delta^p_{36} - \Delta^p_{37} - \Delta^p_{38}) + \cdots + 2^{-\lambda}(\quad)]. \tag{19}$$

Since $\Sigma_1^\infty \beta_n < \infty$, it follows from (b) of Lemma 1.1 and the existence of the τ–limit that defines Δ^t_{ni}, that this limit still exists for each (n,i) if differences in P are replaced by the corresponding differences in Q.

To show that (i) is satisfied, observe that

$$t_{mj} - t_{11} = \tau\text{–}\lim_{\lambda \in S_m}[(\text{average over k of } \{p_{\lambda k}\colon (\lambda,k) \succ (m,j)\})$$
$$- (\text{average over i of } \{p_{\lambda i}\colon (\lambda,i) \succ (1,1)\})].$$

Since $p_{\lambda k}$ and $p_{\lambda i}$ can be replaced by $p_{\lambda k} - p_{11}$ and $p_{\lambda i} - p_{11}$, respectively, the right member is changed in norm by less than $2\Sigma_1^\infty \beta_n < r_+ - \rho_+$ if $q_{\lambda j}$ is used in place of $p_{\lambda j}$ and $q_{\lambda k}$ in place of $p_{\lambda k}$. But with these changes, it follows from (12) that the norm of the bracketed expression in the right member is less than ρ_+. Since F is norming, $\|t_{mi} - t_{11}\| < r_+$.

To show that (ii) is satisfied, let $\Delta^t_{m\beta}$ be the first of the two differences immediately following t_{mj}. Similarly to (19), there are integers $k, \cdots$ for which

$$\Delta^t_{m\beta} = -\Delta^t_{m,\beta+1} = t_{m+1,\beta} - t_{mj} = \tau\text{–}\lim_{\lambda \in S_m}[\tfrac{1}{2}(\Delta^p_{m\beta} - \Delta^p_{m,\beta+1})$$

$$+ \tfrac{1}{4}(\Delta^p_{m+1,k} + \Delta^p_{m+1,k+1} - \Delta^p_{m+1,k+2} - \Delta^p_{m+1,k+3}) + \cdots + 2^{-\lambda}(\quad)]. \tag{20}$$

If we replace p by q everywhere in the third member of (20), then the expression in brackets changes in norm by less than $\Sigma_1^\infty \beta_n < \rho_- - r_-$. It follows from our construction that the new expression in brackets has norm greater than $\frac{1}{2}\rho_{-}$, by virtue of there being a member $f_{m\beta}$ of F that "nearly" norms this expression for all λ. Therefore, $\|\Delta_{m\beta}^{t}\| > \frac{1}{2}r_{-}$.

We will show now that $\{\Delta_{11}^{t},\Delta_{21}^{t},\Delta_{23}^{t},\Delta_{31}^{t},\Delta_{33}^{t},\Delta_{35}^{t},\Delta_{37}^{t},\Delta_{41}^{t},\cdots\}$ is a "nearly" monotone basic sequence. If we replace p by q in equalities such as (17)–(19) for these differences, then the change in norm for a difference of order n in T is less than $\Sigma_{n}^{\infty}\beta_{i}$ and there are 2^{n} such differences of order n. Because of our construction, the new expressions give a "nearly" monotone basic sequence. The sum of all changes in norm is less than $\Sigma_{1}^{\infty}(2^{n}\cdot\Sigma_{n}^{\infty}\beta_{i}) < \Sigma_{1}^{\infty}\eta_{n}$. If $\Sigma_{1}^{\infty}\eta_{n}$ is small enough, the change in "being nearly monotone" will be very small. Moreover, if x^{q} belongs to the linear span of the first n basis elements when p has been replaced by q, and y^{q} belongs to the linear span of the remainder, then there is an f in F for which $f(x^{q}) \doteq \|x^{q}\|$ and $f(y^{q}) = 0$. With $\Sigma_{1}^{\infty}\eta_{i}$ small enough, we can change q back to p and have $f(x^{p}) \doteq \|x^{p}\|$ and $f(y^{p}) \doteq 0$ unless y^{p} is very large, which we can take to mean that $\|y^{p}\| > 3\|x^{p}\|$.

THEOREM 3.2. Let F be a countable norming set of continuous linear functionals on a Banach space X and let τ be the weakest topology for which each f in F is continuous. Let C be a bounded subset of X that is convex and τ–compact. Then RNP and KMP are equivalent for C. Moreover, if B is a bush in C, then τ–cl[con(B)] contains a closed convex set without extreme points.

COROLLARY 3.3. Let X be a separable Banach space and C be a w^{*}–compact convex subset of X^{*}. Then RNP and KMP are equivalent for C. If C fails RNP, then C contains a "nearly" monotone basic 2–tree T for which w^{*}–cl[con(T)] has a closed convex subset without extreme points.

Proof. Let F be a countable dense subset of the unit sphere of X. Then F is norming for X^*. Let τ be the weakest topology for which each $x \in F$ is continuous on X^*. If C is a w^*–compact convex subset of X^*, then C is bounded and τ–compact. It follows from Theorem 3.1 that C contains a "nearly" monotone basic 2–tree. Now use Theorem 3.2.

Before proving Theorem 3.2, we establish the following lemma. A set of nonnegative numbers $\{\lambda_i^n\}$ corresponding to members of a bush $\{b_{ni}\}$ is said to be *spreading* if $\lambda_1^1 = 1$ and $\lambda_i^n = \Sigma \lambda_j^{n+1}$, where the sum is over all j for which $(n+1,j) \succ (n,i)$. A *τ–wedge intersection* for a branch β is the intersection of the t–closures of wedges on β.

LEMMA 3.4. Let F, X, τ and C be as in Theorem 3.2. For some positive $\Theta < 1$, let T be a basic 2–tree in C as described in Theorem 3.1. Then τ–cl[con(T)] is the set of all x which have representations as

$$x = \Sigma_{i=1}^{d(n)} \lambda_i^n [\tau\text{–}\lim_{r\to\infty} x_{ni}(r)] \text{ for each } n, \tag{21}$$

where d(n) is the number of wedges that start at column n, $\{\lambda_i^n\}$ is spreading, and $x_{ni}(r)$ is in W_{ni} for each i. If C_0 is the set of all x with such a representation for which

$$\lim_{n\to\infty} \sup\{\lambda_i^n \colon 1 \le i \le d(n)\} = 0, \tag{22}$$

then C_0 is norm–closed and each representation of a member of C_0 as in (21) satisfies (22). Also, C_0 is bounded, convex, nonempty, and no τ–wedge intersection of B contains a point of C_0.

Proof of the Lemma. Let $x \in \tau$–cl[con(T)]. Since F is countable and total, there is a sequence $\{\xi(k)\}$ in con(T) that converges to x (and only x) in the τ–topology. For each n and k, $\xi(k)$ can be represented as

$$\xi(k) = \Sigma_{i=1}^{d(n)} \lambda_i^n(k) \xi_{ni}(k), \quad \xi_{ni}(k) \in W_{ni},$$

where $\Sigma_i \lambda_i^n(k) = 1$ for each n and each $\lambda_i^n(k) \geq 0$. By using subsequences, the dependence of $\lambda_i^n(k)$ on k can be eliminated and $\{\lambda_i^n\}$ will be spreading as well. A further subsequence of $\{\xi(k)\}$ has the property that, for each (n,i), $\{\xi_{ni}(k)\}$ converges in the τ–topology. That is, we have (21).

To prove that C_0 is norm–closed, we need only show that if $\bar{x} \in$ norm–cl(C_0) and $\bar{x}$ is represented as in (21), then (22) also is satisfied. If there is a representation of $\bar{x}$ as in (21) for which

$$\lim_{n\to\infty} \sup\{\lambda_i^n:\ 1 \leq i \leq d(n)\} \neq 0,$$

then there is a branch β along which the values of $\{\lambda_i^n\}$ converge to a number $\lambda \neq 0$. Choose $x \in C_0$ for which

$$\|\bar{x}-x\| < \tfrac{1}{2}\,\Theta^2\lambda\delta, \tag{23}$$

where δ is a separation constant for T. For an integer $m > 0$, let b_{mi} be the point of order m on β and let Δ and $-\Delta$ be the two differences of order m that follow b_{mi}, where Δ is the difference on branch β. If m is sufficiently great, then the sum of the norms of these differences, when each is multiplied by the corresponding λ_j^{m+1} for the representation of x as in (21), is less than $\frac{1}{4}\lambda\delta$. Also, we can have m great enough that the contribution to $\bar{x}$ from $-\Delta$ has norm less than $\frac{1}{4}\lambda\delta$, but the contribution to $\bar{x}$ from Δ is at least $\lambda\|\Delta\| > \lambda\delta$. Let $\bar{x}_m$ and x_m be the "expansions" of $\bar{x}$ and x using basis elements having order less than or equal to m as differences in T, that is, with coefficients determined by $\{\lambda_i^n\colon n \le m+1\}$ in representations of $\bar{x}$ and x as in (21). Then

$$\|\bar{x}_m - x_m\| > \Theta(\lambda\delta - \tfrac{1}{4}\lambda\delta - \tfrac{1}{4}\lambda\delta) = \tfrac{1}{2}\Theta\lambda\delta.$$

By choosing m so that $\|\bar{x}_m - x_m\| > \frac{3}{4}\limsup_{k\to\infty}\|\bar{x}_k - x_k\|$, we can also have

$$\|\bar{x}_r - x_r\| < 2\|\bar{x}_m - x_m\| \tag{24}$$

if $r > m$. Now use (iii) of Theorem 3.1 to choose an $f \in F$ for which $r > m$ implies either $\|(\bar{x}_r - x_r) - (\bar{x}_m - x_m)\| > 3\|\bar{x}_m - x_m\|$ or

$$|f(\bar{x}_r - x_r)| \ge \Theta\|\bar{x}_m - x_m\|. \tag{25}$$

Since the first inequality contradicts (24), (25) is true for all $r > m$. This implies that $|f[\tau\text{–}\lim(\bar{x}_r - x_r)]| \ge \frac{1}{2}\Theta^2\lambda\delta$, which contradicts (23).

Since $T \subset C_0$, C_0 is not empty. If x and y are in C_0, then for any μ in $(0,1)$, the representations of x and y that satisfy (23) can be combined to obtain such a representation for $\mu x + (1-\mu)y$. Thus C_0 is convex.

To see that no τ–wedge intersection of T contains a point of C_0, we need merely observe that if x is in a τ–wedge intersection for a branch β, then x has a representation as in (21) such that, for each n, $\lambda_i^n = 1$ if (n,i) is on β.

Proof of Theorem 3.2. Since RNP $\Rightarrow$ KMP and C contains a bush if C fails RNP, we only need to prove that if B is a bush in C, then τ–cl[con(B)] contains a closed convex set without extreme points. Let T be a 2–tree in τ–cl[con(B)] as described in Theorem 3.1. Let C_0 be the set given by Lemma 3.4. If x is an extreme point of C_0, then for any representation of x as in (21) and for each n, x belongs to each of the τ–wedges of order n for which $\lambda_i^n \neq 0$. But this would imply that x belongs to at least one τ–wedge intersection of T, none of which contians a point of C_0.

4. UNCONDITIONAL BASIC CONDITIONS

Although it is satisfied by many classical spaces, the assumption that X can be embedded in a Banach space with an unconditional basis is very severe. For example, it implies that X is reflexive unless X contains c_0 or ℓ_1 [1], and therefore that X is reflexive if X has CPCP and X^* is separable. In the fall of 1984, Schachermayer gave a proof that X has CPCP if X has KMP and is isomorphic to subspace of a space Y with an unconditional basis. It was announced by Rosenthal and Wessel in the summer of 1985 that a bounded closed convex subset K of Y fails KMP if K fails CPCP and Y has an unconditional basis [16]. Schachermayer also gave a proof of this that is a modification of his proof for the case K is a subspace of Y. We will prove that if K is a bounded closed convex subset of a Banach space Y with a UBFDD, then:

(i) K failing PCP implies K fails KMP because K contains a complemented bush with empty wedge–intersections, each branch of which spans a space isomorphic with c_0 (Theorem 4.5). (In general, CPCP and KMP imply RNP [18, Theorem 2.1], so KMP $\Longleftrightarrow$ RNP for all such K).

(ii) For K, RNP $\Longrightarrow$ KMP $\Longrightarrow$ CPCP. If Y has no subspace isomorphic with c_0, then K has RNP, KMP, and PCP (Theorems 4.6 and 4.7).

(iii) If a Banach space X has a UBFDD, then (for X) each of RNP, KMP, PCP, and CPCP is equivalent to not having a subspace isomorphic with c_0 (Theorem 4.8).

The next two lemmas were proved by Bourgain for the case $\beta = 1$ and K failing CPCP [2, Lemmas 9–10]. The proofs for general $\beta > 0$ and K failing PCP are not essentially different, but because of their importance in what follows we give proofs. It is advantageous to use Helly's Condition: "For members $\{f_i\colon 1 \le i \le n\}$ of X^* and numbers $\{c_1, \cdots, c_n; M\}$, $|\Sigma_1^n a_i c_i| \le M\|\Sigma_1^n a_i f_i\|$ for all $\{a_i\}$ is equivalent to the existence, for each $\epsilon > 0$, of an x in B such that $\|x\| < M + \epsilon$ and $f_i(x) = c_i$ if $1 \le i \le n$." The following rather easy result also will be used (a similar fact is given on page 138 of [2]). "If a bounded closed nonempty set $K \subset X$ fails PCP, then there is a closed nonempty set $A \subset K$ and a positive number $\epsilon(A)$ such that

$$\operatorname{diam}(U \cap A) > \epsilon(A) \text{ if } U \cap A \neq \emptyset \text{ and } U \text{ is w–open."} \tag{26}$$

<u>LEMMA 4.1</u>. Suppose $A \subset K \subset X$ and $\epsilon(A)$ are as in (26). Then for any $\beta > 0$, any $\epsilon < \frac{1}{2}\epsilon(A)$, any $x \in \mathscr{A} = A + N(X,\beta)$, and any subspace E of X with finite codimension, there exists Δ such that

$$x+\Delta \in \mathscr{A}, \ \Delta \in E, \ \|\Delta\| > \epsilon. \tag{27}$$

<u>Proof</u>. Given $\beta > 0$, $\epsilon < \frac{1}{2}\epsilon(A)$, $x \in \mathscr{A}$, and a subspace E with finite codimension, choose a number $\alpha < \beta$ for which there exist an x_a in A with $\|x-x_a\| < \alpha$. Let $E = \cap_{i=1}^n \ker(f_i)$, where the f_i's are linearly independent. For $\delta < \min\{\beta-\alpha, \frac{1}{2}\epsilon(A)-\epsilon\}$, choose $c > 0$ so that, for all numbers $\{a_i\}$,

$$|\Sigma_1^n a_i c_i| \leq \tfrac{1}{2}\delta\|\Sigma_1^n a_i f_i\| \text{ if } |c_i| \leq c \text{ for each } i.$$

If $U = \{u: |f_i(u-x_a)| < c, 1 \leq i \leq n\}$, then it follows from (26) that there exist y_a and z_a in $U \cap A$ for which $\|y_a-z_a\| > \epsilon(A)$. Let $f_i(y_a-x_a) = c_i$ for each i. Since $y_a \in U$, $|c_i| < c$ and we can use Helly's Condition to find $\eta \in X$ such that $\|\eta\| < \delta$ and $f_i(\eta) = -c_i$ for each i. Let

$$y = y_a + (x-x_a) + \eta = x + (y_a-x_a+\eta).$$

Since $f_i(y_a-x_a+\eta) = 0$ for each i, $y_a-x_a+\eta \in E$. Also, it follows from $y_a \in A$ and $\|(x-x_a)+\eta\| < \alpha+\delta < \beta$ that $y \in \mathscr{A}$. Similarly, there exists $\zeta \in X$ such that $\|\zeta\| < \delta$ and, if

$$z = z_a + (x-x_a) + \zeta = x + (z_a-x_a+\zeta),$$

then $z_a-x_a+\zeta \in E$ and $z \in \mathscr{A}$. Since

$$\|(y_a-x_a+\eta) - (z_a-x_a+\zeta)\| > \|y_a-z_a\| - 2\delta > \epsilon(A) - 2\delta > 2\epsilon,$$

at least one of $y_a-x_a+\eta$ or $z_a-x_a+\zeta$ can be the Δ needed in (27).

LEMMA 4.2. Suppose A and $\epsilon(A)$ are as in Lemma 4.1. Then for any $\beta > 0$, any $\epsilon < \frac{1}{2}\epsilon(A)$, any $x \in \mathscr{A} = A + N(X,\beta)$, and any subspace E of X with finite codimension,

$$X \in \overline{\text{con}}\{x+\Delta: x + \Delta \in \mathscr{A}, \Delta \in E, \|\Delta\| > \epsilon\}.$$

Proof. Suppose $x \notin \overline{\mathrm{con}}\{x+\Delta\colon x+\Delta \in \mathcal{A},\ \Delta \in E, ||\Delta|| > \epsilon\} = D$. Then there is an f in X^* such that $f(x)$ is greater than the sup of f on D. Let $F = E \cap \ker(f)$. It follows from Lemma 4.1 that there exists Δ_0 in F for which $x+\Delta_0 \in \mathcal{A}$ and $||\Delta_0|| > \epsilon$. Then $f(x+\Delta_0) = f(x)$, but $x+\Delta_0 \in D$ and therefore $f(x+\Delta_0) < f(x)$.

THEOREM 4.3. Let K be a bounded closed convex subset of a Banach space X for which $\{\Omega_n\}$ is an FDD. If K fails PCP, then for any sequence of positive numbers $\{\beta_n\}$ there is a bush $B \subset K$ and an increasing sequence of positive integers $\{q(p)\}$ such that, if

$$\Phi_p = \mathrm{lin}(\cup_{q(p)}^{q(p+1)-1}\Omega_i) \text{ for } p \geq 1,$$

and if (n,i) has position p in the lexicographic ordering, then

$$\Delta_{n+1,j} \in \Phi_p + N(X,\beta_p) \text{ if } (n+1,j) \succ (n,i). \tag{28}$$

Proof. First construct an approximate bush B^a as follows. Let $A \subset K$ and $\epsilon(A)$ be as in Lemma 4.2. Let δ be any positive number less than $\epsilon(A)$. Let b_{11}^a be any member of A, and let $q(1) = 1$. For an arbitrary sequence of positive numbers $\{\beta_n\}$, let $\{\eta_n\}$ be a sequence of positive numbers such that $4\Sigma_n^\infty\eta_i < \max\{\frac{1}{2}\beta_n, \frac{1}{2}\delta\}$ and $\delta+\eta_n < \epsilon(A)$ for each n. Use Lemma 4.2 to choose finitely many members $\{b_{11}^a+u_{1i}\colon 1 \leq i \leq d(1)\}$ of $A+N(X,\eta_1)$ such that $||u_{1i}|| > \delta+\eta_1$ for each i, and

$$\mathrm{dist}[b_{11}^a, \mathrm{con}\{b_{11}^a+u_{1i}\colon 1 \leq i \leq d(1)\}] < \eta_1.$$

Now replace u_{1i} by Δ_{1i}^a for which $b_{11}^a+\Delta_{1i}^a \in A$ and $||u_{1i}-\Delta_{1i}^a|| < \eta_1$. Then $||\Delta_{1i}^a|| > \delta$.

Let $b^a_{11}+\Delta^a_{1i} = b^a_{2i}$ for each i. Then

$$\delta_1 = \text{dist}[b^a_{11}, \text{con}\{b^a_{2i}\colon 1 \le i \le d(1)\}] < 2\eta_1, \tag{29}$$

where δ_1 is the first "error in averaging".

Choose $q(2)$ so that

$$\text{dist}[\Delta^a_{1i}, \text{lin}(\cup_{q(1)}^{q(2)-1}\Omega_i)] < \tfrac{1}{2}\beta_1 \text{ for each } i. \tag{30}$$

The preceding process can be continued inductively, first working successively with $b^a_{21}, b^a_{22}, \cdots, b^a_{2,d(1)}$ to produce $\{b^a_{3i}\}$, then working successively with $b^a_{31}, b^a_{32}, \cdots$, etc. If (n,i) occurs in the pth position in lexicographic ordering, then (29) and (30) become:

$$\delta_p = \text{dist}[b^a_{ni}, \text{con}\{b^a_{n+1,j}\colon (n+1,j) \succ (n,i)\}] < 2\eta_p; \tag{31}$$

$$\text{dist}[\Delta^a_{ni}, \text{lin}(\cup_{q(p)}^{q(p+1)-1}\Omega_i)] < \tfrac{1}{2}\beta_p. \tag{32}$$

We now have an approximate bush $B^a = \{b^a_{ni}\}$. Let $B = \{b_{ni}\}$ be the bush obtained from B^a by "averaging back". That is,

$$b_{ni} = \lim_{\lambda\to\infty}[\text{wt. average}\{b^a_{\lambda j}\colon (\lambda,j) \succ (n,i)\}],$$

where the weights are chosen in the obvious way. For each n, it follows from (31) that "averaging back" changes b^a_{ni} by less than $2\Sigma_p^\infty\eta_i$ if (n,i) is in the pth position in lexicographic ordering. Therefore,

$$||\Delta^a_{ni}-\Delta_{ni}|| < 4\Sigma^\infty_p \eta_i < \frac{1}{2}\beta_p$$

if Δ_{ni} is the new difference corresponding to Δ^a_{ni}. This and (32) imply that (28) is satisfied. Since $||\Delta^a_{ni}|| > \delta$ for each (n,i), we have $||\Delta_{ni}|| > \delta - 4\Sigma^\infty_p\eta_i > \frac{1}{2}\delta$. Thus $\frac{1}{2}\delta$ is a separation constant for B. Since K is closed and convex, $B \subset K$. However, B need not be a subset of A.

COROLLARY 4.4. If a bounded closed convex subset K of a Banach space X fails PCP, then K contains a nearly monotone basic bush.

Proof. Since $\overline{\text{lin}}(K) \subset C[0,1]$ and C[0,1] has a monotone basis, this corollary follows from Theorem 4.3.

THEOREM 4.5. If K is a bounded closed convex subset of a Banach space with an unconditional basis and if K fails PCP, then K fails KMP because K contains a complemented bush with empty wedge–intersections. Moreover, each branch of this bush has the property that the closed linear space of differences on that branch is isomorphic with c_0.

Proof. Suppose $K \subset X$ and $\{\omega_n\}$ is an unconditional basis for X. Choose a sequence $\{\beta_n\}$ such that if B is as described in Theorem 4.3, and if D_{ni} is the linear span of all $\Delta_{n+1,j}$ that are differences between b_{ni} and a successor $b_{n+1,j}$ of b_{ni}, then it follows from (28) that $\{D_{ni}\}$ is a UBFDD of $\overline{\text{lin}}\{D_{ni}\}$. Since B is basic, all wedge–intersections of B are empty. It follows from unconditionality of $\{D_{ni}\}$ that B is complemented. Therefore $\overline{\text{con}}(B)$ has no extreme points [8, Theorem A, pg. 354]. Since partial sums of differences along a branch are bounded, it follows from unconditionality of $\{D_{ni}\}$ that all sums of differences on the same branch are bounded. Therefore, for any branch β, the closed linear span of differences on β is isomorphic with c_0.

It is interesting, as has been noted by N. Ghoussoub (see the proof of Corollary 2.11 in [18]), that if K is the unit ball of a Banach space X that is contained in a space with an unconditional basis, then Theorem 4.5 can be proved easily without using any of the preceding results except Lemma 4.1. In this case, it follows easily from Lemma 4.1 that there is an $\epsilon > 0$ and a sequence $\{\Delta_i\}$ of members of K for which $\|\Delta_i\| > \epsilon$, the Δ_i's are "nearly" in consecutive disjoint blocks of the unconditional basis, and $\Sigma_1^n \Delta_i \in K$ for each n. From unconditionality, it follows that $\{\Delta_n\}$ spans a space isomorphic to c_0. Therefore, K contains a space isomorphic to c_0 and contains complemented trees of all orders.

THEOREM 4.6. If K is a bounded closed convex subset of a Banach space with an unconditional basis, then for K we have RNP $\Longleftrightarrow$ KMP $\Longrightarrow$ PCP.

Proof. If K has PCP, then RNP $\Longleftrightarrow$ KMP [18, Theorem 2.1]; if K fails PCP, then K fails RNP and KMP (see Theorem 4.5).

Remark. The next theorem has some relation to the known fact that if X is a Banach lattice with no subspace isomorphic with c_0, then RNP $\Longleftrightarrow$ CPCP for all bounded closed convex separable subsets of X [7, Theorem 2].

THEOREM 4.7. If K is a bounded closed convex subset of a Banach space X which has a UBFDD and if no subspace of X is isomorphic with c_0, then K has RNP, KMP, and PCP.

Proof. Because of Theorem 4.6, it is sufficient to prove that X has a subspace isomorphic with c_0 if K fails RNP. Suppose K fails RNP. Then it follows from Lemma 1.2 that there is a positive number δ such that, for any sequence of positive numbers $\{\beta_n\}$, there is a bush B in K and a UBFDD $\{\Phi_n\}$ of X such that B has separation constant δ and, if Δ is a difference of order n for B, then

$$\Delta \in \text{lin}(\Phi_{n-1},\Phi_n) + N(X,\beta_n),$$

where $\Phi_0 = \{0\}$. Therefore, a sum of differences along a branch is of type

$$\varphi_1 + (\varphi_1'+\varphi_2) + (\varphi_2'+\varphi_3) + \cdots + (\varphi_{r-1}'+\varphi_r) + (\varphi_r'+\varphi_{r+1}), \tag{33}$$

where φ_n and φ_n' belong to $\Phi_n + N(X,\beta_n)$. Let Θ be the basis constant for $\{\Phi_n\}$ and assume that each $\beta_n < \delta$ and that $\Sigma_1^\infty \beta_n$ is small enough that $\{x_n\}$ is unconditionally basic if $\|\omega_n - x_n\| < \beta_n$ for each n and the members of $\{\omega_n\}$ are in disjoint blocks of $\{\Omega_n\}$, the unconditional basis for X from which $\{\Phi_n\}$ was derived by using Lemma 1.2.

Consider the case B has the property:

(S). For each wedge of B and each positive number ϵ, there is a difference $\varphi_{n-1}'+\varphi_n$ and a successor $\varphi_n'+\varphi_{n+1}$ such that $\|\varphi_n\| < \epsilon$ and $\|\varphi_n'\| < \epsilon$.

Let φ_1 be any difference of order 1 for B and let $\{\phi_1'(i)+\phi_2(i)\colon 1 \le i \le d(1)\}$ be the successors of φ_1. If $\|\varphi_1+\phi_1'(i)\| \le \delta$ for each i, then it follows from the average of $\phi_1'(i)$ being 0 that $\|\varphi_1\| \le \delta$. Since $\|\varphi_1\| > \delta$, we can assume for (33) that $\|\varphi_1+\varphi_1'\| > \delta$. Also, r in (33) can be chosen so that $\|\varphi_r\| < \frac{1}{2}\beta_r$ and $\|\varphi_r'\| < \frac{1}{2}\beta_r$. Let $\phi_{r+1}'(i) + \phi_{r+2}(i)$ be the successors of $\varphi_r'+\varphi_{r+1}$ in B. As before, if $\|\varphi_{r+1}+\phi_{r+1}'(i)\| \le \frac{1}{2}\delta$ for each i, then $\|\varphi_{r+1}\| \le \frac{1}{2}\delta$. But this is not possible, since then

$$\|\varphi_r'+\varphi_{r+1}\| < \tfrac{1}{2}\beta_r + \tfrac{1}{2}\delta < \delta.$$

By choosing φ_{r+1}' as a suitable $\phi_{r+1}(i)$, it follows from $\{\Phi_n\}$ having basis constant Θ that we have

$$[(\varphi_1+\varphi_1') + \cdots + (\varphi_{r-1}+\varphi_{r-1}')] + (\varphi_r+\varphi_r') + [(\varphi_{r+1}+\varphi_{r+1}') + \cdots]. \tag{34}$$

where the sum in the first set of brackets has norm greater than $\Theta\|\varphi_1+\varphi_1'\| > \Theta\delta$, any sum beginning with $\varphi_{r+1}+\varphi_{r+1}'$ has norm greater than $\frac{1}{2}\Theta\delta$, and $\|\varphi_r+\varphi_r'\| < \beta_r$. This process can be continued. If we discard all such terms $\varphi_r+\varphi_r'$, we have a series of bracketed terms with norms greater than $\frac{1}{2}\Theta\delta$ whose partial sums are bounded. Because of the choice of $\Sigma_1^\infty\beta_n$, these remaining terms are unconditionally basic with bounded partial sums. Therefore, they span a space isomorphic with c_0.

Now consider the case B does not have property (S). That is, there is a wedge W of B and a positive number ϵ such that, if

$$(\varphi_s'+\varphi_{s+1}) + (\varphi_{s+1}'+\varphi_{s+2}) + \cdots + (\varphi_t'+\varphi_{t+1})$$

$$= \varphi_s' + (\varphi_{s+1}+\varphi_{s+1}') + (\varphi_{s+2}+\varphi_{s+2}') + \cdots + (\varphi_t+\varphi_t') + \varphi_{t+1}$$

is any sum along a branch in W, then for each i at least one of $\|\varphi_{s+i}\|$ or $\|\varphi_{s+i}'\|$ is greater than ϵ. Suppose $s < k < t+1$. If $\|\varphi_k+\phi_k'(i)\| \le \frac{1}{2}\epsilon$ for each successor $\phi_k'(i)+\phi_{k+1}(i)$ of $\varphi_{k-1}'+\varphi_k$ in B, then as before we have $\|\varphi_k\| \le \frac{1}{2}\epsilon$. Since $\|\phi_k'(i)\| \le \|\varphi_k+\phi_k'(i)\| + \|\varphi_k\| \le \epsilon$ for each i, we have the contradiction (for each i) that neither $\|\varphi_k\|$ nor $\|\phi_k'(i)\|$ is greater than ϵ. Thus it is possible to chooses a branch in W along which

$$(\varphi_{s+1}+\varphi'_{s+1}) + (\varphi_{s+2}+\varphi'_{s+2}) + \cdots + (\varphi_t+\varphi'_t)$$

is bounded for each t and each term has norm greater than $\frac{1}{2}\epsilon$. As for (34), the sequence $\{\varphi_{s+i}+\varphi'_{s+i}\colon i \geq 1\}$ is unconditionally basic and spans a space isomorphic with c_0.

Remark. There may exist a Banach space X which has an unconditional basis and a subspace isomorphic with c_0, but has a bounded closed convex subset K for which RNP, KMP, and PCP are not equivalent. However, it follows from Theorem 4.6 that such a K must fail RNP, but have PCP. It then must also fail KMP [18]. It is known that there are Banach spaces that have CPCP and fail both RNP and KMP, but are not contained in Banach spaces with an unconditional basis (see [4] and [6, remark following Corollary II.6]).

THEOREM 4.8. If X has a UBFDD, then each of the properties RNP, KMP, PCP, and CPCP is equivalent to X not having a subspace isomorphic with c_0.

Proof. If X has no subspace isomorphic with c_0, then it follows from Theorem 4.7 that the unit ball of X, and therefore X itself, has RNP, KMP, and PCP. If X has a subspace isomorophic with c_0, then X contains trees and bushes and fails RNP, X fails KMP because the unit ball of c_0 has no extreme points, and X fails CPCP because the unit ball of c_0 has no points at which the weak and norm topologies coincide.

REFERENCES

1. C. Bessaga and A. Pelczyński, *On subspaces of a space with an absolute basis*, Bull. Acad. Sci. Pol. 6 (1958), 313–315.

2. J. Bourgain, *Dentability and finite–dimensional decompositions*, Studia Math. 67 (1980), 136–148.

3. J. Bourgain and H. Rosenthal, *Martingales valued in certain subspaces of* L^1, Isreal J. Math. 37 (1980), 54–75.

4. ________ , *Geometrical implications of certain finite–dimensional decompositions*, Bull. Soc. Math. Belg. 32 (1980), 57–82.

5. R.D. Bourgin, "Geometric Aspects of Convex Sets with the Radon–Nikodým Property", Springer–Verlag No. 993 (1983).

6. N. Ghoussoub and B. Maurey, *G_δ–embeddings in Hilbert space*, J. Funct. Anal. 61 (1985), 72–97.

7. ________ , *On the Radon–Nikodým property in function spaces*, Seminaire d'Analyse–Paris VII (1983–1984).

8. A. Ho, *The Krein–Milman property and complemented bushes in Banach spaces*, Pacific J. Math. 98 (1982), 347–363.

9. R.E. Huff and P.D. Morris, *Dual spaces with the Krein–Milman property have the Radon–Nikodým property*, Proc. Amer. Math. Soc. 49 (1975), 104–108.

10. R.C. James, *Structure of Banach spaces: Radon–Nikodým and other properties*, General Topology and Modern Analysis, Academic Press, 1980.

11. ________ , *Subbushes and extreme points in Banach spaces*, Proceedings of Research Workshop on Banach Space Theory, Univ. Iowa, 1981, edited by Bor–Luh Lin.

12. J. Lindenstrauss and L. Tzafriri, "Classical Banach Spaces I", Springer–Verlag (New York), 1977.

13. R.R. Phelps, *Dentability and extreme points in Banach spaces*, J. Funct. Anal. 17 (1974), 78–90.

14. M.A. Rieffel, *Dentable subsets of Banach spaces with applications to a Radon–Nikodým theorem in functional analysis*, Functional Analysis (Proc. Conf., Irvine, CA, 1966), Academic Press (London), Thompson (Washington, D.C.), pages 71–77.

15. H. Rosenthal, *On the structure of non–dentable closed bounded convex sets*, Advances in Math., to appear.

16. H. Rosenthal and A. Wessel, *The Krein–Milman property and a martingale coordinatization of certain non–dentable convex sets*, to appear in Pac. J. Math.

17. W. Schachermayer, *For a Banach space isomorphic to its square, the Radon–Nikodým property and the Krein–Milman property are equivalent*, Studia Math. 81 (1985), 329–339.

18. ________ , *The Radon–Nikodým property and the Krein–Milman property are equivalent for strongly regular sets*, to appear in Trans. Amer. Math. Soc.

19. C. Stegall, *The Radon–Nikodým property in conjugate Banach spaces*, Trans. Amer. Math. Soc. 206 (1975), 213–223.

Claremont Graduate School, Emeritus
[14385 Clear Creek Place, Grass Valley, CA 95949]

Contemporary Mathematics
Volume **85**, 1989

Independence in separable Banach spaces

N. J. KALTON[1]

Abstract. *We answer a question of Fremlin and Sersouri concerning independence of uncountable sets in separable Banach spaces.*

Recently Fremlin and Sersouri [1] proved the following theorem:

THEOREM 1. *Let X be a separable Banach space and let G be an uncountable subset of X. Then, for any $1 < p < \infty$, there exists a sequence (g_n) of distinct points of G and a sequence of real numbers (a_n), not all zero, such that $\sum |a_n|^p < \infty$ and*

$$\sum_{n=1}^{\infty} a_n g_n = 0.$$

This theorem answered a question of Lipecki. They note that one cannot allow $p = 1$ in the above theorem, and ask if instead one can require $(a_n) \in \ell(1,\infty)$ where $\ell(1,\infty)$ denotes the space weak ℓ_1 of all sequences (a_n) such that the decreasing rearrangement (a_n^*) of $(|a_n|)$ satisfies $\sup n a_n^* < \infty$.

We show in this note that this question has an affirmative answer and, in fact, a rather stronger result is true. For convenience, we restrict attention to real Banach spaces; some obvious rewording is necessary in the complex case. Our theorem is:

[1] Research supported by NSF-grant DMS-8601401

AMS Classification: 46B15

Key words: independence, Banach spaces

THEOREM 2. *Let X be a separable Banach space and let G be an uncountable subset of X. Suppose (a_n) is a sequence of real numbers satisfying $\sum |a_n| = \infty$ and $\lim a_n = 0$. Then there is a sequence of distinct points (g_n) of G and a sequence of signs $\epsilon_n = \pm 1$ so that*

$$\sum_{n=1}^{\infty} \epsilon_n a_n g_n = 0.$$

As in [1] the first step is to reduce the problem to the case when G is dense in itself, and one may then suppose that X is the closed linear span of G. Thus Theorem 2 follows from Theorem 3 below.

THEOREM 3. *Let X be an arbitrary Banach space and let G be a subset of X. Suppose H is the set of accumulation points of G, and that X is the closed linear span of H. Then given any $x \in X$ and any sequence of real numbers (a_n) with $\sum |a_n| = \infty$ and $\lim a_n = 0$ we may find a sequence of signs ϵ_n, and a sequence of distinct elements $g_n \in G$ so that*

$$x = \sum_{n=1}^{\infty} \epsilon_n a_n g_n .$$

PROOF: We may suppose (a_n) is a sequence of nonnegative numbers satisfying the conditions of the theorem. For convenience let $b_n = \max_{i>n} |a_i|$. We shall require the following lemma:

LEMMA 4. *Suppose $\alpha \in \mathbf{R}$ and $m \in \mathbf{N}$. Then we may choose signs ϵ_i, $(i \geq m+1)$ so that if $s_m = \alpha$ and $s_k = \alpha + \sum_{i=m+1}^{k} \epsilon_i a_i$ then we have $\lim s_k = 0$ and $\sup |s_k| \leq \max(b_m, |\alpha|)$.*

PROOF OF THE LEMMA: Define ϵ_k so that $\epsilon_k s_{k-1} \leq 0$. Then the sequence s_k must change signs infinitely often since otherwise ϵ_k is eventually constant and this would imply that $\sum a_n < \infty$. If s_k and s_{k-1} have the same signs then $|s_k| \leq |s_{k-1}|$. If s_k and s_{k-1} have opposite signs then $|s_k| \leq |a_k|$. Thus we have $|s_k| \leq \max(|s_{k-1}|, |a_k|)$ and this implies the second assertion of the lemma. Since $\lim a_k = 0$ and s_k changes sign infinitely often it also implies the first assertion.

RESUMPTION OF THE PROOF OF THE THEOREM: Let us define $F(N,\delta)$, for $N \in \mathbf{N}$, $\delta > 0$, to be the subset of X of all x with the property that for any $m \geq N$ we can find $n > m$ and $\epsilon_i = \pm 1$, $h_i \in H$, $(m+1 \leq i \leq n)$ such that

$$\|x + \sum_{i=m+1}^{n} \epsilon_i a_i h_i\| < \delta,$$

and

$$\|x + \sum_{i=m+1}^{k} \epsilon_i a_i h_i\| \leq \|x\| + \delta \qquad m+1 \leq k \leq n.$$

Note that the h_i are not required to be distinct.

Next we define $F = \cap_{\delta>0} \cup_{N\in\mathbf{N}} F(N,\delta)$. We note first that F is easily seen to be closed. Define $E = \{x : \alpha x \in F \ \forall \alpha \in \mathbf{R}\}$; E is also closed.

We show E contains H. In fact suppose $h \in H$, and $\alpha \in \mathbf{R}$. For arbitrary $\delta > 0$ we pick N so large that $b_N \|h\| < |\alpha|\|h\| + \delta$. If $m \geq N$, we pick ϵ_i according to the Lemma, applied to α and $m = N$, stopping at n where $|s_n|\|h\| < \delta$. Letting $h_i = h$, $(m+1 \leq i \leq n)$ we see that $H \subset E$.

Next we claim that E is a linear subspace. In fact, it is only necessary to show that if $x \in E$ and $y \in E$ then $x+y \in F$. Suppose $\delta > 0$. Let $M = \max(\|x\|, \|y\|)$ and then choose an integer s so large that $6M < s\delta$. Next choose N so that $s^{-1}x, s^{-1}y \in F(N, \delta/(4s))$. Now suppose $m \geq N$. Set $p_0 = m$; then we may inductively define q_k, $(1 \leq k \leq s)$, p_k, $(1 \leq k \leq s)$, ϵ_i, $(p_0 + 1 \leq i \leq p_s)$ and $h_i \in H$, $(p_0 + 1 \leq i \leq p_s)$ so that $p_{k-1} < q_k < p_k$, $(1 \leq k \leq s)$,

$$\|\sum_{p_{k-1}+1}^{q_k} \epsilon_i a_i h_i + s^{-1}x\| < \frac{\delta}{4s}$$

$$\|\sum_{p_{k-1}+1}^{j} \epsilon_i a_i h_i + s^{-1}x\| < \frac{(4\|x\| + \delta)}{4s},$$

for $1 \leq k \leq s$ and $p_{k-1} + 1 \leq j \leq q_k$, and

$$\|\sum_{q_k+1}^{p_k} \epsilon_i a_i h_i + s^{-1}y\| < \frac{\delta}{4s}$$

$$\|\sum_{q_k+1}^{j} \epsilon_i a_i h_i + s^{-1}y\| < \frac{(4\|y\| + \delta)}{4s},$$

for $1 \leq k \leq s$ and $q_k + 1 \leq j \leq p_k$.

Then

$$\|x + y + \sum_{m+1}^{p_s} \epsilon_i a_i h_i\| < \delta.$$

If $p_{k-1} + 1 \leq j \leq q_k$ then

$$\|x + y + \sum_{m+1}^{j} \epsilon_i a_i h_i\| < \frac{s-k+1}{s}\|x+y\| + \frac{(k-1)\delta}{2s} + \frac{(4\|x\| + \delta)}{4s}.$$

If $q_k + 1 \leq j \leq p_k$ then

$$\|x + y + \sum_{m+1}^{j} \epsilon_i a_i h_i\| < \frac{s-k+1}{s}\|x+y\| + \frac{(k-1)\delta}{2s} + \frac{(4\|x\| + \delta)}{4s} + \frac{(4\|y\| + \delta)}{4s}.$$

In either case we conclude that

$$\|x + y + \sum_{m+1}^{j} \epsilon_i h_i\| < \|x+y\| + \delta$$

so that $x + y \in F(N, \delta)$.

It now follows immediately that $E = X$. Now fix any $h_0 \in H$ and let $\gamma = \|h_0\|$. Then, we claim (*) that for any $x \in X$, $m \in \mathbf{N}$ and $\delta > 0$ we can find $n > m$, $h_i \in H$, $m+1 \leq i \leq n$ and $\epsilon_i = \pm 1$, $m+1 \leq i \leq n$ so that

$$\|x + \sum_{m+1}^{n} \epsilon_i a_i h_i\| < \delta$$

and, for $m + 1 \leq j \leq n$,

$$\|x + \sum_{m+1}^{j} \epsilon_i a_i h_i\| < \gamma b_m + \|x\| + \delta.$$

In fact there exists N so that $x \in F(N, \delta/2)$. By the Lemma, applied to $\alpha = 0$ and m, we may find ϵ_i, $m + 1 \leq i \leq k$ where $k \geq N$ so that

$$|\sum_{m+1}^{k} \epsilon_i a_i| < \frac{\delta}{2\gamma}$$

and

$$|\sum_{m+1}^{j} \epsilon_i a_i| < b_m$$

for $m+1 \leq j \leq k$. Now choose $n > k$ and $h_i \in H$, $\epsilon_i = \pm 1$, $(k+1 \leq i \leq n)$ so that

$$\|\sum_{k+1}^{n} \epsilon_i a_i h_i + x\| < \frac{\delta}{2}$$

and

$$\|\sum_{k+1}^{j} \epsilon_i a_i h_i + x\| < \|x\| + \frac{\delta}{2}$$

for $k+1 \leq j \leq n$. We now put $h_i = h_0$ for $m+1 \leq i \leq k$ and our claim is substantiated.

We now may complete the proof of Theorem 3. Suppose $x \in X$ is fixed. Let $p_0 = 0$. Since H is the set of accumulation points of G, we may inductively choose signs ϵ_i and $g_i \in G$, $(p_{k-1}+1 \leq i \leq p_k)$ so that g_i, $(1 \leq i \leq p_k)$ are distinct,

$$\|\sum_{i=1}^{p_k} \epsilon_i a_i g_i - x\| < 2^{-k}$$

for $k \geq 1$ and if $p_{k-1}+1 \leq j \leq p_k$,

$$\begin{aligned}\|\sum_{i=1}^{j} \epsilon_i a_i g_i - x\| &< \|\sum_{i=1}^{p_{k-1}} \epsilon_i a_i g_i - x\| + 2^{-k} + \gamma b_{p_{k-1}} \\ &< 4.2^{-k} + \gamma b_{p_{k-1}}\end{aligned}$$

for $k \geq 1$. The series constructed in this way converges to x and the proof is complete.

References.

1. D. H. Fremlin and A. Sersouri, On ω-independence in separable Banach spaces, Quart. J. Math. to appear.

Department of Mathematics

University of Missouri-Columbia

Columbia

Missouri 65211.

Contemporary Mathematics
Volume **85**, 1989

ON THE WEAK*–FIXED POINT PROPERTY

M.A. Khamsi

Abstract. We establish the Karlovitz's lemma for a contraction map T (i.e. $||T(x) - T(y)|| \leq ||x - y||$) in a dual Banach space X^*, when X has a shrinking strongly monotone basis. This gives a proof which shows that the dual of Banach space, with a shrinking 1–unconditional basis, has weak*–fixed point property.

INTRODUCTION.

Let X be a Banach space, and C a nonempty convex subset of X. We say that C has the fixed point property (f.p.p) if and only if every contraction map T, which leaves C invariant, has a non–empty fixed point set, and we say that X has the weak–f.p.p. (weak*–f.p.p.) if and only if every weakly compact (weak* compact convex subset has f.p.p.

I would like to thank B. Baillom, G. Godefroy and P. Turpin for some useful discussion regarding this paper.

1980 Mathematics Subject Classification (1985 Revision): Primary 47H110, Secondary 46B05

Key Words and Phrases: Nonexpansive mappings, fixed point property.

Suppose X fails to have the weak–f.p.p. Then there exists a weakly compact convex subset C and a contraction T, which leaves C invariant, without a fixed point. Using classical Zorn's lemma, there exists non–empty closed convex subset K of C which is invariant and minimal under T. This convex minimal K has two very nice properties. The first is due to W.A. Kirk. [5]

Lemma (Ki): The minimal convex set K is abnormal in the sense:

$$\sup_{y\in K} \|x - y\| = \mathrm{diam}(K) \quad \text{for every } x \text{ in } K$$

The second property of the minimal convex K is well known as Karlovitz' lemma. It seems that this property was known and proved by Goebel [11].

Lemma (Ka): Let (x_n) be an approximate fixed point sequence for T (i.e. $\lim_{n\to\infty} \|x_n - Tx_n\| = 0$). Then for all $n \in K$, there holds:

$$\lim_{n\to\infty} \|x_n - x\| = \mathrm{diam}(K).$$

Let us recall, that if C is any closed convex subset of X and $T\colon C \longrightarrow C$ is nonexpansive then there exists a sequence (x_n) in C, which is approximate fixed point sequence. Karlovitz proved this fact and used it to show that the R.C. James' renorming of l_2 has f.p.p.

Recent developments of the theory of f.p.p., which includes B. Maurey's result [10] and P.K. Lin's theorems [8], shows that the Karlovitz' lemma is fundamental. Our purpose here is to study the validity of Karlovitz' lemma in the dual case.

THE WEAK*–PROPERTIES OF THE MINIMAL CONVEX K

We shall suppose, without loss of generality, that X is a separable dual space which we denote X^*.

Suppose that X^* fails to have the weak*–f.p.p. As in the case for the weak–f.p.p., there exists a non empty weak*–compact convex subset K, which is minimal for a contraction map.

Lemma 1. The minimal weak*–compact convex set K satisfies the following properties:

1) $\overline{\text{conv}}^*(T(K)) = K$;

2) $\sup_{y \in K} \|x - y\| = \text{diam}(K)$ for every x in K.

The proof is the same as in the weak case.

In view of lemma 1, it is natural to ask if Karlovitz' lemma is valid in a dual space.

While it is not known if this is true in general case, we have positive results in two settings.

TWO IMPORTANT CASES

First Case. Recall that X is a stable [6] Banach space if and only if for all bounded sequences (x_n), (y_n) and all ultrafilters $\mathcal{U}, \mathcal{V}$ on N, we have:

$$\lim_{n,\mathcal{U}} \lim_{m,\mathcal{V}} \|x_n+y_m\| = \lim_{m,\mathcal{V}} \lim_{n,\mathcal{U}} \|x_n + y_m\|.$$

A mapping σ: $X \longrightarrow \mathbb{R}$ σ is said to be a type [6] on X if and only if there exist bounded sequence (x_n) and an ultrafilter $\mathcal{U}$ on N such that:

$$\sigma(x) = \lim_{\mathcal{U}} \|x_n+x\|.$$

Sublemma. Suppose X^* is stable and let σ be a type on X^*. Then σ is weak*–lower semi continuous.

Proof. Let $\beta > \inf_{x \in K} \sigma(x)$. We have to prove that

$$C_\beta = \{x;\ \sigma(x) \le \beta\} \text{ is weak}^*\text{–closed.}$$

Let (y_n) be a weak*–convergent sequence (to y) with $y_n \in C_\beta$, and let $\mathcal{V}$ be an ultrafilter on N. Then

$$\lim_{n,\mathcal{V}} \sigma(y_n) = \lim_{n,\mathcal{V}} \lim_{m,\mathcal{U}} \|x_m+y_n\|.$$

The stability of X^* implies

$$(*) \qquad \lim_{n,\mathcal{V}} \sigma(y_n) = \lim_{m,\mathcal{U}} \lim_{n,\mathcal{V}} \|x_m+y_n\|.$$

Since (y_n) is weak*–convergent to y, then

$$\|x_m+y\| \leq \lim_{n,\mathcal{V}} \|x_m+y_n\|.$$

Thus, using (*), we have $\sigma(y) \leq \lim_{n,\mathcal{V}} \sigma(y_n)$. Since $\sigma(y_n) \leq \beta$ for every n, then $\sigma(y) \leq \beta$. We deduce that $y \in C_\beta$.

From the sublemma, we obtain Karlovitz' lemma in stable Banach spaces. Indeed, suppose that X^* is stable and consider the convex K, which is minimal for T and satisfies the hypothesis of lemma 1. Let (x_n) be an a.f.p.s. for T and define $\sigma(x) = \lim_{\mathcal{U}} \|x_n-x\|$, where $\mathcal{U}$ is an ultrafilter over N. For $\epsilon > 0$, consider $C_\epsilon = \{x; \sigma(x) \leq \inf \sigma + \epsilon\}$. Since C_ϵ is a non void convex subset of K, the sublemma implies that C_ϵ is weak*–closed. Since (x_n) is an a.f.p.s., then $\sigma(Tx) \leq \sigma(x)$. This implies that C_ϵ is invariant under T. By minimality of K, we have $K = C_\epsilon$ for every $\epsilon > 0$. We deduce that σ is constant on K. Since K is weak*–compact, the weak*–limit x of (x_n) over $\mathcal{U}$ exists. For every y in X^* we have: $\|x-y\| \leq \lim_{\mathcal{U}} \|x_n-y\|$. Then $\sup_{y \in K} \|x-y\| \leq \sigma$ and by the lemma 1 $\sigma(x) = \operatorname{diam} K$ for every x in K. And since $\mathcal{U}$ is arbitrary, we obtain:

Proposition 1. Suppose that X^* is stable and fails to have the weak*–f.p.p. Let K be a minimal weak*–compact convex subset of X^* for a contraction map T. Then, for any x in K and every a.f.p.s.,

$$\lim_{n\to\infty} \|x_n-x\| = \operatorname{diam} K.$$

Remark. Aldous [1] has shown, that every subspace of L_1 is stable and B. Maurey proved in [10] that every reflexive subspace of L_1 has the weak–f.p.p. We do not know if the dual subspace of L_1, with unique predual, has the weak*–f.p.p.

Second case. Let X be a Banach space with a Schauder basis (e_i). We suppose that (e_i) is shrinking (if this is not the case, we may replace X^* by the subspace generated by (e_i^*), the biorthogonal system of (e_i)). Let F be an interval of N, and define P_F to be the natural projection associated to F (i.e. $P_F\left[\sum \beta_i e_i\right] = \sum_{i\in F} \beta_i e_i$). We shall say that (e_i) is strongly monotone if and only if:

$$||P_F|| = ||I-P_F|| = 1 \text{ for every } F.$$

It is easy to see that if (e_i) is strongly monotone, then (e_i^*) is also strongly monotone. Suppose that, in this case, X^* fails to have the weak*–f.p.p. Let K be a minimal weak*–compact convex set for a contraction map T and let (x_n) be an a.f.p.s. for T.

Let x be in K, and s a cluster point of $(||x_n - x||)$. Then there exists a subsequence $(||x_{n'} - x||)$ such that $\lim_{n' +\infty} ||x_{n'} - x|| = s$.

Let $\mathcal{U}$ be an ultrafilter over N and x_0 the weak*–limit, over $\mathcal{U}$, of the sequence $(x_{n'})$. Since (e_i^*) is a basis of X^*, a classical argument shows that there exists a sequence (u_n) of successive blocks (related to (e_i^*)) and a subsequence $(x_{n''})$ such that:

$$(**) \qquad \lim_{n'',\mathcal{U}} ||x_{n''} - u_{n''} - x_0|| = 0.$$

We define the function r on K by:

$$r(y) = \lim_{n'',\mathcal{U}} \|x_{n''}-y\| = \lim_{n'',\mathcal{U}} \|u_{n''}+x_0-y\|.$$

Let $\beta > 0$ and consider the set: $C_\beta = \{y \in K;\ r(y) \leq \inf r + \beta\}$. C_β is non–empty, convex and invariant under T (since $(x_{n''})$ is an a.f.p.s. for T we have $r(Ty) \leq r(y)$ for every y in K).

We now prove that C_β is weak*–closed. Consider a weak*–convergent sequence (y_n), $y_n \in C_\beta$, with limit y. By a classical argument, there exists a sequence (v_i) of successive blocks and a subsequence (y_{n_i}) such that:

$$\lim_{i\to\infty} \|y_{n_i}-y-v_i\| = 0.$$

Since r is continuous, we may suppose, without loss of generality, that y and x_0 have finite support. Since the basis (e_i^*) is strongly monotone, for j'' large enough, we have:

$$\|x_0+u_{j''}-y\| \leq \|x_0+u_{j''}-y-v_i\|.$$

We deduce that $\lim_{j'',\mathcal{U}} \|x_0+u_{j''}-y\| \leq \lim_{j'',\mathcal{U}} \|x_0+u_{j''}-y-v_i\|$. By (**), we have $r(y) \leq e(y_{n_i}) + \|y+v_i-y_{n_i}\|$. Since $y_{n_i} \in C_\beta$, if i tends to ∞, we obtain:

$$r(y) \leq \inf r + \beta, \text{ i.e. } y \in C_\beta.$$

Then we conclude, by minimality of K, that $C_\beta = K$ (for every $\beta > 0$) and thus r is constant on K.

Since $(x_{n''})$ converges weak* to x_0 over $\mathcal{U}$, we have for every y in K:

$$\|y-x_0\| \le \lim_{n'',\mathcal{U}} \|y-x_{n''}\| = r(y) = r.$$

Using the lemma 1, we deduce that $r = \text{diam}(K) = r(y)$ for every $y \in K$. But $r(x) = s$, so diam(K) is the unique cluster point of $(\|x_n-x\|)$. Thus we obtain:

Proposition 2. Suppose that X has shrinking strongly monotone basis, and that X^* fails to have the weak*–fixed point property. Let K be a minimal weak*–compact convex subset (of X^*) for a contraction map T, and let (x_n) be an a.f.p.s. for T (in K), Then for all x in K:

$$\lim_{n\to\infty} \|x_n-x\| = \text{diam}(K).$$

Translation into ultrapower language.

Let X be a Banach space and $\mathcal{U}$ be a free ultrafilter over N. The ultrapower space $\tilde{X}$ of X is the quotient space $\ell_\infty(X)/\mathcal{N}$, where

$$l_\infty(X) = \{(x_n);\ x_n \in X \text{ for all } n \text{ and } \|(x_n)\|_{l_\infty(X)} = \text{Sup}\, \|x_n\| < \infty\}$$

and $\mathcal{N} = \{(x_n) \in l_\infty(X);\ \lim_{n,\mathcal{U}} \|x_n\| = 0\}$.

We shall not distinguish between $(x_n) \in l_\infty(X)$ and the coset $(x_n) + \mathcal{N} \in \tilde{X}$. Clearly $\|(x_n)\|_{\tilde{X}} = \lim_{\mathcal{U}} \|x_n\|_X$. It is also clear that X is isometric to a subspace of $\tilde{X}$ by the mapping: $x \longrightarrow (x,x,\ldots)$. Hence, we may assume that X is a subspace of $\tilde{X}$. We will write $\tilde{x}, \tilde{y}, \tilde{z}$ for the general elements of $\tilde{X}$ and x, y, z for the general elements of X. The ultrapower space $\tilde{X}$ is quite useful because it has a nice extension property: let

A be a subset of X and T: $A \longrightarrow X$. We define the subset $\tilde{A}$ and the map $\tilde{T}: \tilde{A} \longrightarrow \tilde{X}$ by:

$$\tilde{A} = \{(x_n) \in \tilde{X};\ x_n \in A \text{ for every } n\} \text{ and } \tilde{T}(x) = \tilde{T}(x_n) = (Tx_n).$$

Many properties, satisfied by A and T, remain satisfied by $\tilde{A}$ and we now translate Karlovitz' lemma into ultrapower language. Let K be a convex weak*–compact subset of X^* which is minimal for a contraction map T, and let $\tilde{K}$ and $\tilde{T}$ be as above. Then clearly $\tilde{K}$ is closed convex subset of $(X^*)^\sim$ with $\operatorname{diam}(\tilde{K}) = \operatorname{diam}(K)$, and $\tilde{T}$ is a contraction map on $\tilde{K}$. Furthermore, $\tilde{T}$ has fixed points in $\tilde{K}$. Indeed, if (x_n) is an a.f.p.s. for T in K, then:

$$\|\tilde{T}(x_n) - (x_n)\|_{(X^*)^\sim} = \lim_{n,\mathcal{U}} \|T(x_n) - x_n\|_{X^*} = 0,$$

and hence, $\tilde{T}(x_n) = (x_n)$. Also, if $\tilde{T}(y_n) = (y_n)$, then some subsequence of (y_n) is an a.f.p.s. for T.

In the ultrapower language, Karlovitz' lemma becomes:

<u>Proposition 3</u>. Let K be a convex weak*–compact subset of X^* which is minimal for a contraction map T. If $\tilde{x}$ is a fixed point of $\tilde{T}$ in $\tilde{K}$, then for every x in K:

$$\|\tilde{x} - x\|_{(X^*)^\sim} = \operatorname{diam}(K).$$

In other words, if $(\tilde{w}_n)$ is an a.f.p.s. of $\tilde{T}$ in $\tilde{K}$, then:

$$\lim_{n\to\infty} \|\tilde{w}_n - x\|_{(X^*)^\sim} = \mathrm{diam}(K) \text{ for every } x \text{ in } K.$$

We show how we prove the second part of the lemma. Let $(\tilde{w}_n)$ be an a.f.p.s. of $\tilde{T}$, and consider $r(x) = \lim_n\text{-inf}\,\|x - \tilde{w}_n\|$. Let $\beta > 0$. Then there exists n_β such that for every $n \geq n_\beta$, $\|\tilde{w} - x\| \leq r(x) + \beta$, and since $\lim_n \|\tilde{w}_n - \tilde{T}(\tilde{w}_n)\| = 0$ there exists N_β such that for every $n \geq N_\beta$: $\|\tilde{w}_n - \tilde{T}(\tilde{w}_n)\| \leq \beta$.

We shall write $\tilde{w}_n = (w_n(m))$. Hence the set of the integers m such that

$$\|w_n - x\| \leq r(x) + 2\beta \text{ and } \|w_n(m) - T(w_n(m))\| \leq 2\beta,$$

is infinite. Thus we may construct a sequence $(w_n(m_n))$ such that:

$$\lim_n\text{-sup}\,\|w_n(m_n) - x\| \leq r(x)$$

and

$$\lim_n \|w_n(m_n) - T(w_n(m_n))\| = 0.$$

Since Karlotivtz′ lemma implies that $\mathrm{diam}(K) \leq r(x)$, we obtain

$$\lim_{n\to\infty} \|\tilde{w}_n - x\|_{\tilde{X}^*} = \mathrm{diam}(K).$$

From this lemma, we deduce the important result which was noticed by P.K. Lin [8].

<u>Sublemma</u>. Under the hypothesis of the preceding lemma, we have the following:

Let $\tilde{W}$ be a closed non–empty convex subset of $\tilde{K}$ which is invariant under $\tilde{T}$. Then for every x in K:

$$\operatorname*{Sup}_{\tilde{w}\in \tilde{W}} \|\tilde{w} - x\| = \operatorname{diam}(K).$$

SOME NEW RESULTS

Since the proofs of the following theorems are the same as the original ones we shall give just the statements of the theorems. Suppose that X has an unconditional basis (e_n). Associated to (e_n) are two constants (see for more details on Schauder basis the excellent book [9]):

$$\lambda = \operatorname*{Sup}_{\substack{\|\Sigma\beta_i e_i\|\le 1 \\ r_i = \pm 1}} \|\Sigma r_i \beta_i e_i\|$$

$$c = \operatorname*{Sup}_{F\subset N} \|P_F\| \quad \text{with } P_F(\Sigma\beta_i e_i) = \sum_{i\in F} \beta_i e_i.$$

We say that the basis (e_i) is 1–unconditional if $\lambda = 1$ and a suppression unconditional basis if $c = 1$.

Our first theorem states:

Theorem 1[7]. Let X be a Banach space with 1–unconditional basis which is shrinking. Then X (resp. X^*) has the weak–f.p.p. (resp. weak*–f.p.p.).

Remark. We deduce from this theorem that l_1 has the weak*–f.p.p.

Theorem 2[4]. Let X be a Banach space with a suppression unconditional basis which is shrinking. Suppose that X^* has the Alternate–Banach–Saks property (A.B.S. property). Then X^* has the weak*–f.p.p.

Recall that a Banach space X has A.B.S. iff for every bounded sequence (x_n) in X, there exists a subsequence (x_{n_k}) such that the sequence $(k^{-1} \sum_{i \leq k} (-1)^i x_{n_i})$ is convergent. B. Beauzamy has shown in [2] that X has A.B.S. property iff X does not have a spreading model which is isomorphic to l_1. Moreover if X is separable, then X^* has a spreading model isomorphic to l_1 iff X has a quotient which has a spreading model isomorphic to c_0. Our last theorem is:

Theorem 3[4]. Let X be a Banach space with a suppression unconditional basis which is shrinking. Suppose that the constant $\lambda > 2$. Then X (resp. X^*) has the weak–f.p.p. (resp. weak*–f.p.p.).

REFERENCES

[1] D. Aldous, Subspaces of L_1 via Random Measure, Trans. Amer. Math. Soc., 258, (1981).

[2] B. Beauzamy and J.T. Lapreste, Modèles étalès des espaces de Banach, Publication of Department of Mathematics, Université Claude Bernard Lyon (France).

[3] L.A. Karlovitz, Existence of fixed point for non–expansive map in space without normal structure, Pacif. J. Math. 66 (1976), 153–159.

[4] M.A. Khamsi, La properiété du point fixe dans les espaces de Banach avec base inconditionnelle. To appear.

[5] W.A. Kirk, A fixed point theorem for mappings which do not increase distance, Amer. Math. Monthly 72 (1965), 1004–1006 MR 32=6436.

[6] J.L. Krivine and B. Maurey, Espaces de Banach stables, Israel Journal of Math. 39 (1981).

[7] P.K. Lin, Unconditional bases and fixed points of nonexpansive mappings, Pacif. J. Math., Vol. 116, No. 1 (1985), 69–76.

[8] P.K. Lin, Remarks on the fixed point problem for nonexpansive maps II, Longhorn Notes, Texas functional Analysis Seminar, (1982–1983), The University of Texas.

[9] J. Lindenstrauss and L. Tzafriri, Classical Banach spaces–I–Sequence spaces, Springer–Verlag, 1977.

[10] B. Maurey, Points fixes des contractions sur un convexe fermé de L_1 Séminarie d'Analyse Fonctionnelle, 80–61, Ecole Polytechnique, Palaiseau.

[11] K. Goebel, In the structure of minimal invariant sets for nonexpansive mappings, Annales Universitat Mariae Curie–Sklodowska. LUBLIN–POLNIA, Vol. XXIX, 9 (1975), 73–77.

M.A. Khamsi
Equipe d'Analyse
Equipe de Recherche associée au
C.N.R.S. No. 294
Universite Paris VI
4, place Jussieu
75230 PARIS CEDEX 05
Tour 46 4ème étage

Current address:
Department of Mathematics
University of Southern California
Los Angeles, CA 90007

Contemporary Mathematics
Volume **85**, 1989

SOME GEOMETRIC AND TOPOLOGICAL PROPERTIES OF THE UNIT SPHERE IN A NORMED LINEAR SPACE*

Bor–Luh Lin, Pei–Kei Lin[1] and S.L. Troyanski[2]

ABSTRACT. Let K be a bounded closed convex set of a normed linear space. An example is given that there is a point x of K which is an extreme point and also is a point of continuity of K but x is not a denting point of K.

Let X be a normed linear space and let $\tilde{X}$ denote the completion of X. For a bounded closed convex set K in X, an element x in K is called a denting point of K if for all $\epsilon > 0$, $x \notin \overline{co}(K \backslash B(x,\epsilon))$ where $B(x,\epsilon) = \{y : y \in X, \|y-x\| < \epsilon\}$. x is a PC (point of continuity) for K if the identity mapping: $(K,\text{weak}) \longrightarrow (K,\|\cdot\|)$ is continuous at x. x is called a very strong extreme point of K if for every sequence $\{x_n\}$ of K–valued Bochner integrable functions on [0,1], the condition $\lim_{n\to\infty} \|\int_0^1 x_n(t)dt-x\| = 0$ implies $\lim_{n\to\infty} \int_0^1 \|x_n(t)-x\|dt = 0$. In [LLT 3], we have proved the following results.

1980 Mathematics Subject Classification (1985 Revision): Primary 46B20.

Key words and phrases: Denting point, extreme point, normed linear space, point of continuity.

*Research partially supported by a grant from NSF to the Research Workshop of Banach Space Theory, University of Iowa, July 5–July 25, 1987.

[1]Research supported by NSF DMS–8514497.

[2]Research supported by the Science Committee of Bulgaria, Contract No. 54/2703, 1987.

__Theorem 1__. Let x be an element in a bounded closed convex set K of a Banach space. Then the following are equivalent:

(i) x is a denting point of K,

(ii) x is a very strong extreme point of K,

(iii) x is a PC for K and x is an extreme point of K.

Now, let x be an element in a bounded closed convex set K of a normed linear space X. Let $\tilde{K}$ be the closure of K in $\tilde{X}$. It is easy to see that x is a denting point (resp. very strong extreme point; PC) of K if and only if x is a denting point (resp. very strong extreme point; PC) of $\tilde{K}$. By Theorem 1, it follows that

__Theorem 2__. Let x be an element in a bounded closed convex set K of a normed linear space X. Then the following are equivalent:

(i) x is a denting point of K,

(ii) x is a very strong extreme point of K,

and each of (i) or (ii) implies

(iii) x is a PC for K and x is an extreme point of K,

In the case when x is an extreme point of $\tilde{K}$ then (iii) implies (i).

__Corollary__. In a normed linear space X, the following are equivalent:

(i) X has property (G), i.e., every point on the unit sphere of X is a denting point of the closed unit ball of X.

(ii) X is p–average locally uniformly rotund (p–ALUR), $i < p < +\infty$, i.e., if for every $x \in X$ and $x_n \in L^p(\lambda,X)$ where λ is the Lebesgue measure on $[0,1]$, such that $\lim_n \|x+x_n\|_p = \|x\|$ and the expectation $Ex_n = 0$, then $\lim_n \|x_n\|_p = 0$,

and each of (i) or (ii) implies

(iii) X is strictly convex and X has property (K), i.e., the norm topology and the weak topology coincide on the unit sphere of X.

In [LLT 1], it was proved that (i), (ii), (iii) are equivalent in Banach spaces. However, if X is a normed linear space which is not complete, then the property (iii) does not imply (i) or (ii). In fact, there exists a strictly convex normed linear space X for which property (K) holds, but X is not even midpoint point locally uniformly convex (MLUR). Recall that a normed linear space X is said to be (MLUR) if for any sequences $\{x_n\}$, $\{y_n\}$, and x in X such that $\|x_n\| \le 1$, $\|y_n\| \le 1$, $\|x\| = 1$, and $\lim_n \|2x-(x_n+y_n)\| = 0$ then $\lim_n \|x_n-y_n\| = 0$.

<u>Example</u>. Let X be the space of all sequences of real numbers with only finitely many nonzero members. Let $\{e_i, f_i\}_{i=1}^{\infty}$ be the unit vector basis of X. Put $g_1 = f_1$, $h_1 = \sup_{i\geq 1} |f_i(x)|$, $g_i = 2f_{i+1}-f_i$, $h_i(x) = \sup_{k\geq i} |g_k(x)|$, $i = 2,3,\cdots$. The norm of X is defined by

$$\|x\| = \left\{\sum_{i=1}^{\infty} 2^{-i}(g_i^2(x)+h_i^2(x)\right\}^{1/2}.$$

Let $x_n = \left[1+\frac{1}{2^{2n+1}}\right]^{-1/2}\left[e_1+\sum_{i=2}^{n} 2^{-i}e_i\right]$ and $y_n = \left[1+\frac{1}{2^{2n+1}}\right]^{-1/2}\left[e_1-\sum_{i=2}^{n} 2^{-i}e_i\right]$.

It is easy to check that $\|x_n\| = \|y_n\| = \|e_1\| = 1$ and $\lim_n \|2e_1-(x_n+y_n)\| = 0$ but $\lim_n \|x_n-y_n\| = 2^{-3/2}$. Hence X is not (MLUR).

Let $\|x\| = \|y\| = \frac{1}{2}\|x+y\|$. Then by convexity argument we obtain $g_i(x-y) = 0$, $i = 1,2,\cdots$. But $f_i(x-y) = 0$ for i sufficiently large, hence $f_i(x-y) = 0$ for all i. Therefore $x = y$, i.e., X is strictly convex.

The idea of the proof that X has property (K) goes back to Kadec [K]. We shall prove that $\tilde{X}$ has property (K). Let $\|x_n\| = \|x\| = 1$ and $\lim_{n\to\infty} f_i(x_n) = f_i(x)$ for all $i \in N$. Then

$$\lim_{n\to\infty} g_i(x_n) = g_i(x), \quad i \in N.$$

So

$$h_i(x) = \sup_{k\geq i} |g_k(x)| = \sup_{k\geq i} |\lim_{n\to\infty} g_k(x_n)|$$

$$= \sup_{k\geq i} \lim_{n\to\infty} |g_k(x_n)| \leq \lim_{n\to\infty} \inf \sup_{k\geq i} |g_k(x_n)|.$$

Since

$$\|x\| = \left\{\sum_{i=1}^{\infty} 2^{-i}(g_i^2(x) + h_i^2(x))\right\}^{1/2} \quad \text{and} \quad \|x_n\| = \|x\|,$$

we conclude that $\lim_{n\to\infty} h_i(x_n) = h_i(x)$ for all $i \in N$. Given $\epsilon > 0$, let k be an integer such that $f_i(x) < \epsilon$ for $i \geq k$. So we have

$$|f_i(x_n - x)| > \epsilon, \quad 1 \leq i \leq k,$$

$$h_k(x) = \sup_{i\geq k} |2f_{i+1}(x) - f_i(x)| \leq 3\epsilon.$$

Since $\lim_{n\to\infty} h_k(x_n)$, there exists N such that if $n > N$ then

$$|h_k(x_n) - h_k(x)| < \epsilon \quad \text{and} \quad h_k(x_n) < 4\epsilon.$$

Then for $i \geq k$ and $n > N$, we have $|g_i(x_n)| < 4\epsilon$. Thus for $j > k$ and $n > N$,

$$|2^{j-k}f_j(x_n) - f_k(x_n)| \leq \sum_{i=k}^{j-1} 2^{i-k}|g_i(x_n)| < 4(2^{j-k}-1)\epsilon.$$

Since $f_k(x) < \epsilon$, hence $|f_k(x_n)| < 4\epsilon$ for $n > N$. Therefore for $j > k$ and $n < N$ we have $|f_j(x_n)| < 4\epsilon$. Hence for $n > N$, $\sup_{i\geq 1}|f_i(x_n-x)| \leq 5\epsilon$. Thus $\|x_n-x\| < 34\epsilon$. So $\tilde{X}$ has property (K).

<u>Remark</u>. Let K be a bounded closed convex set of a normal linear space. It is easy to see that an extreme point of K may not be an extreme point of $\tilde{K}$. The example above shows that there exists a strictly convex normed linear space X such that $\tilde{X}$ is not strictly convex. However, we don't know whether if X is a normed linear space possessing the property (G) (resp. (K)), is $\tilde{X}$ also possessing the property (G) (resp. (K))?

REFERENCES

[K] M.I. Kadec, On spaces isomorphic to locally uniformly rotund spaces, Izvestia vysših učebnyh zavedenii, 6(1959), 51–59 (Russian).

[LLT 1] Bor–Luh Lin, Pei–Kee Lin and S.L. Troyanski, Some geometric and topological properties of the unit sphere in a Banach space, Math. Ann. 274(1986), 613–616.

[LLT 2] Bor–Luh Lin, Pei–Kee Lin, and S.L. Troyanski, A characterization of denting points of a closed convex set, Longhorn Notes, The University of Texas–Austin, 1985–86, pp. 99–101.

[LLT 3] Bor–Luh Lin, Pei–Kee Lin and S.L. Troyanski, Characterizations of denting points, Proc. Amer. Math. Soc. 102(1988), 526–528.

BOR–LUH LIN
DEPARTMENT OF MATHEMATICS
THE UNIVERSITY OF IOWA
IOWA CITY, IA 52242

PEI–KEE LIN
DEPARTMENT OF MATHEMATICS
MEMPHIS STATE UNIVERSITY
MEMPHIS TN 38152

and

S. L. TROYANSKI
DEPARTMENT OF MATHEMATICS
UNIVERSITY OF SOFIA
5 BUL. A. IVANOV
BG1126–SOFIA, BULGARIA

Contemporary Mathematics
Volume **85**, 1989

THE λ-FUNCTION IN BANACH SPACES*

ROBERT H. LOHMAN†

Abstract If X is a Banach space with the λ-property such that the λ-function is locally bounded away from 0 on the unit sphere at each of its points, then each member of the closed unit ball B_X is expressible as a convex series of members of $\text{ext}(B_X)$. A flatness condition which guarantees that the λ-function is a Lipschitz mapping is introduced. It is also shown that the λ-function does not necessarily characterize extreme points of B_X when X is of infinite dimension.

If X is a normed space and x is in the closed unit ball B_X of X, a triple (e, y, λ) is said to be amenable to x in case $e \in \text{ext}(B_X)$, $y \in B_X$, $0 < \lambda \leq 1$ and $x = \lambda e + (1-\lambda)y$. In this case, the number $\lambda(x)$ is defined by

$$\lambda(x) = \sup\{\lambda : (e, y, \lambda) \text{ is amenable to } x\}. \tag{1}$$

X is said to have the λ-property if each $x \in B_X$ admits an amenable triple. If X has the λ-property and $\inf\{\lambda(x) : x \in B_X\} > 0$, X is said to have the uniform λ-property.

These ideas were introduced in [1], where several types of classical spaces were shown to have the λ-property or uniform λ-property and the λ-functions defined by (1) were explicitly calculated for these spaces. These properties have recently been considered for Lorentz sequence spaces, $C(K)$ spaces and other function and sequence spaces ([2],[3],[5]). Consequently, many fundamental normed spaces are now known to have the λ-property or uniform λ-property.

These properties have useful geometric implications. Namely, it was shown in [1] that X has the uniform λ-property if and only if each $x \in B_X$ is expressible as a convex series of members of $\text{ext}(B_X)$ (we'll call this the convex series representation property) and the sequence of partial sums of these convex series converge uniformly on B_X. In looking at analogous results for the λ-property, it has been conjectured that the λ-property is equivalent to the convex series representation property. It is clear that the convex series representation property implies the λ-property and it was shown in [1] that if a Banach space X has the λ-property, then X is the closed convex hull of $\text{ext}(B_X)$. In this note, we prove (Corollary 1.3) that the λ-property implies the convex series representation property if the λ-function is locally bounded away from 0 at each point of the unit sphere S_X of X.

If X has the λ-property and $0 < r < 1$, it was shown in [1] that the λ-function is a Lipschitz mapping on rB_X. Although the λ-functions for such classical spaces as ℓ_1, ℓ_∞ are Lipschitz mappings on their closed unit balls, λ-functions are not, in general, continuous at each point in B_X. Another purpose of this note is to introduce a flatness condition which guarantees that the λ-function is a Lipschitz mapping on B_X (Theorem 2.3). As a consequence, we will see, from a general point of view, why the specific λ-functions for ℓ_1, ℓ_∞, calculated in [1], have Lipschitz constants $\frac{1}{2}$. Another consequence of this flatness condition is the fact that the λ-function in a Banach space whose unit ball is a polyhedron is a Lipschitz mapping. This significantly improves Theorem 2.13 of [1].

*AMS 1980 Mathematics Subject Classification: Primary 46B20; Secondary 46B99.

Key Words and phrases: extreme point, λ-property, convex series, strict convexity, Lipschitz

†Research supported by a Kent State University Summer Research Appointment.

Finally, we show that $\{x \in B_X : \lambda(x) = 1\}$ does not always coincide with $\text{ext}(B_X)$ when X is an infinite-dimensional space. This differs markedly from the situation in a finite-dimensional space (see Theorem 2.10 of [1]) and is related to a potential characterization of strictly convex spaces by means of the λ-function.

O. Notation

If $A \subset B_X$, the λ-function is said to be locally bounded away from 0 on A at each of its points if for each $x \in A$, there is a $\delta > 0$ such that $\inf\{\lambda(z) : z \in A, \|x - z\| < \delta\} > 0$. The convex hull of A is denoted by $co(A)$. The sequence with 1 in the nth coordinate and zeros elsewhere is denoted by e_n. We will also need three auxiliary functions introduced in [1]. Namely, if $u \in S_X$, we let

$$\lambda(u,x) = \sup\{\lambda : 0 \leq \lambda \leq 1 \text{ and } x = \lambda u + (1-\lambda)y \text{ for some } y \in B_X\}, \qquad x \in B_X,$$

$$\alpha(u,x) = \sup\{\alpha : \alpha \geq 0, \|x + \alpha(x-u)\| = 1\}, \qquad x \in B_X\backslash\{u\},$$

$$y(u,x) = x + \alpha(u,x)\,(x-u)\,, \qquad x \in B_X\backslash\{u\}.$$

1. Convex Series Representations

Theorem 1.1 Let X be a Banach space with the λ-property and let F be a closed face of S_X. If the λ-function is locally bounded away from 0 on F at each of its points, then each $x_0 \in F$ is expressible as a convex series of members of $\text{ext}(B_X)$.

Proof. By repeated use of the λ-property, we obtain sequences $(e_k) \subset \text{ext}(B_X)$, $(x_k) \subset S_X$ and (λ_k) such that for $k = 0, 1, 2, \ldots$

$$\text{(a)} \quad x_k = \lambda_{k+1}\, e_{k+1} + (1 - \lambda_{k+1})x_{k+1},$$
$$\text{(b)} \quad 0 < \tfrac{1}{2}\,\lambda(x_k) < \lambda_{k+1} \leq 1.$$

Let $P_0 = 1$ and for $n \geq 1$, write $P_n = \prod_{k=1}^{n} (1 - \lambda_k)$.

From (a), we obtain

$$x_0 = \sum_{k=1}^{n} \lambda_k\, P_{k-1}\, e_k + P_n x_n \tag{2}$$

Since $\sum_{k=1}^{n} \lambda_k P_{k-1} + P_n = 1$ for all n, the proof will be completed by showing that $P_n \to 0$. Now (P_n) is a nonnegative, non-increasing sequence so that $P = \lim_n P_n$ exists. We may assume $P_n > 0$ for all n and hence $\lambda_k < 1$ for all k. Since F is a support of S_X, we have $(x_k) \subset F$. Also, if $x_n \in \text{ext}(B_X)$ for some n, then by (2), we have $x_0 \in \text{co}(\text{ext}(B_X))$ and we are done. Consequently, we may also assume no x_n is an extreme point of B_X. In particular, $e_{k+1} \neq x_{k+1}$ for all k.

Assume, to the contrary, that $P > 0$. Since $P = \prod_{k=1}^{\infty}(1-\lambda_k)$, we have $\sum_{k=1}^{\infty}\lambda_k < \infty$. But for all k,

$$\lambda_{k+1} = \frac{\|x_k - x_{k+1}\|}{\|e_{k+1} - x_{k+1}\|}.$$

Therefore, $\sum_{k=1}^{\infty}\|x_k - x_{k+1}\| \leq 2\sum_{k=1}^{\infty}\lambda_k < \infty$. It follows that (x_k) is a Cauchy sequence in F and hence converges to some point $x \in F$. Since the λ-function is locally bounded away from 0 on F at x, it is the case that $\inf_k \lambda(x_k) > 0$. But $0 < \lambda(x_k) < 2\lambda_{k+1}$ and $\lambda_k \to 0$. The contradiction shows that we have $P = 0$.

Lemma 1.2 Let X be a normed space and let D be a subset of B_X such that $\text{co}(D)$ is dense in B_X. If $\|x\| < 1$, then x is expressible as a convex series of members of D.

Proof. If $z \in B_X$ and $\epsilon > 0$, a standard argument (see Proposition I.2.13 of [4]) shows that we can write $z = \sum_{k=1}^{\infty}\lambda'_k z_k$, where $z_k \in \text{co}(D)$, $\lambda'_k \geq 0$, $\sum_{k=1}^{\infty}\lambda'_k \leq 1+\epsilon$. Write each z_k as a convex combination of members of D and ungroup these terms (by absolute convergence) to obtain $\left(\sum_{k=1}^{\infty}\lambda'_k\right)^{-1} z = \sum_{k=1}^{\infty}\lambda_k x_k$, where $x_k \in D$, $\lambda_k \geq 0$, $\sum_{k=1}^{\infty}\lambda_k = 1$, completing the proof.

Corollary 1.3 Let X be a Banach space with the λ-property. If the λ-function is locally bounded away from 0 on S_X at each of its points, then B_X has the convex series representation property.

Proof. As noted in [1], $\text{co}(\text{ext}(B_X))$ is dense in B_X. By Lemma 1.2, each $x \in B_X$ with $\|x\| < 1$ has the indicated representation. If $\|x\| = 1$, choose $f \in X^*$ such that $\|f\| = f(x) = 1$. Then $F = f^{-1}(1)\bigcap B_X$ is a closed face of S_X containing x. Theorem 1.1 completes the proof.

Corollary 1.4 If X is a Banach space with the λ-property and $\lambda|_{S_X}$ is continuous, then B_X has the convex series representation property.

2. Lipschitz λ-Functions

We now consider a geometric property which will be shown to be sufficient to guarantee that a λ-function is a Lipschitz mapping.

Definition 2.1 Let X be a normed space and let $\delta > 0$. X is said to have the δ-inclination property if for each $e \in \text{ext}(B_X)$ and $x \in B_X$ with $x \neq e$, there exists $f \in X^*$ such that

$\|f\| = f(y(e,x)) = 1$ and $f(e) \leq 1 - \delta$.

Intuitively, the δ-inclination property guarantees that the line segments from extreme points e of B_X to points $y(e,x)$ in B_X are of length at least δ and are uniformly inclined away from a supporting hyperplane for $y(e,x)$. In particular, if X has the δ-inclination property and $e, e' \in \operatorname{ext}(B_X)$ with $e \neq e'$, then $y(e,e') = e'$ and we have $\|e - e'\| \geq \delta$. Thus, X must be a real normed space. In spite of these restrictions, the next result shows that spaces with this property are quite common.

Proposition 2.2

(a) ℓ_1 over the reals has the $2-$ inclination property
(b) ℓ_∞ over the reals has the $2-$ inclination property
(c) If B_X is a polyhedron, then X has the $\delta-$ inclination property for some $\delta > 0$.

Proof. (a) In ℓ_1, let $e = e_1$ and let $x = (x_n) \in B_X$, where $x \neq e$.
Then

$$\begin{aligned} y(e,x) &= x + \alpha(e,x)(x-e) \\ &= (x_1 + \alpha(e,x)(x_1 - 1),\ (1+\alpha(e,x))x_2,\ (1+\alpha(e,x))x_3, \ldots) \end{aligned}$$

so that

$$1 = \|y(e,x)\| = |x_1 + \alpha(e,x)(x_1 - 1)| + (1+\alpha(e,x))(\|x\| - |x_1|) \tag{3}$$

If we can show that $x_1 + \alpha(e,x)(x_1 - 1) \leq 0$, then the functional $f = (-1, sgn\, x_2, sgn\, x_3, \cdots)$ satisfies $\|f\| = f(y(e,x)) = 1$ and $f(e) = -1$, establishing our assertion.

Case I. $x_1 \leq 0$

We clearly have $x_1 + \alpha(e,x)(x_1 - 1) \leq 0$.

Case II. $0 < x_1 < 1$

Suppose $x_1 + \alpha(e,x)(x_1 - 1) > 0$. Then from (3) we obtain $\alpha(e,x)(\|x\| - 1) = 1 - \|x\|$. If $\|x\| < 1$, we contradict the nonnegativity of $\alpha(e,x)$. Thus, $\|x\| = 1$ and we may choose $\eta > 0$ such that $x_1 + (\alpha(e,x) + \eta)(x_1 - 1) > 0$. If we let

$$y_\eta = x + (\alpha(e,x) + \eta)(x - e),$$

then

$$\|y_\eta\| = x_1 + (\alpha(e,x) + \eta)(x_1 - 1) + (1 + \alpha(e,x) + \eta)(1 - x_1) = 1,$$

which contradicts the definition of $\alpha(e,x)$. Therefore, $x_1 + \alpha(e,x)(x_1 - 1) \leq 0$.
The proof for an arbitrary member (i.e., $\pm e_n$) of $\operatorname{ext}(B_{\ell_1})$ is similar.

(b) For the space ℓ_∞, let $e = (1, 1, \ldots)$ and let $x = (x_n) \in B_{\ell_\infty}$, where $x \neq e$. Then $y(e,x) = x + \alpha(e,x)(x-e)$ implies

$$1 = \|y(e,x)\| = \sup_n |x_n + \alpha(e,x)(x_n - 1)|.$$

Write $y(e,x) = (y_n)$. We claim that $\inf_n y_n = -1$. Suppose, to the contrary, that $\inf_n y_n > -1$. then $\eta = \frac{1+\inf_n y_n}{\|x-e\|} > 0$ and for each n, we have

$$1 \geq y_n \geq y_n + \eta(x_n - 1) \geq y_n - (1 + \inf_n y_n) \geq -1.$$

This implies $\|x + (\alpha(e,x) + \eta)(x-e)\| \leq 1$, contradicting the definition of $\alpha(e,x)$ and establishing our claim.

If $y_N = -1$ for some $N, f = -e_N$ satisfies $\|f\| = f(y(e,x)) = 1$ and $f(e) = -1$. Otherwise, there is a subsequence (y_{n_k}) of (y_n) converging to -1. In this case, let $f_k = -e_{n_k}$ for all k. By the weak*-compactness of the unit ball of $\ell_\infty^*, (f_k)$ has a weak*-cluster point f which satisfies $\|f\| = f(y(e,x)) = 1$ and $f(e) = -1$.

That the same conclusion holds for an arbitrary member of $\text{ext}(B_{\ell_\infty})$ follows from the fact that the mapping $(x_n) \to (\varepsilon_n x_n)$ is a linear isometry of ℓ_∞ onto ℓ_∞ for any sequence (ε_n) of ± 1's.

(c) Let B_X be a polyhedron, $e \in \text{ext}(B_X)$ and $x \in B_X$ with $x \neq e$. Then $y(e,x)$ lies on a face F of B_X that does not contain e (see Remark 2.12 of [1]). Therefore, there exists $f \in X^*$, depending on e and F, such that $\|f\| = 1$, $f(F) = \{1\}$ and $f(e) < 1$. Let $\delta = \min(1 - f(e))$, where this minimum is taken over all such pairs (e, F). Since the number of such pairs is finite, $\delta > 0$ and X has the δ-inclination property.

The usefulness of the δ-inclination property lies in the following result whose essential motivation comes from Theorem 2.6 of [1].

Theorem 2.3 Let X be a normed space with the λ-property. If X has the δ-inclination property, then the λ-function is a Lipschitz mapping satisfying

$$|\lambda(x) - \lambda(z)| \leq \delta^{-1}\|x - z\|\,, \qquad x, z \in B_X.$$

Proof. Fix $e \in \text{ext}(B_X)$ and assume $x, z \in B_X$, where $x \neq e \neq z$. Choose f as in Definition 2.1 for e and x. Then

$$\delta \leq 1 - f(e) = f(y(e,x) - e) = (1 + \alpha(e,x))f(x - e)$$

so that $f(x-e) \geq \frac{\delta}{1+\alpha(e,x)}$. Write

$$y(e,z) = x + \alpha(e,z)(x-e) + (1 + \alpha(e,z))(z - x)$$

and obtain

$$\begin{aligned} |\,\|y(e,x)\| - \|x + \alpha(e,z)(x-e)\|\,| &= |\,\|y(e,z)\| - \|x + \alpha(e,z)(x-e)\|\,| \\ &\leq (1 + \alpha(e,z))\|x - z\|. \end{aligned}$$

Then

$$\begin{aligned}(\alpha(e,z)-\alpha(e,x))\,f(x-e) &= f(x+\alpha(e,z)(x-e))-f(x+\alpha(e,x)(x-e))\\ &= f(x+\alpha(e,z)(x-e))-\|y(e,x)\|\\ &\le \|x+\alpha(e,z)(x-e)\|-\|y(e,x)\|\\ &\le (1+\alpha(e,z))\|x-z\|.\end{aligned}$$

Consequently,

$$\alpha(e,z)-\alpha(e,x)\le \delta^{-1}(1+\alpha(e,x))(1+\alpha(e,z))\|x-z\|. \tag{4}$$

Interchanging the roles of x, z and dividing by $(1+\alpha(e,x))(1+\alpha(e,z))$ in (4), we obtain

$$|\lambda(e,x)-\lambda(e,z)|\le \delta^{-1}\|x-z\|. \tag{5}$$

If $z = e$ but $x \neq e$, let $0 < r < 1$. By the proof of Lemma 2.1 of [1], $\lim\limits_{r\to 1^-} \lambda(e, re) = 1$. From (5), we have

$$|\lambda(e,x)-\lambda(e,re)|\le \delta^{-1}\|x-re\|$$

which, by letting $r \to 1^-$, implies

$$|\lambda(e,x)-1|\le \delta^{-1}\|x-e\| = \delta^{-1}\|x-z\| \tag{6}$$

From (5), (6) and the definition of the λ-function, it is straightforward to show

$$|\lambda(x)-\lambda(z)|\le \delta^{-1}\|x-z\|, \qquad x, z \in B_X,$$

completing the proof.

Corollary 2.4 If X is a normed space with the λ-property and satisfies the 2-inclination property, then the λ-function has Lipschitz constant equal to $1/2$.

Proof. By Theorem 2.3, the Lipschitz constant is at most $1/2$. On the other hand, if $e \in \operatorname{ext}(B_X)$, then $|\lambda(e) - \lambda(0)| = 1/2 = \|e - 0\|/2$. Thus, the Lipschitz constant is at least $1/2$.

Remarks 2.5 The formulas for the λ-functions on ℓ_1 and ℓ_∞ over the reals were explicitly calculated in [1]. A direct calculation with these formulas shows that their Lipschitz constants equal $1/2$. That this should have been expected from general principles follows from Proposition 2.2 and Corollary 2.4. In addition, Proposition 2.2 and Theorem 2.3 yield the following strengthening of Theorem 2.13 of [1]:

Corollary 2.6 If X is a finite-dimensional normed space such that B_X is a polyhedron, then the λ-function is a Lipschitz mapping on B_X.

Remark 2.7 We finally wish to observe that the λ-property and δ-inclination property are independent geometric properties. For example, let $X = \left(\oplus \sum\limits_{n=1}^{\infty} \ell_1^n\right)_{\ell_\infty}$. Then X fails to have the λ-property (see [5]), yet X can be shown to have the 2-inclination property. On the other hand, every strictly convex space has the uniform λ-property but fails to have the δ-inclination property for any $\delta > 0$.

3. Strict Convexity and the λ-Function

Let X be a finite-dimensional normed space and let $x \in B_X$. By Theorem 2.10 of [1], we have

$$x \in \text{ext}(B_X) \text{ if and only if } \lambda(x) = 1 \tag{7}$$

Consequently, the values of the λ-function characterize the extreme points of B_X in the finite-dimensional case. It has been asked whether (7) holds in general. If the answer were yes, then we would be able to characterize strictly convex normed spaces in terms of the λ-function. We will show here that (7) fails in general. Thus, individual extreme points cannot be determined by the λ-function in a space with the λ-property.

Example 3.1 In c_0 over the reals, let $E_0 = \{\pm m^{-1}\, e_m \;:\; m \in \mathbb{N}\}$ and $B = \overline{\text{co}}\,(E_0)$. Note that B is compact and $B = \left\{\sum_{m=1}^{\infty} \beta_m\, m^{-1}\, e_m \;:\; \sum_{m=1}^{\infty} |\beta_m| \leq 1\right\}$. Define an equivalent norm on ℓ_∞ by letting the new unit ball be $B_{\ell_\infty} + B$. We let X denote ℓ_∞ with this equivalent norm and now consider some properties of X.

Lemma 3.2 $\text{ext}(B_X) = \{e + \text{ sgn } e(m)\, m^{-1}\, e_m \;:\; e \in \text{ext}(B_{\ell_\infty}),\; m \in \mathbb{N}\}$.

Proof. Write $E_\infty = \text{ext}(B_{\ell_\infty})$. Then the sets B_X, $E_\infty + (E_0 \cup \{0\})$ are weak*-compact and the weak*-closed convex hull of $E_\infty + (E_0 \cup \{0\})$ equals B_X. By Milman's theorem, $\text{ext}(B_X) \subset E_\infty + (E_0 \cup \{0\})$. It is routine to check that if $e \in E_\infty$, then $e + \text{ sgn } e(m) m^{-1}\, e_m \in \text{ext}(B_X)$. On the other hand, if $y \in \text{ext}(B_X)$, we can write $y = e + \varepsilon\, m^{-1}\, e_m$, where $e \in E_\infty$, $m \in \mathbb{N}$ and $\varepsilon = 0$ or ± 1. If $\varepsilon = 0$, it is clear that $y \notin \text{ext}(B_X)$. Similarly, if $\varepsilon \neq \text{ sgn } e(m)$, then $|y(m)| < 1$ and $|y(n)| = 1$ if $n \neq m$. This easily implies $y \notin \text{ext}(B_X)$, completing the proof.

Lemma 3.3 If $e \in E_\infty$ and $\varepsilon > 0$, there is a triple (e', y, λ) amenable to e in X with $\lambda > 1 - \varepsilon$.

Proof. The vector e is of the form $e = ((-1)^{k_m})$, where $k_m = 0$ or 1 for all m. If $m \geq 2$, write $e = (1 - m^{-1})e' + m^{-1}y$, where $e' \in \text{ext}(B_X)$ and $y \in B_X$ are given, respectively, by

$$\left((-1)^{k_1}, \ldots, (-1)^{k_{m-1}},\; 1 + m^{-1},\; (-1)^{k_{m+1}}, \ldots\right)$$

$$\left((-1)^{k_1}, \ldots, (-1)^{k_{m-1}},\; m^{-1},\; (-1)^{k_{m+1}}, \ldots\right)$$

if $k_m = 0$. A similar argument holds if $k_m = 1$. It follows that $(e', y, 1 - m^{-1})$ is amenable to e in X.

Lemma 3.4 X has the λ-property.

Proof. Let $z \in B_X$ and write $z = x + \sum_{m=1}^{\infty} \beta_m\, m^{-1}\, e_m$, where $x \in B_{\ell_\infty}$ and $\sum_{m=1}^{\infty} |\beta_m| \le 1$. Let $M = \{m : |x(m) + \beta_m\, m^{-1}\,| \le 1\}$, $x_M = x + \sum_{m \in M} \beta_m\, m^{-1}\, e_m$ and $M' = \mathbb{N}\backslash M$. By rewriting $z = x_M + \sum_{m\in M'} \beta_m\, m^{-1}\, e_m$, we may assume z has the form $z = x + \sum_{m\in M'} \beta_m\, m^{-1}\, e_m$, where $x \in B_{\ell_\infty}$, $\sum_{m\in M'} |\beta_m| \le 1$ and $m \in M'$ implies $|x(m) + \beta_m\, m^{-1}| > 1$. Thus, $m \in M'$ implies $x(m) \ne 0 \ne \beta_m$ and sgn $x(m) =$ sgn β_m.

Case I. $M' = \phi$

In this case, $z = x \in B_{\ell_\infty}$. If $x(m) = \pm\, 1$ for all m, Lemma 3.3 shows there are triples (e, y, λ) amenable to x in X with λ arbitrarily close to 1. Hence, we may assume $|x_{m_0}| < 1$ for some integer m_0. We can write $x = \frac{1}{2}(e+y)$, where $e \in E_\infty$ and $y \in B_{\ell_\infty}$ are given by

$$\begin{array}{ll} e(m) = y(m) = x(m) & ,\ \text{if}\ \ x(m) = \pm\, 1 \\ e(m) = 1,\ y(m) = -1 & ,\ \text{if}\ \ x(m) = 0 \\ e(m) = 1,\ y(m) = 2x(m) - 1 & ,\ \text{if}\ \ 0 < x(m) < 1 \\ e(m) = -1,\ y(m) = 2x(m) + 1 & ,\ \text{if}\ \ -1 < x(m) < 0. \end{array}$$

Then $z = \frac{1}{2}(e' + y')$, where $e' \in \text{ext}(B_X)$ and $y' \in B_X$ are given by

$$e' = e + \text{ sgn } e(m_0) m_0^{-1}\, e_{m_0}\ ,\ y' = y - \text{ sgn } e(m_0) m_0^{-1}\, e_{m_0}.$$

Therefore, the triple $(e', y', 1/2)$ is amenable to z.

Case II. $M' \ne \phi$

Choose $m_0 \in M'$ such that $|\beta_{m_0}| = \max_{m\in M'} |\beta_m|$. By considering $-z$ instead of z, we may assume $\beta_{m_0} > 0$ and so sgn $x(m_0) = 1$.

If $\beta_{m_0} \ge 1/2$, write $x = \frac{1}{2}(e+y)$, where e, y are as in Case I. Then

$$\begin{aligned} z &= x + \sum_{m\in M'} \beta_m\, m^{-1}\, e_m \\ &= \tfrac{1}{2}\,(e+y) + \tfrac{1}{2}\, m_0^{-1}\, e_{m_0} + (\beta_{m_0} - \tfrac{1}{2})\, m_0^{-1}\, e_{m_0} + \sum_{\substack{m\in M' \\ m\ne m_0}} \beta_m\, m^{-1}\, e_m \\ &= \tfrac{1}{2}\, e' + \tfrac{1}{2}\, y', \end{aligned}$$

where $e' = e + m_0^{-1}\, e_{m_0}, y' = y + (2\beta_{m_0} - 1)\, m_0^{-1}\, e_{m_0} + 2 \sum_{\substack{m\in M' \\ m\ne m_0}} \beta_m\, m^{-1}\, e_m$. Since *sgn* $x(m_0) =$ *sgn* $e(m_0) = 1$, we have $e' \in \text{ext}(B_X)$. Also, $|2\beta_{m_0} - 1| + 2 \sum_{\substack{m\in M' \\ m\ne m_0}} |\beta_m| \le 1$ implies $y' \in B_X$. This shows that $(e', y', 1/2)$ is amenable to z.

Finally, assume $0 < \beta_{m_0} < 1/2$. Write $x = \frac{1}{2}(e+y)$ as in Case I and observe that there is a y' on line segment from y to x such that $x = \beta_{m_0}\, e + (1-\beta_{m_0})y'$ (see (c) of Proposition 1.2 of [1]). Then

$$\begin{aligned} z &= x + \sum_{m \in M'} \beta_m\, m^{-1}\, e_m \\ &= \beta_{m_0} e + (1-\beta_{m_0})y' + \beta_{m_0}\, m_0^{-1}\, e_{m_0} + \sum_{\substack{m \in M' \\ m \neq m_0}} \beta_m\, m^{-1}\, e_m \\ &= \beta_{m_0}\, e' + (1-\beta_{m_0})y'', \end{aligned}$$

where $e' = e + m_0^{-1}\, e_{m_0} \in \operatorname{ext}(B_X)$ and $y'' = y' + \sum_{\substack{m \in M' \\ m \neq m_0}} \dfrac{\beta_m}{1-\beta_{m_0}}\, m^{-1}\, e_m \in B_X$.

Consequently, (e', y'', β_{m_0}) is amenable to z.

We see that (7) fails because Example 3.2 shows $\lambda(e) = 1$ and $e \notin \operatorname{ext}(B_X)$ for each $e \in E_\infty$. With a little additional work, it is possible to show that

$$\{x \in B_X \,:\, \lambda(x) = 1\} = E_\infty \bigcup \operatorname{ext}(B_X).$$

Although the preceding example shows that attainment locally of the value 1 by the λ-function does not characterize individual members of $\operatorname{ext}(B_X)$, it is not known whether the corresponding global property characterizes strictly convex spaces. We state this as the

Strict Convexity Problem: Let X be a normed space with the λ-property. If $\lambda(x) = 1$ for all $x \in S_X$, is X strictly convex?

References

[1] Richard M. Aron and Robert H. Lohman, A geometric function determined by extreme points of the unit ball of a normed space, Pacific J. Math., 127 (1987), 209-231.

[2] Antonio S. Granero, On the Aron-Lohman λ-property, preprint.

[3] Antonio S. Granero, the λ-function in the spaces $\left(\oplus \sum_{i\in I} X_i\right)_p$ and $L_p(\mu, X)$, $1 \leq p \leq \infty$, preprint.

[4] Joram Lindenstrauss and Lior Tzafriri, Classical Banach Spaces, Lecture Notes in Mathematics, 338, Springer-Verlag, New York, 1973.

[5] Robert H. Lohman and Thaddeus J. Shura, Calculation of the λ-function for several classes of normed linear spaces, Nonlinear and Convex Analysis: Proceedings in Honor of Ky Fan, Marcel Dekker Lecture Notes in Pure and Applied Mathematics, 1987, 167-174.

Robert H. Lohman
Department of Mathematical Sciences
Kent State University
Kent, Ohio 44242

Contemporary Mathematics
Volume **85**, 1989

SMOOTH FUNCTIONS IN ORLICZ SPACES

R.P. Maleev* S.L. Troyanski**

Department of Mathematics
University of Sofia, 5 bd A. Ivanov
1126 Sofia, Bulgaria

Abstract. The best order of smoothness of bump functions in Orlicz sequence spaces is found.

1. Introduction.

In many problems of the nonlinear analysis on Banach spaces it is of importance the existence of bump functions with prescribed order of smoothness. Our aim is to find upper bound for the order of smoothness of bump functions in Orlicz sequence spaces.

We begin with some notations and definitions. If X is a Banach space, then $B^j(X)$ will denote the space of all continuous symmetric j–linear functionals T on X. In the

---

* Research supported partially by Bulgarian Committee of Science, contract No. 50/25.03.87.

** Research supported partially by Bulgarian Committee of Science, contract No. 54/25.03.87.

1980 Mathematics Subject Classification (1985 Revision): Primary 46B20, 46B25.

Key Words: Orlicz sequences space, bump function, smoothness.

next

$$T(x,x,\dots,x) = T(x^{(j)}).$$

A function $f:X \longrightarrow \mathbb{R}$, the reals, with bounded nonempty support is called bump function or simply bump. If a bump is k–times Frechet differentiable in X $(f \in F^k(X))$, i.e. for every $x \in X$

$$f(x + ty) = f(x) + \sum_{i=1}^{k} t^i T_x^i(y^{(i)}) + o_x(|t|^k) \tag{1}$$

uniformly on $y \in S(X) = \{x \in X; \|x\| = 1\}$, it is called k–smooth bump. The functional $i!T_x^i$ is called i–th Frechet derivative of f at x and is denoted $D^i f(x;\cdot)$.

It is easy to verify that if X admits an equivalent norm, k–times Frechet differentiable on $X\backslash\{0\}$, then there is in X a k–smooth bump. Let us observe also that the existence of a k–smooth bump f in X implies the existence of a k–smooth bump in every subspace of X.

Further we recall the definition of Orlicz sequence spaces and some related notions. An even, continuous, nondecreasing in $(0,\infty)$, convex function M is called Orlicz function. The space of all sequences $x = \{x_i\}_{i=1}^{\infty}$ such that

$$\tilde{M}(x/\lambda) = \sum_{i=1}^{\infty} M(|x_i|/\lambda) < \infty$$

for some $\lambda > 0$, equipped with the norm

$$\|x\| = \inf\{\lambda > 0;\ \tilde{M}(x/\lambda) \le 1\}$$

is called Orlicz sequence space, generated by M and is denoted ℓ_M.

The subspace of ℓ_M consisting of all $x \in \ell_M$ such that $\tilde{M}(x/\lambda) < \infty$ for all $\lambda > 0$ is denoted h_M.

The following numbers give information on the behavior of the Orlicz function M near 0 and play a special role in the theory of isomorphic embedding of Orlicz sequence spaces into Orlicz sequence spaces:

$$\alpha_M = \sup\{q;\ \sup\{M(uv)/u^q M(v);\ u,v \in (0,1]\} < \infty\},$$
$$\beta_M = \inf\{q;\ \inf\{M(uv)/u^q M(v); u,v \in (0,1]\} > 0\}.$$

Here we mention only that always $\alpha_M \ge 1$ and that $\beta_M < \infty$ implies $\ell_M = h_M$. If $\beta_m = \infty$ then ℓ_M contains isomorphic copy of ℓ_∞ (see e.g. [LT], Prop. 4.a.4) and therefore there is no even Gateau smooth bump in ℓ_M in this case.

2. Preliminary results.

In [BF] the existence of smooth bumps in ℓ_p, $p \ge 1$, is discussed. We shall use in the next the following upper bound for the order of smoothness of bumps in ℓ_p, proved there.

PROPOSITION 2.1 ([BF]). In ℓ_p, $p \ge 1$ not an even integer, there is no $E(p)+1$–smooth bump, where

$$E(p) = \begin{cases} p-1, & p \text{ integer} \\ [p], & p \text{ not integer.} \end{cases}$$

It turns out that the idea of the proof of this result can be generalized to a method for determination of upper bound for the order of smoothness of bumps, usable for other spaces. This method is given by the following

THEOREM 2.2. Let X be Banach space and $f:X \longrightarrow \mathbb{R}$ be boundedly supported continuous function. Let $g \in X \longrightarrow [0,\infty)$, $G:[0,\infty) \longrightarrow [0,\infty)$ be functions with the following properties: g is uniformly continuous on bounded sets and the family $\{z \in X;\ g(z-x) < 1/n\}_{n=0}^{\infty}$ forms a base of neighborhoods of x in the norm topology; G is continuous, $G(0) = 0$, $G(t) > 0$ for $t > 0$.

If for every $x \in X$ and $\epsilon,\gamma > 0$ there exist $\delta \in (0,\gamma)$ and sequence $\{y_n\} \subset X$ such that:

a) $\|y_n\| = \delta$;

b) $\lim_{n\to\infty} g(z+y_n) = g(z) + G(\delta)$ for every $z \in X$;

c) $|f(x+y_n) - f(x)| < \epsilon G(\delta)$, $n = 1,2,\ldots$,

then $f \equiv 0$.

Proof. Without loss of generality we may assume that supp f is in the open unit ball of X and $f(0) \neq 0$. Let $x \in X$ and

$$0 < \epsilon < |f(0)|/2B, \qquad B = \sup\{|g(z)|;\ \|z\| \leq 1\}.$$

We define inductively sets $A_n \subset X$, numbers $\lambda_n \geq 0$, y_n, $x_n \in X$, $\|x_n\| < 1$. Let $A_0 = \{0\}$, $\lambda_0 = 0$, $y_0 = x_0 = 0$. Suppose A_n, λ_n, x_n, y_n are already chosen. Let

A_{n+1} be the set of all y which satisfy the conditions

1. $\|y\| < 1 - \|x_n\|$;
2. $|f(x_n+y) - f(x_n)| < \epsilon G(\|y\|)/2$;
3. $g(x_n+y) \geq g(x_n) + G(\|y\|)/2$;
4. $g(x_n-x_j+y) \leq g(x_n-x_j) + 2G(\|y\|)$, $j = 0,1,\dots,n$.

Let $\lambda_{n+1} = \sup_{y\in A_{n+1}} \|y\|$. From the assumptions a), b), c) it follows that $\lambda_{n+1} > 0$. Choose now $y_{n+1} \in A_{n+1}$ with $\|y_{n+1}\| \geq \lambda_{n+1}/2$ and put $x_{n+1} = x_n + y_{n+1}$. Note that a new step is always possible provided $\|x_n\| < 1$.

Obviously $\|x_{n+1}\| \leq \|x_n\| + \|y_{n+1}\| < 1$ and $g(x_{n+1}) \geq g(x_n) + G(\|y_{n+1}\|/2$. Therefore

$$\frac{1}{2}\sum_{i=1}^{n+1} G(\|y_i\|) \leq g(x_{n+1}) \leq B.$$

Thus $\lim_{n\to\infty} y_n = 0$ which implies $\lim_{n\to\infty} \lambda_n = 0$. Now from 4. we find

$$g(y_{n+1} + \cdots + y_{k+1}) \leq 2 \sum_{i=k+1}^{n+1} G(\|y_i\|)$$

and according to the assumption on g, $\{x_n\}$ is a Cauchy sequence. Let $\lim_{n\to\infty} \|x_n-x\| = 0$. From 2. it is readily seen that

$$|f(x_n)-f(0)| \leq (\epsilon/2) \sum_{i=1}^{n} G(\|y_i\|) \leq f(0)/2.$$

Then $|f(x)-f(0)| \leq f(0)/2$. If $||x|| = 1$ then $|f(0)| \leq |f(0)|/2$ which is a contradiction. Therefore $||x|| < 1$. Choose $0 < \delta < (1-||x||)/2$ and a sequence $\{y'_n\} \in X$, $||y'_n|| = \delta$ with

$$\lim_{n\to\infty} g(z+y'_n) = g(z) + G(\delta) \quad \text{for every } z \in X,$$

$$|f(x+y'_n)-f(x)| \leq \epsilon G(\delta)/8, \; n = 1,2,\ldots$$

Find $\eta > 0$ such that

$$|G(s) - G(t)| < G(\delta)/10, \; s \geq 0, \; 0 \leq t \leq 2, \; |t-s| < \eta;$$

$$|g(v) - g(w)| < G(\delta)/10, \; ||v|| \leq 2, \; ||v-w|| < \eta.$$

Fix m large enough to ensure

$$||x-x_m|| \leq \min(\delta/4,\eta), \; \delta < (1-||x_m||)/2,$$

$$|f(x)-f(x_m)| \leq \epsilon G(\delta)/8, \; \lambda_{m+1} \leq \delta/2.$$

There is $y = y'_i$ (i large enough) such that $||y|| = \delta$ and

$$|f(x+y)-f(x)| \leq \epsilon G(\delta)/8, \; g(x+y) \geq g(x) + 3G(\delta)/4,$$

$$g(x-x_j+y) \leq g(x-x_j) + 4G(\delta)/3, \; j = 0,1,\ldots,m.$$

Put $u = x - x_m + y$. From

(2) $$\big|\|u\| - \delta\big| \le \|x - x_m\| \le \min(\delta/4, \eta)$$

it follows that $\|u\| > 3\delta/4 > \lambda_{m+1}$, i.e. $u \notin A_{m+1}$. (2) implies

(3) $$|G(\|u\|) - G(\delta)| < G(\delta)/10.$$

Now we shall obtain a contradiction showing that $u \in A_{m+1}$. Indeed, from (2) it follows

1. $\|u\| \le 5\delta/4 < 1 - \|x_m\|$;

From (3) we verify

2. $|f(x_m+u) - f(x_m)| \le |f(x+y) - f(x)| + |f(x) - f(x_m)|$
$\le \epsilon G(\delta)/4 \le 5G(\|u\|)/18$;

3. $g(x_m+u) = g(x+y) \ge g(x) + 3G(\delta)/4$
$\ge g(x_m) - G(\delta)/10 + 3G(\delta)/4 \ge g(x_m) + 13G(\|u\|)/22$;

4. $g(x_m-x_j+u) = g(x-x_j+y) \le g(x-x_j) + 4G(\delta)/3$
$= g(x-x_m+x_m-x_j) + 4G(\delta)/3 \le g(x_m-x_j) + G(\delta)/10 + 4G(\delta)/3$
$\le g(x_m-x_j) + (1/10 + 4/3)\, 10G(\|u\|)/9$
$< g(x_m-x_j) + 2G(\|u\|)$, $j = 0,1,\dots,m$.

Thus Theorem 2.2 is proved.

We need for the sequel some auxiliary results concerning polynomials in Orlicz sequence spaces. In all of them M is Orlicz function.

LEMMA 2.3. Let X be Banach space with symmetric basis $\{e_n\}$ and $\lambda_n = \|\sum_{i=1}^{n} e_i\|$. If $\lambda_{n_i}^m/n_i \to 0$ for some increasing sequence of naturals $\{n_i\}$ then

$$\lim_{n\to\infty} P_k(e_n) = P_k(0)$$

for every polynomial P_k of degree $k \leq m$.

We recall that a polynomial in X is a function of the form

$$P_k(x) = y + \sum_{i=1}^{k} T^i(x^{(i)}), \text{ where } y \in X, \ T^i \in B^i(X).$$

Proof. We make induction on k. Our assertion is obvious for $k = 1$. Suppose it is true for every polynomial of degree less than m. What we have to prove is that for every monomial P of degree m

$$\lim_{k\to\infty} P(e_k^{(m)}) = 0. \tag{4}$$

Suppose the contrary. There is a monomial P of degree m such that $P(e_{k_j}^{(m)}) \geq d > 0$ for $j = 1,2,\dots$. According to the inductive assumption for every i we can find naturals $j_1 < j_2 < \dots < j_{n_i}$ such that

$$P((x_i)^{(m)}) \geq n_i d/2 \quad \text{for } x_i = \sum_{r=1}^{n_i} e_{k_{j_r}}.$$

Then

$$P((x_i/||x_i||)^{(m)}) \geq n_i d/2\lambda_{n_i}^m$$

which contradicts the boundedness of P. Thus Lemma 2.3 is proved.

<u>COROLLARY 2.4</u>. Let $M(t_n)/t_n^m \longrightarrow 0$ for some sequence $t_n \longrightarrow 0$, $t_n > 0$. Then

$$\lim_{n\to\infty} P_k(\ell_n) = P_k(0)$$

for every polynomial P_k on h_M of degree $k \le m$.

<u>Proof</u>. Put $\lambda_n = \|\sum_{i=1}^n e_i\|$. Suppose that for some $c > 0$ $\lambda_n \ge (cn)^{1/m}$ for every n. Then from $nM(1/\lambda_n) = 1$ it follows

$$1 \le nM((cn)^{-1/m}). \tag{7}$$

Find integer n_i with

$$(cn_i)^{-1/m} \le t_i \le (c(n_i-1))^{-1/m}, \quad i = 1,2,\ldots$$

Then $c(n_i-1) \le t_i^{-m}$ and from (7) we obtain

$$1 \le n_i M((cn_i)^{-1/m}) \le (n_i/c(n_i-1))M(t_i)/t_i^m \le (2/c)\, M(t_i)/t_i^m, \quad i = 1,2,\ldots,$$

which is a contradiction. This implies

$$\lim_{n\to\infty} \lambda_n^m/n = 0$$

and the Corollary follows from Lemma 2.3.

COROLLARY 2.5. Let $M(t_n)/t_n^k \longrightarrow 0$ for some sequence $t_n \longrightarrow 0$, $t_n > 0$, and $P \in B^k(h_M)$. Then for every $\epsilon > 0$, naturals m, r, and reals $\alpha_1, \alpha_2, \ldots, \alpha_r$ there exist positive integers $n_1, n_2, \ldots, n_r$ such that

$$|P((\sum_{j=1}^{r} \alpha_j e_{n_j})^{(k)})| < \epsilon. \tag{8}$$

Proof. A simple proof of this Corollary can be given by induction on r, using Corollary 2.4 and the representation for $P \in B^k(h_M)$

$$P((x+y)^{(k)}) = P(x^{(k)}) + Q(x,y) + P(y^{(k)}),$$

where Q is a polynomial of degree less than k with respect to y and $Q(x,0) = 0$.

LEMMA 2.6. $\lim_{n\to\infty} (\tilde{M}(x+y_n) - \tilde{M}(y_n)) = \tilde{M}(x)$ for every $x \in h_M$ and every sequence $\{y_n\} \subset h_M$, $\|y_n\| \leq 1/2$ which satisfies $w\text{-}\lim_{n\to\infty} y_n = 0$.

The proof is straightforward.

LEMMA 2.7. Let $\alpha_M = k$, $\lim_{n\to\infty} M(t_n)/t_n^k = 0$ for some sequence $t_n \longrightarrow 0$, $t_n > 0$ and $P \in B^k(h_M)$. Then for every $\delta \in (0,1)$ there exists a sequence $\{x_n\} \subset h_M$ such that:

i) x_n have disjoint finite supports;

ii) $\|x_n\| = \delta$;

iii) $\lim_{n\to\infty} \tilde{M}(x_n) = \delta^k$;

iv) $\lim_{n\to\infty} P(x_n) = 0$.

Proof. It is known (see e.g. [LT], Th. 4.a.9) that $t^k \in \bigcap_{\Lambda>0} C_{M,\Lambda}$ where $C_{M,\Lambda} = \overline{\mathrm{conv}}\{M(\lambda t)/M(t);\ 0 < \lambda < \Lambda\}$ (the closure is taken in the norm topology of $C(0,T)$, T arbitrary in $(0,1)$). Choose $T > \delta$. According to a deep theorem of Lindenstrauss and Tzafriri ℓ_k is isomorphic to a subspace of h_M. Following the proof of this theorem for every n we pick $\tau_n \in (1-2^{-n-2},1)$ and choose $\Lambda_n \in (0,2^{-n-2}(1-\tau_n))$ and a probability measure μ_n, supported by the interval $(0,\Lambda_n)$ such that

$$|t^k - \int_0^{\Lambda_n} M(\lambda t)d\mu_n/M(\lambda)| < 2^{-n-2}, \quad t \in [0,T].$$

Set $\alpha_{j,n} = \int_{\tau_n^j\Lambda_n}^{\tau_n^{j-1}\Lambda_n} d\mu_n(\lambda)/M(\lambda)$, $k_{j,n} = [\alpha_{j,n}]$,

$$\phi(t) = \int_0^{\Lambda_n} M(\lambda t)d\mu_n(\lambda)/M(\lambda).$$

For sufficiently big k_n

$$F_n(\tau_n t) - 2^{-n-1} \le t^k \le F_n(t) + M(t)\Lambda_n/(1-\tau_n) + 2^{-n-1}, \quad t \in [0,T], \tag{9}$$

(10) $$F_n(\tau_n t) \le \phi(t) \le F_n(t) + M(t)\Lambda_n/(1-\tau_n), \quad t \in [0,1],$$

where $F_n(t) = \sum_{j=1}^{k_n} k_{j,n} M(\tau_n^{j-1}\Lambda_n t)$.

Denote $c_{j,n} = \tau_n^{j-1}\Lambda_n$ and consider $x_n = \alpha_n \delta \sum_{j=1}^{k_n} c_{j,n} \sum_{i\in A_{j,n}} e_i$ ($\{e_n\}$ the unit vector basis in h_M), where $A_{j,n}$ are sets of positive integers with $A_{r,n} \cap A_{s,n} = \phi$, $r \ne s$, $|A_{j,n}| = k_{j,n}$, while α_n satisfies

$$\sum_{j=1}^{k_n} k_{j,n} M(c_{j,n}\alpha_n) = 1.$$

Obviously i) and ii) are satisfied. For α_n the estimates hold

(11) $$1 - 2^{-n-2} \le \alpha_n \le 1/(1-2^{-n-2}).$$

Indeed, if $\alpha_n \le 1$, from (10)

$$1 = F_n(\alpha_n) = \phi(1) \ge F_n(\tau_n),$$

which implies $\alpha_n \ge \tau_n$. On the other hand, if $\alpha_n > 1$ from (10) follows

$$1 = F_n(\alpha_n) \ge \alpha_n F_n(1) \ge \alpha_n(\phi(1) - \Lambda_n/(1-\tau_n)) \ge \alpha_n(1-2^{-n-2}).$$

Now we are ready to prove iii). From (9) we deduce for $t \in [0,T]$

$$F_n(\tau_n t) \leq t^k + 2^{-n-1}, \quad t^k - 2^{-n} \leq F_n(t).$$

As $\alpha_n \delta < T$ for sufficiently big n, putting $t = \alpha_n \delta / \tau_n$ in the first inequality and $t = \alpha_n \delta$ in the second we obtain

$$\alpha_n^k \delta^k - 2^{-n} \leq F_n(\alpha_n \delta) = \tilde{M}(x_n) \leq (\alpha_n 2^{n+2}/(2^{n+2}-1))^k \delta^k + 2^{-n-1},$$

which gives, according to (11), iii).

Finally, we observe that the sets $A_{j,n}$ of integers appearing in the definition of x_n have to satisfy only the condition $|A_{j,n}| = k_{j,n}$ and therefore the sequence $\{x_n\}$ can be situated over groups of unit vectors from $\{e_n\}$ corresponding to P such that according to Corollary 2.5 to have iv).

Lemma 2.7 is proved.

3. Main result.

THEOREM 3.1. Let M be Orlicz function, $M \not\sim t^k$, k even, at 0 and f be k–smooth bump in h_M. Then

$$k \leq E(\alpha_M). \tag{12}$$

Before the proof we recall that two Orlicz functions M and N are called equivalent at 0 $(M \sim N)$ if there exist positive constants c, C and t_0 such that

$$C^{-1}M(c^{-1}t) \le N(t) \le CM(ct), \quad t \in [0,t_0].$$

It is well known that from the equivalence of M and N in 0 follows ℓ_M isomorphic to ℓ_N.

Proof of Theorem 3.1. As ℓ_{α_M} is isomorphic to a subspace of h_M the estimate (12) follows immediately from Proposition 2.1 if α_M is not even.

Let $\alpha_M = k$, k even, and $M \not\sim t^k$ at 0. We consider separately the cases:

A. $\lim\limits_{n\to\infty} M(t_n)/t_n^k = 0$ for some sequence $t_n \longrightarrow 0$, $t_n > 0$;

B. $\lim\limits_{n\to\infty} M(t_n)/t_n^k = \infty$ for some sequence $t_n \longrightarrow 0$, $t_n > 0$.

Case A. Suppose f is k–smooth bump in h_M. For arbitrary $x \in h_M$ and $\epsilon,\gamma > 0$ there exists δ, $0 < \delta < \min(\gamma,1/2)$, such that

$$|f(x+y) - f(x) - \sum_{i=1}^{k} (1/i!)D^i f(x;y^{(i)})| \le \epsilon\|y\|^k/2 \tag{13}$$

for every y, $\|y\| \le \delta$.

From (13), Corollary 2.4, Corollary 2.5 and Lemma 2.7 it follows the existence of a sequence $\{y_n\} \subset h_M$ with the properties:

a) $\|y_n\| = \delta$;

b) $\lim\limits_{n\to\infty} \tilde{M}(z+y_n) = \tilde{M}(z) + \delta^k$ for every $z \in h_M$;

c) $|f(x+y_n) - f(x)| < \epsilon\delta^k$, $n = 1,2,\ldots$

Now Theorem 2.2 with $g = \tilde{M}$ and $G(\delta) = \delta^k$ implies $f \equiv 0$ which is a contradiction.

Case B. Let $f \in F^k(h_M)$. For every $x \in h_M$ and arbitrary $\epsilon, \gamma > 0$ we can find $\delta \in (0, \min(\gamma, 1/2))$ and a sequence $\{y_n\} \subset h_M$ with:

a) $\|y_n\| = \delta$;

b) $\lim_{n\to\infty} \tilde{M}(z+y_n) = \tilde{M}(z) + M(\delta)$ for every $z \in h_M$;

c) $|f(x+y_n) - f(x)| \leq \epsilon M(\delta)$, $n = 1,2,\ldots$

Indeed, we have for $\|y\| \leq \delta$ (δ sufficiently small)

$$f(x+y) - f(x) - \sum_{i=1}^{k-1} (1/i!)D^i f(x; y^{(i)}) = R(x,y), \tag{14}$$

where $\|R(x,y)\| \leq C\|y\|^k$, C positive constant.

Choose m such that $t_m^k/M(t_m) < C\epsilon/2$ and put $\delta = t_m$. Consider the sequence $h_n = \delta e_n$. Using Corollary 2.4 we can find n_0 such that

$$\left| \sum_{i=1}^{k-1} (1/i!)D^{k-1} f(x; h_n^{(i)}) \right| < \epsilon M(\delta)/2, \quad n > n_0.$$

Then for the sequence $y_n = h_{n_0+n}$ obviously

$$|f(x+y_n) - f(x)| < C\delta^k + \epsilon M(\delta)/2 = (C\delta^k/M(\delta) + \epsilon/2)M(\delta) < \epsilon M(\delta)$$

for every n, i.e. c).

a) is obviously satisfied, while b) follows from Lemma 2.6.

But we can now use one more time Theorem 2.2 with $g = \tilde{M}$ and $G(\delta) = M(\delta)$ to obtain $f \equiv 0$. The Theorem is proved.

REMARK 3.2. From a result of Sundaresan [S] it follows that the usual norm in ℓ_p, p even, is infinitely many times Frechet differentiable in $\ell_p \setminus \{0\}$.

REMARK 3.3. For $\alpha_M = 2$ the estimate (12) follows from a result of Makarov [M] and the result of Lindenstrauss and Tzafriri used above.

REMARK 3.4. In [MT] it is proved that in h_M, $\alpha_M > 1$ there is an equivalent norm, which is $E(\alpha_M)$–times Frechet differentiable. This fact shows that the estimate (12) is exact.

REFERENCES

[BF] N. Bionic and J. Frampton, Smooth functions on Banach Manifolds, J. Math. Mechanics 15 (1966), 877–898.

[LT] J. Lindenstrauss, L. Tzafriri, Classical Banach Spaces I, Sequence spaces, Springer Verlag, 1977.

[M] B.M. Makarov, One characteristic of Hilbert spaces, Mat. Zamet. 26 (1979), 739–746 (Russian).

[MT] R.P. Maleev, S.L. Troyanski, Differentiability of the norm in Orlicz spaces, Proceedings International Conf. Geometry of Banach spaces and related topics, Mons, Belgium, 1987, to appear.

[S] K. Sundaresan, Smooth Banach spaces, Math. Ann. 173 (1967), 191–199.

Contemporary Mathematics
Volume **85**, 1989

FEFFERMAN SPACES AND C*–ALGEBRAS

Paul S. Muhly*

ABSTRACT

It is shown that the Banach space dual of a C^*–algebra is a Fefferman space. This answers a question of Blasco and Pelczynski.

§1. Let $\mathcal{X}$ be a Banach space and let $H^1_{\mathcal{X}}$ be the set of all Bochner integrable, $\mathcal{X}$–valued functions f on the circle $\mathbb{T}$ such that the Fourier coefficients of f, $\hat{f}(n)$, vanish when $n < 0$. Equivalently, a function $f \in L^1_{\mathcal{X}}$ lies in $H^1_{\mathcal{X}}$ if and only if f is the limit in the $L^1_{\mathcal{X}}$ norm of a sequence of functions of the form $\sum_{k=0}^{n} a_k z^k$ where $a_k \in \mathcal{X}$; i.e. a sequence of analytic, $\mathcal{X}$–valued, trigonometric polynomials. We write $\ell^1_{\mathcal{X}}$ for the set of all sequences $\{X_k\}_{k=0}^{\infty}$, with $X_k \in \mathcal{X}$, such that $\sum_{k=0}^{\infty} \|X_k\|_{\mathcal{X}} < \infty$. A sequence $m = (m_k)_{k=0}^{\infty}$ of complex numbers is called an $H^1_{\mathcal{X}}$–$\ell^1_{\mathcal{X}}$–*multiplier* if whenever $f \in H^1_{\mathcal{X}}$, $\{m_k \hat{f}(k)\}_{k=0}^{\infty}$ is in $\ell^1_{\mathcal{X}}$. We write $(H^1_{\mathcal{X}}, \ell^1_{\mathcal{X}})$ for the collection of all such multipliers. It

---

*Supported, in part, by a grant from the National Science Foundation.

1980 *Mathematics Subject Classification (*1985 *Revision).* Primary 42A45, 46B20. Secondary 46L05, 46L10, 47D25, 47D30. *Key words and phrases.* Fefferman space, Banach dual space of a C^*–algebra, multipliers.

will be convenient to think of a sequence of scalars, $m = \{m_n\}_{n=0}^{\infty}$, as an operator, also denoted m, defined initially on the analytic trigonometric polynomials by the formula $m \cdot f = \{m_n \hat{f}(n)\}_{n=0}^{\infty}$. Then m belongs $(H^1_{\mathscr{X}}, \ell^1_{\mathscr{X}})$ if and only if m extends to a bounded operator defined on all of $H^1_{\mathscr{X}}$. In this case, we write $\|m\|$ or $\|m\|_{\mathscr{X}}$ for the norm of this operator. If $\mathscr{X}$ is the scalars $\mathbb{C}$, we usually drop the subscript $\mathscr{X}$ from our notation and simply write H^1, ℓ^1, and (H^1,ℓ^1) for $H^1_{\mathscr{X}}$, $\ell^1_{\mathscr{X}}$, and $(H^1_{\mathscr{X}},\ell^1_{\mathscr{X}})$, respectively. By considering functions of the form $f(z) = \varphi(z)X$ where X is a fixed non–zero vector in $\mathscr{X}$ and $\varphi \in H^1$, one sees immediately that $(H^1_{\mathscr{X}},\ell^1_{\mathscr{X}})$ is contained in (H^1,ℓ^1) for any Banach space $\mathscr{X}$ and that inequality $\|m\|_{\mathbb{C}} \le \|m\|_{\mathscr{X}}$ is always satisfied. However, the inclusion may be proper [BP].

Definition ([BP]). A Banach space $\mathscr{X}$ is called a *Fefferman space* if $(H_{\mathscr{X}},\ell^1_{\mathscr{X}}) = (H^1,\ell^1)$.

In [BP], the authors prove that the space of trace class operators on a Hilbert space $\mathscr{H}$, $S^1(\mathscr{H})$, is a Fefferman space and they prove that the dual of the algebra, $B(\mathscr{H})$, of all bounded linear operators on $\mathscr{H}$ is also a Fefferman space. Since $S^1(\mathscr{H})$ is the dual of the algebra of compact operators on $\mathscr{H}$–a C^*–algebra, as is $B(\mathscr{H})$–Blasco and Pelczynski were led to ask if the dual of every C^*–algebra is a Fefferman space. Our primary objective here is to show that indeed this is the case.

Theorem 1.1. If $\mathscr{X}$ is the dual of a C^*–algebra, then $\mathscr{X}$ is a Fefferman space. Moreover, $\|m\|_{\mathscr{X}} = \|m\|_{\mathbb{C}}$.

Proof. Let A be a C^*–algebra such that $\mathscr{X} = A^*$. Then the dual of $\mathscr{X}$ is A^{**}, which is a von Neumann algebra under the Arens multiplication. This von Neumann algebra is huge and acts on a highly non–separable Hilbert space. Our first reduction will be to show that A^{**} can be replaced by a countably decomposable von Neumann algebra M with $\mathscr{X}$ replaced by M_*. (Note: In this replacement, M_* need not be the dual of anything.) Then we argue, using Haagerup's theory of noncommutative L^p–spaces [H1,2], that we may assume that M is *finite* and countably decomposable. We then prove that if M is such a von Neumann algebra, then M_* is a Fefferman space. The proof, here, follows the outline of the proof of Theorem 2.2 in [BP]. A critical role will be played by Arveson's theory of subdiagonal algebras [A] and an extension of a factorization theorem proved by Sarason [Sar].

First Reduction. We may replace A^{**} by a countably decomposable von Neumann algebra M and $\mathscr{X}$ by M_*.

We want to show that a given $m \in (H^1,\ell^1)$ lies in $(H^1_{\mathscr{X}},\ell^1_{\mathscr{X}})$. To do this, we need to test m on an arbitrary function $f \in H^1_{\mathscr{X}}$. So fix f and let p be the support projection in A^{**} of all the elements $\hat{f}(n) \in \mathscr{X}$, $n = 0,1,\cdots$. That is, p is the smallest projection in A^{**} such that $p \cdot \hat{f}(n) = \hat{f}(n) \cdot p = \hat{f}(n)$ for all n. (Recall the notation that if φ is a linear functional on a C^*–algebra A and if $a \in A$, then $a \cdot \varphi$ and $\varphi \cdot a$ are the linear functionals on A defined by the formulae $a \cdot \varphi(x) = \varphi(xa)$ and $\varphi \cdot a(x) = \varphi(ax)$. Recall, too, that if A is a von Neumann algebra and if φ comes from its pre–dual, then $a \cdot \varphi$ and $\varphi \cdot a$ are also in the predual.) Let $M = pA^{**}p$. Then, evidently, f takes its values in $M_* = p \cdot \mathscr{X} \cdot p$ which is isometrically imbedded in $\mathscr{X}$ as the collection of all $\varphi \in \mathscr{X}$ such that $p \cdot \varphi = \varphi \cdot p = \varphi$. Thus the sequence $\{m_n\hat{f}(n)\}_{n=0}^{\infty}$ lies in $\ell^1_{\mathscr{X}}$ if and only if it lies in $\ell^1_{M_*}$. It remains to show that M is countably decomposable. We may assume, without loss of generality, that M acts on a Hilbert space $\mathscr{H}$ in such a way that the functionals $\hat{f}(n)$ are given by vectors; i.e., we may assume that there are vectors ξ_n, $\eta_n \in H$, $n = 0,1,\cdots$, such that $\hat{f}(n)(x) = (x\xi_n,\eta_n)$ for all $x \in M$. Suppose that $\{q_\alpha\}_{\alpha\in \mathbf{A}}$ is an orthogonal family of non–zero projections in M. Then since $\|\xi_n\|^2 \geq \sum_{\alpha\in\mathbf{A}} \|q_\alpha \xi_n\|^2$ for each n, and likewise for the η_n, there is a *countable* set $\mathbf{A}' \subseteq \mathbf{A}$ such that for $\alpha \notin \mathbf{A}'$, $q_\alpha \xi_n = q_\alpha \eta_n = 0$ for all n. For such an α, then, $q_\alpha \cdot \hat{f}(n) = \hat{f}(n) \cdot q_\alpha = 0$. So for $\alpha \notin \mathbf{A}'$, $p - q_\alpha$ is a strictly smaller projection than p which supports all the $\hat{f}(n)$. This contradicts the definition of p and show that $\mathbf{A}' = \mathbf{A}$; i.e., M is countably decomposable.

Second Reduction. We may assume that M is finite.

The fact that M is countably decomposable enables us to assert that M admits a faithful normal state φ and that modular theory is available to us. According to Haagerup [H1,2], we may identify M_* with his space $L^1(M,\varphi)$. Also, by [H1,2], we may find a Banach space $\mathscr{Y}$, a sequence $\{(M_n,\varphi_n)\}_{n=1}^\infty$ of *finite* von Neumann algebras M_n with faithful, normal, *finite* traces φ_n, and a sequence $\{j_n\}_{n=1}^\infty$ of *isometric* imbeddings mapping $L^1(M_n,\varphi_n)$ into $\mathscr{Y}$ such that

(a) $\bigcup_{n=1}^\infty j_n(L^1(M_n,\varphi_n))$ is dense in $\mathscr{Y}$;

(b) $j_n(L^1(M_n,\varphi_n)) \subseteq j_m(L^1(M_m,\varphi_m))$, if $n \leq m$; and

c) $L^1(M,\varphi)$ is isometrically isomorphic to a (complemented) subspace of $\mathscr{Y}$.

So, to complete the proof of Theorem 1.1, it suffices to prove that if M is a finite von Neumann algebra with faithful normal finite trace φ, then $L^1(M,\varphi)$ is a Fefferman space and $\|m\|_{L^1(M,\varphi)} = \|m\|_{\mathfrak{C}}$ for all m in (H^1,ℓ^1).

Henceforth, M will denote such a von Neumann algebra and φ will denote such a trace. The L^p–spaces associated with M and φ, $L^p(M,\varphi)$, may be viewed as spaces of (generally unbounded) operators affiliated with M. The reader should consult [N] for a lucid exposition of these spaces. The Banach space $L^1_{L^1(M,\varphi)}$ may be viewed as the predual of the von Neumann algebra tensor product, $L^\infty \otimes M$ [Sak, §1.22]. This algebra is a finite von Neumann algebra which may be viewed as the von Neumann algebra of all essentially bounded, M–valued, measurable functions on the circle with trace, $\mu \otimes \varphi$, given by the formula $\mu \otimes \varphi(f) = \int_{\mathbb{T}} \varphi(f(z))d\mu(z)$, where μ is Lebesgue measure on the circle. When this is done, the non–self–adjoint subalgebra, $H^\infty \otimes M$, is the set of functions f in $L^\infty \otimes M$ such that $\hat{f}(n) = 0$, $n < 0$ and is a maximal, finite, subdiagonal subalgebra of $L^\infty \otimes M$ in the sense of Arveson [A]. The closure of $H^\infty \otimes M$ in the L^1–norm determined by $\mu \otimes \varphi$ is simply $H^1_{L^1(M,\varphi)}$. The following variant of Sarason's theorem [Sar, Theorem 4] will be proved in the next section.

<u>Lemma</u> 1.2. Given $f \in H^1_{L^1(M,\varphi)}$, there exist functions g and h in $H^2_{L^2(M,\varphi)}$ such that f is the pointwise product of g and h, and $||f||_1 = ||g||_2||h||_2$ where the subscript 1 denotes the norm taken in $H^1_{L^1(M,\varphi)}$ and the subscript 2 denotes the norm taken in $H^2_{L^2(M,\varphi)}$.

Now the proof of Theorem 1.1 follows the argument for the proof of Theorem 2.2 in [BP] explicitly. The only change necessary is to replace S^p by $L^p(M,\varphi)$, $p = 1,2$. Indeed, given $f \in H^1_{L^1(M,\varphi)}$, apply Lemma 1.2 to write $f(z) = g(z)h(z)$, a.e. μ, for functions $g, h \in H^2_{L^2(M,\varphi)}$. Set

$$\Phi(z) = \sum_{j=0}^{\infty} \sum_{k=0}^{j} \|\hat{g}(k)\|_2 \|\hat{h}(j-k)\|_2 z^j$$

and note that $\Phi \in H^1$ because $\Phi = G \cdot H$ where $G(z) = \sum_{k-0}^{\infty} \|\hat{g}(k)\|_2 z^k$ and $H(z) = \sum_{k=0}^{\infty} \|\hat{h}(k)\|_2 z^k$. These functions, in turn, are in H^2 because $\sum_{k=0}^{\infty} \|\hat{g}(k)\|_2^2$ and $\sum_{k=0}^{\infty} \|\hat{h}(k)\|_2^2$ are finite; indeed, these sums are the norms squared of the functions g and h. Since $\|A \cdot B\|_{L^1(M,\varphi)} \le \|A\|_{L^2(M,\varphi)} \|B\|_{L^2(M,\varphi)}$, we conclude that $|\hat{\Phi}(j)| \ge \|\hat{f}(j)\|_{L^1(M,\varphi)}$, as in [BP]. On the other hand, since $\|f\|_1 = \|g\|_2 \|h\|_2$, there results the inequality

$$\begin{aligned} \|\Phi\|_{H^1} &\le \|G\|_{H^2} \|H\|_{H^2} \\ &= \|g\|_2 \|h\|_2 = \|f\|_1, \end{aligned}$$

and so, by Corollary 2.3 in [BP], $L^1(M,\varphi)$ is a Fefferman space. Moreover, because we have produced an H^1 function Φ such that $|\hat{\Phi}(k)| \ge \|\hat{f}(k)\|_{L^1(M,\varphi)}$ and $\|\Phi\|_{H^1} \le \|f\|_1$, the norm of m as an element of (H^1,ℓ^1) *equals* that of m as an element of $(H^1_{L^1(M,\varphi)},\ell^1_{L^1(M,\varphi)})$. Indeed, $\|m \cdot f\| = \sum_{n=0}^{\infty} \|m_n \hat{f}(n)\| \le \sum_{n=0}^{\infty} |m_n|\, |\hat{\Phi}(n)|$ $= \sum_{n=0}^{\infty} |m_n \hat{\Phi}(n)| = \|m \cdot \Phi\| \le \|m\|_{\mathbb{C}} \|\Phi\|_{H^1} \le \|m\|_{\mathbb{C}} \|f\|_1$. This shows that

$||m||_{L^1(M,\varphi)} \leq ||m||_{\mathbb{C}}$. Since the reverse inequality is obvious, as was noted earlier, we conclude that $||m||_{L^1(M,\varphi)} = ||m||_{\mathbb{C}}$. This completes the proof of Theorem 1.1.

When M is an arbitrary von Neumann algebra with a faithful, normal, semifinite *weight* φ, there are a variety of definitions for the L^p–spaces, $L^p(M,\varphi)$, $1 \leq p \leq \infty$. However, for a fixed M,φ, and p any two definitions give isometrically isomorphic spaces. Moreover, thanks to Haagerup's analysis [H1,2], the study of the Banach space properties of $L^p(M,\varphi)$ frequently can be reduced to the case when M is finite and φ is a trace. This is because for each p, $L^p(M,\varphi)$ can be imbedded as a complemented subspace of an "inductive limit" of L^p–spaces associated with finite von Neumann algebras as we observed in the proof of Theorem 1.1, where we used the case when $p = 1$. As a result, we are able to prove the following result which is a generalization of Corollary 2.2 and the second half of Theorem 2.2 in [BP].

Theorem 1.3. Let M be a countably decomposable von Neumann algebra and let φ be a faithful, normal, semifinite weight on M. Then for each p, $1 \leq p \leq 2$, $L^p(M,\varphi)$ is a Fefferman space.

Proof. As just noted, following the proof of Theorem 1.1, we may assume that M is finite and that φ is a faithful normal finite trace on M. But then the proof follows the argument for the proof of the second half of Theorem 2.2 word for word. The only special observations one needs are:

a) $L^p(M,\varphi)$, $1 < p < \infty$, is a UMD space; and

b) $L^p(M,\varphi)$, $1 \leq p \leq 2$ is an interpolation scale.

Observation a) is proved in [B,G,M1], although it should be noted that in [B,G,M2] it is shown that all non–commutative L^p–spaces, $1 < p < \infty$, are UMD spaces so that the assumption of semifiniteness (which includes finiteness) made in [B,G,M1] is not necessary. Observation b) is well known in the finite case, but we refer the reader to [K] for an exposition which covers L^p–spaces associated with a faithful normal state. In fact Kosaki *defines* his L^p–spaces through interpolation. In any event, the proof of Theorem 1.3 requires no more detail.

§2 In this section, we prove Lemma 1.2. In fact, we prove a slightly stronger assertion which is an analogue of Sarason's Theorem 4 in [Sar]. Throughout this section, M will be a finite von Neumann algebra with faithful normal finite trace φ. To simplify notation, we write $\mathbb{L}^\infty$ for $L^\infty \otimes M$ and $\mathbb{H}^\infty$ for $H^\infty \otimes M$. As noted earlier, $\mathbb{L}^\infty$ is a finite von Neumann algebra with trace $\mu \otimes \varphi$. The L^p–spaces associated with $\mu \otimes \varphi$ will be denoted $\mathbb{L}^p$, $1 \leq p \leq \infty$. The closure of $\mathbb{H}^\infty$ in $\mathbb{L}^p$ will be denoted by $\mathbb{H}^p$. It is easily seen that $\mathbb{H}^1 = \mathbb{H}^1_{L^1(M,\varphi)}$ in the notation of Lemma 1.2. The algebras $\mathbb{L}^\infty$ and $\mathbb{H}^\infty$ act on the left and on the right of each $\mathbb{L}^p$. We write L and R for these actions; i.e. $L(x)\xi = x\xi$ and $R(x)\xi = \xi x$ for all $x \in \mathbb{L}^\infty$ and $\xi \in \mathbb{L}^p$. Elements of $\mathbb{L}^p$ may be viewed either as functions or as operators which are generally unbounded. We extend L and R to these spaces of operators in the obvious way. Because M and hence $\mathbb{L}^\infty$ is finite, there are no real problems with the arithmetic of the operators L(f) and R(f), $f \in \mathbb{L}^p$.

A partial isometry Θ in $\mathbb{L}^\infty$ is called an *inner operator* if $\Theta \in \mathbb{H}^\infty$. In this case, as is well–known, the initial projection of Θ is constant, i.e., if Θ is viewed as a partial isometry–valued function on $\mathbb{T}$, then $\Theta^*\Theta$ is a constant function. In general a partial isometry–valued function is called *rigid* if its initial projection is a constant function. The terminology is due to Halmos [Hal] who showed that inner operators are rigid. An element $g \in \mathbb{H}^q$, $1 \leq q < \infty$, is called *left–outer* if $[g\mathbb{H}^\infty]_q = L(p)\mathbb{H}^q$ for some projection $p \in M$. (Again, if p is viewed as an M–valued function, then the assertion is that the function is constant.) Right–outer elements of $\mathbb{H}^q$ are defined similarly. The notion of a left or right outer element in $\mathbb{H}^\infty$ may be defined as well, but the topology is the σ–weak topology.

The algebra $\mathbb{H}^\infty$ is an example of an analytic crossed product (formerly called a non–self–adjoint crossed product) as studied in [MMS 1,2]. In particular, see Example 2.4 in [MMS1] from which it is easy to deduce that $\mathbb{H}^\infty$ is the analytic crossed product determined by the *trivial* automorphism of M. By Theorem 3.2 of [MMS2], we conclude that a complete analogue of the Beurling–Lax–Halmos Theorem is valid for subspaces of $\mathbb{L}^p$ that are invariant under left (or right) multiplication by elements in $\mathbb{H}^\infty$. To state it, we require some definitions. A subspace $\mathcal{M}$ of $\mathbb{L}^p$ (that is norm closed if $p<\infty$ and σ–weakly closed if $p = \infty$) is called *left–invariant* if $L(x)\mathcal{M} \subseteq \mathcal{M}$ for all $x \in \mathbb{H}^\infty$. We say that $\mathcal{M}$ is *left–pure* if it contains no subspace that is invariant under $L(x)$ for every $x \in \mathbb{L}^\infty$. Right–hand notions of invariance and purity are defined similarly. We state the version of the Beurling–Lax–Halmos theorem that we need as the following lemma. The proof may be found in [MMS2, Theorem 3.2]. We note for the record that in [MMS2, Theorem 3.2] it is not asserted that the partial isometry produced there is rigid. However, the proof shows that it is.

<u>Lemma</u> 2.1. Let $\mathcal{M}$ be a right–pure, right–invariant subspace of $\mathbb{L}^p$. Then there is a rigid partial isometry $\Theta \in \mathbb{L}^\infty$ such that $\mathcal{M} = L(\Theta)\mathbb{H}^p$. Moreover, Θ is uniquely determined by $\mathcal{M}$ up to right multiples by partial isometries in M. Left–pure, left invariant subspaces have a similar representation as $R(\Theta)\mathbb{H}^p$ for an essentially unique partial isometry Θ in $\mathbb{L}^\infty$.

The following corollary is not really needed for the sequel, but we include it for completeness.

Corollary 2.2. Every (nonzero) element $f \in \mathbb{H}^p$, $1 \leq p \leq \infty$, can be written as $f = \Theta g$ where Θ is inner and $g \in \mathbb{H}^p$ is left–outer.

Proof. We consider the case when $p < \infty$; $p = \infty$ is handled similarly. The space $[f\,\mathbb{H}^\infty]_p$ is right–invariant and is pure because it is contained in $\mathbb{H}^p$. By Lemma 2.1, we may write this space as $L(\Theta)\mathbb{H}^p$ for a suitable partial isometry $\Theta \in \mathbb{L}^\infty$. Since the space is in $\mathbb{H}^p$, Θ is an inner operator. Thus we may write $f = \Theta g$, $g \in \mathbb{H}^p$. Since Θ is inner and therefore rigid the initial projection of $L(\Theta)$ is $L(q)$ for a constant projection function q. It follows that $[g\mathbb{H}^\infty]_p = L(q')\mathbb{H}^p$ for a (perhaps smaller) projection $q' \in M$. Thus g is outer.

Theorem 2.3. Let f be a non–zero function in $\mathbb{H}^1$. Then $f = g\,h$ where g and h lie in $\mathbb{H}^2$, h is outer, $h^*h = (f^*f)^{1/2} := |f|$, and $g^*g = hh^*$.

Proof. Consider $[|f|^{1/2}\mathbb{H}^\infty]_2$ – a right–invariant subspace of $\mathbb{L}^2$. If this space is not right–pure it contains a subspace of the form $L(p)\mathbb{L}^2$ for some projection in $\mathbb{L}^\infty$. But then, $[|f|\mathbb{H}^\infty]_1 \supseteq [|f|^{1/2}p\mathbb{L}^\infty]_1$ which is clearly right–reducing; i.e., this latter space is of the form $L(q)\mathbb{L}^1$ for some $q \in \mathbb{L}^\infty$. Write the polar decomposition of f as $f = v|f|$ and observe that

$$\begin{aligned} \mathbb{H}^1 \supseteq [f\mathbb{H}^\infty]_1 &= L(v)[|f|\mathbb{H}^\infty]_1 \\ &\supseteq L(v)L(q)\mathbb{L}^1. \end{aligned}$$

This last space is a right–reducing subspace of $\mathbb{L}^1$ and so must be zero since it is contained in $\mathbb{H}^1$. This implies that $vq = 0$ which in turn implies that q is zero, since $L(q)$ is contained in the final space of the (possibly unbounded) operator $L(|f|)$, and this is the initial space of $L(v)$. Hence $[|f|^{1/2}\mathbb{H}^\infty]_2$ is a right–pure right invariant subspace and so by Lemma 2.1, is of the form $L(\psi)\mathbb{H}^2$ for a rigid partial isometry in $\mathbb{L}^\infty$. It follows that $|f|^{1/2} = \psi h$ for a left outer function h with $[h\mathbb{H}^\infty]_2 = L(\psi^*\psi)\mathbb{H}^2$ and $|f| = |f|^{1/2}|f|^{1/2} = h^*\psi^*\psi h = h^*h$. Now set $g = vh^*$ and observe that $g^*g = hv^*vh^* = hh^*$, while $gh = vh^*h = v|f| = f$. So it remains to show $g \in \mathbb{H}^2$. For this we use Sarason's argument with only slight modifications. Recall from [A] that we need to show that $\langle g,\xi\rangle = 0$ for all elements $\xi \in \mathbb{L}^2$ such that $\xi^* \in \mathbb{H}^2_0$. Here, $\mathbb{H}^2_0 = \{f \in \mathbb{L}^2 \mid \hat{f}(n) = 0 \text{ for all } n \le 0\}$. Alternatively, we may write $\mathbb{H}^2_0 = R(\zeta)\mathbb{H}^2 = L(\zeta)\mathbb{H}^2$ where $\zeta(z) = zI_M$, a central element of $\mathbb{L}^\infty$. Since h is outer, $[h\mathbb{H}^\infty]_2 = L(p)\mathbb{H}^2$ for a constant projection p. This tells us that $L(p)$ is the final projection of the (possibly unbounded) operator $L(h)$ and this, in turn, contains the initial projection of $L(g)$ by the definition of g. Thus, since

$$\begin{aligned}\langle g,\xi\rangle &= \int \varphi(\xi^*(z)g(z))d\mu(z)\\ &= \int \varphi(g(z)\xi^*(z))d\mu\\ &= \int \varphi(g(z)p\xi^*(z))d\mu,\end{aligned}$$

we may assume that $\xi^* \in L(p)\mathbb{H}^2_0 = L(p)R(\zeta)\mathbb{H}^2$. But then $R(\zeta)^*\xi^* \in L(p)\mathbb{H}^2$ and there is a sequence of elements $\{\eta_n\}_{n=1}^\infty$ in $\mathbb{H}^\infty$ such that $h\eta_n \longrightarrow R(\zeta)^*\xi^*$. It follows that $R(\zeta)h\eta_n \longrightarrow \xi^*$ and $R(\zeta)h\eta_n = hR(\zeta)\eta_n \in \mathbb{H}^2_0$. We have, then, that

$$\begin{aligned}\langle g,\xi\rangle &= \int \varphi(g(z)p\xi^*(z))d\mu \\ &= \lim \int \varphi(g(z)ph(z)R(\zeta)\eta_n(z))d\mu \\ &= \lim \int \varphi(f(z)z \cdot \eta_n(z))d\mu(z).\end{aligned}$$

But the functions $z \cdot \eta_n(z) = (R(\zeta)\eta_n)(z)$ lie in $\mathbb{H}_0^\infty$ and since $f \in \mathbb{H}^1$, $\int \varphi(f(z)z \cdot \eta_n(z))d\mu(z) = 0$ for every n. This proves that $g \in \mathbb{H}^2$ and completes the proof of Theorem 2.3.

Remark 2.4. We structured the proof of Theorem 2.3 so it carries over with no essential changes to the context of those analytic crossed products in which the Beurling–Lax–Halmos theorem is valid. These have been identified in [MMS2].

The proof of Lemma 1.2 is now immediate. Given $f \in H^1_{L^1(M,\varphi)} = \mathbb{H}^1$, use Theorem 2.3 to write $f = gh$, $g,h \in H^2_{L^2(M,\varphi)} = \mathbb{H}^2$, with $|f| = h^*h$ and $gg^* = h^*h$. Then

$$\begin{aligned}\|f\|_1 &= \int \varphi(|f|(z))d\mu(z) \\ &= \int \varphi(h^*(z)h(z))d\mu(z) = \|h\|_2^2 \\ &= \|g\|_2^2, \text{ and so } \|f\|_1 = \|g\|_1\|h\|_1.\end{aligned}$$

POSTSCRIPT When this paper was circulating as a preprint, we received a preprint from Uffe Haagerup and Gilles Pisier, Factorization of Analytic Functions with Values in Non–Commutative L_1–Spaces and Applications, in which they prove a far–reaching generalization of Lemma 1.2., among many other interesting things.

REFERENCES

[A] Wm. Arveson, Analyticity in operator algebras, Amer. J. Math. 89 (1967), 578–642.

[B,G,M1] E. Berkson, T.A. Gillespie, and P.S. Muhly, Abstract spectral decompositions guaranteed by the Hilbert transform, Proc. London Math. Soc. (3) 53 (1986), 489–517.

[B,G,M2] ———, ———, and ———, A generalization of Macaev's theorem to non–commutative L^p–spaces, Integral Equations and Operator Theory, 10 (1987), 164–186.

[BP] O. Blasco and A. Pelczynski, Theorems of Hardy and Paley for vector–valued analytic functions and related classes of Banach spaces, to appear in Trans. Amer. Math. Soc.

[H_1] U. Haagerup, L^p–spaces associated with an arbitrary von Neumann algebra, Algèbres d'Opérateurs et Leurs Applications en Physique Mathématique (Colloques Internationaux du CNRS, No. 274, Marseilles, 20–24 juin 1977), 175–184; Éditions du CNRS, Paris 1979.

[H_2] ———, Non–commutative integration theory, lecture given at the Symposium in Pure Mathematics of the American Mathematical Society, Queens University, Kingston, Ontario, 1980 (also circulated as an unpublished note).

[Hal] P. Halmos, Shifts on Hilbert spaces, J. Reine Angew. Math. 208 (1961), 102–112.

[K] H. Kosaki, Application of the complex interpolation method to a von Neumann algebras: non–commutative L^p–spaces, J. Functional Analysis 56 (1984), 29–78.

[N] E. Nelson, Notes on non–commutative integration, J. Functional Analysis 15 (1974), 103–116.

[MMS1] M. McAsey, P.S. Muhly, and K.–S. Saito, Non–self–adjoint crossed products I (Invariant subspaces and maximality), Trans. Amer. Math. Soc. 248 (1979), 381–409.

[MMS2] ———, ———, and ———, Non–self–adjoint crossed products II, J. Math. Soc. Japan 33 (1981), 485–495.

[Sak] S. Sakai, *C^*–algebras and W^*–algebras*, Erg. Math. Und Ihrer Grenz., Springer Verlag, New York, Heidelberg, Berlin, 1971.

[Sar] D. Sarason, Generalized interpolation in H^∞, Trans. Amer. Math. Soc., 127 (1967), 179–203.

[Tak] M. Takesaki, *Theory of Operator Algebras I*, Springer Verlag, New York, Heidelberg, Berlin, 1979.

Department of Mathematics
University of Iowa
Iowa City, Iowa 52242

Contemporary Mathematics
Volume **85**, 1989

JH^* has the PCP

E. ODELL[1] AND C.S. SCHUMACHER[2]

Abstract. We prove that JH^*, the dual space of a tree space constructed by J. Hagler, has the point of continuity property.

Key Words: point of continuity property, boundedly complete skipped blocking property

AMS Subject Classification Number: 46B20

1. Introduction.

The study of the geometry of the closed bounded convex subsets of a Banach space has, within the last ten years, led to the formulation and investigation of a number of important new properties. Perhaps the most extensively studied of these new properties is the point of continuity property (PCP).

The PCP was introduced in **[B-R]** where it was shown that the PCP is strictly weaker than the RNP. This was accomplished by showing first that the PCP is implied by the existence of a boundedly complete skipped blocking finite dimensional decomposition or BCSBD, in short. Secondly it was observed that if X is either JT (the James Tree space **[J]**) or JH (a tree space constructed by Halger **[H]**) then X^* contains a separable subspace E (spanned by the biorthogonal functionals to the natural basis of X) which has a BCSBD and yet fails the RNP.

Ghoussoub and Maurey **[G-M]** have shown that a separable Banach space has the PCP iff X has a BCSBD. (Other results concerning skipped blocking decompositions are given in **[B-R]**, **[R]** and **[G-M]** for finite dimensional decompositions and **[G-M-S]** for infinite dimensional decompositions.)

Edgar and Wheeler **[E-W]** have shown that JT^* has the PCP, providing an example of an entire dual space with the PCP failing the RNP. One way of seeing

[1] This author's research was partially supported by NSF Grant #DMS-8601752.
[2] This research constitutes part of this author's Ph.D. dissertation prepared at The University of Texas at Austin under the supervision of E. Odell.

this is as follows. If Y is a separable subspace of X and both Y and X/Y have the PCP then so does X (the case where X is separable is proved in [**G-M**], Proposition II.2, and the nonseparable case appears in [**R**], Theorem 3.1.5). To apply this to JT^*, one merely lets $Y = E$ and recalls that JT^*/E is isometric to $\ell_2(\Gamma)$.

It is thus a natural question to ask whether JH^* has the PCP. The "short proof" for JT^* does not apply since JH^*/E is isometric to $c_0(\Gamma)$ which fails the PCP. However in this paper, with analysis which is similar to but more complicated than that used in [**B-R**], we prove that JH^* has the PCP. W. Schachermayer has communicated to us that he has independently found a different proof of this result.

Before we proceed, we would like to express sincere thanks to H. Rosenthal for allowing us to include an elegant principle which he formulated and which has application in a number of James Tree type spaces (see Lemma 2.1).

2. Terminology and Preliminary Results.

Throughout this section X will be a Banach Space. Subspace will mean closed, linear subspace.

For the record, we recall that X has the PCP if every non-empty, closed, bounded subset of X has a point at which the relative weak and norm topologies coincide. Equivalently, X has the PCP if every non-empty, closed bounded subset of X has a relatively weakly open subset of arbitrarily small diameter.

However, we will not need this definition. As discussed in the introduction, we will use instead the concept of a boundedly complete skipped blocking decomposition which we now define.

We will abide by the following notational conventions: Suppose (G_i) is a sequence of finite dimensional subspaces of a Banach space X, then

1) If $1 \leq n < \infty$, $[G_i]_{i=n}^{\infty}$ and $G[n, \infty)$ will represent the closed linear span of $\bigcup_{i=n}^{\infty} G_i$.
2) If $1 \leq n \leq m < \infty$, $G[n, m]$ will represent the linear span of $\bigcup_{i=n}^{m} G_i$.
3) More generally, if $A \subset \mathbb{N}$, $[G_j]_{j \in A}$ will be the closed linear span of $\bigcup_{j \in A} G_j$.

Definition. A sequence (G_i) of finite dimensional subspaces of a Banach space X is a BCSBD of X provided that the following conditions hold:

a) $X = [G_i]_{i=1}^{\infty}$,

b) $G_i \cap [G_j]_{j \neq i} = \{0\}$ for all i,

c) If (m_k) and (n_k) are sequences of positive integers with $m_k < n_k + 1 < m_{k+1}$, then the sequence $(G[m_k, n_k])_{k=1}^{\infty}$ is a boundedly complete finite dimensional decomposition (FDD) for its closed linear span.

We now present the necessary terms and results pertaining to the James Hagler space JH. Let T be the dyadic tree, $T = \bigcup_{n=0}^{\infty}\{0,1\}^n$. The elements of T are called *nodes*. If φ is a node of the form $\varphi = (\varepsilon_i)_{i=1}^{n}$, we say that φ has *length* n and denote this by $|\varphi| = n$. If $\varphi = (\varepsilon_1, \varepsilon_2, \ldots, \varepsilon_n)$ and $\psi = (\delta_1, \delta_2, \ldots, \delta_m)$ are nodes we say that $\varphi \le \psi$ if $n \le m$ and $\varepsilon_i = \delta_i$ for all $i \le n$.

A *segment* S in T is a subset of the form $S = \{\psi : \alpha \le \psi \le \varphi\} = [\alpha, \varphi]$ where α and φ are fixed nodes. A *branch* is a maximal, linearly ordered subset of T. We denote the set of all branches by Γ. Suppose that φ is a node and γ is a branch. If $\varphi \in \gamma$ we say that "*φ belongs to γ*" or that "*γ goes through φ.*" Similarly, if S is a segment and $\varphi \in S$, we say that "*φ belongs to S*" or that "*S goes through φ.*"

A pairwise disjoint family of segments $S_1, S_2, \ldots, S_n$ with $S_i = [\varphi_i, \psi_i]$ is called *admissible* if there exists $m_1 \le m_2$ so that $|\varphi_i| = m_1$ and $|\psi_i| = m_2$ for all $i = 1, \ldots, n$.

JH is the completion of the set of all finitely supported functions $x : T \to \mathbb{R}$ under the norm:

$$\|x\| = \sup\left\{\sum_{i=1}^{n} |S_i^*(x)| : S_1, S_2, \ldots, S_n \text{ is an admissible family of segments}\right\}$$

where for a segment S, $S^*(x) = \sum_{\alpha \in S} x(\alpha)$. Similarly, if $\gamma \in \Gamma$, we define $\gamma^*(x) = \sum_{\alpha \in \gamma} x(\alpha)$. These are well defined norm-1 functionals on JH. We call them segment functionals and branch functionals, respectively. The set of all branch functionals we call Γ^*, and the set of all segment functionals we call Σ^*.

If $\alpha \in T$, let $e_\alpha \in JH$ be defined by $e_\alpha(\varphi) = \delta_{\alpha\varphi}$ for all $\varphi \in T$. Ordered lexicographically, the set $(e_\alpha)_{\alpha \in T}$ is a basis for JH which we call the *node ba-*

sis. Its biorthogonal functionals, $(e^*_\alpha)_{\alpha \in T}$, are called *node functionals*. The set of biorthogonal functionals we call F; its closed linear span we will denote by E.

It is known that $JH^* = [\Gamma^* \cup F]$. A similar formulation holds in JT^*. H. Rosenthal devised the following lemma which encapsulates the principle behind this fact in JT, JH, and spaces like them. It also gives a quick way to see that they do not contain ℓ^1. The origin of the lemma lies in an argument in [**O**].

Lemma 2.1. (Rosenthal) *Let B be a separable Banach Space. Suppose that $W_0 \subset W \subset B^*$ with W_0 countable. For each $b \in B$, let $\hat{b}$ be the canonical inclusion of b into B^{**}. If for all bounded sequences (b_n) in B:*

i) $\hat{b}_n \to 0$ pointwise on W implies that $b_n \to 0$ weakly, and

ii) $\hat{b}_n \to 0$ pointwise on W_0 implies that some subsequence of $(\hat{b}_n)$ converges pointwise on W,

then $\ell^1 \not\hookrightarrow B$ and $B^* = [W]$.

Proof. Suppose $\ell^1 \hookrightarrow B$ (*i.e.*, suppose ℓ_1 embeds into B). Let (x_n) be a sequence in B which is equivalent to the unit vector basis of ℓ^1. Since W_0 is countable we may assume (by extracting a subsequence if necessary) that $(\hat{x}_n)$ converges pointwise on W_0. Define $b_n = x_{2n} - x_{2n-1}$ for all n. Then $(\hat{b}_n)$ goes to zero pointwise on W_0. Extract a subsequence (b'_n) of (b_n) that converges pointwise on W. Since (b'_n) is bounded, (b'_n) is weak Cauchy by (i). But this cannot be since (b'_n) is equivalent to the ℓ^1 basis. We conclude that $\ell^1 \not\hookrightarrow B$.

We must now show that $B^* = [W]$. Choose $g \in B^{**}$ such that $g(w) = 0$ for all $w \in W$. Since $\ell^1 \not\hookrightarrow B$, we can choose a sequence (x_n) in B so that $\hat{x}_n \overset{w^*}{\to} g$ [**O-R**]. But if $w \in W$, then $\hat{x}_n(w) \to g(w) = 0$. Then (i) says that $(x_n) \to 0$ weakly. Hence g is the zero functional, as needed. ∎

Lemma 2.2. *JH satisfies the hypotheses of lemma 2.1 with $W_0 = F$ and $W = \Gamma^* \cup F$. Hence $\ell^1 \not\hookrightarrow JH$, and $JH^* = [\Gamma^* \cup F]$.*

Proof. Let (b_n) be a bounded sequence in JH with $\hat{b}_n(w) \to 0$ for all $w \in W$. If (b_n) is not weakly null we may assume (by passing to a subsequence that there exists

$x^* \in JH^*$, $\|x^*\| = 1$ and $\delta > 0$ so that $\hat{b}_n(x^*) > \delta$ for all n. In particular $\|d\| > \delta$ if d is a convex combination of (b_n).

$\Sigma^* \cup \Gamma^*$ is a weak* compact subset of Ball(JH^*) and thus the map $b \to \hat{b}|_{\Sigma^* \cup \Gamma^*}$ is a norm one mapping of JH into the Banach space $C(\Sigma^* \cup \Gamma^*)$. Since $(\hat{b}_n)$ is pointwise null on W, $(\hat{b}_n|_{\Sigma^* \cup \Gamma^*})$ is weakly null in $C(\Sigma^* \cup \Gamma^*)$. Thus by replacing (b_n) by a convex block subsequence of itself we may, in addition, suppose that

$$\sup\{|S^*(b_n)| : S^* \in \Sigma^* \cup \Gamma^*\} \to 0 \quad \text{as} \quad n \to \infty .$$

But then an easy calculation shows that a subsequence of (b_n) is equivalent to the unit vector basis of c_0, a contradiction. Indeed we may assume, by passing to a subsequence, that there exists $p_1 < q_1 < p_2 < q_2 < \cdots$ such that (b_n) is a seminormalized block basis of (e^*_ϕ) satisfying for all i,

a) $b_i \in$ span $(F^{q_i}_{p_i})$,

b) $\|\hat{b}_i|_{\Sigma^* \cup \Gamma^*}\| \le \varepsilon_i$,

c) $2^{q_i} \sum_{j=i+1}^{\infty} \varepsilon_j \le 1$,

where $F^j_i = \{e^*_\phi : i \le |\phi| \le j\}$. Let $x = \sum a_j b_j$ and let $(S_i)_{i=1}^k$ be an admissible family of segments with $m_1 = \inf\{|\phi| : \phi \in S_1\}$ and $m_2 = \sup\{|\phi| : \phi \in S_1\}$. Note that $k \le 2^{m_1}$. We claim that

$$\sum_{i=1}^{k} |S_i^*(x)| \le \max_j \|a_j b_j\| + \max |a_j|$$

and hence (b_n) is a c_0 basis.

To see the claim observe that we may assume (by altering $(S_i)_1^k$ if necessary) that there exists $j_1 \in \mathbb{N}$ with $p_{j_1} \le m_1 \le q_{j_1}$. Thus

$$\sum_{i=1}^{k} |S_i^*(x)| \leq \sum_{i=1}^{k} |S_i^*(a_{j_1} b_{j_1})| + \sum_{i=1}^{k} \left| S_i^* \left(\sum_{j=j_1+1}^{j_2} a_j b_j \right) \right|$$

$$\leq \|a_{j_1} b_{j_1}\| + k \max_j |a_j| \sum_{j=j_1+1}^{j_2} \|\hat{b}_j|_{\Sigma^* \cup \Gamma^*}\|$$

$$\leq \max_j \|a_j b_j\| + \max_j |a_j| \left(2^{m_1} \sum_{j=j_1+1}^{\infty} \varepsilon_j \right) .$$

Since $2^{m_1} \leq 2^{q_{j_1}}$, the claim follows from c). This shows that JH satisfies hypothesis (i) of lemma 2.1.

To see that JH satisfies hypothesis (ii) of lemma 2.1, we suppose that (b_n) is bounded and that $(\hat{b}_n)$ goes to zero on the node functionals. It can be easily seen that for all $\varepsilon > 0$, and each subsequence (b'_n) of (b_n), there exists a further subsequence (b''_n) of (b'_n) so that $\overline{\lim}|\gamma^*(b''_n)| \leq \varepsilon$ for all but finitely many $\gamma \in \Gamma$. Thus there exists K, a countable subset of Γ, and (b'_n) a subsequence of (b_n) so that $0 = \overline{\lim}|\gamma^*(b'_n)| = \lim \gamma^*(b'_n)$ for all $\gamma \notin K$. Extract a further subsequence (b''_n) of (b'_n) so that (b''_n) converges pointwise on K. Then (b''_n) converges pointwise on all of W, as needed. ■

Remark. Of course in **[H]** it is shown with a more complicated argument that if (b_n) is a normalized sequence satisfying (i) then (b_n) has a subsequence equivalent to the unit vector basis of c_0. From this it follows that $JH^* = [\Gamma^* \cup F]$.

The following norm-1 projections on JH and their adjoints will be needed in the sequel. Fix $x \in JH$, φ and $\psi \in T$, $\gamma \in \Gamma$, and $n \leq m \in \mathbb{N}$. Define:

$$\Pi_\varphi(x)(\psi) = x(\psi) \text{ if } \psi \geq \varphi;\ 0 \text{ otherwise.}$$

$$\Pi_n(x)(\psi) = x(\psi) \text{ if } |\psi| = n;\ 0 \text{ otherwise.}$$

$$\Pi_n^m(x)(\psi) = x(\psi) \text{ if } n \leq |\psi| \leq m;\ 0 \text{ otherwise.}$$

$$\Pi_n^\infty(x)(\psi) = x(\psi) \text{ if } |\psi| \geq n;\ 0 \text{ otherwise.}$$

$$\Pi_\gamma(x)(\psi) = x(\psi) \text{ if } \psi \in \gamma;\ 0 \text{ otherwise.}$$

Recall that a sequence of nodes $(\varphi_n)_{n\in\mathbb{N}}$ is said to be *stongly incomparable* if any two of the nodes are incomparable and if each admissible family of segments contains at most two elements in the sequence.

A simple argument [**H**] shows that any sequence of distinct nodes in T has a subsequence which is either strongly incomparable or all of whose elements lie on a branch.

It is easy to see that if $(\varphi_n)_{n\in\mathbb{N}}$ is a strongly incomparable sequence of nodes then $(e_{\varphi_n})_{n\in\mathbb{N}}$ is equivalent to the unit vector basis of c_0. As we mentioned before, Hagler [**H**] proved the more general fact that every normalized, weakly null sequence in JH has a subsequence which is equivalent to the unit vector basis of c_0. The arguments used to prove this theorem will play a key role in our proof.

If $S_1, S_2, \ldots, S_n$ are pairwise disjoint, infinite segments all beginning at the kth level, it is straightforward to see that $(S_1^*, S_2^*, \ldots, S_n^*)$ is isometrically equivalent to the unit vector basis of ℓ_∞^n. That is, if $y^* = \sum_{i=1}^n a_i S_i^*$, then $\|y^*\| = \max\{a_1, a_2, \ldots, a_n\} = |a_j|$ for some $j \leq n$. Elements y^* of this form will be called elements of *type A*. It is clear that given any such y^* there exists j with $1 \leq j \leq n$ such that for any integer $m \geq k$, $\|y^*\| = |y^*(e_\varphi)|$ where φ is the node belonging to S_j with $|\varphi| = m$. S_j is called a *norming component for* y^*.[1]

If $y \in JH$, the *support of* y is $\text{supp}(y) = \{\varphi \in T : y(\varphi) \neq 0\}$. If $y^* \in JH^*$, the *support of* y^* is $\text{supp}(y^*) = \{\varphi \in T : y^*(e_\varphi) \neq 0\}$.

---

[1] This analysis also applies to $S_1^*, S_2^*, \ldots, S_n^*$ if $S_1, S_2, \ldots, S_n$ is an admissible family of segments whose support lies between the kth and ℓ^{th} levels where $k \leq \ell \in \mathbb{N}$. In this case, y^* is called an element of *finite type A*, and given any $m \in \mathbb{N}$ with $k \leq m \leq \ell$ there is a $j \leq n$ and a node $\varphi \in S_j$ such that $|\varphi| = m$ and $\|y^*\| = |y^*(e_\varphi)|$.

3. JH^* has the PCP.

Recall that the PCP is a property which is separably determined [**B-R**]; that is, a Banach space X has the PCP if every separable subspace of X has the PCP. Thus it is sufficient to show that every separable subspace of JH^* has the PCP. It was noted in section 2 that $JH^* = [\Gamma^* \cup F]$. Hence if Y is a separable subspace of JH^* there is a countable subset Γ_0^* of Γ^* so that $Y \subset [\Gamma_0^* \cup F]$. This, together with the fact that the PCP is an hereditary property (every subspace of a Banach space with the PCP has the PCP), means that it is enough to show that given a countable subset Γ_0^* of Γ^*, there is a BCSBD of $[\Gamma_0^* \cup F]$. We now fix the subset Γ_0^*. Write $Z = [\Gamma_0^* \cup F]$.

Enumerate Γ_0^*; $\Gamma_0^* = \{\gamma_1^*, \gamma_2^*, \gamma_3^*, \ldots\}$. For each $n \in \mathbb{N}$ let $d(n)$ be the smallest integer such that given any two distinct positive integers i and j with $i \leq n$ and $j \leq n$, $\left(\text{supp}(\Pi_{d(n)}^*(\gamma_i^*))\right) \cap \left(\text{supp}(\Pi_{d(n)}^*(\gamma_j^*))\right) = \emptyset$. (That is, $d(n)$ is the smallest integer so that the supports of the first n branches in Γ_0^* are disjoint at the $d(n)$ level.)

If $n \leq m$, define $F_n^m = \{e_\varphi^* : n \leq |\varphi| \leq m\}$.

We now construct the desired sequence (G_i) of finite dimensional subspaces of Z. We proceed by induction.

Let $n_1 = 0$. Choose $m_1 > n_1$ arbitrarily. Define $\alpha_1^* = [\Pi_{m_1+1}^\infty]^*(\gamma_1^*)$. Let $G_1 = [F_{n_1}^{m_1} \cup \{\alpha_1^*\}]$. Let $n_2 = m_1 + 1$. Choose $m_2 > \max\{n_2, d(2)\}$. Define $\alpha_2^* = [\Pi_{m_2+1}^\infty]^*(\gamma_2^*)$ and $G_2 = [F_{n_2}^{m_2} \cup \{\alpha_2^*\}]$. Next let $n_3 = m_2 + 1$, choose $m_3 > \max\{n_3, d(3)\}$, and let $\alpha_3^* = [\Pi_{m_3+1}^\infty]^*(\gamma_3^*)$. Define $G_2 = [F_{n_3}^{m_3} \cup \{\alpha_3^*\}]$. Continue this process inductively.

We must now show that the sequence $(G_i)_{i=1}^\infty$ is a BCSBD of Z. Our first few lemmas are quite elementary; however, it is worth going through them in order to get a feel for the norm in JH^*. It is clear that $Z = [G_i]_{i=1}^\infty$. The following lemma shows that $G_i \cap [G_j]_{j \neq i} = \{0\}$ for all i.

Lemma 3.1. *Given $z^* \in G_i$, $\|z^*\| = 1$, and $w^* \in \text{span}(\cup_{j\neq i} G_j)$, $\|z^* - w^*\| \geq 1/4$. Thus $z^* \notin [G_j]_{j \neq i}$.*

Proof. Write $z^* = x^* + y^*$ where $x^* \in \text{span}(F_{n_i}^{m_i})$ and $y^* = a\alpha_i^*$. If $\|y^*\| \geq 1/4$, then for n sufficiently large

$$\|z^* - w^*\| \geq \|[\Pi_n^\infty]^*[\Pi_{\alpha_i^*}]^*(z^* - w^*)\| = \|[\Pi_n^\infty]^* y^*\| = \|y^*\| \geq 1/4 \,.$$

Similarly, since for n sufficiently large $[\Pi_n^\infty]^* w^*$ is an element of type A whose support is disjoint from that of y^*, if $\lim_{n\to\infty} \|[\Pi_n^\infty]^* w^*\| \geq 1/4$, we have that $\|z^* - w^*\| \geq 1/4$.

Thus we may assume that $\|y^*\| < 1/4$ and $\lim_{n\to\infty} \|[\Pi_n^\infty]^* w^*\| < 1/4$. Note then that $\|x^*\| \geq 3/4$. Also note that $[\Pi_{n_i}^{m_i}]^* w^*$ is an element of finite type A with the property that $\|[\Pi_{n_i}^{m_i}]^* w^*\| \leq \lim_{n\to\infty} \|[\Pi_n^\infty]^* w^*\|$. Thus

$$\|z^* - w^*\| \geq \|[\Pi_{n_i}^{m_i}]^*(z^* - w^*)\| = \|x^* - [\Pi_{n_i}^{m_i}]^* w^*\| \geq \|x^*\| - \|[\Pi_{n_i}^{m_i}]^* w^*\|$$
$$> 3/4 - 1/4 = 1/2 \,. \qquad \blacksquare$$

Let us now fix a skipped blocking of the sequence (G_i). Choose sequences (r_k) and (t_k) of natural numbers so that $r_k < t_k + 1 < t_{k+1}$. We must show that $(G[r_k, t_k])_{k=1}^\infty$ is a boundedly complete FDD. To simplify the notation a bit let us take $p_i = n_{r_i}$ and $q_i = m_{t_i}$ for all i. Thus $G[r_k, t_k] = \left[F_{p_k}^{q_k} \cup \{\alpha_i^*\}_{i=r_k}^{t_k}\right]$.

Lemma 3.2. *For all sequences (w_i^*) with $w_i^* \in G[r_i, t_i]$, for all sequences of scalars (a_i), and for all pairs of positive integers $r \leq s$:*

$$\left\|\sum_{i=1}^{r} a_i w_i^*\right\| \leq 2\left\|\sum_{i=1}^{s} a_i w_i^*\right\| \,.$$

Thus $(G[r_k, t_k])_{k=1}^\infty$ is an FDD.

Proof. Fix $z_1^* = \sum_{i=1}^r a_i w_i^*$ and $z_2^* = \sum_{i=r+1}^s a_i w_i^*$. Define: $x_1^* = [\Pi_0^{q_r}]^*(z_1^*)$ and $y_1^* = z_1^* - x_1^*$.

We distinguish two cases:

<u>Case A</u>: Suppose that $\|z_1^*\| \leq 2\|x_1^*\|$. Then

$$\|z_1^*\| \leq 2\|x_1^*\| = 2\|[\Pi_0^{q_r}]^*(z_1^* + z_2^*)\| \leq 2\|z_1^* + z_2^*\|$$

Case B: Suppose that $\|z_1^*\| \le 2\|y_1^*\|$. Note that y_1^* is an element of type A. Let φ be a node with $|\varphi| = q_r + 1$ such that $\|y_1^*\| = |y_1^*(e_\varphi)|$. Since the $(q_r + 1)$st level was skipped, $|z_2^*(e_\varphi)| = |x_1^*(e_\varphi)| = 0$ so we have that

$$\|z_1^*\| \le 2\|y_1^*\| = 2|y_1^*(e_\varphi)| = 2|z_1^*(e_\varphi) + z_2^*(e_\varphi)| \le 2\|z_1^* + z_2^*\| . \qquad \blacksquare$$

Finally, we must show that the FDD, $(G[r_k, t_k])$, is boundedly complete. Let us fix a sequence (z_i^*) with $z_i^* \in G[r_i, t_i]$ for all i. Assuming that $\sum z_i^*$ does not converge, we must show that the partial sums are unbounded. By passing to further blocks and relabeling if necessary we may assume that (z_i^*) is a bounded sequence with $\|z_i^*\| > \eta > 0$ for all i.

The following simple lemma will be used many times below. We include it for the sake of completeness and for reference.

Lemma 3.3. *Suppose that (z_j^*) is a sequence in JH^* with $z_j^* = v_j^* + w_j^*$ for all j. Let $M = \{m_1, m_2, \ldots\}$ be an increasing sequence of positive integers, $(b_{m_i})_{i \in \mathbb{N}}$ a sequence in JH which is equivalent to the unit vector basis of c_0 and fix $\alpha > 0$. If*

1) $|\langle v_{m_i}^*, b_{m_i}\rangle| > \alpha$ for all $i \in \mathbb{N}$ and $\langle v_k^*, b_{m_i}\rangle = 0$ for all $k \ne m_i$, and

2) (k_i) is a sequence of positive integers such that $\sum_{i=1}^\infty \|w_{k_i}^*\| < \infty$, and $\langle w_j^*, b_{m_i}\rangle = 0$ for all $j \ne k_i$, or

2′) $\langle w_k^*, b_{m_i}\rangle = 0$ for all k and $i \in \mathbb{N}$,

then $\sum z_j^*$ has unbounded partial sums.

Proof. We will prove the lemma for conditions 1 and 2; the proof is similar, though simpler, when 1 and 2′ hold.

Let $K \in \mathbb{R}$ be such that $\|\sum_{i=1}^r \varepsilon_i b_{m_i}\| \le K$ for all $r \in \mathbb{N}$ and for all sequences (ε_i) with $\varepsilon_i = 1$ or -1; and such that $\sum_{i=1}^\infty \|w_{k_i}^*\| \le K$. Fix $n \in \mathbb{N}$. For each $i \in \mathbb{N}$, let $c_{m_i} = \operatorname{sgn}\langle v^*_{m_i}, b_{m_i}\rangle b_{m_i}$. Let s be the largest index so that $m_s \leq n$. We have:

$$\begin{aligned}
\Big\|\sum_{j=1}^n z_j^*\Big\| &\geq K^{-1}\Big|\Big\langle \sum_{j=1}^n z_j^* \,,\, \sum_{i=1}^n c_{m_i}\Big\rangle\Big| \\
&= K^{-1}\Big|\Big\langle \sum_{j=1}^n v_j^* + \sum_{j=1}^n w_j^* \,,\, \sum_{i=1}^n c_{m_i}\Big\rangle\Big| \\
&\geq K^{-1}\sum_{i=1}^s \langle v^*_{m_i}, c_{m_i}\rangle - K^{-1}\sum_{i=1}^n |\langle w^*_{k_i}, c_{m_i}\rangle| \\
&\geq K^{-1}\sum_{i=1}^s |\langle v^*_{m_i}, b_{m_i}\rangle| - \sum_{i=1}^\infty \|w^*_{k_i}\| \\
&\geq \alpha K^{-1}s - K
\end{aligned}$$

This last quantity goes to ∞ as n does. ∎

The rest of the proof consists of 4 lemmas which together encompass all possible cases. Before we proceed, let us define $y_i^* = [\Pi^\infty_{q_i+1}]^*(z_i^*)$ and $x_i^* = z_i^* - y_i^*$ for all i. This decomposition will be understood throughout the rest of the proof. Before we proceed any further, we observe several things about it. Note that y_i^* is an element of type A whose support begins that the (q_i+1)st level, that the y_j^*'s are disjointly supported, and that $\operatorname{supp}([\Pi_{q_i+1}]^*(y_i^*)) \cap \operatorname{supp}(x_j^*) = \emptyset$ for all j. Note also that $x_i^* \in [F^{q_i}_{p_i}]$ for all i.

Lemma 3.4. *If* $\overline{\lim}\, \|y_i^*\| > \delta > 0$, *then the partial sums of* $\sum z_i^*$ *are unbounded.*

Proof. For each i choose a node φ_i such that $|\varphi_i| = q_i + 1$ and $\|y_i^*\| = |\langle y_i^*, e_{\varphi_i}\rangle|$.

There is an infinite subset M of $\mathbb{N}$ so that $\|y_i^*\| > \delta$ for all $i \in M$ and such that either $(\varphi_i)_{i\in M}$ is strongly incomparable or φ_i lies on a branch γ for all $i \in M$.

Case A. Suppose that $(\varphi_i)_{i\in M}$ is strongly incomparable. $|\langle y_i^*, e_{\varphi_i}\rangle| > \delta$ for each $i \in M$. Note that if $j \in M$, $\langle x_i^*, e_{\varphi_j}\rangle = 0$ for all $i \in \mathbb{N}$ and that $\langle y_i^*, e_{\varphi_j}\rangle = 0$ for all $i \in \mathbb{N}$ with $i \neq j$. Since $(e_{\varphi_i})_{i\in M}$ is equivalent to the unit vector basis of c_0, taking $y_j^* = v_j^*$, $x_j^* = w_j^*$ ($j \in \mathbb{N}$), and $e_{\varphi_i} = b_i$ ($i \in M$) in lemma 3.3$'$, we conclude that $\sum z_j^*$ has unbounded partial sums.

Case B. Suppose then that $\varphi_i \in \gamma$ for all $i \in M$. Let α_i be the norming component of y_i^* containing φ_i. Enumerate M in increasing order: $M = \{m_1, m_2, m_3, \ldots\}$. Let ψ_{m_i} be the node on α_{m_i} with $|\psi_{m_i}| = |\varphi_{m_{i+1}}|$. Note that $\psi_{m_i} \neq \varphi_{m_{i+1}}$, since $\alpha_{m_i} \cap \alpha_{m_{i+1}} = \emptyset$. Clearly, the sequence (ψ_{m_j}) is strongly incomparable. Hence $(e_{\psi_{m_j}})$ is equivalent to the unit vector basis of c_0. We have that $|\langle y_{m_i}, \psi_{m_i}\rangle| = \|y_{m_i}^*\| > \delta$ for each $i \in \mathbb{N}$. Note that $\langle x_i^*, \psi_{m_j}\rangle = 0$ for all i and j and that $\langle y_i^*, \psi_{m_j}\rangle = 0$ for all $i \neq m_j$. As before, the conditions of lemma 3.3′ are satisfied with $y_i^* = v_i^*$, $x_i^* = w_i^*$ and $\psi_{m_i} = b_{m_i}$ for all i. ■

From now on we assume that $\lim_{i\to\infty} \|y_i^*\| = 0$; given this, we may also assume that $\|x_i^*\| > 1$ (by deleting a finite number of terms in the sequence, multiplying by $1/\eta$, and relabeling). The following will be fixed for the remainder of the proof. Let (β_i) be a sequence in JH such that for all i:

i) $\beta_i \in [e_\varphi : p_i \le |\varphi| \le q_i]$,

ii) $\|\beta_i\| = 1$,

iii) $\langle x_i^*, \beta_i\rangle > 1$.

We make use of the following decomposition technique which was employed by Hagler [H]. For $\varepsilon > 0$ and $k \in \mathbb{N}$, define:

$$F_{k,\varepsilon} = \{\ \varphi \in T : |\varphi| = p_k \ \text{ and } \ |\gamma^*(\beta_k)| > \varepsilon \ \text{ for some } \ \gamma \in \Gamma \text{ passing through } \varphi\ \}$$

Let $\beta_{k,\varepsilon} = \sum_{\varphi \in F_{k,\varepsilon}} \Pi_\varphi(\beta_k)$ and $\omega_{k,\varepsilon} = \beta_k - \beta_{k,\varepsilon}$. Note that $^{\#}F_{k,\varepsilon} < 1/\varepsilon$, since $\|\beta_k\| = 1$.

Lemma 3.5. *If for all $\varepsilon > 0$, $\overline{\lim}\langle x_k^*, \omega_{k,\varepsilon}\rangle \ge 1/2$, then the partial sums of $\sum z_i^*$ are unbounded.*

Proof. For all $\varepsilon > 0$ there are infinitely many $k \in \mathbb{N}$ such that $\langle x_k^*, \omega_{k,\varepsilon}\rangle > 1/4$. Let (ε_i) be a sequence of positive numbers decreasing to zero. Let (k_i) be an increasing sequence of positive integers so that $\langle x_{k_i}^*, \omega_{k_i,\varepsilon_i}\rangle > 1/4$ for each $i \in \mathbb{N}$. Note that $\langle x_n^*, \omega_{k_i,\varepsilon_i}\rangle = 0$ for all $n \neq k_i$.

We may write $\omega_{k_i,\varepsilon_i} = \sum_{j=1}^{m_i}(\Pi_{\varphi_{ij}}(\beta_{k_i}))$ where $|\varphi_{ij}| = p_{k_i}$ for $1 \le j \le m_i$.

Let us adopt the notational convention that if φ is a node, φ^- is its (unique) immediate predecessor in the tree T.

For each node φ_{ij} we will define an element v_{ij} in JH. Fix φ_{ij}.

Case i: If $\varphi_{ij} \notin \operatorname{supp}(y_k^*)$ for any k, then let $v_{ij} = \Pi_{\varphi_{ij}}(\beta_{k_i})$. Note that $\langle y_k^*, v_{ij}\rangle = 0$ for all $k \in \mathbb{N}$.

Case ii: If $\varphi_{ij} \in \operatorname{supp}(y_k^*)$ for some $k \in \mathbb{N}$ we take a different tack. First we note that k is unique, since the y_j^*'s are disjointly supported. Write $\Pi^*_{\varphi_{ij}}(y_k^*) = a_{ij}\alpha^*_{ij}$. Thus $\langle y_k^*, \Pi_{\varphi_{ij}}(\beta_{k_i})\rangle = a_{ij}\langle \alpha^*_{ij}, \beta_{k_i}\rangle$. Define

$$v_{ij} = -\langle \alpha^*_{ij}, \beta_{k_i}\rangle e_{\varphi^-_{ij}} + \Pi_{\varphi_{ij}}(\beta_{k_i}) \ .$$

Since the y_j^*'s are disjointly supported and $\operatorname{supp}(\Pi^*_{p_{k_i}-1}(y_k^*)) \neq \emptyset$, k is the unique integer for which $\varphi^-_{ij} \in \operatorname{supp}(y_k^*)$. Thus $\langle y_k^*, e_{\varphi^-_{ij}}\rangle = a_{ij}$ and $\langle y_n^*, e_{\varphi^-_{ij}}\rangle = 0$ for $n \neq k$. We observe then that $\langle y_m^*, v_{ij}\rangle = 0$ for all $m \in \mathbb{N}$.

For each $i \in \mathbb{N}$, define $v_{k_i} = \sum_{j=1}^{m_i} v_{ij}$.

Since (by the skipping) $\varphi^-_{ij} \notin \operatorname{supp}(x_m)$ for any $m \in \mathbb{N}$, $\langle x_{k_i}, v_{k_i}\rangle = \langle x_{k_i}, \omega_{k_i,\varepsilon_i}\rangle > 1/4$ and for $n \neq k_i$, $\langle x_n, v_{k_i}\rangle = \langle x_n, \omega_{k_i,\varepsilon_i}\rangle = 0$.

Fix $\gamma \in \Gamma$ and $i \in \mathbb{N}$. If $\varphi_{ij} \notin \gamma$ for any $j \leq m_i$, then either

$$|\langle \gamma^*, v_{k_i}\rangle| = 0 \ , \quad \text{or}$$

for some $j \leq m_i$, $|\langle \gamma^*, v_{k_i}\rangle| = |-\langle \alpha^*_{ij}, \beta_{k_i}\rangle| < \varepsilon_i$, depending on whether or not $\varphi^-_{ij} \in \gamma$.

On the other hand, if $\varphi_{ij} \in \gamma$ for some $j \leq m_i$ then either

$$\begin{aligned}
|\langle \gamma^*, v_{k_i}\rangle| &= |\langle \gamma^*, \omega_{k_i,\varepsilon_i}\rangle| \leq \varepsilon_i \ , \quad \text{or} \\
|\langle \gamma^*, v_{k_i}\rangle| &= |-\langle \alpha^*_{ij}, \beta_{k_i}\rangle \gamma^*(e_{\varphi^-_{ij}}) + \langle \gamma^*, \Pi_{\varphi_{ij}}(\beta_{k_i})\rangle| \\
&\leq |-\langle \alpha^*_{ij}, \beta_{k_i}\rangle \gamma^*(e_{\varphi^-_{ij}})| + |\langle \gamma^*, \Pi_{\varphi_{ij}}(\beta_{k_i})\rangle| \\
&\leq |-\langle \alpha^*_{ij}, \omega_{k_i,\varepsilon_i}\rangle| + |\langle \gamma^*, \omega_{k_i,\varepsilon_i}\rangle| \leq 2\varepsilon_i \ ,
\end{aligned}$$

depending on whether φ_{ij} belongs in case i or case ii.

We conclude that $\lim_{i\to\infty}\langle \gamma^*, v_{k_i}\rangle = 0$ for all $\gamma \in \Gamma$. Since (v_{k_i}) is a bounded block basis of $(e_\varphi)_{\varphi\in T}$ and $JH^* = [\Gamma^* \cup F]$, v_{k_i} goes to zero weakly. By extracting a subsequence if necessary we may assume that (v_{k_i}) is equivalent to the unit vector basis of c_0 [H].

We observe that the conditions of lemma 3.3′ are satisfied with $x_i^* = v_i^*$, $y_i^* = w_i^*$, and $v_{k_i} = b_{k_i}$ where $M = \{k_1, k_2, k_3, \ldots\}$. We conclude that the partial sums of $\sum z_i^*$ are unbounded. ∎

Recall our standing assumption that $\lim_{i\to\infty} \|y_i^*\| = 0$. In addition, we can now assume that there is an $\varepsilon > 0$ such that $\overline{\lim}\langle x_k^*, \beta_{k,\varepsilon}\rangle > 1/2$. Let M be an infinite subset of $\mathbb{N}$ so that $\langle x_k^*, \beta_{k,\varepsilon}\rangle > 1/2$ for all $k \in M$. Since $\#F_{k,\varepsilon} < 1/\varepsilon$, for each $k \in M$ there exists $\varphi_k \in F_{k,\varepsilon}$ so that $\langle x_k^*, \Pi_{\varphi_k}(\beta_k)\rangle > \varepsilon/2$. By extracting a subsequence if necessary we may assume either that $(\varphi_k)_{k\in M}$ is a strongly incomparable sequence of nodes or that there is some $\gamma \in \Gamma$ with $\varphi_k \in \gamma$ for all $k \in M$. All of these things shall remain fixed in the lemmas that follow.

Lemma 3.6. *If the sequence φ_k is strongly incomparable, then the partial sums of $\sum z_i^*$ are unbounded.*

Proof. Since (φ_k) is strongly incomparable, $(\Pi_{\varphi_k}(\beta_k))_{k\in M}$ is equivalent to the unit vector basis of c_0.

Case A: Suppose there are infinitely many $k \in M$ such that $\varphi_k \notin \mathrm{supp}(y_i^*)$ for any $i \in \mathbb{N}$. By extracting a subsequence once more we may assume that this condition holds for all $k \in M$. Hence

$$\mathrm{supp}\big(\Pi_{\varphi_k}(\beta_k)\big) \cap \mathrm{supp}(y_i^*) = \emptyset$$

for all $k \in M$ and all $i \in \mathbb{N}$. In particular,

i) $\langle y_i^*, \Pi_{\varphi_k}(\beta_k)\rangle = 0$ for all $i \in \mathbb{N}$ and all $k \in M$,

ii) $\langle x_k^*, \Pi_{\varphi_k}(\beta_k)\rangle > \varepsilon/2$ for all $k \in M$, and

iii) $\langle x_i^*, \Pi_{\varphi_k}(\beta_k)\rangle = 0$ for all $k \in M$ and all $i \neq k$.

Thus the hypotheses of lemma 3.3′ are satisfied with $x_i^* = v_i^*$, $y_i^* = w_i^*$, $(i \in \mathbb{N})$, and $\Pi_{\varphi_k}(\beta_k) = b_k$ $(k \in M)$, and so the partial sums of $\sum z_i^*$ are unbounded.

Case B: Suppose that for all but finitely many $k \in M$, $\varphi_k \in \mathrm{supp}(y_i^*)$ for some $i \in \mathbb{N}$. We may once more extract a subsequence and assume that this condition holds for all $k \in M$. Since (φ_k) is strongly incomparable, we know that given $j \in \mathbb{N}$ there is at most one $k \in M$ with $\varphi_k \in \text{supp}(y_j^*)$. Enumerate M in increasing order: $M = \{m_1, m_2, m_3, \ldots\}$. For each i, let k_i be such that $\varphi_{m_i} \in \text{supp}(y_{k_i}^*)$. Our assumptions above assure that k_i is well-defined and that if $i \neq j$ then $k_i \neq k_j$. Hence:

i) $\langle y_j^*, \Pi_{\varphi_{m_i}}(\beta_{m_i})\rangle = 0$ if $j \neq k_i$,

ii) $\langle x_{m_i}^*, \Pi_{\varphi_{m_i}}(\beta_{m_i})\rangle > \varepsilon/2$ for all $i \in \mathbb{N}$, and

iii) $\langle x_k^*, \Pi_{\varphi_{m_i}}(\beta_{m_i})\rangle = 0$ for all $k \neq m_i$.

Extracting a subsequence for one final time, we ensure that $\sum_{i=1}^{\infty} \|y_{k_i}^*\| < 1$ and thus that our sequences (x_j^*), (y_j^*), and $(\Pi_{\varphi_{m_i}}(\beta_{m_i}))$ will satisfy the hypotheses of lemma 3.3. We conclude that the partial sums of $\sum z_i^*$ are unbounded, as needed. ■

Lemma 3.7. *If the sequence $(\varphi_k)_{k\in M}$ all lies on the branch γ, then the partial sums of $\sum z_i^*$ are unbounded.*

Proof. Before proceeding, let us establish some notation and make a definition. As before, if φ is a node we will take φ^- to be its immediate predecessor in the tree. For each $i \in M$ define:

$$u_i = -\langle \gamma^*, \beta_i\rangle e_{\varphi_i^-} + \Pi_{\varphi_i}(\beta_i)\,.$$

Since $|\varphi_i| = p_i$, $|\varphi_i^-| = p_i - 1 > q_{i-1}$. We observe that if $i \in M$:

i) $\langle x_j^*, u_i\rangle = 0$ for all $j \in \mathbb{N}$ with $j \neq i$, and

ii) $\langle x_i^*, u_i\rangle = \langle x_i, \Pi_{\varphi_i}(\beta_i)\rangle > \varepsilon/2$.

Since $\langle \gamma^*, u_k\rangle = 0$ for all $k \in M$, and given any other branch κ, $\langle \kappa^*, u_k\rangle = 0$ for all k sufficiently large, we observe that $u_k \to 0$ weakly (again, u_k is a bounded block basis of $(e_\varphi)_{\varphi\in T}$). We could thus extract a subsequence in order to assure that $(u_k)_{k\in M}$ is equivalent to the unit vector basis of c_0. (Actually, it is easy to see directly that (u_k) is already equivalent to the unit vector basis of c_0.)

We now proceed to distinguish three cases:

Case A: Suppose that there are infinitely many $k \in M$ for which $\varphi_k^- \notin \operatorname{supp}(y_i^*)$ for any $i \in \mathbb{N}$. By extracting a subsequence if necessary we may assume that this condition holds for all $k \in M$. Then $\langle y_i^*, u_k \rangle = 0$ for all $i \in \mathbb{N}$ and for all $k \in M$. Hence the sequences $(x_j^*)_{j \in \mathbb{N}}$, $(y_j^*)_{j \in \mathbb{N}}$, and $(u_k)_{k \in M}$ satisfy the conditions of lemma 3.3$'$, and we are done.

We may thus assume (by extracting a subsequence if necessary) that given any $k \in M$, $\varphi_k^- \in \operatorname{supp}(y_i)$ for some $i \in \mathbb{N}$. This can happen in two ways (giving us our two remaining cases).

Case B: Suppose that there is some $j \in \mathbb{N}$ such that there exist infinitely many $k \in M$ with $\varphi_k^- \in \operatorname{supp}(y_j^*)$. Once more we extract a subsequence if necessary in order to assume that for all $k \in M$, φ_k^- lies in the support of y_j^*. The disjointness of support of the y_i^*'s implies that if $k \in M$ and $i \neq j$, $\varphi_k^- \notin \operatorname{supp}(y_i^*)$ for any i. Hence if $i \neq j$, $\langle y_i^*, u_k \rangle = 0$ for all $k \in M$.

Thus the sequence $(z_{j+1}^*, z_{j+2}^*, z_{j+3}^*, \ldots)$ satisfies the conditions which held in case A, and $\sum_{i=j+1}^{\infty} z_i^*$ has unbounded partial sums. We conclude that $\sum z_i^*$ has unbounded partial sums, as well.

Case C: Suppose that given $j \in \mathbb{N}$ there are at most finitely many $k \in M$ with $\varphi_k^- \in \operatorname{supp}(y_j^*)$. We know that for all $k \in M$, $\varphi_k^- \in \operatorname{supp}(y_i^*)$ for some i. Since $\{i : \text{there exists some } k \text{ such that } \varphi_k^- \in \operatorname{supp}(y_i^*)\}$ is infinite, we may assume also, (by extracting a subsequence of M) that given $i \in \mathbb{N}$ there is at most one $k \in M$ with $\varphi_k^- \in \operatorname{supp}(y_i^*)$. Enumerate M in increasing order: $M = \{m_1, m_2, m_3, \ldots\}$. Let k_i be the index so that $\varphi_{m_i}^- \in \operatorname{supp}(y_{k_i}^*)$. (Then if $j \neq k_i$, $\langle y_j^*, u_{m_i} \rangle = 0$ for all $i \in \mathbb{N}$.)

By extracting a subsequence if necessary we may also assume that $\sum_{i=1}^{\infty} \|y_{k_i}^*\| < 1$. Then, taking $x_i^* = v_i^*$, $y_i^* = w_i^*$, and $u_{m_i} = b_{m_i}$ for all i, we see that the hypotheses of lemma 3.3 are satisfied. This concludes case C and the proof of lemma 3.7. ∎

We have just proved that if the series $\sum z_j^*$ does not converge then the corresponding partial sums are unbounded. This yields the result that:

Theorem 3.8. *JH^* has the* PCP.

Remark. Let us also note that this technique, with minor modifications, works in the space JT^*, yielding yet another proof that JT^* has the PCP.

References

[B-R] J. Bourgain and H.P. Rosenthal, *Geometrical implications of certain finite dimensional decompositions*, Bull. Soc. Math. Belg. **32** (1980), 57–82.

[E-W] G.A. Edgar and R.H. Wheeler, *Topological properties of Banach spaces*, Pacific J. Math. **115** (1984), 317–350.

[G-M] N. Ghoussoub and B. Maurey, *G_δ-embeddings in Hilbert space*, Journal of Functional Analysis **61** (1985), 72–79.

[G-M-S] N. Ghoussoub, B. Maurey, and W. Schachermayer, *Geometrical implications of certain infinite dimensional decompositions*, (to appear).

[H] J. Hagler, *A Counterexample to several questions about Banach spaces*, Studia Math. **60** (1977), 289–308.

[J] R.C. James, *A separable somewhat reflexive Banach space with nonseparable dual*, Bull. Amer. Math. Soc. **80** (1974), 738–743.

[O] E. Odell, *A normalized weakly null sequence with no shrinking subsequence in a Banach Space not containing ℓ^1*, Compositio Math. **41** (1980), 287–295.

[O-R] E. Odell and H.P. Rosenthal, *A double-dual characterization of separable spaces containing ℓ^1*, Israel J. Math. **20** (1975), 375–384.

[R] H.P. Rosenthal, *Weak*-Polish Banach spaces*, J. Funct. Anal. **76** (1988), 267–316.

Contemporary Mathematics
Volume **85**, 1989

AN ANALOGUE OF THE F. AND M. RIESZ THEOREM FOR SPACES OF DIFFERENTIABLE FUNCTIONS

A. Pełczyński*)

Abstract. The first dual of the space of k–times continuously differentiable functions on $\mathbb{R}^n$ vanishing at infinity is isomorphic to a Cartesian product of the space of regular Borel measures on [0,1] by a separable Banach space.

Introduction. The first duals of several Banach spaces naturally appearing in Analysis are of the form $M \oplus F$ where M is as a Banach space isomorphic to the dual of C[0,1] and F is a separable Banach space. The decomposition in the case of the Disc Algebra is a simple consequence of the F. and M. Riesz Theorem. G.M. Henkin [He] has discovered an analogous result for the ball algebra and more generally for the algebra of

---

*)The author would like to express his gratitude to the Ida Beam distinguished visiting program which made it possible for him to participate in the Workshop on Banach spaces at the University of Iowa, July 5–July 25, 1987.

1980 Mathematics subject classification (1985 Revision). Primary: 46E15, 46E27.

Key words and phrases. Banach spaces of differentiable functions, spaces of measures, separable distortion of a space of measures, Banach module over spaces of continuous functions.

uniformly continuous analytic functions in several variables on a strictly pseudoconvex domain. We refer to [R] and [P] for simplified proofs of these facts and for further references.

Around 1970, G.M. Henkin communicated to the author that a similar result holds for the space of k–times continuously differentiable functions on the n–torus; the same fact was also mentioned by Bourgain [B]. To the best of our knowledge no proofs exist in the literature.

The main result of the present paper is Theorem A stated in Section 0. It gives a general criterion for anisotropic spaces of continuously differentiable functions to have the first duals of the form $M \oplus F$. Our result immediately yields the Henkin–Bourgain observation (Corollary B). Surprisingly, the argument for the spaces of differentiable functions is much "softer" than that for analytic functions. It does not require any estimates of singular integrals. Still the proof of the main result presented in Section 2 is an adaptation of Henkin's method for analytic functions. It is based on the concept of a vector–valued Henkin measure with respect to a given set of partial derivatives. It appears that the space of these Henkin measures is a module over the space of scalar–valued continuous functions. This fact enables us to apply the result of Section 1 that moduli over space of continuous functions consisting of Hilbert space valued measures are complemented in the whole space of Hilbert space valued measures.

0. Preliminaries and formulation of the main result

Denote by $C_c(X)$ the space of scalar–valued continuous functions with compact supports on a locally compact Hausdorff space X, and by $C_0(X)$ the closure of $C_c(X)$ in the norm

$$||f||_\infty = \sup_{x \in X} |f(x)|.$$

$M(X)$ stands for the space of regular scalar–valued (finite) Borel measures on X with the norm of total variation.

If $a = (a(j))_{1 \leq j \leq n} \in Z_+^n$ (= the nth Cartesian power of the nonnegative integers) then

$$\partial^a = \frac{\partial^{a(1)+a(2)+\cdots+a(n)}}{\partial_1^{a(1)} \partial_2^{a(2)} \cdots \partial_3^{a(n)}}$$

denotes the operator of the partial derivatives corresponding to a. The identity operator corresponds to $0 = (0,0,\cdots,0)$.

Given a finite nonempty set $S \subset Z_+^n$ the symbol $C_0^S(\mathbb{R}^n)$ stands for the Banach space of scalar–valued functions f on $\mathbb{R}^n$ such that $\partial^a f \in C_0(\mathbb{R}^n)$ for $a \in S$. We admit

$$||f||_{C_0^S} = \max_{a \in S} ||\partial^a f||_\infty.$$

An n–dimensional smoothness is a finite nonempty set $S \subset Z_+^n$ which satisfies the "saturation" condition

(s) if $a \in S$ then $b \in S$ for $0 \leq b \leq a$.

For $a = (a(j)) \in Z_+^n$ and $b = (b(j)) \in Z_+^n$ we write

$$b \leq a \text{ iff } b(j) \leq a(j) \text{ for } j = 1,2,\ldots,n; \quad b < a \text{ iff } b \leq a \text{ and } b \neq a.$$

Given $S \subset Z_+^n$, we put

$$S_* = \{b \in S : \exists\, a \in S \text{ with } b < a\}.$$

Clearly, if $\{0\} \neq S$ and S satisfies (s) then S_* is nonempty; the formal identity map $C_0^S \longrightarrow C_0^{S_*}$ is called the natural embedding; it is a contraction, i.e., $\|f\|_{C_0^{S_*}} \le \|f\|_{C_0^S}$.

Finally (cf. [Se]), recall that the dual space $[C_0(X)]^*$ can be identified with $M(X)$. Moreover if X is a locally compact metric space of power continuum then $M(X)$ is isometrically isomorphic to the space

$$M = (\ell^1(\Gamma, L^1) \oplus \ell^1(\Gamma))_1$$

where Γ is a set of power continuum and $\ell^1(\Gamma, L^1)$ is the sum in the sense of $\ell^1(\Gamma)$ of Γ copies of the usual Lebesgue space $L^1 = L^1[0, 1]$. Since L^1 is isomorphic to its Cartesian product with the one–dimensional space, the space M is isomorphic as a Banach space to $\ell^1(\Gamma, L^1)$.

A Banach space E is said to be a separable distortion of M provided E is isomorphic to the direct sum $M \oplus F$ with F separable.

Now we are ready to state the main result of the paper.

Theorem A. *Assume that the natural embedding* $C_0^S \longrightarrow C_0^{S_*}$ *is compact* ($S \subset Z_+^n$ *a finite nonempty set satisfying* (s)). *Then the dual space* $[C_0^S(\mathbb{R}^n)]^*$ *is a separable distortion of* M.

Corollary B. *The space* $[C_0^{(k)}(\mathbb{R}^n)]^*$ *is a separable distortion of* M.

Here for $k = 0,1,\ldots;\ n = 1,2,\ldots$ we put

$$(k;\ n) = (k) = \{a \in Z_+^n : a(1)+a(2)+\cdots+a(n) \leq k\}.$$

Clearly if $k \geq 1$ then $(k)_* = (k-1)$, and it is classical that the natural embedding $C^{(k)}(\mathbb{R}^n) \longrightarrow C^{(k-1)}(\mathbb{R}^n)$ is compact.

1. Projections onto $C_c(X)$–moduli of Hilbert space–valued measures

Let (X, Σ, λ) be a probability measure space. Let H be a Hilbert space with the scalar product $\langle\cdot,\cdot\rangle_H$. By $L^p(\lambda;\ H)$ we denote the space of H–valued functions on X which are Bochner λ–integrable in power p $(1 \leq p < \infty)$. The symbol $L^p(\lambda)$ (for $1 \leq p \leq \infty$) stands for usual Lebesgue spaces of scalar–valued functions.

A (closed linear) subspace F of $L^p(\lambda;\ H)$ is an $L^\infty(\lambda)$–module provided

(m) if $f \in F$ and $m \in L^\infty(\lambda)$ then $mf \in F$.

Let E be a subspace of $L^p(\lambda;H)$. A projection $P : L^p(\lambda;\ H) \xrightarrow[\text{onto}]{} E$ is called pointwise orthogonal provided

For every $h \in L^p(\lambda;\ H)$ and every $f \in E$,

$$\langle (h-P(h))(x), f(x)\rangle_H = 0 \quad \text{for} \quad x \in X \quad \lambda\text{–almost everywhere.}$$

It is easy to verify that the pointwise orthogonal projection, if it exists, is unique and contractive.

Our first result is probably well known.

Proposition 1.1. *For every* $L^\infty(\lambda)$*–module* $F \subset L^p(\lambda; H)$ *there exists the pointwise orthogonal projection from* $L^p(\lambda; H)$ *onto* F.

Proof. Call a $g \in L^p(\lambda; H)$ normalized provided $\|g(x)\|^2 = \|g(x)\|$ for $x \in X$, λ–a.e. Observe that if F is a nonzero $L^\infty(\lambda)$–module then there is a nonzero normalized element in F. Indeed, if $0 \neq f \in F$ then for some $c > 0$, $\lambda\{x : \|f(x)\| > c\} > 0$. Put $m(x) = \|f(x)\|^{-1}$ whenever $\|f(x)\| > c$ and $m(x) = 0$ otherwise. Then $m \in L^\infty(\lambda)$ and by (m), mf is the desired normalized element of F.

Now by a standard transfinite inductive procedure we construct a maximal family $(g_\alpha)_{\alpha\in\mathfrak{A}}$ of normalized nonzero mutually pointwise orthogonal elements. (We call functions g and h pointwise orthogonal provided $\langle g(x), h(x)\rangle_H = 0$ for $x \in X$, λ–a.e.) The construction uses the observation that if B is any nonempty subset of an $L^\infty(\lambda)$-module F then the set of all functions in F which are pointwise orthogonal to all elements of B is again an $L^\infty(\lambda)$–module. The maximality of the family $(g_\alpha)_{\alpha\in\mathfrak{A}}$ yields that if some f in F is pointwise orthogonal to all g_α for $\alpha \in \mathfrak{A}$ then $f = 0$.

Let $\mathcal{F}$ be the family of all finite nonempty subsets of $\mathfrak{A}$. We consider $\mathcal{F}$ to be directed by inclusion of sets. For $\Phi \in \mathcal{F}$ we put

$$P_\Phi(h) = \sum_{\alpha\in\Phi} a_\alpha^h g_\alpha \qquad \text{for } h \in L^p(\lambda; H),$$

where $a_\alpha^h(x) = \langle h(x), g_\alpha(x)\rangle$ for $x \in X$.

Clearly $a_\alpha^h \in L^p(\lambda)$ because $h \in L^p(\lambda; H)$ and the g_α's are normalized. Let

$$F_\Phi = \{f \in L^p(\lambda; H) : f = \sum_{\alpha \in \Phi} m_\alpha g_\alpha\}.$$

It follows from (m) and the closedness of F that F_Φ is a closed subspace of F which is the range of the pointwise orthogonal projection P_Φ.

We shall show that the net $\{P_\Phi\}_{\Phi \in \mathcal{F}}$ consisting of contractive projections is strongly convergent and its limit is the desired pointwise orthogonal projection onto F.

Case 1°: $p = 2$. Then $\{P_\Phi\}_{\Phi \in \mathcal{F}}$ is the net of orthogonal projections in the Hilbert space $L^2(\lambda; H)$. Regarding each P_Φ as a selfadjoint operator in this Hilbert space we have

$$\text{if } \phi_1 \subset \phi_2 \text{ then } P_{\Phi_1} \leq P_{\Phi_2}.$$

Thus the net $\{P_\Phi\}_{\Phi \in \mathcal{F}}$ converges strongly to some orthogonal projection, say P.

Case 2°: $1 \leq p < 2$. Then $L^2(\lambda; H)$ can be regarded as a norm dense subset of $L^p(\lambda; H)$ (because λ is a probability measure); moreover the norm convergence in $L^2(\lambda; H)$ yields the norm convergence in $L^p(\lambda; H)$. Hence $\{P_\Phi\}_{\Phi \in \mathcal{F}}$ can be regarded as a net of contractive projections in $L^p(\lambda; H)$ which strongly converges on a dense subset of $L^p(\lambda; H)$. Hence the net converges strongly on the whole space $L^p(\lambda; H)$.

Case 3°: $p > 2$. Then the unit ball of $L^p(\lambda;H)$ regarded as a subset of $L^2(\lambda; H)$

is closed. For every $f \in L^p(\lambda; H)$, $\lim_{\phi \in \mathscr{F}} \|P(f)-P_\phi(f)\|_{L^2(\lambda;H)} = 0$ and $\|P_\phi(f)\|_{L^p(\lambda;H)} \le \|f\|_{L^p(\lambda;H)}$ for every $\phi \in \mathscr{F}$. Hence $\|P(f)\|_{L^p(\lambda;H)} \le \|f\|_{L^p(\lambda;H)}$. Thus the pointwise orthogonal projection P is $L^p(\lambda; H)$–bounded.

Finally we show that the range of the limit projection P coincides with F. For, pick $h \in L^p(\lambda; H)$. Since $P(h)$ is the strong limit of the net $\{P_\Phi(h)\}_{\Phi \in \mathscr{F}}$ and since $P_\Phi(h) \in F_\Phi \subset F$ for $\Phi \in \mathscr{F}$, we infer that $P(h) \in F$. On the other hand, if $f \in F$ then $f-P(f) \in F$ is pointwise othogonal to every g_α for $\alpha \in \mathfrak{A}$. Thus $P(f) = f$. ■

Denote by $M(X; H)$ the Banach space of regular Borel measures of bounded variation with values in a Hilbert space H defined on Borel sets of locally compact topological Hausdorff space X.

A closed linear subspace G of $M(X; H)$ is a $C_c(X)$–module provided for every $\mu \in G$ and every $f \in C_c(X)$, $f\mu \in G$ (the measure $f\mu$ is defined by $f\mu(A) = \int_A f(x)d\mu$ for every Borel set A of X).

Now we are ready to formulate the main result of the present section.

Proposition 1.2. *If* G *is a closed linear subspace of* $M(X; H)$ *which is a* $C_c(X)$*–module then* G *is complemlented in* $M(X; H)$.

Before proving the result we recall (cf. [D–S], Chapt. IV, §10, Lemma 5; [D–U], Chapt. I) that for every $\mu \in M(X; H)$ there exists a regular probability Borel measure λ on X which controls μ, i.e.,

$$\lim_n \lambda(A_n) = 0 \text{ implies } \lim_n \|\mu\|(A_n) = 0$$

for every sequence (A_n) of Borel subsets of X,

where $\|\mu\|(\cdot)$ denotes the semivariation of μ (cf. [D–S] or [D–U] for the definition).

Given a regular probability Borel measure λ on X we put

$$M_\lambda = \{\nu \in M(X; H) : \lambda \text{ controls } \nu\}.$$

Since H has the Radon–Nikodym property (cf. [D–U]), there exists an isometric isomorphism U_λ from M_λ onto $L^1(\lambda; H)$. We shall need the following fact.

If λ_1 is another regular Borel probability measure such that λ is absolutely continuous with respect to λ_1 then for every $\nu \in M_\lambda$ one has

$$U_{\lambda_1}(\nu) = bU_\lambda(\nu) \tag{1}$$

where $b \in L^1(\lambda_1)$ is the Radon–Nikodym derivative of λ with respect to λ_1 and (1) is understood as the pointwise equality of functions λ_1–a.e.

Proof of Proposition 1.2. Fix $\mu \in M(X; H)$. Let λ be any regular Borel probability measure on X which controls μ. Put $F^\lambda = U_\lambda(M_\lambda \cap G)$. Taking into account (1), the assumption that G is a $C_c(X)$–module and the density of $C_c(X)$ in $L^\infty(\lambda)$ (because λ is regular) we infer that $F^\lambda \subset L^1(\lambda, H)$ is an $L^\infty(\lambda)$–module. Thus, by Proposition 1.1, there exists the pointwise orthogonal projection, say P^λ, from $L^1(\lambda; H)$ onto F^λ. We put

$$Q(\mu) = U_\lambda^{-1} P^\lambda U_\lambda(\mu).$$

Next we show that $Q(\mu)$ is independent of the choice of the measure λ which controls μ. First consider a regular Borel probability measure λ_1 such that λ is absolutely continuous with respect to λ_1. Note that if b is the Radon–Nikodym derivative of λ with respect to λ_1 then

$$bP^{\lambda}(h) = P^{\lambda_1}(bh) \quad \text{for} \quad h \in L^1(\lambda). \tag{2}$$

Thus, combining (2) with (1), we get

$$\begin{aligned} U_{\lambda_1}^{-1} P^{\lambda_1} U_{\lambda_1}(\mu) &= U_{\lambda_1}^{-1} P^{\lambda_1}(bU_{\lambda}(\mu)) \\ &= U_{\lambda_1}^{-1}(bP^{\lambda} U_{\lambda}(\mu)) \\ &= U_{\lambda}^{-1} P^{\lambda} U_{\lambda}(\mu) \\ &= Q(\mu). \end{aligned}$$

Now if λ' is an arbitrary regular Borel probability measure which controls μ then $\lambda_1 = 2^{-1}(\lambda+\lambda')$ is a probability measure which has the property that both λ and λ' are absolutely continuous with respect to λ_1. Hence from the previous observation

$$U_{\lambda'}^{-1} P^{\lambda'} U_{\lambda'}(\mu) = U_{\lambda_1}^{-1} P^{\lambda_1} U_{\lambda_1}(\mu) = Q(\mu).$$

Now it is clear that Q is the desired projection from M(X; H) onto G. For instance the linearity of Q follows from the fact that given μ and ν in M(X; H) there exists a regular Borel probability measure on X which simultaneously controls both μ and ν. ■

Remark. It follows from the analysis of the proofs of Propositions 1.1 and 1.2 that the projection Q constructed above has the property that ker Q is also a $C_c(X)$–module.

2. Proof of the main result

Preliminary remarks. Let $\bigoplus_S C_0(\mathbb{R}^n)$ and $\bigoplus_S M(\mathbb{R}^n)$ denote the Cartesian products of #S (= the number of elements of S) copies of $C_0(\mathbb{R}^n)$ and $M(\mathbb{R}^n)$, respectively, equipped with the norms: maximum of the norms of the coordinates and sum of the norms of the coordinates, respectively. The dual $[\bigoplus_S C_0(\mathbb{R}^n)]^*$ is isometrically isomorphic to $\bigoplus_S M(\mathbb{R}^n)$; the duality is given by

$$(f,\mu) \longrightarrow \mu(f) = \sum_{a\in S} \int_{\mathbb{R}^n} f_a d\mu_a$$

for $f = (f_a)_{a\in S} \in \bigoplus_S C_0(\mathbb{R}^n)$ and $\mu = (\mu_a)_{a\in S} \in \bigoplus_S M(\mathbb{R}^n)$. There is the canonical isometric isomorphism $\mathcal{I}_S : C_0^S \longrightarrow C_0(\mathbb{R}^n)$ defined by $\mathcal{I}_S(f) = (\partial^a f)_{a\in S}$ for $f \in C_0^S$ which induces the isometric isomorphism from the dual $(C_0^S)^*$ onto the quotient space $\bigoplus_S M(\mathbb{R}^n)/(C_0^S)^\perp$, where

$$(C_0^S)^\perp = \{\mu \in \bigoplus_S M(\mathbb{R}^n) : \mu(\mathcal{I}_S(f)) = 0 \text{ for } f \in C_0^S\}.$$

The following concept plays an important role in the proof of Theorem A. Given a sequence $(f_m) \subset C_0^S$, we write $f_m \xrightarrow[A]{} 0$ provided

(3) $$\sup_m \|f_m\|_{C_0^S} < \infty,$$

(4) $$\lim_m (\partial^a f_m)(x) = 0 \quad \text{for } x \in \mathbb{R}^n \text{ and for } a \in S_*.$$

Call a $\mu = (\mu_a)_{a \in S} \in \bigoplus_S M(\mathbb{R}^n)$ a *Henkin measure with respect to* S provided

(5) $$\lim_m \sum_{a \in S} \int_{\mathbb{R}^n} \partial^a f_m \, d\mu_a = 0 \quad \text{whenever } f_m \xrightarrow[A]{} 0.$$

We denote by G_S the set of all Henkin measures with respect to S.

Proposition 2.1. (i) $G_S \supset (C_0^S)^\perp$;

(ii) G_S *is a norm closed subspace of* $\bigoplus_S M(\mathbb{R}^n)$;

(iii) G_S *is a* $C_c(\mathbb{R}^n)$*–module.*

Proof. (i) is obvious. (ii) is routine and easy. We show (iii). Note that every continuous function on $\mathbb{R}^n$ with compact support is a uniform limit of infinitely differentiable functions on $\mathbb{R}^n$ with compact supports (denoted by $C_c(\mathbb{R}^n) \cap C^\infty(\mathbb{R}^n)$). Thus taking into account (ii) it suffices to establish

$$\text{if } \mu = (\mu_a)_{a \in S} \in G_S \text{ and } g \in C_c(\mathbb{R}^n) \cap C^\infty(\mathbb{R}^n) \text{ then } g\mu = (g\mu_a)_{a \in S} \in G_S;$$

equivalently,

(6) $$\lim_m \sum_{a\in S} \int_{\mathbb{R}^n} (\partial^a f_m) g\, d\mu_a = 0 \text{ whenever } f_m \xrightarrow[A]{} 0$$

for $\mu = (\mu_a)_{a\in S} \in G_S$ and for $g \in C_c(\mathbb{R}^n) \cap C^\infty(\mathbb{R}^n)$.

To this end, first observe

(7) if $f_m \xrightarrow[A]{} 0$ and $g \in C_c(\mathbb{R}^n) \cap C^\infty(\mathbb{R}^n)$ then $gf_m \xrightarrow[A]{} 0$.

Indeed, given $a \in S$ we have

(8) $$\partial^a(gf_m) = \sum_{0\le c\le a} \begin{bmatrix} a \\ c \end{bmatrix} \partial^c f_m \cdot \partial^{a-c} g$$

where $\begin{bmatrix} a \\ c \end{bmatrix} = \prod_{j=1}^{n} \frac{a(j)!}{c(j)!(a(j)-c(j))!}$ for $a = (a(j))$, $c = (c(j))$ in Z_+^n with $c \le a$. Thus $\|\partial^a(gf_m)\|_\infty \le k^k \|f_m\|_{C_0^S} \sum_{b\in S} \|\partial^b g\|_\infty$ where $k = \#S$. Hence if (f_m) satisfies (3) so does (gf_m). Now assume that $a \in S_*$. Then $c \in S_*$ for $0 \le c \le a$. Thus it follows from (8) that if (f_m) satisfies (4) so does (gf_m). This proves (7).

Next it follows from (8) that for every $a \in S$,

$$g\partial^a f_m = \partial^a(gf_m) - \sum_{0\le c<a} \partial^c f_m\, \partial^{a-c} g.$$

Note that if $a \in S$ and $c < a$ then $c \in S_*$. Thus taking into account that $f_m \xrightarrow[A]{} 0$ and using the Lebesgue Domination Convergence Theorem we infer that for every $\nu \in M(\mathbb{R}^n)$,

$\lim_m \int_{\mathbb{R}^n} \partial^c f_m \, \partial^{a-c} g d\nu = 0$ whenever $0 \le c < a$. Consequently,

$$\lim_m \int_{\mathbb{R}^n} [\partial^a(gf_m) - g\partial^a f_m] d\mu_a = \lim_m \sum_{0 \le c < a} \partial^c f_m \, \partial^{a-c} g d\mu_a = 0.$$

Hence, summing over $a \in S$,

$$\lim_m \Big[\sum_{a \in S} \int_{\mathbb{R}^n} g\partial^a f_m \, d\mu_a - \sum_{a \in S} \int_{\mathbb{R}^n} \partial^a(gf_m) d\mu_a \Big] = 0. \tag{9}$$

Combining (9) with (7) and taking into account that $\mu = (\mu_a)_{a \in S} \in G_S$. we get (6). ■

Proof of Theorem A. For $a \in S$ and for $x \in \mathbb{R}^n$ let $\delta_{x,a}(f) = f_a(x)$ for $f = (f_a)_{a \in S} \in \oplus_S C_0(\mathbb{R}^n)$. Let $\delta^*_{x,a}$ be the corresponding functional on C_0^S, i.e.,

$$\delta^*_{x,a}(f) = (\partial^a f)(x) \quad \text{for} \quad f \in C_0^S.$$

Let

$$E = \{\text{the closed subspace of } (C_0^S)^* \text{ generated by } \delta^*_{x,a} \text{ for } x \in \mathbb{R}^n, a \in S_*\}.$$

Then

$$E \text{ is separable;} \tag{10}$$

$$E \supset G_S/(C_0^S)^\perp. \tag{11}$$

For (10) consider the sets $Z_r = \bigcup_{a \in S_*} \bigcup_{\|x\|_2 \le r} \delta^*_{x,a}$ for $r > 0$. The assumption that the natural embedding $C_0^S \longrightarrow C_0^{S_*}$ is compact yields the compactnes of the adjoint operator $(C_0^{S_*})^* \longrightarrow (C_0^S)^*$. Clearly each Z_r is the image under the adjoint operator of a subset of the ball of $(C_0^{S_*})^*$ centered at zero and of radius r because each $\delta^*_{x,a}$ regarded as a functional on $C_0^{S_*}$ is of norm $\leq \|x\|_2 \leq r$ (remember that $a \in S_*$). Thus Z_r is a norm totally bounded subset of $(C_0^S)^*$. Hence each Z_r for $r > 0$, and $\bigcup_{r>0} Z_r$, and E are norm separable subsets of $(C_0^S)^*$.

Assume to the contrary that (9) is false. Then there would exist a $\mu^* \in (C_0^S)^* \setminus E$ such that $\mu^* \circ J_S$ extends to a Henkin measure with respect to S. In other words, there would exist a $\mu = (\mu_a)_{a \in S} \in \bigoplus_S M(\mathbb{R}^n)$ such that

(12) $$\mu \in G_S; \text{ there exists } \mu^* \in (C_0^S)^* \setminus E \text{ with}$$
$$\mu^*(f) = \sum_{a \in S} \int_{\mathbb{R}^n} \partial^a f \, d\mu_a \quad \text{for } f \in C_0^S.$$

Clearly $\mu^* \neq 0$. Thus by the Hahn–Banach Extension Principle there would exist an $F^{**} \in (C_0^S)^{**}$ such that

$$\|F^{**}\| = 1; \; F^{**}(\mu^*) \neq 0; \; F^{**}(e^*) = 0 \text{ for } e^* \in E.$$

In particular, $F^{**}(\delta^*_{x,a}) = 0$ for $x \in \mathbb{R}^n$ and for $a \in S_*$. Let (x_p) be a sequence whose elements form a dense subset in $\mathbb{R}^n$. By the Goldstine Theorem ([D–S], Chapt. V, §4, Theorem 5) there would exist a sequence (f_m) in C_0^S such that

(13) $$\|f_m\| \leq 2 \quad \text{for } m = 1,2\dots,$$

(14) $$\mu^*(f_m) = F^{**}(\mu^*) \neq 0 \quad \text{for } m = 1,2,\dots,$$

(15) $$(\partial^a f_m)(x_p) = \delta^*_{x_p,a}(f_m) = F^{**}(\delta^*_{x_p,a}) = 0$$

$$\text{for } a \in S_*,\ p = 1,2,\ldots,m;\ m = 1,2\ldots .$$

Since the natural embedding $C_0^S \longrightarrow C_0^{S_*}$ is compact, it follows from (13), (15) and density of the sequence (x_p) in $\mathbb{R}^n$ that there would exist a subsequence (f'_m) of the sequence (f_m) such that $\lim_m \partial^a f'_m(x) = 0$ for all $x \in \mathbb{R}^n$ and for $a \in S_*$. The latter property again combined with (13) yields that $f'_m \xrightarrow[A]{} 0$. Thus, by (12), $\lim_m \mu^*(f'_m) = 0$, which contradicts (14).

We complete the proof using a routine decomposition argument. The space $\bigoplus_S M(\mathbb{R}^n)$ is naturally isomorphic to $M(\mathbb{R}^n; \ell_k^2)$ where $k = \#S$; the isomorphism carries $C_c(\mathbb{R}^n)$ moduli into $C_c(\mathbb{R}^n)$ moduli. Thus Proposition 2.1(iii) combined with Proposition 1.2 yields a decomposition: $\bigoplus_S M(\mathbb{R}^n) = V \oplus G_S$ for some closed subspace V of $M(\mathbb{R}^n)$. Thus, in view of Proposition 2.1(i), we get

$$(C_0^S)^* = \bigoplus_S M(\mathbb{R}^n)/(C_0^S)^\perp = V \oplus F$$

where $F = G_S/(C_0^S)^\perp$. By (10) and (11), F is separable and, by Remark after the proof of Proposition 2.1, V can be chosen to be a $C_c(\mathbb{R}^n)$–module.

Finally we show that $V \approx M$ (the sumbol "$\approx$" reads "is isomorphic to"). Pick $a \in S$ so that $a(1) = \max\{b(1) : b \in S\}$. Define the operator $U : M(\mathbb{R}) \longrightarrow (C_0^S)^*$ by

$$U(\nu) = \nu^* \text{ where } \nu^*(f) = \int_{\mathbb{R}^n} (\partial^a f)(t,0,0,\ldots,0)dt \text{ for } f \in C_0^S.$$

It can be easily shown that U isomorphically maps $M(\mathbb{R})$ into $(C_0^S)^*$. Since $(C_0^S)^* = V \oplus F$ with F separable, it follows that there is another subspace of $(C_0^S)^*$, say M_1, such that $M_1 \approx M$ and $M_1 \subset V$. Now we construct by transfinite induction a maximal family $\{\mu_\alpha\}_{\alpha\in\mathfrak{A}}$ of measures in M_1 such that the semivariations $\|\mu_\alpha\|(\cdot)$ are mutually orthogonal (i.e., for every Borel set $B \subset \mathbb{R}^n$ and every $\alpha \neq \beta$, $\|\mu_\alpha + \mu_\beta\|(B) = \|\mu_\alpha\|(B)+\|\mu_\beta\|(B)$) and $\|\mu_\alpha\|(\mathbb{R}^n) = 1$ for $\alpha \in \mathfrak{A}$. Since each μ_α is a measure with values in the k–dimensional Hilbert space, there exists a regular Borel probability measure λ_α on $\mathbb{R}^n$ such that

$$\lambda_\alpha(B) \leq \|\mu_\alpha\|(B) \leq \sqrt{k}\,\lambda_\alpha(B) \quad \text{for Borel } B \subset \mathbb{R}^n.$$

Let L be the sum in the sense of $\ell^1(\mathfrak{A})$ of the spaces $L^1(\lambda_\alpha)$. The orthogonality of the semivariations allows extension of the correspondence $\lambda_\alpha \longrightarrow \mu_\alpha$ to an isomorphism from L onto the minimal $C_c(\mathbb{R}^n)$–module V_1 which contains the family $\{\mu_\alpha\}_{\alpha\in\mathfrak{A}}$. Thus $V_1 \subset V$ because V is a $C_c(\mathbb{R}^n)$–module. Moreover, V_1 is complemented in V because V_1 being a $C_c(\mathbb{R}^n)$–module is complemented in $\bigoplus_S M(\mathbb{R}^n)$. The maximality of the family $\{\mu_\alpha\}_{\alpha\in\mathfrak{A}} \subset M_1 \approx M$ yields that $L \approx M$. Thus V contains a complemented subspace isomorphic to M. On the other hand, V being a $C_c(\mathbb{R}^n)$–module is (by Proposition 2.1) complemented in $\bigoplus_S M(\mathbb{R}^n)$. Note that $\bigoplus_S M(\mathbb{R}^n) \approx M^k \approx M$ (because $k = \#S < \infty$). Hence by the decomposition technique, $V \approx M$. ∎

3. Final remarks

3.1. There are many examples of smoothness S such that $(C_0^S)^*$ is not a separable distortion of M. The simplest example seems to be the 3–dimensional smoothness

$$(1;2)\times(0;1) = \{(0,0,0),(1,0,0),(0,1,0)\}.$$

(Then we have $C_0^{(1;2)\times(2;1)}(\mathbb{R}^3) = C_0(\mathbb{R}; C^{(1)}(\mathbb{R}^2))$.)

More generally we have

Proposition 3.1. *Let* $S = S' \times S''$ *be a smoothness which is a Cartesian product of smoothnesses* S' *and* S''. *Assume that* S *is not an interval. Then* $(C_0^S)^*$ *is not a separable distortion of* M.

Recall that S is an interval provided $S\backslash S_*$ is a one point set, equivalently if $S = \{b \in Z_+^n : 0 \le b \le a\}$ for some $a \in Z_+^n$.

To prove Proposition 3.1, we adopt the argument of [P], Theorem 11.5, and we use the result obtained independently in [K–Si] and [P–S] that if S' is not an interval then $(C_0^{S'})^*$ is not isomorphic to any subspace of M.

3.2. By a theorem of Borsuk (cf., e.g., [P–S]), if S is a 1–dimensional smoothness then C_0^S is isomorphic to $C_0(\mathbb{R})$; hence $(C_0^S)^*$ is isomorphic to M. For 2–dimensional smoothnesses we have

Propostion 3.2. *If* S *is a* 2–*dimensional smoothness then* $[C_0^S(\mathbb{R}^2)]^*$ *is a separable distortion of* M.

The proof of Proposition 3.2 combines Theorem A with the following facts.

Lemma 3.1 (due to T. Serbinowski). *Let a* 2–*dimensional smoothness satisfy*

(+) *there are in* S *exactly one element which maximizes the first coordinate and exactly one element which maximizes the second coordinate.*

Then the natural embedding $C_0^S \longrightarrow C_0^{S_*}$ *is compact.*

Lemma 3.2. *Let* S *be an arbitrary* 2–*dimensional smoothness. Then there are unique elements* a_1 *and* a_2 *in* $S \backslash S_*$ *which maximize the first and the second coordinate respectively. Let*

$$S_1 = \{b \in Z_+^2 : b + (a_2(1), a_1(2)) \in S\}.$$

Then S_1 *is a smoothness satisfying* (+). *Moreover the map* $f \longrightarrow \partial_1^{a_2(1)} \partial_2^{a_1(2)} f$ *is an isomorphism from* C_0^S *onto* $C_0^{S_1}$.

3.3. Theorem A and Corollary B remain valid if the space $C_0^S(\mathbb{R}^n)$ is replaced by the analogous space of differentiable functions on a bounded domain in $\mathbb{R}^n$ as well as of periodic functions on the n–torus. The proof requires minor changes.

3.4. Given μ and ν in $\underset{S}{\oplus} M(\mathbb{R}^n)$ write $\nu << \mu$ provided each coordinate of ν is absolutely continuous with respect to the total variation of the corresponding coordinate of μ. Contrary to Henkin's result for analytic functions (cf., e.g., [R], Theorem 9.3.1) in general G_S does not have the property that if $\mu \in G_S$ and $\nu << \mu$ then $\nu \in G_S$. A simple example (due to S. Kwapien) is the following.

Let $S = (1; 2)$. Define $\mu \in \underset{S}{\oplus} M(\mathbb{R}^2)$ by $\mu((f_{(0)}, f_{(1)}, f_{(2)})) = \int_0^1 f_{(1)}(t,t)dt + \int_0^1 f_{(2)}(t, t)dt$ for $(f_{(0)}, f_{(1)}, f_{(2)}) \in \underset{S}{\oplus} C_0(\mathbb{R}^2)$. Clearly $\mu \in G_S$ because for $f \in C_0^{(1)}(\mathbb{R}^2)$,

$$\mu((f, \partial_1 f, \partial_2 f)) = \int_0^1 (\partial_1 f)(t, t)dt + \int_0^1 (\partial_2 f)(t, t)dt$$
$$= f(1, 1) - f(0, 0).$$

Now define $\nu \in \underset{S}{\oplus} M(\mathbb{R}^2)$ by

$$\nu((f_{(0)}, f_{(1)}, f_{(2)})) = \int_0^1 f_{(1)}(t, t)dt.$$

Clearly $\nu << \mu$. One can easily show that $\nu \notin G_S$ by constructing a sequence $(f_m) \subset C^{(1)}(\mathbb{R}^2)$ such that $(\partial_1 f_m)(t, t) = ||\partial_1 f_m||_\infty = 1$ for $0 \le t \le 1$; $||f_m||_\infty \le \frac{4}{m}$, $||\partial_2 f_m||_\infty \le \frac{4}{m}$ for $m = 1,2,\dots$.

References

[B] J. Bourgain, The Dunford–Pettis Property for the ball–algebras, the polydisc–algebras and the Sobolev spaces, Studia Math. 27(1986), 245–253.

[D–S] N. Dunford and J.T. Schwartz, Linear Operators I, Interscience Publishers, New York–London 1958.

[D–U] J. Diestel and J.J. Uhl, Jr., Vector Measures, Amer. Math. Soc., Providence, R.I. 1977.

[He] G.M. Henkin, Banach spaces of analytic functions on the ball and on the bicilinder are not isomorphic, Funkc. Anal. i Priložen. 4(1968), 82–91 (Russian).

[K–Si] S.V. Kisliakov and N.G. Sidorenko, Anisotropic spaces of smooth functions without local unconditional structure, LOMI preprint E–2–86, Leningrad 1986.

[P] A. Pełczyński, Banach spaces of analytic functions and absolutely summing operators, CBMS regional conference series in mathematics No 30, Amer. Math. Soc., Providence, R.I.

[P–S] A. Pełczyński and K. Senator, On isomorphism of anistropic Sobolev spaces with "classical Banach spaces" and a Sobolev type embedding theorem, Studia Math. 84(1986), 169–215.

[R] W. Rudin, Function Theory in the Unit Ball of C^n, Springer–Verlag, Berlin-Heidelberg–New York 1980.

[Se] Z. Semadeni, Banach spaces of continuous functions, Monografie Matematyczne, Warszawa 1971.

Institute of Mathematics
Polish Academy of Sciences
Sniadeckich 8, Ip
00950 Warszawa
Poland

Contemporary Mathematics
Volume **85**, 1989

ALMOST ISOMETRIC METHODS IN SOME ISOMORPHIC EMBEDDING PROBLEMS

Yves RAYNAUD

ABSTRACT: We study the finite representability of $\ell_p(\ell_q)$ in some spaces whose ultrapowers are well known, as L_s, $L_r(L_s)$, L_φ, by reducing to several isometric lemmas in low dimension related to this question, and give for these spaces an analogue of the Guerre–Lévy theorem. We sketch a proof of an extension of Kalton's theorem, concerning embeddings of $\ell_p(X)$ in L_0, to Orlicz function spaces.

INTRODUCTION: We will examine here how some isomorphic finite representability problems can be solved by methods of isometric or partially isometric nature.

The prototype of these problems is the following: given a Banach space X, for which $p \in [1,\infty]$ is the space ℓ_p crudely finitely representable in X? (Recall that a Banach space Y is crudely finitely representable in X iff there exists a real C such that each finite dimensional subspace F of Y is $(C+\epsilon)$–isomorphic to a subspace G of X; which is equivalent to saying that Y is C–isomorphic to a subspace of an ultrapower $\tilde{X}$ of X – see [6] for the definition of ultrapowers). In the case of $Y = \ell_p$, by a well known

---

1980 Mathematics Subject Classification (1985 Revision): Primary 46B20, 46E30.

Key words and phrases: Finite representability, ultrapowers, Lebesgue and Orlicz spaces, random probabilities.

theorem of Krivine ([14], th. 12.4), crude finite representability implies finite representability in the ordinary sense $(C = 1)$, and hence isometric embeddability in some ultrapower $\tilde{X}$.

If the ultrapowers $\tilde{X}$ of X are sufficiently well known it may be easy to decide if $\ell_p \subset \tilde{X}$ isometrically. Generally an obstruction to isomorphic finite representability has a counterpart in low dimension. For example, for $p > 2$, it can be shown that ℓ_p^3 does not isometrically embed in L_1; and ℓ_p^2 does not embed in L_q, $1 < q \le 2$.

In §I below we apply this approach to the finite representability of $\ell_p(\ell_q)$ in some spaces whose ultrapowers are well known, such as L_1, L_r (L_s), L_φ, and give several isometric lemma (in low dimension) related to this question. This allows us to give an analogue of the Guerre–Levy theorem ([5]) for these spaces.

In §II we are interested in the finite representability of $\ell_p(X)$ in L_φ. We show that (except for $p = 2$) this implies that X embeds in L_p; this is a kind of generalization of a result of N. Kalton ([7]). The proof is related to the theory of "types" defined by Krivine and Maurey ([8]), which is of almost isometric nature. However we avoid formulating it in the language of [8].

I. FINITE REPRESENTABILITY OF $\ell_p(\ell_q)$ IN SOME CLASSICAL BANACH SPACES.

1) Refinement of crude finite representability

PROPOSITION 1. *If the space $\ell_p(\ell_q)$ is crudely finitely representable in a normed space X, then it is finitely representable in X.*

Proof. Let $T_1 : \ell_p(\ell_q) \hookrightarrow X$ be a C–embedding of $\ell_p(\ell_q)$ in an ultrapower $X_{(1)}$ of X.

We will denote by $(e_i)_{i=1}^{\infty}$, resp. $(f_j)_{j=1}^{\infty}$, the natural basis of ℓ_p, resp. ℓ_q; $(e_i \otimes f_j)_{i,j=1}^{\infty}$ is then the basis of $\ell_p(\ell_q)$. By an easy reformulation of the proof of Krivine's theorem given by H. Lemberg (see [2] or [14]) we see that there exists a sequence $(b_i)_{i=1}^{\infty}$ of disjoint normalized blocks on the sequence $(e_1 \otimes f_n)_{n=1}^{\infty}$, whose image belongs to a u–extension of $X_{(1)}$ (in the language of [18]) and form a "ℓ_q–space over $X_{(1)}$", i.e.:

$$\forall x \in X_{(1)},\ \forall (\lambda_i)_i \in \mathbb{R}^{(\mathbb{N})},\ \|x + \sum_i \lambda_i T_1 b_i\| = \|x + (\Sigma|\lambda_i|^q)^{\frac{1}{q}} T_1 b_1\|$$

Put $F_1 = \overline{\mathrm{Span}}[Tb_i]_{i=1}^{\infty}$, $X_{(2)} = X_{(1)} \otimes F_1$ and let T_2 be the operator $\ell_p(\ell_q) \hookrightarrow X_{(2)}$ defined by:

$$\begin{cases} T_2(e_1 \otimes f_n) = T_1(b_n) \\ T_2(e_i \otimes f_n) = T_1(e_i \otimes f_n) \quad \forall i > 1. \end{cases}$$

Then T_2 is a C–embedding of $\ell_p(\ell_q)$ into $X_{(2)}$. By iterating this procedure, we obtain an increasing sequence of spaces $(X_{(n)})_n$, with $X_{(n)} = X_{(n-1)} \oplus F_{n-1}$, $X_{(n)}$ being a u–extension of $X_{(n-1)}$ (and thus f.r. in X) and of C–embeddings $T_n : \ell_p(\ell_q) \longrightarrow X_{(n)}$ such that:

$$\cdot\ T_{n+1}\Big|_{[e_i \otimes f_j]_{i \neq n+1}} = T_n\Big|_{[e_i \otimes f_j]_{i \neq n+1}}$$

$\cdot\ (T_n(e_k \otimes f_j))_{j=1}^{\infty}$ forms, for each $k \le n$, a ℓ_q space over the space $Y_k^n = X_{(k-1)} \oplus F_k \oplus \dots \oplus F_{n-1}$. Let us consider the inductive limit $\hat{X}$ (resp. $\hat{Y}_k$) of the spaces $X_{(n)}$, $n \geq 1$ (resp. Y^n_k, $n \geq 1$). We obtain a C–embedding $\hat{T} : \ell_p(\ell_q) \to \hat{X}$ such that, for each $k \geq 1$, the sequence $(\hat{T}(e_k \otimes f_j))_{j=1}^{\infty}$ forms an ℓ_q–space over $\hat{Y}_k$. In particular:

$$\forall (\lambda_{ij}) \in \mathbb{R}^{(\mathbb{N}^2)},\ \forall x \in X : \|x + \sum_{ij} \lambda_{ij} \hat{T}(e_i \otimes f_j)\| = \|x + \sum_i (\sum_j |\lambda_{ij}|^q)^{\frac{1}{q}} \hat{T}(e_i \otimes f_1)\|$$

Now we can find blocks $g^n_k = \sum_{i \in I_{k,n}} a^{k,n}_i e_i$, $k \leq n$, such that $(\hat{T} g^n_k \otimes f_1)^n_{k=1}$ is $\left(1 + \frac{1}{n}\right)$–isomorphic to the basis of ℓ^n_p. We obtain an embedding $\tilde{T}$ of $\ell_p(\ell_q)$ in an ultrapower $\tilde{X}$ of $\hat{X}$ ($\tilde{T}(e_k \otimes f_i)$ being defined by the sequence $(\hat{T}\, g^n_k \otimes f_i)_{n=1}^{\infty}$) and $\tilde{T}$ is clearly an isometric embedding. ∎

2) <u>Case where</u> $X = L_1$.

It is well known that $\ell_p(\ell_q)$ isometrically embeds in L_1 for $p \leq q \leq 2$, and not for $p > q$. In fact in this case there is an obstruction in low dimension, for we have:

<u>LEMMA 2</u>. *If* $\infty \neq p > q \neq 2$ *then the space* $(\ell^2_q \oplus \mathbb{R})_{\ell_p}$ *does not isometrically embed into* L_1.

Here the notation $(\ell^2_p \oplus \mathbb{R})_{\ell_p}$ means that the direct sum $\ell^2_q \oplus \mathbb{R}$ is equipped with the norm $\|(x,y)\| = (\|x\|^p + |y|^p)^{1/p}$.

As ℓ^2_q does not isometrically embed in L_p of $2 \neq q < p \neq \infty$, (as an easy consequence of Beckner's inequalities, see [13], 1.e. 14 and 1.e. 15), Lemma 2 is a corollary of the following result.

PROPOSITION 3. *If the normed space* Y *is such that* $(Y \oplus \mathbb{R})_{\ell_p}$ *isometrically embeds in* L_1, *then* Y *isometrically embeds into* L_p *(when* $1 \le p \le 2$*); resp. is isometric to a Hilbert space (when* $2 \le p < \infty$*).*(*)

Proof. a) Let us first suppose $1 < p \le 2$.
If $x,y \in L_1(\Omega,\mathcal{A},\mathbb{P})$ are such that $\|\alpha x + \beta y\|_1 = (|\alpha|^p + |\beta|^p)^{1/p}$ $(\forall \alpha,\beta \in \mathbb{R})$, it is not hard to see that these elements have the same support. If $\mu = |y| \cdot \mathbb{P}$ and $z = \frac{x}{y}$, we have:

$$\|\alpha z + \beta\|_{L_1(\mu)} = (|\alpha|^p + |\beta|^p)^{1/p}.$$

By Rudin's equimeasurability theorem (in the real case, see [10]) this equality determines uniquely the probability distribution of the random variable z. Thus $z \sim Z$ in distribution, where Z is the quotient of two independent p–stable random variables X, Y (over $[0,1]$ with Lebesgue measure λ), the underlying probability being $|Y|.\lambda$.
Therefore

$$\forall r < p \quad \mathbb{E}|z|^r = \mathbb{E}\left[|\tfrac{X}{Y}|^r \cdot |Y|\right] = \mathbb{E}(|X|^r\, |Y|^{r-1}) = K_r < \infty$$

(because Y has moments of negative orders $\rho > -1$).

Now suppose that $(Y \oplus \mathbb{R})_{\ell_p}$ embeds in L_1. By a change of density we may suppose the existence of $X \subset L_1$ isometric to Y such that:

(*) Proposition 3 was obtained already by L. Dor ([20]) as I became aware after writing these notes.

$$\forall \alpha, \beta \in \mathbb{R}, \ \forall x \in X, \ \|\alpha x + \beta\|_1 = (|\alpha|^p \|x\|^p + |\beta|^p)^{1/p}.$$

Thus: $\|x\|_r = K_r \|x\|_1$, i.e. X embeds isometrically in L_r for each $r < p$, and thus for $r = p$.

If now $p \geq 2$ then, as ℓ_2^2 is a sublattice of $L_1(\ell_p^2)$ (see [15]) we have:

$$(Y \oplus \mathbb{R})_{\ell_p} \underset{1}{\overset{\subset}{\sim}} L_1 \Longrightarrow (Y \oplus \mathbb{R})_{\ell_2} \underset{1}{\overset{\subset}{\sim}} L_1$$

hence Y is Hilbertian. ∎

3) Case where $X = L_p(L_q)$, $1 \leq p \leq q < \infty$.

Recall that ultrapowers of $L_p(L_q)$ are isometrically embeddable (as sublattices) in spaces $L_p(\tilde{\Omega})(L_q(\tilde{\Omega}'))$. (See [9]). As a consequence of this fact and of Proposition 9 we may observe that if $\ell_r(\ell_s)$ is (crudely) finitely representable in $L_p(L_q)$ (a fortiori if it embeds isomorphically) then it embeds isometrically in $L_p(L_q)$.

We restrict here ourselves to the case $p \leq q$ (the structure of $L_p(L_q)$, $p > q$ is by far simpler, see [15]).

Let us first give a definition. We say that the space $(\bigoplus_i X_i)_{\ell_p}$ embeds in a Banach lattice L "with disjoint fibers" if there is an embedding $T : (\bigoplus X_i)_{\ell_p} \hookrightarrow L$ such that the images TX_i are pairwise disjoint in L; we say that $\ell_p(X)$ is "finitely representable with disjoint fibers" in L if it embeds with disjoint fibers in an ultrapower $\tilde{L}$ of L. Recall the following lemma of [16].

LEMMA 4. *If* $\ell_p(\ell_q)$ *is finitely representable in a Banach lattice* L *then either it embeds in* L_1 *or it is finitely representable with disjoint fibers in* L.

PROPOSITION 5. *If* $(X \oplus \mathbb{R})_{\ell_r}$ *embeds isometrically in* $L_p(L_q)$, $p \le q < \infty$, *with disjoint fibers, then* X *embeds isometrically in* $L_r(L_q)$.

Proof. It is analogous to the proof of Proposition 3 (see [15]), using an analogue on $\mathbb{R}_+$ of Rudin's equimeasurability theorem ([15] lemma [18] or [11] lemma 2). ■

LEMMA 6. *Let* $1 < r < q < \infty$. *If* ℓ_s^2 *isometrically embeds in* $L_r(L_q)$ *then* $r \le s \le q \vee 2$ or $s = 2$.

This is the low dimensional isometric version of [15], Proposition 1.

Proof. We can suppose $q > 2$ (otherwise $L_r(L_q) \subset L_r$ isometrically). Using Beckner's inequalities it is easy to verify that $L_r(L_q)$ is $2 \wedge r$ – smooth, in the following sense:

$$\exists C, \forall f,g \in L_r(L_q), \quad \left[\frac{\|f+g\|^r + \|f-g\|^r}{2}\right]^{1/r} \le \left(\|f\|^{2\wedge r} + C\|g\|^{2\wedge r}\right)^{\frac{1}{2\wedge r}}$$

It follows easily, taking $f = f_0$ and $g = t\, g_0$ where (f_0,g_0) is a ℓ_s^2–basis in $L_r(L_q)$ and letting $t \longrightarrow 0$, that $s \ge 2 \wedge r$. Moreover if $s > 2$, using Beckner's inequalities again, we see that $\dfrac{\|(f_0^2 + t^2 g_0^2)^{1/2}\| - 1}{t^2} \xrightarrow[t\to 0]{} 0$ and thus f_0 and g_0 are disjoint.

But $L_r(L_q)$ is r–smooth on disjoint vectors, hence $s \ge r$. On the other hand, by using Beckner's inequalities and Clarkson's inequalities (in L_q, $q > 2$) we obtain:

$$\frac{\|f+g\|^r + \|f-g\|^r}{2} \geq \int (\|f\|_q^q + \gamma^q \|g\|_q^q)^{r/q} \quad \left[\text{were } \gamma = \sqrt{\frac{r-1}{q-1}}\right]$$

and thus by the inverse Minkowski inequality:

$$\geq (\|f\|^q + \gamma^q \|g\|^q)^{r/q} \quad \text{as } r/q < 1.$$

It follows easily that $s \leq q$. ■

COROLLARY 7. *If* $r > s \neq 2$ *then* $(\ell_s^2 \oplus \mathbb{R})_{\ell_r}$ *does not embed isometrically with disjoint fibers in* $L_p(L_q)$, $1 \leq p \leq q < \infty$.

COROLLARY 8. *If* $r > s \neq 2$ *then* $\ell_r(\ell_s)$ *is not finitely representable in any* $L_p(L_q)$, $1 \leq p \leq q < \infty$.

Corollary 7 improves Proposition 15 of [15]; Corollary 8 was obtained in [15].

4) Case where $X = L_\varphi$.

Recall that an Orlicz function φ is said to be moderate if $\sup_{t>0} \frac{t\varphi'(t)}{\varphi(t)} < \infty$. Ultrapowers of the associate spaces $L_\varphi(\Omega,\mathcal{A},\mu)$ can be conveniently described: (see e.g. [16]).

Fact 9. *Ultrapowers of* L_φ, φ *being moderate, are Musielak–Orlicz "modular spaces", associated to uniformly moderate Musielak–Orlicz functions.*

Recall that a Musielak–Orlicz function on the space Ω is a measurable

$\psi : \Omega \times \mathbb{R}_+ \longrightarrow \mathbb{R}_+$ such that the partial functions $\psi_\omega = \psi(\omega,.)$ are Orlicz functions; it is uniformly moderate if $\operatorname{Ess\,sup}_{\omega} \sup_{t>0} \frac{t\,\psi'(\omega,t)}{\psi(t)} = q < \infty$, where $\psi'(\omega,t)$ is the right derivative w.r.t. to t. We define the "modular":

$$\Psi(f) = \int_\Omega \psi(\omega, |f(\omega)|)d\mu(\omega)$$

then $\|f\|_\psi = \inf\{\lambda > 0 : \Psi(\frac{f}{\psi}) \le 1\}$ and $L_\psi = \{f \in L^\circ(\Omega)/\|f\|_\psi < \infty\}$.

PROPOSITION 10. *If* $(X \oplus \mathbb{R})_{\ell_r}$ *isometrically embeds with disjoint fibers in* L_ψ *then* X *embeds (isometrically) in* L_r.

We will sketch the proof of Proposition 9. Suppose there are $e_0 \in L_\psi$ and a subspace X of L_ψ such that:

$$\forall x \in X,\ \forall \beta \in \mathbb{R},\ \Psi\left[\frac{\beta x + e_0}{(1+\beta^r\|x\|^r)^{1/r}}\right] = 1.$$

Differentiating this with respect to β^r (at the point $\beta = 0$) we obtain:

$$\lim_{\beta \to 0} \frac{1}{\beta^r}\,\Psi(\beta x) = K\|x\|^r$$

where $K = \frac{1}{r}\int_\Omega e_0 \cdot \psi'(\omega,e_0)d\mu(\omega)$ does not depend on x. Proposition 10 is thus a consequence of the following lemma:

LEMMA 10. *Let* $\mathcal{G} \subset L_\psi$ *be the order–ideal consisting of the elements* g *for which*

$\overline{\lim\limits_{\beta\to 0}}\frac{\Psi(\beta g)}{\beta^r} < \infty$. *There exists a lattice homomorphism* u *from* $\mathcal{G}$ *into an abstract* L_r *space such that*

$$\|u(g)\|_r^r = \lim_{\beta\to 0}\frac{\Psi(\beta g)}{\beta^r}$$

whenever the latter limit exists.

Proof. For $g \in \mathcal{G}$ set $\theta(g) = \lim\limits_{\Lambda,\mathcal{U}} \frac{1}{2\log\Lambda}\int_{\frac{1}{\Lambda}}^{\Lambda} \lim\limits_{\beta,\mathcal{V}} \frac{1}{\beta^r}\psi(\lambda\beta g)\frac{d\lambda}{\lambda^{r+1}}$ where $\mathcal{U}$ (resp. $\mathcal{V}$) is an ultrafilter finer than the filter of neighborhoods of $+\infty$ (resp. 0) in $\mathbb{R}_+$. Then θ is clearly disjointly additive; an easy calculation shows that θ is homogeneous of degree r, i.e.:

$$\forall\lambda \in \mathbb{R},\ \forall g \in \mathcal{G},\ \theta(\lambda g) = |\lambda|^r\theta(g).$$

Moreover it is non–decreasing. Then $\theta(g)^{1/r}$ is easily shown to define an abstract L_r–semi–norm on $\mathcal{G}$. (See the proof of [7], lemma 2.1). Thus there exists a lattice homomorphism u from $\mathcal{G}$ into an abstract L_r space (the Hausdorff completion of $\mathcal{G}$ equipped with this semi–norm) such that: $\forall g \in \mathcal{G}$, $\theta(g) = \|u(g)\|_r^r$. We have $\theta(g) = \lim\limits_{\beta\to 0}\frac{1}{\beta^r}\Psi(\beta g)$ when this limit exists: for then $\forall\lambda > 0$, $\lim\limits_{\beta\to 0}\frac{1}{\beta^r}\Psi(\lambda\beta g) = \lambda^r \lim\limits_{\beta\to 0}\frac{1}{\beta^r}\Psi(\beta g)$. ■

As a consequence we obtain:

PROPOSITION 12. *If* $\ell_p(\ell_q)$ *is crudely finitely representable in an Orlicz space* L_φ *not containing* c_0, *then* ℓ_q *embeds in* L_p (*i.e.* $p \leq q \leq 2$ *or* $q \in \{2,p\}$).

5) An analogue to Guerre–Lévy theorem.

The results of this section were announced in [16].

If X is a Banach space, let us note:

$$p_X = \operatorname{Sup}\{p \geq 1 : X \text{ is of type } p\}$$

$$q_X = \operatorname{Inf}\{q \leq \infty : X \text{ is of cotype } q\}.$$

Recall that by Maurey–Pisier theorem (see [14], Theorem 13.2) ℓ_{p_X} and ℓ_{q_X} are finitely representable in X.

Now if X is an infinite dimensional subspace of L_1, Guerre–Lévy theorem ([5]) states that ℓ_{p_X} embeds (almost isometrically) in X.

We will give an analogous result for subspace of Orlicz spaces.

Recall first the following fact due to Krivine and Maurey (see [15] for a proof).

Fact 13. *If* ℓ_q *is finitely representable in a stable Banach space* (*see* [8] *for a definition*) *then there exists a* $p \in [1,\infty]$ *such that* $(\bigoplus_{n=1}^{\infty} \ell_q^n)_{\ell_p}$ *embeds* (*almost isometrically*) *in* X.

Note that, for example, $L_p(L_q)$ and Orlicz space L_φ (not containing c_0) are stable.

In the following we suppose that $L_\varphi \not\supset c_0$.

COROLLARY 14. *Let* $q > 2$. *If* ℓ_q *is finitely representable in an infinite dimensional subspace* X *of* L_φ *then it embeds in* X.

Proof. By Fact 13 there exists p such that $(\oplus \ell_q^n)_{\ell_p}$ embeds in L_φ; by Proposition 12 we have $q \in \{2,p\}$. So $q = p$. ∎

COROLLARY 15. *Let* X *be an infinite dimensional subspace of* L_φ. *Then* ℓ_p *embeds (almost isometrically) in* X *for* $p \in \{p_X, q_X\} \setminus \{2\}$. *If* $p_X = q_X = 2$ *then* ℓ_2 *embeds a.i. into* X.

Proof. If $q_X > 2$, then ℓ_{q_X} embeds in X by Corollary 14.

If $p_X < 2$, then ℓ_{p_X} is f.r. in X and thus $(\overset{\infty}{\underset{n=1}{\oplus}} \ell_{p_X}^n)_{\ell_p}$ embeds a.i. in X by Fact 13. Then $p \le p_X$ by Proposition 11, and on the other hand $p_X \le p$ as ℓ_p embeds in X.

If $p_X = q_X = 2$, then X contains a ℓ_p subspace (by Aldous Theorem [1]) and clearly $p = 2$. ∎

Recall also the following result for spaces $L_p(L_q)$, consequence of Corollary 8 (and Fact 12), which was first derived in ([15], Theorem 10):

COROLLARY 16. *Let* X *be a infinite dimensional subspace of* $L_p(L_q)$, $1 \le p \le q < \infty$. *If* $p_X < 2$ *then* X *contains* ℓ_{p_X} *(almost isometrically).*

II. FINITE REPRESENTABILITY OF $\ell_p(X)$ IN ORLICZ SPACES.

N. Kalton proved the following theorem ([7]):

THEOREM. *Let be* $1 \leq p < 2$; *if* X *is a Banach space such that* $\ell_p(X)$ *embeds in* L_0 *then* X *embeds into* L_p.

Later B. Maurey removed the condition $p \neq 2$ (unpublished). We will give an extension of this theorem in the setting of Orlicz space. (We remain in the frame of normed spaces theory, but could extend to non–locally convex Orlicz spaces, at least in the locally bounded case).

THEOREM 17. *Let* L_φ *be an Orlicz space not containing* c_0, *and let* $1 < p \neq 2 < \infty$. *If* $\ell_p(X)$ *is crudely finitely representable in* L_φ, *then* X *embeds (isomorphically) in* L_p. *If* $p = 2$, *the same is true if we suppose* φ *to be* 2*–concave or* 2*–convex.*

The method we follow here is a mixture of Maurey's ideas for the L_1 case and the ideas used in the $p > 2$ case treated in [16]. We will only sketch the proof of Theorem 17 here, and refer the interested reader to [17] for more details.

We may suppose X to be separable. In view of Fact 9, we will take as hypothesis that a $\ell_p(X)$ C–embeds into a Musielak–Orlicz space (with uniformly moderate ψ), on a probability space (after a suitable change of density).

1) The $p > 2$ case.

If $T : \ell_p(X) \hookrightarrow L_\psi$ is the given embedding, then for each $x \in X$ the sequence $(T(e_n \otimes x))_{n=1}^\infty$ is a ℓ_p sequence in L_ψ. Let us first examine the behavior of such sequences in L_ψ.

Every ℓ_p–sequence $(y_n)_{n=1}^\infty$ is almost disjoint, by a standard argument due to Kadec and Pelczynski (see [13], 1.c.8). Thus, after extracting if necessary a subsequence, we have:

$$(*)\quad \begin{cases} \forall\lambda\in\mathbb{R},\ \forall x\in L_\psi,\ \lim\limits_{n\to\infty}\Psi(\lambda(x+y_n)) = \Psi(\lambda x)+\lim\limits_{n\to\infty}\Psi(\lambda y_n) \\ \qquad\qquad = \Psi(\lambda x)\ +\ h(\lambda) \end{cases}$$

where h is some Orlicz function (moderate, with the same constant as ψ). As a consequence we have:

$$\lim_{n_1\to\infty}\ \dots\ \lim_{n_k\to\infty}\ \Psi\Big[\lambda\Big[\sum_{j=1}^{k}\alpha_j y_{n_j}\Big]\Big] = \sum_{j=1}^{k} h(\lambda\alpha_j)$$

By a standard argument ([12], Theorem 4.a.9 and 4.a.8) we can find blocks:

$$z_k = \sum_{j=1}^{N_k}\alpha_j^{(k)}\,y_{n_j}^{(k)},\quad n_1<n_2<\dots<n_{N_k},\quad n_1\longrightarrow\infty$$

such that: $\lim\limits_{k\to\infty}\Psi(\lambda z_k) = a^q\lambda^q$ for some $a>0$ and $q\in[1,\infty]$. (i.e. the Orlicz function h associated to the sequence $(z_k)_k$ is $h(\lambda)=a^q\lambda^q$). Then $\overline{\mathrm{span}}[z_k]_k \supseteq \ell_q$ (see [19]), thus $q=p$.

Coming back to our embedding $T:\ell_p(X)\hookrightarrow L_\psi$, and having chosen a dense sequence $(x_i)_{i=1}^\infty$ in X, we apply the preceding procedure to the ℓ_p–sequence $(T(e_n\otimes x_1))_{n=1}^\infty$, obtaining blocks $(T(b_n^{(1)}\otimes x_1))_{n=1}^\infty$; apply it now to $(T(b_n^{(1)}\otimes x_2))_{n=2}^\infty$, and so on; a diagonal argument enables us to find a sequence $(b_n)_{n=1}^\infty$ of normalized disjoint blocks (on the ℓ_p–basis) such that:

$$\lim_{n\to\infty} \Psi(\lambda T(b_n \otimes x)) = a(x)^p \lambda^p$$

for each $x = x_i$, $i \in \mathbb{N}$ and therefore (by density) for each $x \in X$.

It is then clear that $\|x\| \sim a(x)$ $(\forall x \in X)$.

On the other hand, interpreting the preceding equality in an ultrapower $\tilde{L}_\psi = L_{\tilde{\psi}}(\tilde{\Omega})$ of $L_\psi(\Omega)$, we find a linear map $u : X \longrightarrow L_{\tilde{\psi}}$ such that

$$\forall x \in X,\ \forall \lambda \in \mathbb{R},\ \tilde{\Psi}(\lambda u(x)) = a(x)^p \lambda^p.$$

Lemma 11 gives us an embedding v from $u(X)$ into a L_p–space such that $a(x) = \|v o u(x)\|_p$ $(\forall x \in X)$. The proof of Theorem 17 is then achieved in this case.

2) The $1 \leq p < 2$ case.

In this case formula $(*)$ has to be replaced by:

$$\lim_{n\to\infty} \Psi(\lambda(x+y_n)) = \mathbb{E}_\omega \int_{\mathbb{R}} \psi(\lambda(x+t))d\nu_\omega(t) + h(\lambda)$$

where ν_ω is a random probability (see [4]).

Similar reasoning to the one of Aldous ([1]) enable us to obtain, when $(y_n)_n$ is a ℓ_p–sequence, a block sequence $(z_k)_k$ for which:

$$\lim_{n\to\infty} \Psi(\lambda(x+z_k)) = \mathbb{E}_\omega \int_{\mathbb{R}} \psi(\lambda(x+t))d\nu_\omega^{(p)}(t) + a^p \lambda^p$$

where $\nu_\omega^{(p)}$ is a p–stable random probability, that is:

$$\forall t \in \mathbb{R}, \int_{\mathbb{R}} e^{iut} d\nu_\omega^p(a) = e^{-A(\omega)^p t^p}.$$

Equivalently this can be written:

$$\lim_{k\to\infty} \Psi(\lambda(x+z_k)) = \Psi(\lambda(x + A \otimes Y)) + h(\lambda)$$

where Y is a fixed p–stable random variable (independent of Ω).

Now coming back to our embedding $T : \ell_p(X) \hookrightarrow L_\psi$, the reiteration procedure exposed in n°1 gives us two maps:

$$A : X \longrightarrow L_0^+ \text{ and } a : X \longrightarrow \mathbb{R}_+$$

and we see that: $\|x\| \sim \|(A(x),a(x))\|_{\bar\psi}$ where $L_{\bar\psi}$ is the Musielak–Orlicz space defined on the extended measure space $\bar\Omega = \Omega \cup \{\bar\omega\}$ by $\bar\psi(\bar\omega,t) = t^p$ and, $\forall \omega \in \Omega$, $\bar\psi(\omega,t) = \mathbb{E}_Y \psi(\omega,tY)$.

If $p < 2$, $\bar\psi$ is necessarily p–concave; if $p = 2$ and $\bar\psi$ is 2 concave (resp. 2–convex), so is $\bar\psi$.

Other properties of these functions are the following:

i) $x \longmapsto A(x)^p$ and $x \longmapsto a(x)^p$ are of negative type.

For the second it results of the reasoning of n°1; and for the first from the following limit property:

$$\forall t, \ \forall x \in X, \exp(iT(b_n \otimes x)(\omega).t) \xrightarrow[n \to \infty]{\sigma(L_\infty, L_1)} \exp(-A(x)(\omega)^p t^p).$$

(ii) If $x_1,\dots,x_N \in X$ then:

$$\|(\sum_i (A(x_i)^p, a(x_i)^p))^{1/p}\|_{\bar{\psi}} \sim (\Sigma\|x_i\|^p)^{1/p}.$$

Using the p–concavity of $L_{\bar{\psi}}$, the latter equivalence and an argument like Krivine's factorization theorem, ([13], 1.d.11) we obtain a L_1–norm $\| \ \|_1$ on the ideal J generated by the $(A(x)^p,a(x)^p)$, $x \in X$, such that:

$$\|x\|^p \sim \|(A(x)^p, a(x)^p)\|_1$$

Therefore on X, the map $\longmapsto \|x\|^p$ is equivalent to a negative type function, which implies that X embeds in L_p (isomorphic version of a theorem of Bretagnolle, Dacunha–Castelle and Krivine, [3]). ■

REFERENCES

[1] ALDOUS D.
Subspaces of L_1 via random measures, Trans. Am. Math. Soc., 267, n°2, 1981, p. 445–463.

[2] BEAUZAMY B., LAPRESTE J.T.
Modèles étalés des espaces de Banach, Publications du Département de Mathématiques de l'Université de Lyon.

[3] BRETAGNOLLE J., DACUNHA–CASTELLE D., KRIVINE J.L.
Lois stables et espaces L^p, Ann. Inst. Henri–Poincaré, Sect. B, 2, 1966, p. 231–259.

[4] GARLING D.J.H.
Stable Banach Spaces, Random Measures and Orlicz functions space, in: Probability Measures on Groups, Lect. Notes in Math. 928, Springer–Verlag, 1982.

[5] GUERRE S., LEVY M.
Espaces ℓ^p dans les sous–espaces de L^1, Trans. Am. Math. Soc., 279, n°2, (1983), p. 611–616.

[6] HEINRICH S.
Ultraproducts in Banach spaces theory, J. Reine Angew. Math. 313, 1980, p. 72–104.

[7] KALTON N.
Banach spaces embedding into L_0, Israel J. Math. 52, 1985, p. 305–319.

[8] KRIVINE J.L., MAUREY B.
Espaces de Banach Stables, Israel J. Math. 39, n°4, 1981, p. 273–295.

[9] LEVY M., RAYNAUD Y.
Ultrapuissances de $L^p(L^q)$, C.R. Acad. Sc. Paris, t 299, Série I, n°3, 1984, p. 81–84.

[10] LINDE W.
Moments of Measures on Banach Spaces, Math. Ann. 258, 1982, p. 277–287.

[11] LINDE W.
Uniqueness theorems for Measures in L_r or $c_0(\Omega)$, Math. Ann. 274, 1986, p. 617–626.

[12] LINDENSTRAUSS J., TZAFRIRI L.
Classical Banach Spaces, I: Sequence Spaces, Springer–Verlag, 1977.

[13] LINDENSTRAUSS J., TZAFRIRI L.
Classical Banach Spaces, II: Function Spaces, Springer–Verlag, 1979.

[14] MILMAN V., SCHECHTMAN G.
Asymptotic Theory of Finite Dimensional Normed Spaces, Lecture Notes in Math. 1200, Springer–Verlag, 1986.

[15] RAYNAUD Y.
Sur les sous–espaces de $L^p(L^q)$, Séminaire de Géométrie des Espaces de Banach, Universités Paris 7 et 6, 1984–85.

[16] RAYNAUD Y.
Finie représentabilité de $\ell_p(X)$ dans les espaces d'Orlicz, C.R. Acad. Sc. Paris, t 304, Série I, n°12, 1987, p. 331–334.

[17] RAYNAUD Y.
Finite representability of $\ell_p(X)$ in Orlicz function spaces, in preparation.

[18] STERN J.
Ultrapowers and local properties of Banach Spaces, Trans. Amer. Math. Soc., 240, 1978, p. 231–252.

[19] WOO J.T.
On Modular Sequence Spaces, Studia Math. 48 n° 3, 1973, p. 271–289.

[20] DOR L.E.
On isometric embeddings in L_1, The Altgeld Book, University of Illinois, 1975–76.

EQUIPE D'ANALYSE
UNIVERSITE PARIS VI
Tour 46 – 4éme Etage
4, Place Jussieu
75252 – PARIS CEDEX 05

Contemporary Mathematics
Volume **85**, 1989

Sub-simplexes of convex sets and some characterizations of simplexes with the RNP †

HASKELL ROSENTHAL
Department of Mathematics
The University of Texas at Austin
Austin, Texas 78712

Abstract. The notion of a sub-simplex is defined and some geometrical equivalences are obtained. The following results are proved for K a separable closed bounded convex subset of a Banach space: K is a simplex with the Radon-Nikodým Property provided K has the Integral Representation Property and $\overline{co}\, A$ is a sub-simplex of K for all compact $A \subset \operatorname{Ext} K$. (The converse is also true and an immediate consequence of known theorems.) If K is a simplex and a face of some compact convex set in a weaker locally convex topology on K, then every point of K is the barycenter of at most one probability measure supported on $\operatorname{Ext} K$. The latter result gives a partial answer to a question of G.A. Edgar.

We introduce here the apparently new notion of a sub-simplex of a convex set, and use this in obtaining a characterization of "reasonable" simplexes with the RNP (Radon-Nikodým Property). Our primary motivation is the study of closed bounded convex non-compact subsets of Banach spaces; specifically whether such sets have the RNP if they enjoy certain weaker properties. We do obtain some information in the compact setting also. For example, it follows from Theorem 1 below that if K is a compact metrizable convex subset of a locally convex space, then K is a simplex if (and only if) $\overline{co}\, A$ is a sub-simplex of K for every compact subset A of $\operatorname{Ext} K$, the extreme points of K. (As usual, $\overline{co}\, A$ denotes the closed convex hull of A.)

We first formulate the relevant definitions. Let X be a real linear space and K be a non-empty convex subset of X. Recall that K is said to be in algebraically

†This note is based on a talk given at the 1987 Workshop on Banach Space Theory at the University of Iowa. The research for it was partially funded by NSF DMS-8601752.
Key Words and Phrases: sub-simplexes of convex sets, Banach spaces, the Radon-Nikodým Property. AMS Classification Numbers: 46A55, 46B20, 46B22.

general position in X if there is a linear functional p on X with $p(k) = 1$ for all $k \in K$. In case X is a linear topological space, K is said to be in general position if p can be chosen as above with p continuous. If K is in algebraically general position in X, we let $C_K = \bigcup_{\lambda \geq 0} \lambda K$, the cone generated by K, and let $\leq_K$ be the order relation induced by C_K. That is, for $x, y \in X$, we define $x \leq_K y$ provided $y - x \in C_K$. (We also refer to $\leq_K$ as the order induced by K.)

Let K be as above, and W be a non-empty subset of K. Recall that W is said to be an extremal subset of K if for all k, k' in K and $0 < \lambda < 1$, if $\lambda k + (1 - \lambda)k'$ belongs to W, then k and k' belong to W. We say that W *is a semi-extremal subset of* K *if for all* k *in* K, w' *in* W *and* $0 < \lambda < 1$, *if* $\lambda k + (1 - \lambda)w'$ *belongs to* W, *then* k *belongs to* W. Following the standard usage, we say that W is a face of K if W is a convex extremal subset of K. Evidently W is a face of K if and only if W is a convex semi-extremal subset with $K \sim W$ convex.

The next simple result gives our main reason for defining semi-extremality.

Elementary Proposition. *Let* W *be a non-empty convex subset of* K *(with* K *a convex subset of* X *a real linear space). The following are equivalent.*

(a) W *is semi-extremal.*

(b) $\mathrm{Aff}(W) \cap K = W$.

(c) Assuming K *is in algebraically general position, then* $\leq_W$ *coincides with* $\leq_K$ *on* span W.

Remark. span W denotes the linear span of W; $\mathrm{Aff}(W)$ denotes the affine span of W; of course if $w_0 \in W$, then $\mathrm{Aff}(W) = w_0 + \mathrm{span}(W - w_0)$. Notice that (b) yields that the convex semi-extremal subsets of K are precisely the sets W of the form $W = A \cap K$, where A is an affine subspace of X. We also note that K is in algebraically general position if and only if $0 \notin \mathrm{Aff}\, K$.

Proof. (b) is invariant under translation, so by replacing K by $K \times \{1\}$ in $X \times \mathbb{R}$ if necessary, we may assume that K is already in algebraically general position; let p be a linear functional on X with $p \mid K = 1$.

Now assume (a). We first show that

(1) $$W = K \cap \mathrm{span}\, W\ .$$

Of course $W \subset K \cap \operatorname{span} W$ and $\operatorname{span} W = C_W - C_W$. So let $w, w' \in W$, $\alpha, \beta \geq 0$, and $k \in K$ with $k = \alpha w - \beta w'$. Then $1 = p(k) = \alpha p(w) - \beta p(w') = \alpha - \beta$. Hence $k = (1+\beta)w - \beta w'$. Thus $w = \frac{1}{1+\beta}k + \frac{\beta}{1+\beta}w'$. Obviously if $\beta = 0$, $k = w \in W$. Otherwise, $0 < \frac{\beta}{1+\beta} < 1$ so by the definition of semi-extremality, $k \in W$, proving (1).

Next we note that (1) implies

$$C_W = C_K \cap \operatorname{span} W \ . \tag{2}$$

Indeed it suffices to show, assuming (1), that if $x \in C_K \cap \operatorname{span} W$, then $x \in C_W$. If $x = 0$, this is trivial. Otherwise, $p(x) > 0$ and $x/p(x) \in K \cap \operatorname{span} W$; hence by (1), $x/p(x) \in W$ so $x \in C_W$.

It is trivial that (2) is equivalent to (c) of the Proposition. Hence (a) $\Rightarrow$ (c). Assuming (c), we have that (2) holds; but then if $x \in K \cap \operatorname{span} W$, $x \in C_W$ and $p(x) = 1$ so $x \in W$. Hence (1) holds. (1) obviously implies (b) since $W \subset \operatorname{Aff}(W) \cap K \subset \operatorname{span} W \cap K$. Thus (c) $\Rightarrow$ (b). Finally assume (b) and let $k \in K$, $w \in W$ and $0 < \lambda < 1$ with $w \overset{\text{df}}{=} \lambda k + (1-\lambda)w' \in W$. Obviously k lies on the line joining w and w', so $k \in \operatorname{Aff}(W)$, hence $k \in W$. Thus (a) holds, completing the proof. ∎

We next recall that if $(V, \leq)$ is an ordered vector space and Y is a linear subspace of V, Y is called a vector sub-lattice of V if for all $x, y \in Y$, the maximum $x \vee y$ of x and y, exists *and* belongs to Y. (That is, there exists a z in Y with $x \leq z$ and $y \leq z$ so that for all v in V with $x \leq v$ and $y \leq v$, $z \leq v$. z is of course unique, and is denoted $x \vee y$.) Of course the definition is equivalent to: for all $x, y \in Y$, the minimum $x \wedge y$ of x and y, exists and belongs to Y.)

Definition. *Let K be a convex subset of a real linear space. A non-empty subset W of K is called a sub-simplex of K provided W is a convex semi-extremal subset and there exists a convex set $\widetilde{K}$ in algebraically general position in a vector space V and an affine bijection $\alpha : K \to \widetilde{K}$ so that $\operatorname{span} \widetilde{W}$ is a vector sub-lattice of $(V, \leq_{\widetilde{K}})$ where $\widetilde{W} = \alpha(W)$.*

Remark. Let W be a semi-extremal convex subset of K with K a convex subset of X a linear space. It is easily seen that if K is in algebraically general position, then

W is a sub-simplex of K if and only if span W is a vector sub-lattice of $(X, \leq_K)$. Moreover in any case, W is a sub-simplex of K if and only if span $\widetilde{W}$ is a vector sub-lattice of $(\widetilde{X}, \leq_{\widetilde{K}})$, where $\widetilde{X} = X \times \mathbb{R}$, $\widetilde{K} = K \times \{1\}$, and $\widetilde{W} = W \times \{1\}$. We also recall (using our terminology) that K is said to be a simplex if K is a sub-simplex of itself.

It follows immediately from Theorem 1 below that if K is a closed bounded convex separable subset of a Banach space so that K has the IRP (the Integral Representation Property), then K is a simplex with the RNP if and only if $\overline{co}\, A$ is a sub-simplex of K for all compact $A \subset \operatorname{Ext} K$. We prefer to phrase things in terms of a non-hereditary version of the IRP. Although our motivation is in the setting of closed bounded convex subsets of separable Banach spaces, we also prefer to state our results in the setting of certain reasonable possibly non-closed subsets of locally convex spaces.

Definition. Let K be a non-empty subset of a locally convex space X. We say that K is *reasonable* provided K is *bounded, line-closed, measure-convex*, and *Souslin.* Recall that K is

line-closed	provided $L \cap K$ is a closed subset of L for all lines $L \subset K$;
measure-convex	provided $\overline{co}\, A$ is a compact subset of K for all compact $A \subset K$;
Souslin	provided K is a continuous image of some separable complete metric space.

(Evidently the closed reasonable subsets of a Banach space are precisely the non-empty sets which are closed bounded convex and separable.)

Suppose K is convex Souslin. Let $\mathcal{P}(K)$ denote the family of all Borel probability measures on K. More generally, if W is a subset of K, we let $\mathcal{P}(W)$ denote the set of all $\mu \in \mathcal{P}(K)$ so that μ is supported on W; that is, there exists a Borel subset L of W with $\mu(L) = 1$. We recall the classical fact that since K is Souslin, every $\mu \in \mathcal{P}(K)$ is supported on a σ-compact subset of K. Moreover we note that compact convex subsets of a locally convex space are metrizable (*e.g.*, this follows immediately from the Lemma at the end of §1 of [**14**]). Finally, we recall that by results of Fremlin and Pryce [8], K is bounded measure-convex if and only if for all $\mu \in \mathcal{P}(K)$, the barycenter $x \stackrel{\text{df}}{=} \int_K k\,d\mu(k)$ exists and belongs to K. (We may interpret the integral here in the Bochner sense, because of the above remarks.) In particular, if K is bounded measure-convex, K is σ-convex.

Now fix K a reasonable subset of a locally convex space X. *We say that* K *has* ERM (Extremal Representing Measures) *if every* $x \in K$ *is represented by some* $\mu \in \mathcal{P}(\operatorname{Ext} K)$. ($x$ is represented by μ means that x is the barycenter of μ.) Using the result in [8] mentioned above, it is easily seen that K has ERM if and only if K is the σ-convex hull of the set of all $x \in K$ for which there is a compact $A \subset \operatorname{Ext} K$ with $x \in \overline{\operatorname{co}}\, A$. Thus if K has ERM, one has a somewhat effective way of expressing every $x \in K$ as belonging to $\overline{\operatorname{co}}\, \operatorname{Ext} K$. The IRP is the hereditary version of ERM; that is, K is said to have the IRP if L has ERM for all relatively closed convex subsets L of K. We say that K has UERM (Unique Representing Measures) if every $x \in K$ is represented by a unique $\mu \in \mathcal{P}(\operatorname{Ext} K)$. Finally, we recall that K is said to have the RNP if for every probability space $(\Omega, \mathcal{S}, \mu)$ and continuous linear operator $T : L^1(\mu) \to X$ with $Tf \in K$ for all $f \in L^1(\mu)$ with $\int f\,d\mu = 1$ and $f \geq 0$, there is a strongly measurable $\varphi : \Omega \to K$ so that $Tf = \int f\varphi\,d\mu$ for all $f \in L^1(\mu)$. (It is known that in this definition, it suffices to take the probability space to be the unit interval endowed with Lebesgue measure on its Borel subsets.)

We can now formulate the main result of this note (part of which is just a summary of known material).

Theorem 1. *Let* K *be a reasonable subset of a locally convex space* X. *The following are equivalent.*

1. K *is a simplex with the* RNP.
2. K *has* UERM.
3. K *has* ERM *and* $\overline{\operatorname{co}}\, A$ *is a sub-simplex of* K *for all non-empty compact* $A \subset \operatorname{Ext} K$.
4. K *has* ERM *and assuming* K *is in general position in* X, *then for all disjoint compact subsets* A *and* B *of* $\operatorname{Ext} K$,

$$x \wedge y = 0 \quad \text{for all} \quad x \in \overline{\operatorname{co}}\, A \quad \text{and} \quad y \in \overline{\operatorname{co}}\, B \tag{3}$$

(where "$\wedge$" denotes the minimum with respect to the order on X induced by K).

The implications $1 \Leftrightarrow 2$ are known. $2 \Rightarrow 3$ and $3 \Rightarrow 4$ follow easily from standard results and the definitions involved. (We note also that the proof of $2 \Rightarrow 3$ yields another equivalent condition: 3′. *K is a simplex with* ERM *so that* $\overline{\text{co}}\, A$ *is a face of K for all non-empty compact* $A \subset \operatorname{Ext} K$.) $4 \Rightarrow 2$ is deduced using the L^1-convexity theorem of the author [**13**]. We have several comments before giving the proofs of $2 \Rightarrow 3 \Rightarrow 4 \Rightarrow 2$.

First concerning $1 \Leftrightarrow 2$: If K is compact, then K is metrizable (as mentioned above). Then K has the RNP and $1 \Leftrightarrow 2$ then becomes the classical seminal result of Choquet ([**4**], [**5**]). Suppose X is a Banach space and K is closed. Then the result that K has the RNP $\Rightarrow$ K has the ERM (and hence K has the IRP) is due to Edgar [**6**]. In this setting, $1 \Rightarrow 2$ was subsequently proved by Bourgin and Edgar [**3**] and Saint Raymond [**15**]. For K and X as in Theorem 1 with K closed, $1 \Rightarrow 2$ is due to Thomas [**16**]. The extension to possibly non-closed K is given by the author in [**14**]. It follows by standard reasoning that if K satisfies 2, then K is affinely equivalent to $\mathcal{P}(\operatorname{Ext} K)$ and hence K is a simplex. The surprising result that K then also has the RNP is due to Edgar (this is a special case of Corollary 2.7 of [**7**]). (We note incidentally that by a result of Kendall [**9**], (see also [**12**, Propostion 3.1]), if K is a simplex, then K is line-compact. Thus $1 \Leftrightarrow 2$ holds provided K satisfies all the conditions in the definition of reasonable sets except the line-closedness condition; the latter follows automatically if either 1 or 2 holds.) The following open problem considerably motivated the formulation of Theorem 1.

Problem. *Let K and X be as in Theorem 1. Assume that K is a simplex with* ERM. *Does K have the* RNP*?*

The main setting of interest in this problem is for K a closed reasonable subset of a Banach space. The problem is then a slight variation of a question posed by Edgar in [**7**] (in virtue of his result that $2 \Rightarrow 1$ in Theorem 1 above): if K is a simplex, is every point of K represented by at most one $\mu \in \mathcal{P}(\operatorname{Ext} K)$? We obtain in Theorem 6 a partial affirmative answer to both questions.

For one last motivation for our definition of sub-simplexes and condition 3 of Theorem 1, we note the following result of M. Rogalski (Théorème 35 of [**10**]): *There* *exists a compact convex metrizable subset K of a locally convex space X so that K is not a simplex, yet for all compact $A \subset \operatorname{Ext} K$, $\overline{co}\, A$ is a simplex and a face of K. (In fact for all compact convex $L \subset K$, L is a face of K provided* $\operatorname{Ext} L \subset \operatorname{Ext} K$.) It follows that if K is in general position and A, B are disjoint compact subsets of $\operatorname{Ext} K$, then if $x \in \overline{co}\, A$ and $y \in \overline{co}\, B$, x and y are "orthogonal" over the positive elements of X. That is, if $0 \leq u$ and $u \leq x$, $u \leq y$, $u = 0$ (where $\leq$ denotes "$\leq_K$"). However by condition 4 of our Theorem 1, there must exist such x and y so that either $x \wedge y$ does not exist or $x \wedge y \neq 0$. (Rogalski also obtains in [**10**] several other interesting counter examples as well as various characterizations of compact (not necessarily metrizable) Choquet simplexes K, several of which involve the facial structure of $\overline{co}\, A$ for compact $A \subset \operatorname{Ext} K$.)

We now give the straightforward proofs of the implications $2 \Rightarrow 3$ and $3 \Rightarrow 4$ of Theorem 1. We first recall a standard elementary fact: *if K is a simplex and F is a face of K, then F is a sub-simplex.* Indeed, assume K is in algebraically general position in a real linear space X, and suppose p is a linear function on X with $p \mid K = 1$; let $\leq$ denote the order relation induced on X by K and assume $X = \operatorname{span} K$. Since $(X, \leq)$ is thus a vector lattice, to show that $\operatorname{span} F$ is a vector sub-lattice it suffices to that if $x \in \operatorname{span} F$, then $x^+ \in \operatorname{span} F$, where $x^+ = x \vee 0$. Choose u and $v \in C_F$ with $x = u - v$. Since then $u \geq x$ and $u \geq 0$, $u \geq x^+$; letting $w = u - x^+$, then $w \geq 0$ and $u = x^+ + w$. Of course if $x^+ = 0$ or $w = 0$, $x^+ \in \operatorname{span} F$, so assume $x^+ \neq 0$ and $w \neq 0$. Then $p(x^+) > 0$ and $p(w) > 0$; we have that

$$\frac{u}{p(u)} = \frac{x^+}{p(u)} + \frac{w}{p(u)} = \frac{p(x^+)}{p(u)} \frac{x^+}{p(x^+)} + \frac{p(w)}{p(u)} \frac{w}{p(w)} .$$

But $\dfrac{u}{p(u)} \in F$ and $\dfrac{x^+}{p(x^+)}, \dfrac{w}{p(w)} \in K$. Hence since F is a face, $\dfrac{u}{p(u)} \in F$ so $u \in \operatorname{span} F$.

Now suppose that 2 holds and suppose A is a non-empty compact subset of $\operatorname{Ext} K$. Evidently to show that 3 holds, it suffices to show (by the above elementary fact) that $\overline{co}\, A$ is a face of K, since 2 implies K is a simplex. Let $x, y \in K$ and $0 < \beta < 1$ so that $z \overset{\text{df}}{=} \beta x + (1 - \beta) y$ belongs to $\overline{co}\, A$. Choose μ and $\nu \in \mathcal{P}(\operatorname{Ext} K)$ representing x and y respectively. Then evidently $\lambda \overset{\text{df}}{=} \beta \mu + (1 - \beta) \nu$ represents z. Since K is reasonable, $\overline{co}\, A$ is compact metrizable. Hence by Choquet's Theorem, there exists a $\gamma \in \mathcal{P}(\text{Ext}\,\overline{co}\,A)$ with γ representing z. Of course $\text{Ext}\,\overline{co}\,A = A$; thus $\gamma \in \mathcal{P}(\text{Ext}\,K)$ and so since K has UERM, $\gamma = \lambda$. Thus $\lambda(A) = 1$, which implies $\mu(A) = \nu(A) = 1$, and hence $x, y \in \overline{co}\,A$. Thus $\overline{co}\,A$ is a face, proving that $2 \Rightarrow 3$. (We have in fact proved that $2 \Rightarrow 3' \Rightarrow 3$.)

$3 \Rightarrow 4$. Let A and B be disjoint compact subsets of $\text{Ext}\,K$ and let $W = \overline{co}(A \cup B)$. Thus since $A \cup B$ is compact, W is a sub-simplex. Thus W is a simplex and is compact metrizable. It follows that letting $M(\text{Ext}\,W)$ denote the family of all signed measures supported on $\text{Ext}\,W$ ($= A \cup B$), then by the Choquet uniqueness theorem, $\mu \to \int_{\text{Ext}\,K} k\,d\mu(k)$ is a lattice isomorphism from $M(\text{Ext}\,W)$ onto $(\text{span}\,W, \leq_W)$ (since K is in general position, so is W). In particular if $x \in \overline{co}\,A$ and $y \in \overline{co}\,B$, there exist $\mu \in \mathcal{P}(A)$ and $\nu \in \mathcal{P}(B)$ so that x and y are represented by μ and ν respectively; since $\mu \wedge \nu = 0$, $x \wedge_W y = 0$ (where "$\wedge_W$" denotes the minimum in $\leq_W$). Since by assumption W is a semi-extremal subset of K, $x \wedge y = 0$ by our Elementary Proposition.

To deal with the remaining implication $4 \Rightarrow 2$ of Theorem 1, we first require some consequences of "L^1-convexity"; that is, of the main result of [**13**].

Lemma 2. *Let K be a σ-convex line-closed subset of a linear topological space X. Assume that K is in general position; let $\leq$ be the order relation on X induced by K and $p \in X^*$ with $p \mid K = 1$. Let (h_j) be a sequence in X with $h_j \leq h_{j+1}$ for all j and $\sup_j p(h_j) < \infty$. Then $h \overset{\text{df}}{=} \lim_{j\to\infty} h_j$ exists in X and $h = \sup_j h_j$.*

(This is given as part of Theorem 1 of [**13**].)

We require the following consequence. (A sequence (x_j) in $(X, \leq)$ an ordered vector space is called increasing if $x_j \leq x_{j+1}$, for all j.)

Corollary 3. *Let K, X, p and $\leq$ be as in Lemma 2. Let (x_j), (y_j) be increasing sequences in X so that $x_j \wedge y_j = 0$ for all j and $\sup p(x_j) < \infty$, $\sup p(y_j) < \infty$. Then $x \wedge y = 0$, where $x = \lim_j x_j$ and $y = \lim_j y_j$.*

Remark. Of course by Lemma 2, we have that x and y are well-defined and moreover $x = \sup_j x_j$, $y = \sup_j y_j$.

Proof of Corollary 3. We first show this for the case where one of the sequences is constant. For the sake of notational clarity, suppose that (h_n) is an increasing sequence in X, and h, z are in X with $h_n \to h$ and $h_n \wedge z = 0$ for all n. We claim that $h \wedge z = 0$. (Since p is continuous and (h_n) is increasing, $p(h) = \lim_n p(h_n) = \sup_n p(h_n) < \infty$.) Let $u \in X$ with $u \leq h$ and $u \leq z$. We must show that $u \leq 0$. Fix n. Snce $u \leq h = h_n + (h - h_n)$, we have that

$$u - (h - h_n) \leq h_n \ . \tag{4}$$

Since $h - h_n \geq 0$, we have

$$u - (h - h_n) \leq u \ . \tag{5}$$

Since $u \leq z$ by assumption, we have by (4) and (5) that

$$u - (h - h_n) \leq h_n \wedge z = 0 \ . \tag{6}$$

Hence by (6), $u \leq h - h_n$ for all n. But since $h = \sup_n h_n$ by Lemma 2, $0 = \inf_n (h - h_n)$, so $u \leq 0$, providing that $h \wedge z = 0$.

Now let (x_j), (y_j), x and y be as in the statement of Corollary 3. Fix j. Then $x_j \wedge y_n = 0$ for all $n \geq j$. Hence applying the above result for $z = x_j$, $h_n = y_n$ for all $n \geq j$, and $h = y$, we obtain that $x_j \wedge y = 0$. Now applying the above result for $z = y$, $h_j = x_j$ for all j, and $h = x$, we obtain that $x \wedge y = 0$. ■

We now pass to the proof of 4 ⇒ 2 of Theorem 1. Let $p \in X^*$ with $p \mid K = 1$ and assume 4 holds. It then suffices to prove

$$\begin{cases} \text{If } \mu \text{ and } \nu \text{ are disjointly supported members of} \\ \mathcal{P}(\operatorname{Ext} K), \text{ then } \mu \text{ and } \nu \text{ have distinct barycenters.} \end{cases} \tag{7}$$

This was observed in the remarks following Corollary 2.6 of [**12**]; for the sake of completeness, we repeat the argument. Suppose (7) holds yet $\underline{\mu}$ and $\underline{\nu} \in \mathcal{P}(\operatorname{Ext} K)$ have the same barycenter and $\underline{\mu} \neq \underline{\nu}$ (*i.e.*, condition 2 fails). Then letting $\gamma = \underline{\mu} - \underline{\nu}$, $\gamma \neq 0$ yet $\int_K k \, d\gamma(k) = 0$. By the Jordan-Hahn decomposition theorem we may choose μ, ν disjointly supported finite measures, each supported on $\operatorname{Ext} K$, with $\gamma = \mu - \nu$. It follows that $\int_K k \, d\mu(k) = \int_K k \, d\nu(k)$. (Since K is reasonable, K is bounded measure convex, so these integrals exist.) Since p is continuous and $p \mid K = 1$,

$$p\Big(\int_K k \, d\mu(k)\Big) = \int_K p(k) \, d\mu(k) = \mu(K) = \nu(K) = p\Big(\int_K k \, d\nu(k)\Big) \ .$$

Since $\gamma \neq 0$, $\lambda \overset{\mathrm{df}}{=} \mu(K) \neq 0$. It follows that $\lambda^{-1}\mu$ and $\lambda^{-1}\nu$ are disjointly supported members of $\mathcal{P}(\operatorname{Ext} K)$ with the same barycenter, contradicting (7).

Finally, we observe that if μ and ν are given members of $\mathcal{P}(\operatorname{Ext} K)$ and A and B are disjoint compact subsets of $\operatorname{Ext} K$, then

$$(8) \qquad u \wedge v = 0\ , \quad \text{where } u = \int_A k\, d\mu(k) \ \text{ and } \ v = \int_B k\, d\nu(k)\ .$$

Indeed, if $\mu(A) = 0$ or $\nu(B) = 0$, this is obvious. Otherwise, letting $\underline{u} = \dfrac{u}{\mu(A)}$ and $\underline{v} = \dfrac{v}{\nu(B)}$, then $\underline{u} \in \overline{co}\, A$ and $\underline{v} \in \overline{co}\, B$ so $\underline{u} \wedge \underline{v} = 0$ by condition 4. It follows that $u \wedge v = (\mu(A)\underline{u}) \wedge v(B)\underline{v} = 0$.

Now let μ and ν be disjointly supported members of $\mathcal{P}(\operatorname{Ext} K)$ with barycenters x and y respectively. Choose A and B disjoint σ-compact subsets of $\operatorname{Ext} K$ with $\mu(A) = \nu(B) = 1$. Choose (A_j) and (B_j) sequences of compact subsets of $\operatorname{Ext} K$ with $A_j \subset A_{j+1}$, $B_j \subset B_{j+1}$ for all j and $A = \bigcup_j A_j$, $B = \bigcup_j B_j$. Let $x_j = \int_{A_j} k\, d\mu(k)$ and $y_j = \int_{B_j} k\, d\mu(k)$ for all j. It follows by (8) that (x_j) and (y_j) satisfy the hypotheses of Lemma 2, and moreover $x_j \to x$ and $y_j \to y$. Since K is reasonable, K is σ-convex and line closed. Hence by Corollary 3, $x \wedge y = 0$, so $x \neq y$, proving (7) and thus completing the proof of Theorem 1. ∎

It seems worth pointing out that we have not used the assumption that K has ERM, in obtaining the uniqueness part of the assertion $4 \Rightarrow 2$. That is, we have the following immediate consequence of the above argument.

Corollary 4. *Let K be a reasonable subset of a locally convex space X, in general position, so that* (3) *holds for all disjoint compact subsets A and B of* Ext *K. Then each point of K is the barycenter of at most one measure belonging to $\mathcal{P}(\operatorname{Ext} K)$.*

We next give a geometric characterization of sub-simplexes. This characterization generalizes Theorem 3 of [**12**]; however, its proof follows immediately from the proof of the latter result, given in section 3 of [**12**].

Theorem 5. *Let K be a line-compact subset of a linear topological space X, in general position. Let W be a convex semi-extremal subset of K. The following are equivalent.*

1. *W is a sub-simplex of K.*
2. *For all $x, y \in \operatorname{span} W$ and $\alpha, \beta > 0$ with $(x + \alpha K) \cap (y + \beta K) \neq \emptyset$, there exist a $z \in \operatorname{span} W$ and a $\gamma \geq 0$ with*

$$(x + \alpha K) \cap (y + \beta K) = z + \gamma K . \tag{9}$$

Assuming K and W are σ-convex, we have a third equivalent condition:

3. *The same as 2, with $\alpha = \beta = 1$.*

Remark. Assume $W = K$. The remarkable equivalence $1 \Leftrightarrow 2$ was stated by Choquet for K a compact convex subset of a locally convex space; it was formulated and proved in general by Kendall [**9**]. The implication $3 \Rightarrow 1$ is proved by the author in [**12**]. For the case where K is a compact convex subset of a finite-dimensional space, $3 \Leftrightarrow 1$ was proved by Rogers and Shephard. (For this and other references, see the remarks following the statement of Theorem 3 in [**12**].)

We only give some comments concerning the proof, referring to [**12**] for the detailed arguments. Let $\leq$ be the order on X induced by K. $1 \Leftrightarrow 2$ follows from the author's localization of arguments of Kendall [**9**], given in Theorem 3.2 of [**12**]. Thus if 1 holds and x, y are as in condition 2, then if $z = x \vee y$, z satisfies (9) for some $\gamma \geq 0$, whenever the left side of (9) is non-empty. Thus $1 \Rightarrow 2$. Similarly, assuming 2 holds and fixing x, y in $\operatorname{span} W$, then the proof of the result in [**12**] cited above, yields the (considerably more delicate) result that there is a unique z so that whenever $\alpha, \beta > 0$, there is a $\gamma \geq 0$ satisfying (9) whenever its left side is non-empty, and then $z = x \vee y$; hence 1 holds. Trivially $2 \Rightarrow 3$; we consider finally the implication $3 \Rightarrow 1$. Assume that condition 3 holds, and let $p \in X^*$ with $p \mid K = 1$. It then follows directly from the argument for Corollary 3.5 of [**12**] that

(10) for x, y in $\operatorname{span} W$ with $p(x) = p(y)$, $x \vee y$ exists and belongs to $\operatorname{span} W$.

Now let f belong to $\operatorname{span} W$ with $p(f) = 1$. To complete the argument, it suffices to prove

$$f^+ = f \vee 0 \text{ exists and belongs to } \operatorname{span} W. \tag{11}$$

Indeed, once this is shown, let $x, y \in \operatorname{span} W$; we must show that $x \vee y$ exists and belongs to $\operatorname{span} W$. In virtue of (10), we may assume $p(x) \neq p(y)$. So suppose $p(x) > p(y)$; then letting $c = p(x) - p(y)$, $f \stackrel{\mathrm{df}}{=} \frac{1}{c}(x - y)$ belongs to $\operatorname{span} W$ and $p(f) = 1$. But then $x \vee y = (p(x) - p(y))f^+ + y$ exists and belongs to $\operatorname{span} W$.

Now let f be as in the statement preceding (11). If $f \geq 0$, there is nothing to prove, so assume $f \not\geq 0$. Fix $u \in W$. Define a sequence (α_n) and a sequence (g_n) with $g_n \in C_W$ for all n, as follows: let $g_1 = f \vee u$. (g_1 is a well-defined element of C_W by (10), since $p(f) = p(u) = 1$ and $f, u \in \operatorname{span} W$.) Having defined g_n, let $\alpha_n = p(g_n)$ and $g_{n+1} = g_n \vee \alpha_n f$. (Again by (10), g_{n+1} is a well-defined element of C_W.) Now the proof of Lemma 3.7 of [**12**] yields that setting $\alpha_0 = 1$, then $g_{n+1} = \alpha_n f \vee u$ and $\alpha_n \leq \alpha_{n+1}$ for all $n \geq 0$; moreover since K is line-closed σ-convex, $\alpha_n \to \infty$ as $n \to \infty$. (This uses the L^1-convexity theorem, phrased above as Lemma 2.) Now define h_n by $h_n = \dfrac{g_{n+1}}{\alpha_n}$ $(= f \vee \dfrac{u}{\alpha_n})$ for all n. It is immediate that $h_n \in \operatorname{span} W$ and $h_n \geq h_{n+1}$ for all n. Now the discussion preceding the proof of Lemma 3.7 in [**12**] yields that $h \stackrel{\mathrm{df}}{=} \lim_{j\to\infty} h_j$ exists in X and in fact $h = \inf_j h_j = f^+$. Since W is σ-convex and semi-extremal, it follows from Lemma 2 (replacing h_j by $-h_j$ for all j) that $h \in \operatorname{span} W$, proving (11). ∎

The remaining part of this note deals with the following partial answer to the above-mentioned uniqueness question of Edgar.

Theorem 6. *Let K be a reasonable subset of some locally convex space with K a simplex. Suppose there exists a compact convex subset W of a locally convex space and a one-one affine continuous map $\alpha : K \to W$ with $\alpha(K)$ a face of $\overline{\alpha(K)}$. Then every point of K is represented by at most one measure belonging to $\mathcal{P}(\operatorname{Ext} K)$.*

Evidently we thus obtain that if K satisfies the assumptions of Theorem 6 and K has ERM, then K has the RNP (using of course Edgar's result in [**7**], stated in our setting as $2 \Rightarrow 1$ of Theorem 1 above). This gives a partial affirmative answer to the question given after the statement of Theorem 1.

Theorem 6 follows from Corollary 4 above and the crystallization of Choquet's proof of his uniqueness theorem given by the author in [**12**]. We first recall some standard elementary facts. Given K a convex subset of a linear space X and $x \in K$, F_x denotes the smallest face of K containing X.

Proposition 7. *Let K and X be as above, and x and y be elements of K.*

(a) $F_x = \{k \in K :$ *there is a* $k' \in K$ *and* $0 < \lambda < 1$ *with* $x = \lambda k + (1-\lambda)k'\}$

(b) *Assuming K is a simplex in algebraically general position, then $x \wedge y = 0$ if and only if $F_x \cap F_y = \emptyset$.*

Here $x \wedge y$ refers to the minimum of x and y in the order induced by K. Proposition 7 follows immediately from Proposition 2.4 of [**12**].

The next result gives the fundamental remaining ingredient for the proof of Theorem 6. (If W is a topological space and $A \subset K \subset W$, a subset V of K is called a neighborhood of A in K if there is an open subset U of W with $A \subset U \cap K \subset V$.)

Lemma 8. *Let W be a compact convex subset of a locally convex space and let K be a face of W. Let A be a compact subset of* Ext K, *V a neighborhood of A in K, $x \in \overline{\text{co}}\, A \cap K$ and $\varepsilon > 0$. There exist n, scalars $\lambda, \lambda_1, \dots, \lambda_n$ and $k, x_1, \dots, x_n$ in K so that*

(a) $0 \leq \lambda < \varepsilon$, $0 \leq \lambda_i$ *for all* i, *and* $\lambda + \sum_{i=1}^n \lambda_i = 1$

(b) $x = \lambda k + \sum_{i=1}^n \lambda_i x_i$

(c) $F_{x_i} \subset V$ *for all* i.

Remark. The faces F_{x_i} in (c) are with respect to K.

Lemma 8 follows directly from the special case when $K = W$; the latter result is stated as Theorem 2.5 of [**12**], and in turn is an immediate consequence of Choquet's original proof of his uniqueness theorem [4]. We refer the reader to the discussion following the statement of 2.5 in [**12**], for a motivating "coordinate-free" reformulation of the conclusion of Lemma 8.

Proof. Choose V' an open neighborhood of A in W with $V' \cap K = V$. By Theorem 2.5 of [**12**], choose n, scalars $\lambda_1, \dots, \lambda_n$, and $k, x_1, \dots, x_n$ in W so that (a),(b) of Lemma 8 hold, and also so that

$$F'_{x_i} \subset V' \quad \text{for all} \quad i \tag{12}$$

where, for $y \in W$, F'_y denotes the smallest face of W containing y. We may obviously assume that $\varepsilon < 1$ and hence we can assume that $\lambda_i > 0$ for all i. But since K is a face of W, it follows immediately that then $x_1, \dots, x_n$ belong to K and moreover

$k \in K$ provided $\lambda > 0$. If $\lambda = 0$, we just replace k by any element of K, so we can assume $k \in K$ also. Then for each i, since $x_i \in K$, we have immediately from Proposition 7(a) that $F_{x_i} \subset F'_{x_i}$ and hence by (12), $F_{x_i} \subset F'_{x_i} \cap K \subset V' \cap K = V$, proving (c). ∎

It is convenient to isolate out the following consequence of Lemma 8.

Corollary 9. *Let K and W be as in Lemma 8, with K a simplex in algebraically general position. Then for all disjoint compact subsets A_1 and A_2 of* Ext K, $y_1 \wedge y_2 = 0$ *for $y_i \in \overline{co}\, A_i \cap K$, $i = 1,2$ (where "$\wedge$" refers to the minimum in the order relation induced by K).*

Proof. Let $\varepsilon > 0$. For $j = 1,2$, let A_j, y_j be as in the statement of the corollary, and choose V_j a neighborhood of A_j in K with $V_1 \cap V_2 = \emptyset$. By Lemma 8, for each $j = 1,2$, choose $n_j, \lambda_{j1}, \ldots, \lambda_{jn_j}$ and $k_j, x_{j1}, \ldots, x_{jn_j}$ in K so that (a) – (c) of Lemma 8 hold (for "x" = "y_i", "λ" = λ_j, "λ_i" = λ_{ji} etc.). Let $\leq$ be the order-relation induced by K and p a linear functional on the ambient space with $p \mid K = 1$. Letting $X = \operatorname{span} K$, then $(X, \leq)$ is a vector lattice. Thus by elementary vector lattice theory,

$$y_1 \wedge y_2 = \Big(\lambda_1 k_1 + \sum_{j=1}^{n_1} \lambda_{1j} x_{1j}\Big) \wedge \Big(\lambda_2 k_2 + \sum_{k=1}^{n_2} \lambda_{2k} x_{2k}\Big)$$

$$\leq \lambda_1 k_1 \wedge y_2 + \Big(\sum_{j=1}^{n_1} \lambda_{1j} x_{1j}\Big) \wedge \lambda_2 k_2 \tag{13}$$

$$+ \sum_{j=1}^{n_1} \sum_{k=1}^{n_2} \lambda_{1j} x_{1j} \wedge \lambda_{2k} x_{2k} .$$

But for all j and k, $x_{1j} \wedge x_{2k} = 0$ by Proposition 7, since $F_{x_{1j}} \cap F_{x_{2k}} \subset V_1 \cap V_2 = \emptyset$. Hence the third term of (13) equals zero. Moreover

$$\Big(\sum_{j=1}^{n_1} \lambda_{1j} x_{1j}\Big) \wedge \lambda_2 k_2 \leq \Big(\lambda_1 k_1 + \sum_{j=1}^{n_1} \lambda_{1j} x_{1j}\Big) \wedge \lambda_2 k_2 = y_1 \wedge \lambda_2 k_2 . \tag{14}$$

We thus obtain by (13) and (14) that

$$y_1 \wedge y_2 \leq \lambda_1 k_1 \wedge y_2 + y_1 \wedge \lambda_2 k_2 . \tag{15}$$

Since $\lambda_i < \varepsilon$ for $i = 1, 2$, it follows from (15) that

$$\begin{aligned} p(y_1 \wedge y_2) &\le p(\lambda_1 k_1 \wedge y_2) + p(y_1 \wedge \lambda_2 k_2) \\ &\le \lambda_1 p(k_1) + \lambda_2 p(k_2) = \lambda_1 + \lambda_2 < 2\varepsilon \ . \end{aligned}$$

Since $\varepsilon > 0$ is arbitrary, $p(y_1 \wedge y_2) \le 0$; since $y_1 \wedge y_2 \ge 0$ and p is "strictly positive", it follows that $y_1 \wedge y_2 = 0$. ■

We may now easily complete the proof of Theorem 6. Let K, W and α be as in its statement and let $K' = \sigma(K)$. By "adding on" one dimension if necessary, we can assme that both K and K' are in general position in their ambient spaces. Letting X (resp. X') denote $\operatorname{span} K$ (resp. K'), then since K is assumed to be a simplex and α is one-one, K' is also a simplex and hence X and X' are vector lattices in the order induced by K and K' respectively. Now let A and B be disjoint compact subsets of $\operatorname{Ext} K$ and x and y be in $\overline{co}\, A$ and $\overline{co}\, B$ respectively. Then since α is affine and continuous, $\alpha(A)$ and $\alpha(B)$ are disjoint compact subsets of $\operatorname{Ext} K'$ and $\alpha(x) \in \overline{co}\, \alpha(A)$, $\alpha(y) \in \overline{co}\, \alpha(B)$. Hence $\alpha(x) \wedge \alpha(y) = 0$ by Corollary 9. Thus by Proposition 7(b), $F_{\alpha(x)} \cap, F_{\alpha(y)} = \emptyset$, so since α is an affine equivalence (*i.e.*, preserves the affine structure of K and K'), $F_x \cap F_y = \emptyset$, so again by Proposition 7(b), $x \wedge y = 0$. Theorem 6 now follows by Corollary 4. ■

Concluding Remarks. Suppose that K is a separable closed bounded convex subset of some Banach space B. We obtain that if K is a simplex, then K satisfies the conclusion of Theorem 6 provided $B = X^*$ for some Banach space X and K is a face of $\overline{K}^*$, $\overline{K}^*$ denoting the weak* closure of K in X^*. (Thus we have an affirmative answer to Edgar's question in this case.) In particular, this holds if K is a face of $\widetilde{K}$ (where we regard $B \subset B^{**}$ and for $W \subset B$, set $\widetilde{W} = \overline{W}^*$, the weak*-closure of W in $B^{**} = (B^*)^*$). For example, if K is a simplex and a closed bounded subset of ${L^1}^+[0,1]$ $(= \{f \in L^1[0,1] : f \ge 0\})$, then K satisfies the conclusion of Theorem 6. Indeed, W is a face of $\widetilde{W}$ where $W = {L^1}^+[0,1]$. But it is easily seen that if W is a closed convex subset of a Banach space so that W is a face of $\widetilde{W}$, then C is also a face of $\widetilde{C}$ for any closed convex $C \subset W$.

The variation on Edgar's problem (given after the statement of Theorem 1) is motivated by the famous problem: if K is a closed bounded convex subset of a

Banach space, does K have the RNP if K has the KMP? (Recall that K as above is said to have the KMP (the Krein-Milman Property) if $L = \overline{co}(\text{Ext } L)$ for all closed convex $L \subset K$.) In this connection, we note the following result, which follows quite simply from known facts.

Proposition. *Let K be a separable closed bounded convex subset of a Banach space. Suppose that K has the* KMP, *and there is a closed convex subset W of some Banach space with W a face of $\widetilde{W}$ and $\alpha : K \to W$ a one-one continuous affine map with $\alpha(K)$ closed. Then K has the* RNP.

Proof. Let K, W and α be as in the statement of the Proposition. Suppose first that C is a closed bounded convex subset of W. If C fails the RNP, then there exists a closed convex (non-empty) subset L of C so that $(\text{Ext } \widetilde{L}) \cap L = \emptyset$. (This result is due to J. Bourgain [**1**]; for a recent exposition, see [**11**].) Since W is a face of $\widetilde{W}$, L is also a face of $\widetilde{L}$, as remarked above. Hence $(\text{Ext } \widetilde{L}) \cap L = \text{Ext } L = \emptyset$, so we have proved that C fails the KMP. Now since K is assumed to have the KMP, then $C \stackrel{\text{df}}{=} \alpha(K)$ has the KMP. Indeed, let L be a closed convex subset of C. Then $K' \stackrel{\text{df}}{=} \alpha^{-1}(L)$ is a closed convex subset of K. Thus $K' = \overline{co} \text{ Ext } K'$ and since α is affine and one-one continuous, $\alpha(\text{Ext } K') = \text{Ext } L$, so

$$L = \alpha(K') = \alpha(\overline{co} \text{ Ext } K') = \overline{co}\, \alpha(\text{Ext } K') = \overline{co} \text{ Ext } L \ .$$

Hence K' has the RNP. It now follows by Proposition 4.1(c) of [**14**] that K has the RNP. (The latter is proved using the same argument as in the proof of Theorem 1.1 of [**2**].) ∎

References

1. J. Bourgain, *A geometric characterization of the Radon-Nikodým property in Banach spaces*, Compositio Math. **36** (1978), 3–6.
2. J. Bourgain and H.P. Rosenthal, *Applications of the theory of semi-embeddings to Banach space theory*, J. Funct. Anal. **52** (1983), 149–188.
3. R.D. Bourgin and G.A. Edgar, *Noncompact simplexes in Banach spaces with the Radon-Nikodým property*, J. Funct. Anal. (2) **23** (1976), 162–176.
4. G. Choquet, *Unicité des représentations intégrales au moyen des points extrémaux dans les cônes convexes réticulés*, C.R. Acad. Sci. Paris **243** (1956), 555–557.
5. G. Choquet, *Existence des représentations intégrales au moyen des points extrémaux dans les cônes convexes réticulés*, C.R. Acad. Sci. Paris **243** (1956), 699–702.

6. G.A. Edgar, *A noncompact Choquet theorem*, Proc. Amer. Math. Soc. **49** (1975), 354–358.
7. G.A. Edgar, *On the Radon-Nikodým property and martingale convergence*, Springer-Verlag Lecture Notes in Mathematics **645** (1978), 62–76.
8. D.H. Fremlin and I. Pryce, *Semiextremal sets and measure representation*, Proc. London Math. Soc. (3) **29** (1974), 502–520.
9. D.G. Kendall, *Simplexes and vector lattices*, J. London Math. Soc. **37** (1962), 365–371.
10. M. Rogalski, *Caracterisation des simplexes par des propriétés portant sur les faces fermées et sur les ensembles compacts de points extrémaux*, Math. Scand. **28** (1971), 159–181.
11. H.P. Rosenthal, *On non-norm-attaining functionals and the equivalence of the weak*-KMP with the RNP*, Longhorn Notes, The University of Texas Functional Analysis Seminar, 1985-86, 1–12.
12. H.P. Rosenthal, *On the Choquet representation theorem*, Longhorn Notes (1986-87), The University of Texas at Austin (to appear).
13. H.P. Rosenthal, L^1*-convexity*, Longhorn Notes (1986-87), The University of Texas at Austin (to appear).
14. H.P. Rosenthal, *Martingale proofs of a general integral representation theorem*, to appear.
15. J. Saint Raymond, *Représentation intégrale dans certains convexes*, Sem. Choquet, 14e Année, University of Paris VI (1974-75), No.2, 11 pp.
16. E.G.F. Thomas, *A converse to Edgar's theorem*, Springer-Verlag Lecture Notes in Mathematics **794** (1979), 497–512.

Contemporary Mathematics
Volume **85**, 1989

SOME MORE REMARKABLE PROPERTIES OF THE JAMES-TREE SPACE

WALTER SCHACHERMAYER

ABSTRACT

We show that the norm of the dual JT^* of the James-tree space JT has the Kadec-Klee-property, i.e. on the unit sphere of JT^* the weak and the norm topologies coincide. We also show that for every w*-compact convex subset $C \subseteq JT^*$ the strongly exposing functionals form a dense G_δ-subset of JT^{**}.

Hence JT^* although a non-separable dual of a separable Banach space, shares in a rather striking way some of the features of separable duals.

We also obtain as a corollary that there is an equivalent norm $|.|$ on JT such that every point of the unit-sphere of $(JT, |.|)^*$ is strongly exposed.

1. Introduction

A classical result of Kadec and Klee states that a Banach space X with separable dual may equivalently be renormed so that the dual norm on X^* is locally uniformly convex (see [D], th. IV.4.1, to which we refer for unexplained definitions). The local uniform convexity implies in particular that every point of the unit-sphere is strongly exposed which in turn implies that the strong and weak topologies coincide on the unit-sphere (the Kadec- Klee property).

On the other hand, if $(X, \|.\|)$ is such that the dual norm on X^* is locally uniformly convex, then the norm of X is Fréchet differentiable and therefore the density character of X^* is equal to that of X (see [D], chapter II).

Hence a separable Banach space X has a separable dual iff there is an equivalent norm on X such that the dual norm on X^* is locally uniformly convex.

The first example of a separable Banach space not containing ℓ^1 and having a non-separable dual was the James-tree space JT [J]. This space turned out to be a rich source

AMS classification: 46B20, 46B22

Key words: Kadec-Klee-property, strongly exposing functionals for many further results (positive ones as well as counterexamples; compare [L-S], [G-M-S], [S-S-W]).

In the present paper we shall see that JT^* shares in a rather striking way some of the features of separable dual spaces.

After some definitions and notations and presenting the notion of a "molecule" of JT^* in section 2 we prove a series of lemmas in section 3: We obtain very detailed information on JT^*; for example lemma 3.8 implies in particular (the well known fact) that JT_*, the predual of JT has a boundedly complete skipped blocking decomposition (remark 3.9) and that on the unit sphere of JT_* the weak and the norm topologies coincide (proposition 3.10).

In section 4 we obtain the main result: The norm of JT^* has the Kadec- Klee property (i.e., weak and norm topology coincide on the unit sphere) (theorem 4.1) and there is an equivalent norm $|.|$ on JT such that every point of the unit sphere of $(JT, |.|)^*$ is strongly exposed (theorem 4.4). Finally we show in theorem 4.8 that for every weak-star compact subset $C \subseteq JT^*$ the strongly exposing functionals form a dense G_δ-subset of JT^{**} (theorem 4.8). However it may happen that the intersection of the strongly exposing functionals with JT is empty (remark 4.9).

In section 5 we show that - contrary to the case of JT - for a separable Banach space X containing ℓ^1 isomorphically there is x_0^* in the sphere of X^*, where the relative weak and norm topologies do not coincide.

It is a pleasure to acknowledge stimulating discussions on the present questions with M. Fabian, G. Godefroy, A. Sersouri and S. Troyanski during the 15th Winter-School on Abstract Analysis of the Cech Academy of Science in Šrni 1987. We also thank G. Godefroy and M. Fabian for some helpful remarks on a first version of this paper.

2. Definitions and Notations:

For $N \in \mathbf{N}$ denote $\Delta_N = \{-1,+1\}^N$ and $\Delta = \Delta_\omega = \{-1,+1\}^{\mathbf{N}}$. Denote $\mathcal{T} = \cup_{n=0}^{\infty} \Delta_n$ the binary tree and $\mathcal{T}_N = \cup_{n=0}^{N} \Delta_n$ the finite binary tree up to level N. $\Delta_0 = \{-1,+1\}^0$ will be the one point set $\{\emptyset\}$ which will be called the origin of the tree $\mathcal{T}$ (resp. $\mathcal{T}_N$). We shall also consider the extended binary tree $\overline{\mathcal{T}} = \mathcal{T} \cup \Delta_\omega$. We equip $\mathcal{T}_N, \mathcal{T}$ and $\overline{\mathcal{T}}$ with their natural order structures (see, e.g. [G-M-S]).

Recall [L-S] the definition of James-tree space JT as the the space of real valued functions on $\mathcal{T}$ such that the norm

$$\|x\|_{JT} = sup\ \{(\sum_{i=1}^{n} |\sum_{t \in S_i} x_t|^2)^{1/2} : S_1, ..., S_n \text{ disjoint finite segments of } \mathcal{T}\}$$

is finite.

A finite seqment S is a subset $\{t_k, t_{k+1}, ..., t_n\}$ of elements of T, where $k \leq n \in \mathbf{N}$, for $k \leq i \leq n$, $t_i \in \Delta_i$ (which we denote by $|t_i| = i$) and $t_k \leq t_{k+1} \leq ... \leq t_n$. An infinite segment S is a subset of $\overline{T}$ similar as above, but allowing n to be ω, i.e. $S = \{t_i\}_{i=k}^{\omega}$, where $k \in \mathbf{N}$, $|t_i| = i$ for every $k \leq i \leq \omega$ and $t_k \leq t_{k+1} \leq ... \leq t_\omega$.

2.1 DEFINITION: A <u>molecule</u> m is an expression of the form

$$m = \sum_{j=1}^{n} \lambda_j \chi_{S_j}$$

where S_j are disjoint segments of $\overline{T}$ (finite or infinite), $\lambda_j \in \mathbf{R}$ and $\sum \lambda_j^2 \leq 1$.

A molecule m defines a linear functional of norm less than or equal to 1 on JT via

$$\langle x, m \rangle = \sum_{j=1}^{n} \lambda_j (\sum_{t \in S_j} x_t).$$

A molecule m also defines in an obvious way a (bounded, finitely valued) function on $\overline{T}$. However, it will be convenient, <u>not</u> to identify two molecules, if they define the same linear functional on JT (equivalently the same function on $\overline{T}$), as the representation is in general not unique.

The term $(\sum_{j=1}^{n} \lambda_j^2)^{1/2}$ will be called the mass of the molecule; note that the mass of a molecule is greater than or equal to the norm of the induced linear functional on JT.

The notion of molecule (compare [G-M-S]) is a useful tool for analysing the space JT^* in view of the following result: (see [S-S-W], lemma 6.3).

2.2 PROPOSITION: The unit ball of JT^* is the norm-closed convex hull of the molecules. □

For $N \in \mathbf{N}$ denote by $\pi_N : JT \to JT$ the restriction map to Δ_N, i.e. $\pi_N((x_t)_{t \in T} = (y_t)_{t \in T}$ where $y_t = x_t$ if $|t| = N$ and $y_t = 0$ otherwise. For $N \leq M$ denote $\pi_{[N,M]} : JT \to JT$ the restriction map to $\cup_{n=N}^{M} \Delta_n$ and $\pi_{[N,\omega]} : JT \to JT$ the restriction map to $\cup_{n=N}^{\infty} \Delta_n$. Clearly these maps are contractive projections. Their duals which are contractive projections on JT^*, will still be denoted by the same letters. We also may define $\pi_\omega : JT^* \to \ell^2(\Delta_\omega) = \ell^2(\{-1, +1\}^{\mathbf{N}})$ to be the restriction of an element of JT^*, identified with a function on $\overline{T}$, to the bottom- line Δ_ω of $\overline{T}$. As was shown by Lindenstrauss and Stegall [L-S], π_ω is a quotient map and the kernel of π_ω is the space $B = JT_*$ predual to JT.

The subsequent result follows easily from the weak-star lower semicontinuity of the norm on JT^* and from proposition 2.2 above and is left as an exercise to the reader (compare [G-M-S], lemma V.2).

2.4 PROPOSITION: For $x \in JT^*$

$$(a) \quad \|x\| = lim_{N\to\infty}\|\pi_{[0,N]}(x)\|$$

$$(b) \quad \|\pi_\omega(x)\| = lim_{N\to\infty}\|\pi_{[N,\omega]}(x)\| = lim_{N\to\infty}\|\pi_N(x)\|.$$

We still need some geometrical concepts: Let C be a closed, convex, bounded subset of a Banach space X. A slice of C will be a set of the form

$$S(f,\alpha) = \{x \in C : \langle f,x\rangle > M_f - \alpha\}$$

where $f \in X^*$, $\|f\| = 1$, $\alpha > 0$ and

$$M_f = sup\{\langle f,x\rangle : x \in C\}.$$

An element $f \in X^*$, $f \neq 0$, strongly exposes C if

$$lim_{\alpha\to 0}\ diam S(f/\|f\|,\alpha) = 0,$$

where diam denotes the diameter of a set, i.e. for $A \subseteq X$

$$diam(A) = sup\{\|x-y\| : x,y \in A\}.$$

Note that for a strongly exposing $f \in X^*$ there is a unique x_0 in the intersection of $(S(f/\|f\|,\alpha))_{\alpha>0}$. We say that f strongly exposes C at the point x_0.

3. The technical Lemmata:

The first general geometrical lemma is a slight refinement of ([S], lemma 2.7):

3.1 LEMMA: Let C be a convex subset of the unit ball of a Banach space X, f, g in X^*, $\|f\| = \|g\| = 1$ and $1 > \delta > 0$, $1 > \alpha > 0$ and $\alpha/3 > \epsilon > 0$ be given. Define

$$c = sup\{\langle g, x\rangle : x \in S(f, \alpha\epsilon\delta)\}.$$

Then there are $f_1 \in X^*$, $\|f_1\| = 1$ and $\beta > 0$ such that

$$(i)\ S(f_1, \beta) \subseteq S(f, \alpha)$$
$$(ii)\ S(f_1, \beta) \subseteq \{x \in C : \langle g, x\rangle > c - \delta\}$$
$$(iii)\ \|f - f_1\| < 3\epsilon.$$

PROOF: Let $f_1 = (f + \epsilon g)/\|f + \epsilon g\|$ and note that

$$\begin{aligned} M_{f+\epsilon g} &= sup\{\langle f + \epsilon g, x\rangle : x \in C\} \\ &\geq sup\{\langle f + \epsilon g, x\rangle : x \in S(f, \alpha \in S)\} \\ &\geq M_f - \alpha.\epsilon.\delta + \epsilon c \\ &= M_f - \alpha.\epsilon.\delta + \epsilon(c - \delta) + \epsilon.\delta \\ &> M_f + \epsilon(c - \delta). \end{aligned}$$

Hence

$$T = \{x \in C : \langle f + \epsilon g, x\rangle > M_f + \epsilon(c - \delta)\}$$

is a slice of C of the form $T = S(f_1, \beta)$ for some $\beta > 0$.

If, for $x \in C$, $\langle f, x\rangle \leq M_f - \alpha$ then

$$\begin{aligned} \langle f + \epsilon g, x\rangle &\leq M_f - \alpha + \epsilon \\ &\leq M_f - 2\epsilon \\ &\leq M_f + \epsilon(c - \delta) \end{aligned}$$

which proves (i).

As regards (ii) note that if, for $x \in C$, $\langle g, x\rangle \leq c - \delta$ then

$$\langle f + \epsilon g, x\rangle \leq M_f + \epsilon(c - \delta)$$

which proves (ii). The verification of (iii) is straightforward. □.

3.2 REMARK: We shall apply the lemma - just as we did in [S] - to produce for a given slice $S(f,\alpha)$ and for $g \in X^*$ a subslice $S(f_1,\beta)$ on which the oscillation of g is small: Indeed in the setting of 3.1 we may suppose for given $\delta > 0$, by choosing $\alpha > 0$ sufficiently small, that

$$sup\{\langle g,x\rangle : x \in S(f,\alpha)\} < sup\{\langle g,x\rangle : x \in S(f,\alpha\eta)\} + \delta$$

for every $0 < \eta < 1$ (obvious proof). Under this additional hypothesis (ii) implies that the values of g on $S(f_1,\beta)$ are contained in the interval $]c-\delta, c+\delta[$. Lemma 3.1 furnishes a more quantitative version of this observation and in particular the fact that f_1 may be chosen close to f.

We now turn to a series of lemmata which analyse the space JT^*. In the next lemma 3.3 we shall use the estimate

$$1 - r^2 \leq (1-r^2)^{1/2} \leq 1 - r^2/2$$

for $0 \leq r \leq 1$ and we shall apply Jensen's inequality to the concave function $f(r) = 1 - r^2/2$ in the form

$$(*) \qquad \sum_{i=1}^{n} \mu_i(1 - r_i^2/2) \leq 1 - (\sum_{i=1}^{n} \mu_i r_i)^2/2,$$

where μ_i and r_i are in $[0,1]$ and $\sum_{i=1}^{n} \mu_i = 1$.

In fact, for μ_i and r_i in $[0,1]$ and $0 < \sum_{i=1}^{n} \mu_i = \mu \leq 1$ we may estimate

$$\begin{aligned}\sum_{i=1}^{n} \mu_i(1 - r_i^2/2) &= \mu[\sum_{i=1}^{n}(\frac{\mu_i}{\mu})(1 - r_i^2/2)]\\ &\leq \mu[1 - (\sum_{i=1}^{n}(\frac{\mu_i}{\mu})r_i)^2/2]\\ &= \mu - \mu^{-1}(\sum_{i=1}^{n}\mu_i r_i)^2/2\\ (**) \qquad &\leq \sum_{i=1}^{n}\mu_i - (\sum_{i=1}^{n}\mu_i r_i)^2/2.\end{aligned}$$

3.3 LEMMA: Let $N \in \mathbf{N}$ and x be a function on $\mathcal{T}_N$ of the form

$$x = \sum_{i=1}^{n} \mu_i m_i \tag{0}$$

where μ_i are positive scalars, $\sum \mu_i \leq 1$ and m_i are molecules on $\mathcal{T}_N$, i.e.

$$m_i = \sum_{r=1}^{s(i)} \lambda_{i,r} \chi_{S_{i,r}} \tag{1}$$

where $(S_{i,r})_{r=1}^{s(i)}$ are disjoint segments and

$$\sum_{r=1}^{s(i)} \lambda_{i,r}^2 \leq 1$$

for $1 \leq i \leq n$.

For $t \in \Delta_N$ define

$$I_1^t = \{i \in \{1, ..., n\} : m_i(t) \geq 0\}$$

$$I_2^t = \{i \in \{1, ..., n\} : m_i(t) < 0\}$$

$$d_i^t = |m_i(t)| \text{ for } 1 \leq i \leq n$$

$$d^t = min(\sum_{i \in I_1^t} \mu_i d_i^t, \sum_{i \in I_2^t} \mu_i d_i^t)$$

and

$$d = (\sum_{t \in \Delta_N} (d^t)^2)^{1/2}.$$

Then there is a representation

$$x = \sum_{i=1}^{p} \mu_i' m_i'$$

where m_i' are molecules on $\mathcal{T}_N$ and μ_i' positive scalars such that

$$\sum_{i=1}^{p} \mu_i' \leq \sum_{i=1}^{n} \mu_i - d^2/2.$$

In particular $\|x\|_{JT^*} \leq \sum_{i=1}^{n} \mu_i - d^2/2$.

3.4 REMARK: The quantity d^t measures the amount of cancellations at $t \in \Delta_N$ in the representation $x = \sum \mu_i m_i$ and d the total amount of cancellations on Δ_N. Roughly speaking the lemma states that if there is a big amount of cancellations on the bottom level Δ_N of $\mathcal{T}_N$ then $\sum \mu_i m_i$ is not a good representation of x and may be replaced by a "more economical" one.

PROOF OF LEMMA 3.3: Fix $t \in \Delta_N$ and suppose that $d^t > 0$. First note that we may group the cancellations into pairs, i.e. we may assume that there is $k \in \mathbf{N}$ and subsets $(i_{1,j})_{j=1}^k$ and $(i_{2,j})_{j=1}^k$ of I_1^t and I_2^t such that, for $1 \leq j \leq k$,

$$\mu_{i_{1,j}} m_{i_{1,j}}(t) = -\mu_{i_{2,j}} m_{i_{2,j}}(t) \neq 0 \tag{2}$$

and

$$\sum_{j=1}^k \mu_{i_{1,j}} m_{i_{1,j}}(t) = -\sum_{j=1}^k \mu_{i_{2,j}} m_{i_{2,j}}(t) = d^t.$$

Indeed, if this is not possible for the original representation (0) of x we easily can make it possible by writing the $m_i's$ as suitable convex combinations of themselves, i.e. by passing to a representation

$$x = \sum_{i=1}^n \mu_i \sum_{q=1}^{p(i)} \nu_{i,q} m_i = \sum_{i=1}^n \sum_{q=1}^{p(i)} (\mu_i \nu_{i,q}) m_i$$

where $\nu_{i,q} \geq 0$ are suitably chosen and satisfy

$$\sum_{q=1}^{p(i)} \nu_{i,q} = 1 \text{ for } 1 \leq i \leq n.$$

For simplicity of notation we maintain the original representation (0). Now fix $j \in \{1, ..., k\}$ and consider the molecules $m_{i_{1,j}}$ and $m_{i_{2,j}}$. Denote $S_{1,j}$ (resp. $S_{2,j}$) the segment of $\mathcal{T}_N$ appearing in the expression (1) of $m_{i_{1,j}}$ (resp. $m_{i_{2,j}}$) containing t. By (2) these segments are well defined.

Let $S_j = S_{1,j} \cap S_{2,j}$ and note that either $S_{1,j} = S_j$ or $S_{2,j} = S_j$ holds true (possibly both). Denote $C_j = \mathcal{T}_N \setminus S_j$ and χ_{C_j} the characteristic function of C_j. Then

$$\begin{aligned} \mu_{i_{1,j}} m_{i_{1,j}} + \mu_{i_{2,j}} m_{i_{2,j}} &= (\mu_{i_{1,j}} m_{i_{1,j}} + \mu_{i_{2,j}} m_{i_{2,j}}) \chi_{C_j} \\ &= (\mu_{i_{1,j}} m_{i_{1,j}}) \chi_{C_j} + (\mu_{i_{2,j}} m_{i_{2,j}}) \chi_{C_j}. \end{aligned}$$

Suppose that $S_{1,j} = S_j$ (the other case is analogous). On the one hand side the function $m_{i_{2,j}} \chi_{C_j}$ may still be represented by a molecule of mass less than or equal to one: In fact,

$$m_{i_{2,j}} \chi_{C_j} = \sum_{r=1}^{s(i_{2,j})} \lambda_{i_{2,j},r} \chi_{S_{i_{2,j}r} \cap C_j}$$

and the latter expression is a molecule.

On the other hand the function $m_{i_{1,j}}\chi_{C_j}$ may be represented by a molecule of mass less than or equal to $(1-(d^t_{i_{1,j}})^2)^{1/2}$ as one may leave out the term $d^t_{i_{1,j}}\chi_{S_j}$ in the representation (1) of $m_{i_{1,j}}$. Hence letting

$$\begin{aligned}\mu'_{i_{1,j}} &= \mu_{i_{1,j}}(1-(d^t_{i_{1,j}})^2/2)\\ m'_{i_{1,j}} &= (\mu_{i_{1,j}}\chi_{C_j})/(1-(d^t_{i_{1,j}})^2/2)\end{aligned} \tag{3}$$

and

$$\begin{aligned}\mu'_{i_{2,j}} &= \mu_{i_{2,j}}\\ m'_{i_{2,j}} &= m_{i_{2,j}}\chi_{C_j}\end{aligned}$$

we have the equality for the functions on $\mathcal{T}_N$

$$\mu_{i_{1,j}}m_{i_{1,j}} + \mu_{i_{2,j}}m_{i_{2,j}} = \mu'_{i_{1,j}}m'_{i_{1,j}} + \mu'_{i_{2,j}}m'_{i_{2,j}},$$

the $m'_{i_{1,j}}$ and $m'_{i_{2,j}}$ still being molecules but $\mu'_{i_{1,j}}$ being smaller than $\mu_{i_{1,j}}$ by (3).

In the case $S_j \neq S_{1,j}$ (and therefore $S_j = S_{2,j}$) do the construction with the roles of 1 and 2 interchanged to arrive at similar equations where (3) is replaced by

$$\mu'_{i_{2,j}} = \mu_{i_{2,j}}(1-(d^t_{i_{2,j}})^2/2). \tag{4}$$

Do this construction for $1 \leq j \leq k$ and for the remaining indices, i.e. for

$$i \in \{1, ..., n\} \setminus \{(i_{1,j})_{j=1}^k \cup (i_{2,j})_{j=1}^k\},$$

relabel μ_i by μ'_i and m_i by m'_i. We thus obtained a new representation of x

$$x = \sum_{i=1}^n \mu'_i m'_i.$$

Denote by $I_0 = (i_j)_{j=1}^k$ those indecizes for which (3) resp. (4) hold true, i.e. $i_j = i_{1,j}$ if $S_{1,j} = S_j$ and $i_j = i_{2,j}$ otherwise. We may estimate

$$\begin{aligned}\sum_{i=1}^n \mu'_i &\leq \sum_{i\notin I_0}\mu_i + \sum_{i\in I_0}\mu'_i\\ &= \sum_{i\notin I_0}\mu_i(1-0^2/2) + \sum_{j=1}^k \mu_{i_j}(1-(d^t_{i_j})^2/2)\\ \text{by } (**) \quad &\leq \sum_{i=1}^n \mu_i - (\sum_{j=1}^k \mu_{i_j} d^t_{i_j})^2/2\\ &= \sum_{i=1}^n \mu_i - (d^t)^2/2.\end{aligned}$$

Here the application of Jensen's inequality was explained before lemma 3.3 and the last line follows from the line after (2).

This was the construction for $t \in \Delta_N$ fixed. Applying it successively to all $t \in \Delta_N$ for which $d^t > 0$ we finally end up with a representation

$$x = \sum_{i=1}^{p} \widetilde{\mu}_i \widetilde{m}_i$$

such that $\widetilde{m}_i$ are molecules on $\mathcal{T}_N$, $\widetilde{\mu}_i \geq 0$ and

$$\sum_{i=1}^{p} \widetilde{\mu}_i \leq \sum_{i=1}^{n} \mu_i - \sum_{t \in \Delta_N} (d^t)^2/2 = \sum_{i=1}^{n} \mu_i - d^2/2. \square$$

3.5 LEMMA; Let $N \in \mathbf{N}$ and $x \in JT^*$ which we identify with the function $(x(t))_{t\in\mathcal{T}} = (\langle e_t, x\rangle)_{t\in\mathcal{T}}$ on $\mathcal{T}$. Suppose that $\pi_{[N,\omega]}(x) \geq 0$ such that, for every $t \in \mathcal{T}_{[N,\omega[} = \cup_{n=N}^{\infty} \Delta_n$

(1) $$x(t) \geq x(t,1) + x(t,-1),$$

where $(t,1)$ and $(t,-1)$ are the two successors of t. Then

$$\|\pi_{[N,\omega]}(x)\| = \|\pi_N(x)\|.$$

PROOF: Clearly $\|\pi_{[N,\omega]}(x)\| \geq \|\pi_N(x)\|$. For the reverse inequality it suffices to show by 2.4 a) that, for every $M \geq N$,

$$\|\pi_{[N,M]}(x)\| \leq \|\pi_N(x)\|.$$

But, for $M \geq N$ fixed, it is an easy exercise to verify that by condition (1) we may write $\pi_{[N,M]}(x)$ as a finite convex combination of molecules m_i of the form

$$m_i = \sum_{t\in\Delta_N} x(t)\chi_{S_t^i}$$

where S_t^i are segments in $\mathcal{T}_{[N,M]} = \cup_{n=N}^{M} \Delta_n$ such that $t \in S_t^i$ for each $t \in \Delta_N$. $\square$

3.6 REMARK: The lemma 3.5 applies in particular to the case where $\mu_i \geq 0$ and m_i are molecules

$$m_i = \sum_{r=1}^{s(i)} \lambda_{i,r}\chi_{S_{i,r}}$$

such that each $S_{i,r}$ starts at level N or before (i.e., $S_{i,r} \cap T_N \neq \emptyset$ for $1 \leq r \leq s(i)$ and $1 \leq i \leq n$) and such that $\lambda_{i,r} \geq 0$ for all (i,r) such that $S_{i,r}$ passes through Δ_N (i.e. $S_{i,r} \cap \Delta_N \neq \emptyset$ implies $\lambda_{i,r} \geq 0$).

In fact we shall replace the positivity condition by the subsequent "same sign condition":

3.7 LEMMA: Let $N \in \mathbf{N}$ and $x \in JT^*$,

$$x = \sum_{i=1}^{n} \mu_i m_i$$

where $\mu_i \geq 0$, $\sum_{i=1}^{n} \mu_i = 1$ and

$$m_i = \sum_{r=1}^{s(i)} \lambda_{i,r} \chi_{S_{r,i}}$$

such that, for every $1 \leq r \leq s(i)$ and $1 \leq i \leq n$,

$$S_{i,r} \cap T_N \neq \emptyset$$

and such that, for every $t \in \Delta_N$,

$$x(t) \geq 0 \Rightarrow \lambda_{i,r} \geq 0 \text{ for all } (i,r) \text{ such that } t \in S_{i,r}$$

$$x(t) \leq 0 \Rightarrow \lambda_{i,r} \leq 0 \text{ for all } (i,r) \text{ such that } t \in S_{i,r}.$$

Then

$$\|\pi_N(x)\| = \|\pi_{[N,\omega]}(x)\|.$$

PROOF: Immediate from lemma 3.5 and remark 3.6, noting that for $t \in \Delta_N$ the operator

$$S_t : \pi_{[N,\omega]} JT^* \longrightarrow \pi_{[N,\omega]} JT^*$$

$$[S_t(x)](s) = \begin{cases} 0 & \text{if } |s| < N \\ x(s) & \text{if } |s| \geq N \text{ and } s \not\geq t \\ -x(s) & \text{if } |s| \geq N \text{ and } s \geq t \end{cases}$$

that changes the sign of $x(s)$ for t and all successors s of t is an isometrical isomorphism on $\pi_{[N,\omega]} JT^*$, which denotes the range of $\pi_{[N,\omega]}$. □.

3.8 LEMMA: Let $N \in \mathbf{N}$, $0 < \epsilon < 1$ and $0 < \delta < \epsilon^3/2^{10}$. If $x \in JT^*$, $\|x\| \leq 1$ is such that

$$\|\pi_{[1,N]}(x)\| > 1 - \delta$$

then

$$\|\pi_{[N,\omega]}(x)\| < \|\pi_N(x)\| + \epsilon.$$

PROOF: By proposition 2.2 it suffices to show the lemma for convex combinations of molecules, i.e.

$$x = \sum_{i=1}^{n} \mu_i m_i$$

and

$$m_i = \sum_{r=1}^{s(i)} \lambda_{i,r} \chi_{S_{i,r}},$$

where $\mu_i \geq 0$, $\sum \mu_i \leq 1$, $(S_{i,r})_{r=1}^{s(i)}$ disjoint segments of $\overline{T}$ and $\sum_{i=1}^{s(i)} \lambda_{i,r}^2 \leq 1$, for $1 \leq i \leq n$.

For $1 \leq i \leq n$ define three index sets

$$\begin{aligned}
J_3^i &= \{r \in \{1, ..., s(i)\} : S_{i,r} \cap T_N = \emptyset\} \\
J_2^i &= \{r \in \{1, ..., s(i)\} : S_{i,r} \cap \Delta_N = \{t\} \neq \emptyset \text{ and } sign(x(t)) \neq sign(\lambda_{i,r})\} \\
J_1^i &= \{1, ..., s(i)\} \setminus (J_2^i \cup J_3^i).
\end{aligned}$$

We let (by abuse of notation) $sign$ $(0) = +1$ above, which will avoid some technical difficulties.

For $j = 1, 2, 3$ and $1 \leq i \leq n$ let

$$m_{j,i} = \sum_{r \in J_j^i} \lambda_{i,r} \chi_{S_{i,r}}$$

and let

$$x_j = \sum_{i=1}^{n} \mu_i m_{j,i}, \qquad j = 1, 2, 3.$$

Note that x_1 is of the form described in lemma 3.7 hence

$$\|\pi_{[N,\omega]}(x_1)\| = \|\pi_N(x_1)\|.$$

We shall therefore be done if we show that

$$\|\pi_{[N,\omega]}(x_2)\| < \epsilon/4 \text{ and } \|\pi_{[N,\omega]}(x_3)\| < \epsilon/4. \tag{1}$$

As regards x_2 this is also an element of the form described in lemma 3.7 hence, by 3.7 and 3.3

$$\|\pi_{[N,\omega]}(x_2)\| = \|\pi_N(x_2)\| = d$$

where d is the number associated to $\pi_{[0,N]}(x)$ by lemma 3.3 above and for which we get in view of

$$\|\pi_{[0,N]}(x)\| > 1 - \delta > 1 - \epsilon^3/2^{10} > 1 - \epsilon^2/32$$

that

$$d < \epsilon/4,$$

which proves the first half of (1).

As regards x_3 first note that

$$\|x_1 + x_2\| = \|x - x_3\| \geq \|\pi_{[0,N]}(x - x_3)\| = \|\pi_{[0,N]}(x)\| > 1 - \delta.$$

Denoting $I_0 = \{i \in \{1, ..., n\} : \|m_{1,i} + m_{2,i}\| < 1 - \epsilon^2/2^7\}$ we obtain

$$1 - \delta < \|x_1 + x_2\| < (1 - \sum_{i \in I_0} \mu_i) + (1 - \epsilon^2/2^7) \sum_{i \in I_0} \mu_i$$

whence

$$\sum_{i \in I_0} \mu_i < \delta/(\epsilon^2/2^7) < \epsilon/8.$$

For $i \notin I_0$ we may use

$$\|m_{3,i}\|^2 + \|m_{1,i} + m_{2,i}\|^2 \leq 1$$

to estimate

$$\|m_{3,i}\| \leq (1 - (1 - \epsilon^2/2^7)^2)^{1/2} < \epsilon/8$$

whence

$$\|x_3\| \leq \sum_{i \in I_0} \mu_i \|m_{3,i}\| + \sum_{i \notin I_0} \mu_i \|m_{3,i}\| < \epsilon/4. \square$$

3.9 REMARK: Note in passing that lemma 3.8 implies in particular the by now well-known fact ([B-R], [G-M]) that $(\pi_n(JT_*))_{n=0}^{\infty}$ forms a boundedly complete skipped blocking decomposition of JT_*. Indeed observing that $(\pi_n(JT_*))_{n=1}^{\infty}$ is a monotone Schauder-decomposition of JT_*, this fact quickly follows from the subsequent observation.

CLAIM: Let $(x_n)_{n=1}^{\infty}$ be elements of $\pi_n(JT_*) = \pi_n(JT^*)$ such that

$$(i) \quad \lim_{N \to \infty} \|\sum_{n=1}^{N} x_n\| < \infty$$

$$(ii) \quad \liminf_{n \to \infty} \|x_n\| = 0.$$

Then $\sum_{n=1}^{\infty} x_n$ converges in norm in JT_*.

Indeed, we may suppose that the limit in (i) is 1; fix $1 > \epsilon > 0$ and $\epsilon^3/2^{10} > \delta > 0$ and find N big enough such that

$$\|\sum_{n=1}^{N} x_n\| > 1 - \delta \text{ and } \|x_N\| < \epsilon.$$

Then lemma 3.8 implies that, for $M \geq N$,

$$\|\sum_{n=N}^{M} x_n\| < \|x_N\| + \epsilon < 2\epsilon,$$

which proves the claim.

In a similar way lemma 3.8 implies the subsequent proposition which takes care of the easier part of theorem 4.1 below.

3.10 PROPOSITION: Let $x \in JT_*$, $\|x\| = 1$. Then x is a point of weak-star to norm continuity of the unit ball of JT^*. In fact, the sequence of relative weak-star neighbourhoods

$$V_n(x) = \{z \in JT^* : \|z\| \leq 1 \text{ and } \|\pi_{[0,n]}(x - z)\| < n^{-1}\}$$

forms a relative norm-neighbourhood base of x in the unit ball of JT^*.

PROOF: Given $0 < \epsilon < 1$, let $0 < \delta < \epsilon^3/2^{10}$, find $n \in \mathbf{N}$, $n > 2/\delta$ such that

$$\|\pi_{[0,n]}(x)\| > 1 - \delta/2 \text{ and } \|\pi_{[n,\omega]}(x)\| < \epsilon,$$

which is possible by 2.4 and the assumption that $x \in JT_*$, i.e. $\pi_\omega(x) = 0$.

For $z \in V_n(x)$ we have $\|\pi_{[0,n]}(x)\| > 1 - \delta$, hence by 3.8

$$\|\pi_{[n,\omega]}(z)\| \leq \|\pi_n(z)\| + \epsilon.$$

As $\|\pi_n(z)\| < \epsilon + n^{-1}$ we may estimate

$$\begin{aligned} \|x - z\| &\leq \|\pi_{[0,n-1]}(x - z)\| + \|\pi_{[n,\omega]}(x)\| + \|pi_{[n,\omega]}(z)\| \\ &< \delta/2 + \epsilon + (2\epsilon + \delta/2) < 4\epsilon. \quad \square \end{aligned}$$

However, in order to establish theorem 4.1 below we shall need the full strength of 3.8. Before we can do so we still have to establish some more technical results.

3.11 LEMMA: Let $N \in \mathbf{N}$, $1 > \epsilon > 0$, $\epsilon^5/2^{26} > \delta > 0$ and let $x \in JT^*$ be such that

$$(i)\ \|\pi_{[N,\omega]}(x)\| \leq 1$$
$$(ii)\ \|\pi_N(x)\| > 1-\delta \text{ and } \|\pi_\omega(x)\| > 1-\delta.$$

Then for $y \in JT^*$ such that

$$(i)\ \|\pi_{[N,\omega]}(y)\| \leq 1$$
$$(ii)\ \|\pi_N(x-y)\| < \delta \text{ and } \|\pi_\omega(x-y)\| < \delta$$

we obtain

$$\|\pi_{[N,\omega]}(x-y)\| < \epsilon.$$

3.12 REMARK: Loosely speaking the lemma states the following: Let x be in the unit ball of $\pi_{[N,\omega]}(JT^*)$ such that $\|\pi_N(x)\|$ as well as $\|\pi_\omega(x)\|$ are close to 1. Then for every y in the unit ball of $\pi_{[N,\omega]}(JT^*)$ such that $\|\pi_N(x-y)\|$ as well as $\|\pi_\omega(x-y)\|$ are small we obtain that $\|\pi_{[N,\omega]}(x-y)\|$ is small.

The proof will show that y (and therefore x) is in fact close to some molecule of a special form (and not only to a convex combination of molecules).

PROOF: STEP 1: We show that there is a molecule m_0 of the form

(1) $$m_0 = \sum_{t\in\Delta_N} \lambda_t \chi_{S_t}$$

where $\sum_{t\in\Delta_N} \lambda_t^2 \leq 1$ and S_t are segments such that t is the first element of S_t and S_t is infinite (i.e., $S_t \cap \Delta_\omega \neq \emptyset$), such that

(2) $$\|\pi_N(x-m_0)\| < \epsilon/8$$

and

(3) $$\|\pi_\omega(x-m_0)\| < \epsilon/8.$$

Indeed, define the linear functional F on JT^* by

$$F(z) = (\frac{\pi_N(z), \pi_N(x)}{\|\pi_N(x)\|}) + (\frac{\pi_\omega(z), \pi_\omega(x)}{\|\pi_\omega(x)\|})$$

where the first inner product is taken in $\ell^2(\Delta_N)$ and the second in $\ell^2(\Delta_\omega)$. Clearly $\|F\| \leq 2$ and

$$F(x) = \|\pi_N(x)\| + \|\pi_\omega(x)\| > 2(1-\delta).$$

Hence the slice T of the unit ball of JT^* defined by

$$T = T(F, \frac{32\delta}{\epsilon}) = \{z \in JT^* : \|z\| \le 1 \text{ and } F(z) > 2 - \frac{32\delta}{\epsilon}\} \tag{4}$$

is non-empty. By 2.2 the slice T contains a molecule m_0'. If m_0' is of the form

$$m_0' = \sum_{j=1}^{k} \lambda_j \chi_{S_j'}$$

with $\sum_{j=1}^k \lambda_j^2 \le 1$ and $(S_j')_{j=1}^k$ disjoint segments in $\overline{T}$, define

$$m_0 = \sum_{j \in J_0} \lambda_j \chi_{S_j}$$

where

$$J_0 = \{j \in \{1, ..., k\} : S_j' \cap \Delta_N \ne \emptyset \text{ and } S_j' \cap \Delta_\omega \ne \emptyset\}$$

and

$$S_j = S_j' \cap \overline{T}_{[N,\omega]}, \qquad \text{for } 1 \le j \le k,$$

where $\overline{T}_{[N,\omega]}$ denotes the subset of $\overline{T}$ consisting of $t \in \overline{T}$ with $N \le |t| \le \omega$. Clearly m_0 is a molecule of the form (1). In order to estimate $\|\pi_{[N,\omega]}(m_0 - m_0')\|$ let

$$J_1 = \{j \in \{1, ..., k\} : S_j \cap \Delta_N = \emptyset\} \text{ and}$$
$$J_2 = \{j \in \{1, ..., k\} : S_j \cap \Delta_N \ne \emptyset \text{ and } S_j \cap \Delta_\omega = \emptyset\}.$$

Note that

$$\|\pi_N(\sum_{j \notin J_1} \lambda_j \chi_{S_j})\| \ge (\sum_{j \notin J_1} \lambda_j \chi_{S_j}, \frac{\pi_N(x)}{\|\pi_N(x)\|}) > 1 - \frac{32\delta}{\epsilon}$$

hence

$$(\sum_{j \in J_1} \lambda_j^2)^{1/2} \le (1 - (1 - \frac{32\delta}{\epsilon})^2)^{1/2} < (\frac{2^6\delta}{\epsilon})^{1/2} < \epsilon^2/2^{10}.$$

The same estimate holding true for J_2 we conclude that

$$\|\pi_{[N,\omega]}(m_0 - m_0')\| < \epsilon^2/2^9. \tag{5}$$

This implies in particular that

$$F(m_0) > F(m_0') - 2\|\pi_{[N,\omega]}(m_0 - m_0')\| > 2 - \epsilon^2/2^7,$$

i.e.

$$m_0 \in T(F, \epsilon^2/2^7) = \{z \in JT^*, \|z\| \leq 1 \text{ and } F(z) > 2 - \epsilon^2/2^7\}.$$

Hence using the fact established in the proof of 3.13 below for $H = \ell^2(\Delta_N)$ and $H = \ell^2(\Delta_\omega)$ respectively we obtain (2) and (3).

STEP 2: If m_1' is another molecule contained in the slice $T(F, 32\delta/\epsilon)$ defined in (4), we may do the same construction to find a molecule m_1 of the form (1) such that (2) and (3) hold true for m_1 in the place of m_0 and, as in (5),

$$\|\pi_{[N,\omega]}(m_1 - m_1')\| < \epsilon^2/2^9.$$

We may conclude from (3) and the special form (1) of m_0 and m_1 that

$$\|\pi_{[N,\omega]}(m_0 - m_1)\| = \|\pi_\omega(m_0 - m_1)\| < \epsilon/4$$

and therefore

(6) $$\|\pi_{[N,\omega]}(m_0' - m_1')\| < \epsilon/2.$$

In other words the diameter of the image under $\pi_{[N,\omega]}$ of the set of molecules contained in $T(F, 32\delta/\epsilon)$ has diameter less than $\epsilon/2$. It is now routine to deduce that

(7) $$diam(\pi_{[N,\omega]}T(F, 4\delta)) < \epsilon,$$

where

$$T(F, 4\delta) = \{z \in JT^* : \|z\| \leq 1 \text{ and } F(z) > 2 - 4\delta\}.$$

Indeed, if

$$z = \sum_{i=1}^{n} \mu_i m_i$$

is a convex combination of molecules such that $z \in T(F, 4\delta)$ then letting

$$I_0 = \{i \in \{1, ..., n\} : m_i \in T(F, 32\delta/\epsilon\}$$

we obtain

(8) $$\sum_{i \in I_0} \mu_i > 1 - \epsilon/8$$

and therefore, for

$$\bar{z} = \sum_{i \in I_0} \mu_i m_i$$

that

$$\|z - \bar{z}\| < \epsilon/8.$$

If w is another convex combination of molecules in $T(F, 4\delta)$ apply the same construction to obtain a combination $\overline{w}$ of molecules in $T(F, 32\delta/\epsilon)$. Hence by (6) and (8)

$$\|\pi_{[N,\omega]}(w - z)\| \leq \|\pi_{[N,\omega]}(\overline{w} - \bar{z})\| + \epsilon/4 \leq (\epsilon/2 + \epsilon/8) + \epsilon/4 < \epsilon.$$

By the density of the convex combinations of molecules in $T(F, 4\delta)$ we have proved (7).

The proof of the lemma is complete by observing that x and y in the statement of the lemma are contained in $T(F, 4\delta)$. □

As an easy consequence of lemma 3.11 we may state

3.13 LEMMA: Let $N \in \mathbf{N}$, $1 > \epsilon > 0$, $\epsilon^5/2^{26} > \delta > 0$ and let $x \in JT^*$ be such that

(i) $\|\pi_{[N,\omega]}(x)\| \leq 1$

(ii) $\|\pi_N(x)\| > 1 - \delta$ and $\|\pi_\omega(x)\| > 1 - \delta^2/2$.

Define the slice S of the unit ball of JT^* by

$$S = \{z \in JT^*, \|z\| \leq 1 \text{ and } (\frac{\pi_\omega(z), \pi_\omega(x)}{\|\pi_\omega(x)\|}) > 1 - \delta^2/2\}$$

and the weak-star neighbourhood V of x

$$V = \{z \in JT^*, \|\pi_N(x - z)\| < \delta\}.$$

Then $U = S \cap V$ contains x and is a relative weak neighbourhood of x in the unit ball of JT^* such that

$$diam(\pi_{[N,\omega]}(U)) < 2\epsilon.$$

PROOF: The lemma immediately follows from 3.11 and the subsequent elementary fact applied to the Hilbert space $\ell^2(\Delta)$, whose proof we leave to the reader (draw a picture):

FACT: Let S be a slice of the unit ball of a Hilbert space H

$$S = S(x, \alpha) = \{z \in H : \|z\| \leq 1 \text{ and } (x, z) > 1 - \alpha\}$$

where $x \in H$, $\|x\| = 1$ and $\alpha > 0$. Then, for every $y \in S$,

$$\|x - z\| < (2\alpha)^{1/2}$$

and

$$diam(S) < 2(2\alpha)^{1/2}. \quad \square.$$

4. The main results

4.1 THEOREM: The norm of JT^* has the Kadec-Klee property. More precisely, defining for $x \in JT^*$, $\|x\| = 1$ and $n \in \mathbf{N}$

$$V_n(x) = \{z \in JT^* : \|z\| \leq 1 \text{ and } \|\pi_{[0,n]}(x - z)\| < n^{-1}\}$$

and

$$S_n(x) = \{z \in JT^* : (\pi_\omega(z), \pi_\omega(x)) > \|\pi_\omega(x)\|^2 - n^{-1}\}$$

then $(V_n \cap S_n)_{n=1}^{\infty}$ forms a relative norm-neighbourhood base of x in the unit ball of JT^*.

4.2 REMARK: It seems worth noting that $V_n(x)$ are weak-star neighbourhoods of x and $S_n(x)$ are (in the non-trivial case $\|\pi_\omega(x)\| \neq 0$ which is taken care of by 3.10) slices determined by a fixed element $G \in JT^{**}$, $\|G\| = 1$

$$G : JT^* \to \mathbf{R}$$
$$z \to \left(\frac{\pi_\omega(z), \pi_\omega(x)}{\|\pi_\omega(x)\|}\right)$$

PROOF: If $\pi_\omega(x) = 0$ then proposition 3.10 applies. In the other case let $c = \|\pi_\omega(x)\|$, fix $1 > \epsilon > 0$ and $\epsilon^5/2^{26} > \delta > 0$. By proposition 2.4 we may find $N \in \mathbf{N}$ such that

$$(i)\ \|\pi_{[N,\omega]}(x)\| < c/(1 - \delta^2/4)$$
$$(ii)\ \|\pi_n(x)\| > c(1 - \delta/2) \text{ for } n \geq N.$$

Apply lemma 3.12 to the element

$$\frac{x}{[c/(1 - \delta^2/4)]}$$

of JT^*. It follows that we may find $n > max(N, \epsilon^{-1})$ such that $S_n(x) \cap V_n(x)$ satisfies

$$diam(\pi_{[n+1,\omega]}(S_n(x) \cap V_n(x))$$
$$\leq diam(\pi_{[N,\omega]}(S_n(x) \cap V_n(x)) < 2\epsilon.$$

Hence

$$\begin{aligned}&diam(S_n(x)\cap V_n(x))\\ \leq& diam(\pi_{[0,n]}(S_n(x)\cap V_n(x))) + diam(\pi_{[n+1,\omega]}(S_n(x)\cap V_n(x)))\\ <&2\epsilon+2\epsilon=4\epsilon.\end{aligned}$$

The theorem is proved. □

4.3 REMARK: The reader should notice that the points of weak-star to norm continuity on the unit sphere of JT^* are precisely those contained in JT_*. In fact, it is an instructive exercise to verify that, for $x \in JT^*$, $\|x\| = 1$

$$\begin{aligned}inf\{diam(V \cap \text{ unit ball } (JT^*)) : V \text{ a w*-neighbourhood of } x\}\\ = \sqrt{2}\|\pi_\omega(x)\|.\end{aligned}$$

Next note that the norm of JT^* fails to be strictly convex. For example

$$x_1(t) = \begin{cases} 1 & \text{if } t = \emptyset \text{ and } t = (+1) \\ 0 & \text{elsewhere} \end{cases}$$

$$x_2(t) = \begin{cases} 1 & \text{if } t = \emptyset \text{ and } t = (-1) \\ 0 & \text{elsewhere.} \end{cases}$$

Then x_1, x_2 are elements of the unit sphere of JT^* and $x = (x_1 + x_2)/2$ also belongs to the unit sphere. Hence x is not an extreme point of the unit ball of JT^*.

However on the basis of theorem 4.1 it is not to difficult to find a better norm on JT, namely to renorm JT with an equivalent norm $|.|$ such that every point of $(JT, |.|)^*$ is a denting point. This application was essentially shown to me by M. Fabian, G. Godefroy and S. Troyanski at the 15. Winterschool of the Čech Academy of Science on Abstract Analysis in Šrni 1987. I also thank M. Fabian for pointing out a mistake in a previous version of this paper.

4.4 THEOREM: There is an equivalent norm $|.|$ on JT such that every point of the unit sphere of $(JT, |.|)^*$ is a denting point. In particular the dual norm $|.|$ on JT^* is strictly convex and has the Kadec-Klee property.

PROOF OF THEOREM 4.4: Define a norm $|.|$ on JT^* by

$$|x| = [\|x\|^2_{JT^*} + \sum_{n=1}^{\infty} 2^{-2n}\langle x_n, x\rangle^2]^{1/2}$$

for $x \in JT^*$, where $(x_n)_{n=1}^\infty$ is a dense sequence in the unit ball of $(JT, \|.\|)$. Clearly $|.|$ is an equivalent dual norm on JT^*, i.e. $(JT^*, |.|) = (JT, |.|)^*$, where $|.|$ on JT^* (by abuse of notation we use the symbol $|.|$ in both cases).

It follows from Cauchy-Schwarz that every point of the unit sphere of $(JT^*, |.|)$ is extreme.

Also note that the norm $|.|$ is Kadec-Klee as for every net $(x_\alpha)_{\alpha \in I}$ such that x_α tends to x_0 weakly and $|x_\alpha| = |x_0| = 1$ for $\alpha \in I$ it also follows from Cauchy-Schwarz that $lim\|x_\alpha\| = \|x_0\|$ whence by 4.1. $(x_\alpha)_{\alpha \in I}$ tends to x_0 in norm. Hence every point of the unit sphere of $(JT^*, |.|)$ is extreme and a point of weak to norm continuity and therefore by a result of Lin, Lin and Troyanski [L-L-T] a denting point. □

4.6 REMARK: It is shown in ([S-S-W], prop. 3.2) how to deduce from theorem 4.1 that every w*-compact convex subset C of JT^* contains a denting point.

We shall show the more general theorem 4.8 below which implies the existence of strongly exposed points of a set C as above. We shall not use the full strength of theorem 4.1 but only the general geometrical lemma 3.1 and a direct construction.

Let us introduce some more notations: For $n \in \mathbf{N}$ denote

$$\begin{aligned} p_n : \Delta = \{-1,+1\}^{\mathbf{N}} &\to \Delta_n = \{-1,+1\}^n \\ (\epsilon_i)_{i=1}^\infty &\to (\epsilon_i)_{i=1}^n \end{aligned}$$

the restriction to the first n coordinates. If $z \in \ell^2(\Delta)$ is of finite support A and $n \in \mathbf{N}$ is big enough such that $p_n|A$ is injective we may define $p_n(z) \in \ell^2(\Delta_n)$ on the elements $s \in \Delta_n$ by

$$p_n(z)(s) = \begin{cases} z(t) & \text{if there is } t \in A,\ p_n(t) = s \\ 0 & \text{otherwise} \end{cases}$$

and we may define $p_{[n,\omega]}(z) \in JT^*$ on the elements $s \in \overline{T}$ by

$$p_{[n,\omega]}(z)(s) = \begin{cases} z(t) & \text{if } n \le |s| \le \omega \\ & \text{and there is } t \in A,\ p_{|s|}(t) = s \\ 0 & \text{otherwise.} \end{cases}$$

Clearly

$$\|z\|_{\ell^2(\Delta_\omega)} = \|p_n(z)\|_{\ell^2(\Delta_n)} = \|p_{[n,\omega]}(z)\|_{JT^*}.$$

Also note that $f \in JT^{**}$ may be represented as a function on $\overline{T}$ by

$$f(t) = \langle f, e_t \rangle \text{ for } t \in T$$

and

$$f(t) = lim_{n\to\infty}\langle f, \chi_{S_t^n}\rangle \text{ for } t \in \Delta_\omega$$

where in the preceeding line S_t^n is the segment starting at level n such that $t \in S_t^n$. The functional f is uniquely determined by its values in $\overline{T}$ $\{f(t) : t \in \overline{T}\}$ and the functionals f with finite support on $\overline{T}$ are norm-dense in JT^{**} (for more details see [S-S-W]).

Given $f \in JT^{**}$ with finite support $A \subseteq \overline{T}$ let n be such that, for $t \in A \cap T$, $|t| < n$ and such that p_n is injective on $A \cap \Delta_\omega$. Then we may define $p_n(f) \in JT$ on the elements $s \in T$ by

$$p_n(f)(s) = \begin{cases} f(s) & \text{if } s \in A \cap T \\ f(t) & \text{if } |s| = n \text{ and there is } t \in A \cap \Delta_\omega \text{ such that } p_n(t) = s \\ 0 & \text{otherwise.} \end{cases}$$

4.7 LEMMA: a) For $f \in JT^{**}$ of finite support let $n_0 \in \mathbf{N}$ such that $p_{n_0}(f)$ is defined. Then $(p_n(f))_{n \geq n_0}$ tends $\sigma(JT^{**}, JT^*)$ to f.

b) For $x \in JT^*$, $\epsilon > 0$ and $z \in \ell^2(\Delta)$ with finite support such that $\|\pi_\omega(x) - z\| < \epsilon$ let $n_0 \in \mathbf{N}$ be such that $p_{[n_0,\omega]}(z)$ is defined. Then

$$\lim_{n\to\infty, n\geq n_0} \|p_{[n,\omega]}(z) - \pi_{[n,\omega]}(x)\| < \epsilon.$$

PROOF: a) It is clear that for a finite combination of molecules $x \in JT^*$ there is $n \geq n_0$ such that

$$\langle f, x\rangle = \langle p_n(f), x\rangle.$$

The assertion follows from the easily verified fact

$$\|f\| = \|p_n(f)\| \qquad n \geq n_0$$

and proposition 2.2.

b) Suppose that y is a convex combination of molecules. Then there is $n \geq n_0$ such that

$$p_{[n,\omega]}(y) = \pi_{[n,\omega]}(y).$$

Now let $\beta = \epsilon - \|\pi_\omega(x) - z\| > 0$ and find a convex combination of molecules x_1 such that $\|x - x_1\| < \beta/2$, whence

$$\|\pi_\omega(x_1) - z\| < \epsilon - \beta/2.$$

For $n \geq n_0$ big enough we have

$$\|p_{[n,\omega]}(z) - \pi_{[n,\omega]}(x_1)\| = \|p_{[n,\omega]}(z)\| - p_{[n,\omega]}(\pi_\omega(x_1))\|$$
$$= \|z - \pi_\omega(x_1)\| < \epsilon - \beta/2.$$

and

$$\|p_{[n,\omega]}(z) - \pi_{[n,\omega]}(x)\| < \epsilon. \quad \square$$

4.8 THEOREM: Let C be a σ^*-compact, convex subset of JT^*. The set $SE = SE(C) \subseteq JT^{**}$ of strongly exposing functionals is a dense G_δ-subset of $(JT^{**}, \|.\|)$.

In particular C is the closed convex hull of its strongly exposed points.

4.9 REMARK: Theorem 4.8 answers negatively a question of [J-Z].

Let us note, however, that it may happen that $SE(C) \cap JT = \emptyset$. (In fact, this follows from C. Stegall's result [St].) Indeed let C be the σ^*-closed convex hull of the indicator functions of branches γ of $\overline{T}$ (i.e., of the infinite segments of $\overline{T}$ starting at the origin $\emptyset$ of $\overline{T}$).

It is easy to verify that every slice of C determined by an element of JT has diameter equal to $\sqrt{/}2$, whence no element of JT can belong to $SE(C)$. Theorem 4.7 tells us however that we may find "many" functionals in JT^{**} which do strongly expose C.

PROOF OF THEOREM 4.8: We may and do suppose that C is contained in the unit ball of JT^*. For $\epsilon > 0$ denote $SE_\epsilon(C)$ those functionals in JT^{**} which determine slices of diameter less than ϵ, i.e.

$$SE_\epsilon(C) = \{f \in JT^{**} : \text{ there is } \alpha > 0$$
$$\text{such that } diam(S(f,\alpha)) < \epsilon\}.$$

It is easy to verify that $SE_\epsilon(C)$ is norm-open and we shall therefore be done (by Baire's theorem), if we show that for $\epsilon > 0$, $SE_\epsilon(C)$ is norm-dense in JT^{**}.

So fix $f \in JT^{**}$, $\|f\| = 1$ and $1 > \epsilon > 0$. We shall construct $g \in JT^{**}$, $\|g\| = 1$, $\|f - g\| < \epsilon$ and $\alpha > 0$ such that

$$diam(S(g,\alpha)) < \epsilon.$$

STEP 1: We first construct $f_0 \in JT^{**}$ of finite support in $\overline{T}$ and $1 > \alpha_0 > 0$ such that

$$(i)\ \|f - f_0\| < \epsilon/3$$
$$(ii)\ diam(\pi_\omega(S(f_0, \alpha_0))) < \epsilon/10.$$

For $0 < \gamma < 1$ define

$$\varphi(\gamma) = sup\{\|\pi_\omega(x)\| : x \in S(f,\gamma)\},$$

and

$$\varphi_0 = lim_{\gamma\to 0}\varphi(\gamma).$$

If $\varphi_0 < \epsilon/10$ we are done by letting $f_0 = f$ and $1 > \alpha_0 > 0$ sufficiently small. Otherwise find $0 < \gamma_0 < 1$ such that

$$\varphi(\gamma_0) < \varphi_0 + \epsilon^2/4800$$

and $x_0 \in S(f,\eta)$ such that

$$\|\pi_\omega(x_0)\| > \varphi_0 - \epsilon^2/4800,$$

where $\eta = \gamma_0[min(\epsilon,\gamma_0/3)]^3/48000$.

Define the element h in the unit ball of JT^{**} by

$$h : JT^* \to \mathbf{R}$$
$$x \to (\pi_\omega(x), \pi_\omega(x_0)/\|\pi_\omega(x_0)\|).$$

Note that

$$sup\{\langle h,x\rangle : x \in S(f,\eta)\} \geq \langle h,x_0\rangle = \|\pi_\omega(x_0)\| > \varphi_0 - \epsilon^2/4800.$$

By lemma 3.1 there is $f_0 \in JT^{**}$, $\|f_0\| = 1$, $\|f - f_0\| < 3\epsilon/10 < \epsilon/3$ and $\alpha_0 > 0$ such that

$$(i)\ S(f_0,\alpha_0) \subseteq S(f,\gamma_0)$$
$$(ii)\ S(f_0,\alpha_0) \subseteq \{x \in C : \langle h,x\rangle > \varphi_0 - 2\epsilon^2/4800\}.$$

As $\pi_\omega(S(f,\gamma_0)) \subseteq (\varphi_0 + \epsilon^2/4800).ball(\ell^2(\Delta))$ we conclude by the "fact" established in the proof of 3.13 that

$$\begin{aligned} diam\pi_\omega(S(f_0,\alpha_0)) &\leq (1+\epsilon^2/4800).2(2\epsilon^2/1600)^{1/2} \\ &< \sqrt{2}.2.\sqrt{2}(\epsilon/40) \\ &= \epsilon/10. \end{aligned}$$

By the density of the functions of finite support (on $\overline{T}$) in JT^{**} we may assume by an obvious perturbation argument that f_0 is of finite support.

STEP 2: We shall find $f_1 \in JT^{**}$, $\|f_1\| = 1$, $\|f_0 - f_1\| < \epsilon/3$, $1 > \alpha_1 > 0$ and $n \in \mathbf{N}$ such that

$$(i)\ S(f_1,\alpha_1) \subseteq S(f_0,\alpha_0)$$
$$(ii)\ diam\ \pi_{[n,\omega]}(S(f_1,\alpha_1)) < \epsilon/2.$$

From step 1(ii) choose $z \in \ell^2(\Delta)$ with finite support $A \subseteq \Delta$ such that, for $x \in S(f_0, \alpha_0)$, $\|\pi_\omega(x) - z\| < \epsilon/10$. Find n_0 big enough such that $p_{n_0}(f_0)$ and $p_{[n_0,\omega]}(z)$ are defined. If the above assertion were false, then, for $n \geq n_0$ and every slice $S(f_1, \alpha_1) \subseteq S(f_0, \alpha_0)$ such that $\|f_0 - f_1\| \leq \epsilon/3$, $\|f_1\| = 1$, we could find $x \in S(f_1, \alpha_1)$ such that

$$\|\pi_{[n,\omega]}(x) - p_{[n,\omega]}(z)\| > \epsilon/6. \tag{1}$$

On the other hand, for $x \in S(f_0, \alpha_0/2)$ fixed, by lemma 4.7.a)

$$lim_{n\to\infty, n\geq n_0}\langle p_n(f_0), x\rangle = \langle f_0, x\rangle > M_{f_0} - \alpha_0/2. \tag{2}$$

We now construct inductively a nested sequence of slices, using (1), (2) and lemma 3.1 in order to obtain a constradiction:

By (1), for $\alpha_0 > \gamma > 0$ there is $x \in S(f_0, \gamma)$ such that

$$\|\pi_{[n_0,\omega]}(x) - p_{[n_0,\omega]}(z)\| > \epsilon/6$$

and therefore we may find $m_1 > n_0$ and $y_1 \in JT$, $\|y_1\| = 1$ and $y_1 = \pi_{[n_0,m_1]}(y_1)$ such that

$$\langle y_1, \pi_{[n_0,\omega]}(x) - p_{[n_0,\omega]}(z)\rangle > \epsilon/6.$$

If we have chosen γ sufficiently small (namely $\gamma < (\alpha/2)(\epsilon/36)(\epsilon/24)$) then by lemma 3.1 we may find $g_1 \in JT^{**}$, $\|g_1\| = 1$, $\|f_0 - g_1\| < \epsilon/12$ and $\beta_1 > 0$ such that

$$S(g_1, \beta_1) \subseteq S(f_0, \alpha_0/2)$$

and

$$S(g_1, \beta_1) \subseteq \{x \in C : \langle y_1, x - p_{[n_0,\omega]}(z)\rangle > \epsilon/8\}$$

whence, in particular

$$S(g_1, \beta_1) \subseteq \{x \in C : \|\pi_{[n_0,m_1]}(x - p_{[n_0,\omega]}(z))\| > \epsilon/8\}. \tag{3}$$

Now let $\beta > 0$ be sufficiently small (namely $\beta < \beta_1(\alpha_0/4)(\epsilon/24)$) and choose some $x \in S(g_1, \beta) \subseteq S(f_0, \alpha_0/2)$. By (2) we may find $n_1 > m_1$ such that

$$\langle p_{n_1}(f_0), x\rangle > M_{f_0} - \alpha_0/2$$

and by lemma 3.1 we may find $h_1 \in JT^{**}$, $\|h_1\| = 1$, $\|h_1 - g_1\| < \epsilon/12$ and $\gamma_1 > 0$ such that

$$S(h_1, \gamma_1) \subseteq S(g_1, \beta_1)$$

and

$$S(h_1,\gamma_1) \subseteq \{x \in C : \langle p_{n_1}(f_0), x\rangle > M_{f_0} - 3\alpha_0/4\}. \tag{4}$$

For the general induction step suppose we have defined slices

$$S(f_0,\alpha_0/2) \supseteq S(g_1,\beta_1) \supseteq S(h_1,\gamma_1) \supseteq S(g_2,\beta_2) \supseteq \ldots \supseteq S(h_k,\gamma_k)$$

such that, for $1 < i \leq k$,

$$\|g_i - h_{i-1}\| < \epsilon/6.2^i \text{ and } \|h_i - g_i\| < \epsilon/6.2^i$$

and natural numbers $n_0 < m_1 < n_1 < m_2 < \ldots < n_k$ such that, for $1 \leq i \leq k$

$$S(g_i,\beta_i) \subseteq \{x \in C : \|\pi_{[n_{i-1},m_i]}(x - p_{[n_0,\omega]}(z))\| \geq \epsilon/8\} \tag{5}$$

and

$$S(h_i,\gamma_i) \subseteq \{x \in C : \langle p_{n_i}(f_0), x\rangle \geq M_{f_0} - 3\alpha_0/4\}. \tag{6}$$

Now from (1) choose $\gamma_k(\epsilon/36)(\epsilon/12.2^k) > \gamma > 0$ and $x \in S(h_k,\gamma)$ such that

$$\|\pi_{[n_k,\omega]}(x) - p_{[n_k,\omega]}(z)\| > \epsilon/6.$$

By the same argument as for $k = 0$ find $g_{k+1} \in JT^{**}$, $\|g_{k+1}\| = 1$, $\|g_{k+1} - h_k\| < \epsilon/6.2^k$, $\beta_{k+1} > 0$ and $m_{k+1} > n_k$ such that

$$S(g_{k+1},\beta_{k+1}) \subseteq S(h_k,\gamma_k)$$

and

$$S(g_{k+1},\beta_{k+1}) \subseteq \{x \in C : \|\pi_{[n_k,m_{k+1}]}(x - p_{[n_0,\omega]}(z))\| > \epsilon/8\}. \tag{7}$$

Next choose $0 < \beta < \beta_{k+1}(\alpha_0/4)(\epsilon/12.2^k)$ and $x \in S(g_{k+1},\beta)$. By (2) find $n_{k+1} > m_{k+1}$ such that

$$\langle p_{n_{k+1}}(f_0), x\rangle > M_{f_0} - \alpha_0/2$$

and apply once again lemma 3.1 to find $h_{k+1} \in JT^{**}$, $\|h_{k+1}\| = 1$, $\|h_{k+1} - g_{k+1}\| < \epsilon/6.2^k$ and $\gamma_{k+1} > 0$ such that

$$S(h_{k+1},\gamma_{k+1}) \subseteq S(g_{k+1},\beta_{k+1})$$

and

$$(8) \qquad S(h_{k+1},\gamma_{k+1}) \subseteq \{x \in C : \langle p_{n_{k+1}}(f_0), x\rangle > M_{f_0} - 3\alpha_0/4\}.$$

This finishes the inductive step. Denoting by $\sim$ the $\sigma(JT^*, JT)$- closure, choose

$$x \in \bigcap_{i=1}^{\infty} \tilde{S}(g_i,\beta_i) = \bigcap_{i=1}^{\infty} \tilde{S}(h_i,\gamma_i).$$

Note that conditions (5) and (6) pass to the σ^*-closure, hence x lies in the sets appearing in the right hand side of (5) and (6) for all $i \in \mathbf{N}$. Hence

$$\langle f_0, x\rangle = lim_{i\to\infty}\langle p_{n_i}(f_0), x\rangle \geq M_{f_0} - 3\alpha_0/4$$

whence $x \in S(f_0,\alpha_0)$ and therefore

$$\|\pi_\omega(x) - z\| < \epsilon/10.$$

By lemma 4.7 we conclude that, for i big enough

$$\|\pi_{[n_i,\omega]}(x) - p_{[n_i,\omega]}(z)\| < \epsilon/10.$$

This contradiction to (5) finishes the proof of step 2.

STEP 3: Let n be given by step 2 and choose a finite subset $(y_j)_{j=1}^{M}$ in the unit ball of JT such that, for $x \in \pi_{[0,n-1]}(JT^*)$, $\|x\| = 1$,

$$sup_{1\leq j\leq M}\langle y_j, x\rangle \geq 1/2.$$

Apply again lemma 3.1 and remark 3.2 successively for $1 \leq j \leq M$ to find slices $S(f_1,\alpha_1) \supseteq S(f_2,\alpha_2) \supseteq \ldots \supseteq S(f_{M+1},\alpha_{M+1})$ such that, for $1 \leq j \leq M$, $\|f_{j+1}\| = 1$, $\|f_{j+1} - f_j\| < \epsilon/3M$ and such that the oscillation of y_j on $S(f_{j+1},\alpha_{j+1})$ is less than $\epsilon/4$.

The desired slice will be

$$S(g,\alpha) = S(f_{M+1},\alpha_{M+1})$$

noting that $\|g - f\| = \|f_{M+1} - f\| < \epsilon$ and

$$\begin{aligned} diam(S(g,\alpha)) &\leq diam(\pi_{[0,n-1]}S(g,\alpha)) + diam(\pi_{[n,\omega]}S(g,\alpha)) \\ &< \epsilon/2 + \epsilon/2 \\ &= \epsilon. \end{aligned}$$

The proof of the theorem is complete. □

5. The case of separable Banach Spaces containing ℓ^1 isomorphically

In this section we shall show that - contrary to the case of JT - for a separable Banach space X containing ℓ^1 there is a point x_0^* on the unit sphere of X^* where the relative weak and norm-topologies on the unit ball of X^* do not coincide. We do not know whether the same result holds true in the non-separable case (obvious conjecture: yes).

In fact theorem 5.2 follows rather directly from a result of M. Talagrand ([S-S-W], theorem 3.6).

Let us fix some notations: If $h : \Delta \to]0, \infty[$ is a continuous function then

$$\begin{aligned} M_h : C(\Delta) &\to C(\Delta) \\ f &\to f.h \end{aligned}$$

defines an isomorphism on $C(\Delta)$.

Let $s \in \mathbf{N}$, $s \geq 3$ and define μ_s to be the "biased coin" probability measure on the Cantor-groups $\Delta = \{-1, +1\}^{\mathbf{N}}$,

$$\mu_s = \otimes_{i=1}^{\infty}[((s-1)/s).\delta_{\{1\}} + (1/s).\delta_{\{-1\}}].$$

Define the convolution operator

$$\begin{aligned} T_s : C(\Delta) &\to C(\Delta) \\ f &\to f * \mu_s. \end{aligned}$$

Note that the dual operator T_s^* also is given by convolution with μ_s, i.e.

$$\begin{aligned} T_s^* : \mathcal{M}(\Delta) &\to \mathcal{M}(\Delta) \\ \mu &\to \mu * \mu_s \end{aligned}$$

and, for $t = (\epsilon_n)_{n=1}^{\infty} \in \Delta$, denoting δ_t the Dirac-measure at t, we have

$$\begin{aligned} T_s^*(\delta_t) &= \delta_t * \mu_s \\ &= \otimes_{i=1}^{\infty}[((s-1)/s)\delta_{\{\epsilon_n\}} + (1/s).\delta_{\{-\epsilon_n\}}]. \end{aligned}$$

Denote K_s the weak-star compact subset of $\mathcal{M}(\Delta)$ given by

$$K_s = T_s^*(P_\Delta)$$

where P_Δ denotes the probability measure on Δ.

Note that

$$K_s = \overline{conv}^{\sigma^*}\{\delta_t * \mu_s : t \in \Delta\}.$$

The subsequent statement is a slight reformation and generalisation of ([S-S-W], theorem 3.6) and is easily obtained by adapting the proof given in [S-S-W]. We leave the details to the reader.

5.1 THEOREM (Talagrand): For $s \in \mathbf{N}$, $s \geq 3$ define

$$d_s = inf\{diam[1/k(S_1 + ... + S_k)] : S_1, ..., S_k \text{ slices of } K_s\}.$$

Then $\liminf_{s\to\infty} d_s \geq 1$.

If $\alpha > 0$ and $h : \Delta \to [\alpha, \infty[$ is a continuous function then

$$K_s^h = T_s^* D_h^*(P_\Delta)$$

is compact convex too and letting

$$d_s^h = inf\{diam[1/k(S_1 + ... + S_k)] : S_1, ..., S_k \text{ slices of } K_s^h\}$$

we have $d_s^h \geq \alpha d_s$. □

5.2 THEOREM: Let X be a separable Banach space containing ℓ^1 isomorphically. Then, for $\epsilon > 0$ there is a point $x_0^* \in X^*$, $\|x_0^*\| = 1$ such that for every weak neighbourhood V of x_0^*

$$diam(V \cap ball(X^*)) > 1 - \epsilon.$$

To reduce theorem 5.2 to Talagrand's result 5.1 we need a variant of a theorem of Pelczynski [P], which follows easily by combining ([P], theorem 4.3) with a well-known result of James on almost isometric embedability of ℓ^1 ([L-T]). Again we leave the details to the reader.

5.3 LEMMA: If a separable Banach space X contains ℓ^1 then, for $\epsilon > 0$, there is a $\sigma(\mathcal{M}(\Delta, C(\Delta)) - \sigma(X^*, X)$ - continuous injection $i : \mathcal{M}(\Delta) \to X^*$ such that, for $\mu \in \mathcal{M}(\Delta)$

$$\|\mu\| \leq \|i(\mu)\| \leq (1 + \epsilon)\|\mu\|. \square$$

We also need a simple elementary topological result:

5.4 LEMMA: Let $\epsilon > 0$ and $F : \Delta \to [1, 1+\epsilon]$ be a lower semicontinuous function. There is a continuous function $h : \Delta \to [1-\epsilon, 1]$ and $t_0 \in \Delta$ such that

$$1 = h(t_0).F(t_0) = sup\{h(t).F(t) : t \in \Delta\}.$$

PROOF: As F is of first Baire-class there is a point of continuity t_0 of F. It is an easy exercise to construct a continuous function $g : \Delta \to [1, 1+\epsilon]$ such that $g \geq F$ and $g(t_0) = F(t_0)$. The function $h(t) = 1/g(t)$ does the required job. □

5.5 LEMMA: For a dual norm $|||.|||$ on $\mathcal{M}(\Delta)$ - with respect to the duality with $C(\Delta)$ - and $\epsilon > 0$ satisfying

$$||.|| \leq |||.||| \leq (1 + \epsilon/2)||.||$$

there is a σ^*-compact, convex subset K of the unit-ball of $\mathcal{M}(\Delta), |||.|||)$ such that

(i) there is an element $\mu_0 \in K$ with $|||\mu_0||| = 1$;

(ii) for every weakly open set V in $\mathcal{M}(\Delta)$, $V \cap K \neq \emptyset$,

$$diam_{|||.|||}(V \cap K) \geq diam_{||.||}(V \cap K) > 1 - \epsilon.$$

PROOF: Fix $s \in \mathbf{N}$, $s \geq 3$ such that, with the notation of 5.1, $d_s > 1 - \epsilon/2$. Define

$$F : \Delta \to [1, 1 + \epsilon/2]$$
$$F(t) = |||T_0^*(\delta_t)|||.$$

The function F is lower semicontinuous and by Lemma 5.4 there is a continuous function $h : \Delta \to [1 - \epsilon/2, 1]$ and $t_0 \in \Delta$ such that

$$1 = h(t_0).F(t_0) = sup\{h(t).F(t) : t \in \Delta\}.$$

The set

$$K_s^h = T_s^* D_h^*(P(\Delta))$$

satisfies the requirements as it is contained in the ball of $(\mathcal{M}(\Delta), |||.|||)$, $\mu_0 = h(t_0).T_s^*(\delta_{t_0})$ satisfies $|||\mu_0||| = 1$ and by theorem 5.1 and a lemma of J. Bourgain ([G-G-M-S]) for every weakly open V in $\mathcal{M}(\Delta)$, $V \cap K_s^h \neq \emptyset$

$$diam_{||.||}(V \cap K_s^h) \geq d_s^h \geq (1 - \epsilon/2)d_s > 1 - \epsilon. \square$$

PROOF OF THEOREM 5.2: If X is separable and contains ℓ^1 then by 5.3 there is a $\sigma^* - \sigma^*$-continuous injection $i : \mathcal{M}(\Delta) \to X^*$ such that

$$\|\mu\| \leq \|i(\mu)\| \leq (1 + \epsilon/2)\|\mu\|.$$

Defining $|||\mu||| = \|i(\mu)\|$ we obtain a dual norm on $\mathcal{M}(\Delta)$ satisfying the assumptions of lemma 5.5. Let K and μ_0 be as in 5.5 and consider $x_0^* = i(\mu_0)$. This is a point of the unit-sphere of X^* such that for every weak neighbourhood V of x_0^*

$$\begin{aligned} diam_{\|.\|_{X^*}}(V \cap ball(X^*)) &\geq diam_{\|.\|_{X^*}}(V \cap i(K)) \\ &= diam_{|||.|||}(i^{-1}(V) \cap K)) > 1 - \epsilon. \quad \square \end{aligned}$$

REFERENCES

[B] J. Bourgain: On dentability and the Bishop-Phelps property, Israel Journal of Math. 28(1977), p. 265 - 271.

[B-R] J. Bourgain, H.P. Rosenthal: Geometrical implications of certain finite- dimensional decompositions, Bull. Soc. Math. Belg. 32(1980), p. 57 - 82.

[D] J. Diestel: Geometry of Banach spaces - Selected Topics, Springer lecture notes 485 (1975).

[G-H] N. Ghoussoub, B. Maurey: G_δ-embeddings in Hilbert spaces, J. Funct. Anal. 61, No. 1 (1985), p. 72 - 97.

[G-G-M-S] N. Ghoussoub, G. Godefroy, B. Maurey, W. Schachermayer: Some geometrical and topological structures in Banach spaces, to appear in Mem. A.M.S.

[G-M-S] N. Ghoussoub, B. Maurey, W. Schachermayer: Geometrical implications of certain infinite-dimensional decompositions, preprint (1987).

[J] R.C. James: A separable somewhat reflexive space with non-separable dual, Bull. A.M.S. 80(1974), p. 738 - 743.

[J-Z] K. John, V. Zizler: On rough norms on Banach Spaces, Comment. Math. Univ. Carolinae 19 (1978), p. 335 - 349.

[Ka] M. I. Kadec: On weak and norm convergence, Dokl. Akad. Nauk SSSR, 122 (1958), p. 13 - 16.

[Kl] V. L. Klee: Mappings into normed linear spaces, Func. Math. 49 (1960), p. 25 - 34.

[L-L-T] Bor-Luh Lin, Pei-Kee Lin, S.L. Troyanski: Characterizations of denting points, Proc. A.M.S. 102 (1988), p. 526 - 528.

[L-S] J. Lindenstrauss, C. Stegall: Examples of separable spaces which do not contain ℓ^1 and whose duals are non-separable, Studia Math. 54, p. 81 - 105.

[L-T] J. Lindenstrauss, L. Tzafriri: Classical Banach spaces I, Springer, (1977).

[P] R. Phelps: Dentability and extreme points in Banach spaces, J. Functional Analysis 16, p. 78 - 90, (1974).

[Pe] A. Pelczynski: On Banach spaces containing $L^1(\mu)$, Studia Math. 30 (1968), p. 231 - 246.

[S] W. Schachermayer: The Radon-Nikodym and the Krein-Milman properties are equivalent for strongly regular sets, Transactions of the A.M.S. 303 (1987), p. 673 - 687.

[S-S-W] W. Schachermayer, A. Sersouri, E. Werner: Some geometrical results concerned with the Radon-Nikodym property in Banach space, preprint (1987).

[St] C. Stegall: The Radon-Nikodym property in conjugate Banach spaces, Trans. A.M.S. 206 (1975), p. 213 - 223.

[T] S. Troyanski: On locally uniformly convex and differentiable norms in certain non-separable Banach spaces, Studia Math. 37 (1971), p. 173 - 180.

W. Schachermayer
Johannes Kepler Universität Linz
Institut für Mathematik
A-4040 Linz - Austria

Contemporary Mathematics
Volume **85**, 1989

A note of the Lavrientiev index for quasi–reflexive Banach spaces

A. Sersouri

ABSTRACT: To every Banach space X, we associate an ordinal index $\beta(X)$ by considering the topological behavior of the elements of $X^{**}\backslash X$ (considered as functions on (Ball (X^*), weak*)). We prove that for quasi–reflexive spaces, that is, $1 \leq \dim X^{**}/X < \infty$, we have $\beta(X) = \omega$ if we suppose that X and X^{**} are isometric. If we merely suppose that X and X^{**} are only isomorphic, the index $\beta(X)$ can be as large as we want (however always less than ω_1). This shows the existence of a quasi–reflexive space that is isomorphic to its biduals but can never be renormed to be isometric.

Introduction.

Let $\beta(X)$ denote the Lavrientiev index of the Banach space X (see definition below). In this note we will give a boundedness result concerning this index. More precisely we will prove:

Theorem A*. *Let* X *be a quasi–reflexive Banach space which is isometric to its second dual* X^{**}, *then* $\beta(X) = \omega$.

---

*Theorem A and Theorem B were both proved during the 1987 Iowa Banach Workshop.

AMS classification: 46B20.

Key words and phrases: quasi–reflexivity, Baire functions, Lavrientiev index.

This theorem cannot be extended to the isomorphic case. Indeed, using the DFJP–interpolation method, the following can be proved:

Theorem B^*. [HOR] *For every ordinal number* $\eta < \omega_1$, *there exists a quasi–reflexive Banach space* X *satisfying* $\dim(X^{**}/X) = 1$ *and* $\beta(X) = \eta$.

Moreover replacing X by $X \oplus \ell^2$ if necessary, we can suppose that X is isomorphic to X^{**}, since $\beta(X) = \beta(X \oplus \ell^2)$ for any Banach space X (see Prop. 2) and $(X \oplus \ell^2)^{**} \approx X \oplus \ell^2$ if X is assumed to be quasi–reflexive.

Since the β–index is an isomorphic invariant (see Prop. 2), we deduce easily from Theorems A and B that there exists a quasi–reflexive space of order 1 (namely the space X of Theorem B with $\eta \geq \omega + 1$) which is isomorphic to its bidual X^{**} but which cannot be renormed to become isometric to X^{**}. The existence of such a space has been observed for the first time in [S].

Acknowledgement. I want to thank G. Godefroy for suggesting this problem about the β–index, as well as R. Haydon, and E. Odell for the valuable discussions I have had with them. My special thanks go to B.L. Lin who invited me to the Iowa Banach Workshop.

Preliminaries.

The definition of the index $\beta(X)$ is intimately related to the class of Baire–1 functions. So we will first recall several facts concerned with Baire–1 functions. For the definition and usual properties we refer to [D].

The Baire–1 functions can be classified by the use of three "natural" (ordinal) indices known respectively as the separation index, the oscillation index, and the convergence index. Even if these indices are not equal they are equivalent in a very strong sense [KL], and we will refer to them as Lavrientiev indices, who first considered the separation index [L].

In this paper we will work with the variation known as the oscillation index. This variation is also used in [HOR], and we refer to [Bo] for a use of the separation index.

We now recall the definition of the oscillation index.

Let K be a topological space, f a function on K, L a subset of K, and $x \in L$. The oscillation of f at the point x relatively to L is defined by:

$$\mathrm{Osc}(f,x;L) = \operatorname*{Inf}_{\substack{V \text{ open} \\ x \in V}} \ \operatorname*{Sup}_{x_1,x_2 \in V \cap L} |f(x_1) - f(x_2)|.$$

Now for $\epsilon > 0$, and $\alpha < \omega_1$, we define inductively the α-derivative of K with respect to f and ϵ by:

$$\begin{cases} K^{(0)}_{(f,\epsilon)} = K \\ K^{(\alpha+1)}_{(f,\epsilon)} = \left\{x \in K^{(\alpha)}_{(f,\epsilon)} : \mathrm{Osc}(f,x;K^{(\alpha)}_{(f,\epsilon)}) \geq \epsilon\right\} \\ K^{(\alpha)}_{(f,\epsilon)} = \bigcap_{\beta<\alpha} K^{(\beta)}_{(f,\epsilon)} \quad \text{if } \alpha \text{ is a limit ordinal.} \end{cases}$$

We set

$$\beta_K(f,\epsilon) = \inf\left[\{\omega_1\} \cup \{\alpha : K^{(\alpha)}_{(f,\epsilon)} = \emptyset\}\right]$$

and

$$\beta_K(f) = \operatorname*{Sup}_{\epsilon>0} \beta_K(f,\epsilon).$$

The ordinal number $\beta_K(f)$ is what is known as the oscillation index of f relative to K.

In the case that K is a Polish space, f is a Baire–1 function if and only if $\beta_K(f) < \omega_1$.

Before going any further, we summarize in the next proposition some of the elementary results we will need.

Proposition 1. Let K be a topological space, f a function on K, then:

(i) $\beta_K(f+g) = \beta_K(f)$ for every continuous function g on K.

(ii) $\beta_L(f) \leq \beta_K(f)$ for every subset L of K.

(iii) If moreover K is a subset of a linear space E, and f a linear function on E, then for every $k > 0$ one has $\beta_{k\cdot K}(f) = \beta_K(f)$.

(iv) If P is a compact space, $\varphi : P \longrightarrow K$ a surjective continuous application, then $\beta_K(f) = \beta_P(f \circ \varphi)$.

Proof. (i) It is easy to check that for every subset L of K, every $x \in L$, one has $\mathrm{Osc}(f+g,x;L) = \mathrm{Osc}(f,x\,;L)$. From this it clearly follows that $K^{(\alpha)}_{(f+g,\epsilon)} = K^{(\alpha)}_{(f,\epsilon)}$ for every $\epsilon > 0$, and $\alpha < \omega_1$. Hence $\beta_K(f+g) = \beta_K(f)$.

(ii) If $x \in L_1 \subset L_2 \subset K$, it is clear that $\mathrm{Osc}(f,x\,;L_1) \leq \mathrm{Osc}(f,x;L_2)$. From this we deduce easily that $L^{(\alpha)}_{(f,\epsilon)} \subset K^{(\alpha)}_{(f,\epsilon)}$ for every $\epsilon > 0$ and $\alpha < \omega_1$ if we assume that $L \subset K$. This proves (ii).

(iii) Observe that for every $k > 0$, every $x \in K$, we have $\mathrm{Osc}(f,kx;kK) = k\,\mathrm{Osc}(f,x;K)$. This implies that $(kK)^{(\alpha)}_{(f,k\epsilon)} = k \cdot K^{(\alpha)}_{(f,\epsilon)}$ for every $\epsilon > 0$, and $\alpha < \omega_1$, which implies (iii).

(iv) Since P is compact, and φ surjective, φ is an open mapping. Hence if $R \subset P$ and $L \subset K$ are such that $\varphi^{-1}(L) = R$, then for every $t \in R$ we have $\mathrm{Osc}(f \circ \varphi,t;R) = \mathrm{Osc}(f,\varphi(t);L)$, which implies that $R'_{(f\circ\varphi,\epsilon)} = \varphi^{-1}(L'_{(f,\epsilon)})$ for every $\epsilon > 0$. From this we deduce that $P^{(\alpha)}_{(f\circ\varphi,\epsilon)} = \varphi(K^{(\alpha)}_{(f,\epsilon)})$ for every $\epsilon > 0$, $\alpha < \omega_1$ and the result follows. ∎

We now introduce the Lavrientiev index for Banach spaces and establish some of its properties.

If X is a Banach space, we define the index $\beta(X)$ by

$$\beta(X) = \operatorname{Sup}_{\xi \in X^{**}} \beta_K(\xi)$$

where $K = (B(X^*),\omega^*)$ is the unit ball of X^* equipped with the ω^*–topology.

Proposition 2.

(i) If X and Y are isomorphic Banach spaces, then $\beta(X) = \beta(Y)$.

(ii) If R is a reflexive Banach space, then $\beta(X \oplus R) = \beta(X)$ for every Banach space X.

(iii) X is reflexive if and only if $\beta(X) = 1$.

(iv) If X is not reflexive then $\beta(X) \geq \omega$.

(v) If X is separable, then $\beta(X) < \omega_1$ if and only if $X \not\supset \ell^1$.

Proof. (i) Let X be a Banach space, $|\cdot|_1$ and $|\cdot|_2$ two equivalent norms on X, and let $K_1 = (B_{|\cdot|_1}(X^*),\omega^*)$, $K_2 = (B_{|\cdot|_2}(X^*),\omega^*)$ and $k > 0$ such that

$$\frac{1}{k} \cdot K_1 \subset K_2 \subset k \cdot K_1$$

By Proposition 1, for every $\xi \in X^{**}$ we have that

$$\beta_{K_1}(\xi) \leq \beta_{k \cdot K_2}(\xi) = \beta_{K_2}(\xi) \leq \beta_{K_1}(\xi).$$

(ii) Let $K = (B(X^*),\omega^*)$ and $P = (B(R),\omega^*)$, then $K \times P$ is the unit ball of

$(X \oplus_1 R)^*$. To prove (ii) it is enough to show that for every $\zeta = (\xi,\rho) \in X^{**} \oplus R^{**}$ one has $\beta_{K \times P}(\zeta) = \beta_K(\xi)$.

Observe first that if L is a subset of K, and $t = (x,r) \in L \times P$, then $\mathrm{Osc}(\zeta,t;L\times P) = \mathrm{Osc}(\xi,x;L)$ (since ρ in continuous on P), and from this deduce that $(K \times P)^{(\alpha)}_{(\zeta,\epsilon)} = K^{(\alpha)}_{(\xi,\epsilon)} \times P$, for every $\epsilon > 0$, and $\alpha < \omega_1$. The desired result then follows.

(iii) Is clear since $\beta_K(f) = 1$ if and only if f is continuous on K.

(iv) Let $\xi \in X^{**}\backslash X$, and as usual $K = (B(X^*),\omega^*)$. It is well known that the function $x \longrightarrow \mathrm{Osc}(\xi,x;K)$ is concave. If we let $M = \mathrm{Osc}(\xi,0;K)$, we have that $\mathrm{Osc}(\xi,x;K) \geq M(1-\|x\|)$. This implies that for every $\epsilon < M$, $K'_{(\xi,\epsilon)} \supset (1-\frac{\epsilon}{M})K$, and by an inductive argument, one can show that $K^{(n)}_{(\xi,\epsilon)} \supset (1-\frac{n\,\epsilon}{M})K$, for every $n \leq \frac{M}{\epsilon}$ (see the proof of prop. 1(iii)). This shows that $\beta_K(\xi,\epsilon) \geq \frac{M}{\epsilon}$.

(v) Suppose that X contains ℓ^1. We can suppose that ℓ^1 is isometric to a subspace of X. (This is possible after a renorming of X.) Let $i : \ell^1 \longrightarrow X$ be the isometric embedding, by the Hahn–Banach theorem we know that $i^* : (B(X^*),\omega^*) \longrightarrow (B(\ell^\infty),\omega^*)$ induces a surjective continuous mapping, and since $(B(X^*),\omega^*)$ is compact, we are in the situation of Proposition 1(iv).

Let $\mathcal{U}$ be a non–trivial ultrafilter over $\mathbb{N}$, the element $e_{\mathcal{U}} \in \ell^{\infty*}$ (*i.e.*, $e_{\mathcal{U}}(x) = \lim_{\mathcal{U}} x_n$) satisfies

$$\mathrm{Osc}(e_{\mathcal{U}},x;B(\ell^\infty)) = 2 \text{ for every } x \in B(\ell^\infty),$$

hence $\beta_{B(X^*)}(e_{\mathcal{U}} \circ i^*) = \beta_{B(\ell^\infty)}(e_{\mathcal{U}}) = \omega_1$.

The proof of the converse is more sophisticated. Let $L_K(f)$ denote the separation index of the function f relative to K (this is the index originally known as the Lavrientiev index [Bo]), and for a Banach space X we define $L(X) = \sup_{\xi \in X^{**}} L_{B(X^*)}(\xi)$.

Bourgain [Bo] proved that if X is a separable Banach space not containing ℓ^1, then $L(X) < \omega_1$. This implies that $\beta(X) < \omega_1$ in view of the following result [KL]:

If K is a compact metric space, then for every bounded function f on K, every $\alpha < \omega_1$ one has $L_K(f) \leq \omega^\alpha$ if and only if $\beta_K(f) \leq \omega^\alpha$. ■

Remarks.

1. The estimation $\beta_K(\xi,\epsilon) \geq \frac{M}{\epsilon}$ appearing in the proof of Proposition 2(iv) is the best possible. For instance for $X = c_0$, $\xi = (1,1,1,...) \in \ell^\infty$ one can check easily that $\mathrm{Osc}(\xi,a;B(\ell^1)) = 2(1 - \|a\|)$ for every $a \in B(\ell^1)$, this implies that $\beta_{B(\ell^1)}(\xi,\epsilon) \leq \frac{2}{\epsilon}$.

2. The proof of Proposition 2(v) shows that if Y is a subspace of X, and $\xi \in Y^{**}$, then $\beta_{B(Y^*)}(\xi) = \beta_{(B)(X^*)}(\xi)$. In particular we have that $\beta(X) \geq \beta(Y)$.

3. The result of the preceding remark implies that if X is a quasi–reflexive Banach space of order 1, then every non–reflexive subspace Y of X satisfies $\beta(Y) = \beta(X)$.

4. The separability assumption is necessary in Proposition 2(v). Indeed, let $(X_\eta)_{\eta<\omega_1}$ be the family of space of Theorem B, and let $X = (\oplus_{\eta<\omega_1} X_\eta)_{\ell^2}$. It is easy to see that X is an Asplund space (*i.e.*, X^* has RNP), and hence $X \not\supset \ell^1$. On the other hand one has $\beta(X) \geq \beta(X_\eta) \geq \eta$ for every $\eta < \omega_1$.

5. If X is a Banach space (not necessarily separable) such that X^{**}/X is separable, then $\beta(X) < \omega_1$. Indeed, by a result of Valdivia [V], $X = S \oplus R$ where R is a reflexive space and S^{**} is separable. So $\beta(X) = \beta(S \oplus R) = \beta(S) < \omega_1$.

Proof of the Main Result.

To prove Theorem A we will make use of the notions of neighborly basic sequences, and spreading models. These notions are defined as follows:

Definition 3. [Be] Let $(e_n)_{n\geq 1}$ be a basic sequence in some Banach space $(e_n)_{n\geq 1}$ is said to be neighborly if the norm of every finite expansion $\sum_{n=1}^{\infty} a_n e_n$ decreases when (some) a_k is replaced by a_{k-1} or a_{k+1} (k is arbitrary, $a_0 = 0$), and all the other coefficients being unchanged.

Definition 4. [BL] Let $(x_n)_{n\geq 1}$ be a bounded sequence in some Banach space, and we suppose that $(x_n)_{n\geq 1}$ has no norm convergent subsequence. If $\mathcal{U}$ is a nontrivial ultrafilter, we define on $\mathbb{R}^{(\mathbb{N})}$ a norm $[\cdot]$, associated with a sequence of vectors $(\varphi_n)_{n\geq 1}$ as follows:

$$\left[\sum_{i=1}^{k} a_i\varphi_i\right] = \lim_{n_1}{}_{\mathcal{U}} \lim_{n_2}{}_{\mathcal{U}} \cdots \lim_{n_k}{}_{\mathcal{U}} \left\|\sum_{i=1}^{k} a_i x_{n_i}\right\|$$

The completion of $(\mathbb{R}^{(\mathbb{N})},[\cdot])$ is called the spreading model associated to $((x_n),\mathcal{U})$ and the sequence $(\varphi_n)_{n\geq 1}$ its fundamental sequence.

Remarks.

1. The neighborly property we consider here was defined by S. Bellenot [Be]. A slightly different notion was first introduced by R. James [J].

2. The definition we take here for the spreading models is due to Beauzamy and Lapresté, and it is shown in [BL] that it is equivalent to the original one given by Brunel and Sucheston [BS].

We are now ready to prove Theorem A. The proof we present here uses two known results (Theorem 5 and Theorem 6 below) and an elementary lemma.

Theorem 5. [HOR] *Let* K *be a Polish space, and* f *a bounded Baire class–1 function on* K. *Let* $(g_n)_{n\geq 1} \subset C(K)$ *be a sequence of uniformly bounded sequence of continuous functions converging point–wise to* f. *Then:*

If $\beta_K(f,\epsilon) \geq \omega$ *for some* $\epsilon > 0$, *then* (g_n) *has a spreading model whose fundamental sequence is equivalent to the* ℓ^1*–basis.*

Theorem 6. [S] *Let* X *be a quasi–reflexive Banach space which is isometric to its second dual* X^{**}. *Then for every* $\xi \in X^{**}\backslash X$, *there exists an* $x \in X$, *a neighborly basic sequence* $(e_n)_{n\geq 1}$ *such that* $\xi = x + \tilde{\xi}$ *where* $\tilde{\xi} = \sum_{n=1}^{\infty} e_n$ *in the* ω^**–topology.*

Lemma 7. Let $(e_n)_{n\geq 1}$ be a neighborly basic sequence and let $s_n = \sum_{i=1}^{n} e_i$. Then the fundamental sequence of every spreading model on $(s_n)_{n\geq 1}$ is $(s_n)_{n\geq 1}$ itself.

Proof. Using the neighborly property of $(e_n)_{n\geq 1}$ we can prove that $(s_n)_{n\geq 1}$ is invariant under spreading in the terminology of [BS], *i.e.*: $\|\sum_{i=1}^{k} a_i s_{n_i}\| = \|\sum_{i=1}^{k} a_i s_i\|$ for every sequence of scalars $a_1,\ldots,a_k$, and every sequence of integers $n_1 < n_2 < \cdots < n_k$. Indeed, let $b_i = \sum_{j=1}^{i} a_j$ $(1 \leq i \leq k)$, then:

$$\|\sum_{i=1}^{k} a_i s_{n_i}\| = \|\sum_{i=1}^{k} b_i(s_{n_i} - s_{n_{i-1}})\|$$

$$= \|b_1(e_1 + \cdots + e_{n_1}) + b_2(e_{1+n_1} + \cdots + e_{n_2}) + \sum_{i=3}^{k} b_i(s_{n_i} - s_{n_{i-1}})\|$$

$$= \|b_1 e_1 + b_2(e_2 + \cdots + e_{n_2}) + \sum_{i=3}^{k} b_i(s_{n_i} - s_{n_{i-1}})\|$$

(move the coefficient b_1 to obtain "$\geq$" and the coefficient b_2 to obtain "$\leq$")

$$= \|\sum_{i=1}^{k} b_i e_i\| = \|\sum_{i=1}^{k} a_i s_i\|.$$

■

Proof of Theorem A. We are going to prove that $\beta_K(\xi) = \omega$ for every $\xi \in X^{**}\backslash X$, where $K = (B(X^*),\omega^*)$.

By Theorem 6 and Proposition 1(i) we have $\beta_K(\xi) = \beta_K(\tilde{\xi})$, and if we suppose that $\beta_K(\tilde{\xi}) > \omega$, then by Theorem 5 and Lemma 7, the sequence $s_n = \sum_{i=1}^{n} e_i$ must be equivalent to the ℓ^1–basis, which is impossible since the space X does not contain ℓ^1 (as X is quasi–reflexive). This shows that $\beta_K(\tilde{\xi}) \leq \omega$, and the proof of Proposition 2(iv) shows that $\beta_K(\tilde{\xi}) \geq \omega$.

This concludes the proof of Theorem A. ■

Remarks.

1. The proof of Theorem B will appear in [HOR]. Let us mention how the construction is done.

Let η be an ordinal number, $f \in B_1(0,1)$ a bounded Baire–1 function such that $\beta_{[0,1]}(f) = \eta$, and let $(g_n)_{n\geq 1} \subset C(0,1)$ be a uniformly bounded sequence point–wise converging to f. The space X of Theorem B is obtained by applying the DFJP–interpolation method to the space C(0,1) and the convex $C = \overline{\text{conv}}\,[\pm g_n] \subset C(0,1)$.

2. A similar construction is done in [GM].

3. One can give the following improvement of Theorem A:

<u>Theorem</u>. *Let* X *be a quasi–reflexive Banach space such that* $X^{(n)}$ *(i.e., then* n^{th} *dual) is isometric to* $X^{(m)}$ *for some* $m > n$. *Then* $\beta(X^{(p)}) = \omega$ *for every* $p \geq 0$.

<u>Proof</u>. Let $Z = (\oplus \sum_{k=0}^{3(m-n)-1} X^{(2n+k)})_{\ell^2}$, it is clear that Z is quasi–reflexive and not difficult to check that $Z^{**} \equiv Z$. On the other hand for every $p \geq 0$, $X^{(p)}$ is a subspace of $Z^{(p)}$, and then for every $\xi \in X^{(p)**}\setminus X^{(p)}$ we have $\beta_{B(X^{(p)**})}(\xi) = \beta_{B(Z^{(p)**})}(\xi) = \omega$ by the proof of Theorem A, and the remarks following Proposition 2. ∎

4. It would be interesting to know if we can remove the quasi–reflexivity assumption in Theorem A.

References

[Be] S. Bellenot: Transfinite duals of quasi–reflexive Banach spaces. Trans. A.M.S., **273** 2 (1982), 551–577.

[Bo] J. Bourgain: On convergent sequences of continuous functions. Bull. Soc. Math. Belgique, **32** (1980), 235–249.

[BS] A. Brunel, L. Sucheston: On B–convex Banach spaces. Math. Systems Theory, **7** (1984), 294–299.

[BL] B. Beauzamy, J. Lapresté: Modèles étalés des espaces de Banach. Travaux en Cours, Herman, Paris (1983).

[D] J. Diestel: Sequences and series in Banach spaces. Graduate Texts in Math., **92**, Springer Verlag (1987).

[DFJP] W. Davis, T. Figiel, W. Johnson, A. Pelczyński: Factoring weakly compact operators. J. Funct. Anal., **17** (174), 311–327.

[GM] N. Ghoussoub, B. Maurey: G_δ–embeddings in Hilbert spaces II. Preprint.

[HOR] R. Haydon, E. Odell, H. Rosenthal: In preparation.

[J] R. James: Banach spaces quasi–reflexive of order one. Studia Math., **60** (1977), 157–177.

[KL] A. Kechris, A. Louveau: A classification of Baire class–1 functions. Preprint.

[L] M. Lavrientiev: Sur les sous classes de la classification de Baire. C.R. Paris, 1925.

[S] A. Sersouri: On James' type spaces. Trans. A.M.S. (to appear).

[V] M. Valdivia: On a class of Banach spaces. Studia Math., **60** (1977), 11–13.

A. Sersouri
Equipe d'Analyse, Université Paris VI
4 place Jussieu
75252 Paris Cedex 05
France

and

Department of Mathematics
Texas A & M University
College Station, Texas 77843
U.S.A.

Current Address:

Department of Mathematics
University of Illinois
Urbana, Illinois 61801

Contemporary Mathematics
Volume **85**, 1989

ω-independence in non separable Banach spaces

ABDERRAZZAK SERSOURI

Abstract: We show that the non separable $\mathcal{C}(K)$ space constructed by Kunen (assuming the continuum hypothesis) has no uncountable ω-independent family.

A family $(f_\alpha)_{\alpha\in\mathcal{F}}$ in a Banach space X is said to be ω-independent if for every sequence $(\alpha_n)_{n\geq 1} \subset \mathcal{F}$ of distinct indices, and every sequence $(a_n)_{n\geq 1}$ of scalars, the series $\sum_{n=1}^{\infty} a_n f_{\alpha_n}$ converges (in norm) to zero if and only if $a_n = 0$ for every $n \geq 1$. Recently it was proved in **[1]**, **[2]**, that no separable Banach space contains an uncountable ω-independent family. The result stated in the abstract shows that even for non separable Banach spaces, one can not expect (in general) to find ω-independent families with uncountable cardinality. This answers negatively a question of **[1]**.

We will use the following property of Kunen's example **[3]**.

Theorem 1. *Assuming the continuum hypothesis, there exists a non separable $\mathcal{C}(K)$ space satisfying the following property:*

$$(\star) \qquad \begin{cases} \text{For every family } (f_\alpha)_{\alpha<\omega_1} \text{ in } \mathcal{C}(K), \text{ there exists } \alpha < \omega_1 \\ \text{such that } f_\alpha \in \overline{cv}\{f_\beta : \beta > \alpha\} \end{cases}$$

The proof of the result stated in the abstract will be done in two lemmas.

Lemma 2. *A Banach space X has property $(\star)$ if and only if it has the following property:*

$$(*) \qquad \begin{cases} \text{Every family } (C_\alpha)_{\alpha<\omega_1} \text{ of closed convex subsets of } X \\ \text{which is decreasing (i.e.: } C_{\alpha+1} \subset C_\alpha) \text{ is stationary} \\ \text{(i.e.: } \exists \alpha_0 < \omega_1 \text{ such that } C_\alpha = C_{\alpha_0} \text{ for every } \alpha \geq \alpha_0) \end{cases}$$

AMS classification: 46B15, 40J05
key words: ω-independence

Proof. $(\bigstar) \Rightarrow (*)$: Suppose that X fails $(*)$, then there exists a decreasing family $(C_\alpha)_{\alpha<\omega_1}$ of closed convex subsets of X such that the (ordinal) set $A = \{\alpha < \omega_1 : C_\alpha \backslash C_{\alpha+1} \neq \phi\}$ is uncountable. Let $(\alpha(\xi))_{\xi<\omega_1}$ be the canonical ordering of A, and for every $\xi < \omega_1$, let $f_\xi \in C_{\alpha(\xi)} \backslash C_{\alpha(\xi)+1}$. By $(\bigstar)$, there exists $\xi < \omega_1$ such that

$$f_\xi \in \overline{cv}\{f_\zeta : \zeta \geq \xi + 1\} \subset C_{\alpha(\xi+1)} \subset C_{\alpha(\xi)+1}$$

This contradiction concludes the proof.

$(*) \Rightarrow (\bigstar)$: If $(f_\alpha)_{\alpha<\omega_1}$ is a given family in X, define $C_\alpha = \overline{\omega}\{f_\beta : \beta \geq \alpha\}$.
By $(*)$, there exists $\alpha < \omega_1$ such that $C_\alpha = C_{\alpha+1}$, which implies in particular that $f_\alpha \in \overline{cv}\{f_\beta : \beta > \alpha\}$. ∎

Lemma 3. *A Banach space X has no uncountable ω-independent family if it has $(*)$.*

Proof. Let $(a_n)_{n\geq 1}$ be a seqence of positive real numbers such that $\lim\limits_{n=\infty} a_n = 0$ and $\sum\limits_{n=1}^{\infty} a_n = \infty$, and let $b_n = \operatorname{Sup}\limits_{m>n} a_m$.

Let $(f_\alpha)_{\alpha<\omega_1}$ be an uncountable family in X. Without loss of generality we can suppose that $(f_\alpha)_{\alpha<\omega_1} \subset$ Ball (X) (since we want to prove that it is not ω-independent).

By $(*)$, we can also suppose that the sets $C_\alpha = \overline{cv}\{f_\beta : \beta \geq \alpha\}$ satisfy $C_\alpha = C_0$ for every $\alpha \geq 0$.

To prove the lemma we will use the following result due to Kalton [**2**]:

$$\left\{\begin{array}{l} \text{For every } x \in \overline{sp}\{f_\alpha : \alpha < \omega_1\}, \text{ every } \delta > 0, \text{and every } n \in \mathbb{N}, \\ \text{there exist } m > n, \text{ a sequence of signs } (\varepsilon_i)_{i=n+1}^m, \\ \text{and a sequence of (not necessarily distinct) ordinals } (\alpha_i)_{i=n+1}^m \text{ such that:} \\ \left\{\begin{array}{l} \|x + \sum\limits_{i=n+1}^{m} a_i \varepsilon_i f_{\alpha_i}\| < \delta + b_n \\ \operatorname{Sup}\limits_{n<k<m} \|x + \sum\limits_{i=n+1}^{k} a_i \varepsilon_i f_{\alpha_i}\| < \|x\| + \delta + b_n \end{array}\right. \end{array}\right.$$

As in [2], and using the fact that $C_\alpha = C_0$ for every $\alpha \geq 0$, we may construct inductively a sequence of signs $(\varepsilon_n)_{n\geq 1}$, a sequence of positive real numbers $(\lambda_p^n)_{n\geq 1, 1\leq p\leq k(n)}$, and a sequence of ordinals $(\alpha_p^n)_{n\geq 1,1\leq p\leq k(n)}$ such that:

$$\begin{cases} \sum_{p=1}^{k(n)} \lambda_p^n = 1 & \text{for every } n \geq 1 \\ \alpha_1^n < \alpha_2^n < \cdots < \alpha_{k(n)}^n < \alpha_1^{n+1} & \text{for every } n \geq 1 \\ \sum_{n=1}^{\infty} a_n \varepsilon_n x_n = 0 \, , & \text{where } x_n = \sum_{p=1}^{k(n)} \lambda_p^n f_{\alpha_p^n} \end{cases}$$

Now since we have supposed that $(f_\alpha)_{\alpha<\omega_1} \subset$ Ball (X), it's easy to see that the series $\sum_{n=1}^{\infty} a_n \varepsilon_n \Big(\sum_{p=1}^{k(n)} \lambda_p^n f_{\alpha_p^n} \Big)$ also converges to zero.

This proves that $(f_\alpha)_{\alpha<\omega_1}$ is not ω-independent. ∎

Remark: The result of Lemma 3 can also be proved by the method used in [1]. For instance, using the notation of [1], we can construct inductively a sequence of positive numbers $(\lambda_p^\sigma)_{\sigma\in S, 1\leq p\leq n_\sigma}$, a sequence of ordinals $(\alpha_p^\sigma)_{\sigma\in S,1\leq p\leq n_\sigma}$ such that

$$\begin{cases} \sum_{p=1}^{n_\sigma} \lambda_p^\sigma = 1 & \text{for every } \sigma \in S \\ \alpha_1^{\psi(k)} < \alpha_2^{\psi(k)} < \cdots < \alpha_{n_{\psi(k)}}^{\psi(k)} < \alpha_1^{\psi(k+1)} & \text{for every } k \in \mathbb{N} \\ \|x_\sigma - x_{g(\sigma)}\| < 3^{-|\sigma|} & \text{where } x_\sigma = \sum_{p=1}^{n_\sigma} \lambda_p^\sigma f_{\alpha_p^\sigma} \end{cases}$$

As in [1], and using the boundedness of $(f_\alpha)_{\alpha<\omega_1}$, we can prove that the series

$$\sum_{k=1}^{\infty} \frac{(-1)^{d(\psi(k))}}{2^{|\psi(k)|}} \Big(\sum_{p=1}^{n_{\psi(k)}} \lambda_p^{\psi(k)} f_{\alpha_p^{\psi(k)}} \Big)$$

converges to zero.

Acknowledgement: I want to thank Professor S. Bellenot for several helpful conversations, and Professor H. P. Rosenthal for suggesting the formulation we present here of lemma 2.

References

[1] D. Fremlin, A. Sersouri: *ω-independence is separable Banach spaces.* To appear: Quar. J. Math.

[2] N. Kalton: *Independence in separable Banach spaces.* To appear: Proc. of the 1987 Iowa Banach workshop.

[3] K. Kunen, J. Vaugham: "Handbook of set theoretic topology", North Holland (1984) (ch. 23).

A. Sersouri
Equipe d'Analyse, Université Paris 6,
4 place Jussieu, 75252 Paris Cedex 05,
FRANCE

and

Department of Mathematics
The University of Texas at Austin
Austin, TX 78712

Contemporary Mathematics
Volume **85**, 1989

The canonical injection from $C([0,1])$ into $L_{2,1}$ is not of cotype 2

Michel Talagrand

Abstract. We prove the result mentioned in the title.

1. Introduction

For an operator T between Banach spaces X,Y we denote by $C_2(T)$ the constant of Gaussian cotype 2 of T, that is the infimum of the (possibly infinite) numbers C for which

$$\left[\sum_{i\leq n} \|T(x_i)\|^2\right]^{1/2} \leq CE\|\sum_{i\leq n} g_i x_i\| \tag{1.1}$$

for all sequences $(x_i)_{i\leq n}$ of X, where $(g_i)_{i\leq n}$ is an independent sequence of $N(0,1)$ random variables.

1980 Mathematics Subject Classification (1985 Revision): Primary 56E30, Secondary 47B10.

Key Words and phrases: Gaussian cotype, Lorentz spaces, Gaussian randomization.

We denote by λ Lebesgue measure on $[0,1]$. On the Lorentz space $L_{2,1} = L_{2,1}([0,1],\lambda)$, the norm is given by

$$\|h\|_{2,1} = \int_0^\infty \sqrt{\lambda(\{|h|\geq t\})}dt . \tag{1.2}$$

Theorem 1. The canonical injection $j : C([0,1]) \longrightarrow L_{2,1}$ does not have Gaussian cotype 2 (i.e., $C_2(j) = \infty$).

The study of j is motivated in particular by the factorization theorems of G. Pisier [1]. More precisely, an operator $T : X \longrightarrow Y$ between Banach spaces, is called (2,1)–summing if there is a constant C such that for all finite sequences $(x_i)_{i\leq n}$ in X we have

$$\Big[\sum_{i\leq n} \|T(x_i)\|^2\Big]^{1/2} \leq C \sup_{\epsilon_i=\pm 1} \|\sum_{i\leq n} \epsilon_i x_i\|.$$

Moreover, the smallest constant C satisfying this is denoted by $\pi_{2,1}(T)$.

Let X be a Banach. Let us denote by I_X the identity operator on X.

It is well known that if X is of cotype 2 in the above sense then it is also of cotype 2 when one replaces the sequence (g_i) by a sequence of i.i.d random variables taking the values $+1$ and -1 with equal probability $1/2$. It is clear that the latter property implies that the identity of X is (2,1)–summing. The converse is still an open problem (originally posed by Maurey in his thesis), namely: if I_X is (2,1)–summing is X necessarily of cotype 2?

It is natural to ask whether the same is true for an operator. We show that the answer to that question is negative. The inclusion $j : C([0,1]) \longrightarrow L_{2,1}$ is easily seen to be

(2,1)–summing. Indeed, if $(x_i)_{i\leq n} \in C([0,1])$, and if $A = \sup_{\epsilon_i=\pm1} \|\sum_{i\leq n} \epsilon_i x_i\| = \sup_t \sum_{i\leq n} |x_i(t)|$, we have

$$\begin{aligned}
\sum_{i\leq n} \|j(x_i)\|_{2,1}^2 &= \sum_{i\leq n} \left[\int_0^\infty \sqrt{\lambda(\{|x_i|\geq t\})}\, dt\right]^2 \\
&= \sum_{i\leq n} \left[\int_0^A \sqrt{\lambda(\{|x_i|\geq t\})}\, dt\right]^2 \\
&\leq \sum_{i\leq n} A \int_0^\infty \lambda(\{|x_i|\geq t\})\, dt \\
&= A \sum_{i\leq n} \int |x_i|\, d\lambda = A \int \sum_{i\leq n} |x_i|\, d\lambda \leq A^2.
\end{aligned}$$

However, j is not of cotype 2 by the theorem above. This counterexample is the "natural" one since by [1] and (2,1)–summing operator on a C(K) space factors through an inclusion $C(K) \longrightarrow L_{2,1}(\lambda)$ for some probability measure λ of K.

The construction used to prove Theorem 1 is motivated by a general principle of independent interest, (due to G. Pisier) that we explain now. We consider the diagonal operator Δ form $c_0 = c_0(\mathbb{N})$ to itself given by $\Delta((x_\ell)_{\ell\geq1}) = (\gamma_\ell x_\ell)_{\ell\geq1}$, where $\gamma_1 = 1, \gamma_\ell = (\log \ell)^{-1/2}$ for $\ell \geq 2$.

For an operator T between X and Y, we denote by H(T) the supremum of the numbers $\pi_2(T \circ U \circ \Delta)$, where U in any norm one operator from a subspace Z of c_0 into X, and where Δ is restricted to $\Delta^{-1}(Z)$.

Theorem 2. For some universal constant K, we have

$$K^{-1}H(T) \le C_2(T) \le KH(T).$$

Acknowledgement. Theorem 2 is due to G. Pisier. Pisier was well aware that once Theorem 2 has been obtained, deciding whether j has cotype 2 must be well within reach. He nonetheless generously suggested to the author to finish off the question. As expected, this turned out to be a very simple matter. Only the thought that the result might be of interest to others convinced the author to write down a result for which he does not deserve the main credit. More thanks are due to G. Pisier for providing the author with the background for the problem.

2. Proof of Theorem 2

We denote by K a universal constant, that may vary at each occurrence. Consider vectors $x_1, \cdots, x_n \in c_0$; let $x_i = (x_{i\ell})_{\ell \ge 1}$. Set

$$W = \sup_{x^* \in \ell^1, \|x^*\| \le 1} \left[\sum_{i \le n} x^*(x_i)^2 \right]^{1/2}.$$

We have

$$\Delta\Big[\sum_{i \le n} g_i x_i\Big] = \Big[\gamma_\ell \sum_{i \le n} g_i x_{i\ell}\Big]_{\ell \ge 1}.$$

The Gaussian r.v. $Y_\ell = \sum_{i \le n} g_i x_{i\ell}$ satisfies $(EY_\ell^2)^{1/2} = \left[\sum_{i \le n} x_{i\ell}^2\right]^{1/2} \le W$. By a routine computation ([2], p. 102) we see that

$$E\|\Delta\Big[\sum_{i\le n} g_i x_i\Big]\| = E\sup_{\ell} |\gamma_\ell Y_\ell| \le KW.$$

Thus, for each operator U from a subspace Z of c_0 to x, whenever $\Delta(x_i) \in U$ for $i \le n$, we have

$$E\|U \circ \Delta\Big[\sum_{i\le n} g_i x_i\Big]\| \le KW.$$

By definition of $C_2(T)$, we get

$$\Big[\sum_{i\le n} \|T \circ U \circ \Delta(x_i)\|^2\Big]^{1/2} \le KC_2(T)W,$$

and by definition of the 2–summing norm, this proves that $\pi_2(T \circ U \circ \Delta) \le KC_2(T)$, and hence $H(T) \le KC_2(T)$.

We turn to the converse. Consider a sequence $(y_i)_{i\le n}$ of X. The main point is that there exists a subspace Z of c_0, a norm–one operator U from Z to X, and a sequence $(x_i)_{i\le n}$ in $\Delta^{-1}(Z)$ such that $y_i = U\circ \Delta(x_i)$, and that

$$W = \sup_{x^*\in\ell_1, \|x^*\|\le 1} \Big[\sum_{i\le n} x^*(x_i)^2\Big]^{1/2} \le KE\|\sum_{i\le n} g_i y_i\|. \tag{2.1}$$

This is essentially, in another language, the content of Theorem 19 of [2], and is proved in the same way. By definition of $H(T)$, we have

$$\Big[\sum_{i\le n}\|T(y_i)\|^2\Big]^{1/2} = \Big[\sum_{i\le n}\|T\circ U\circ\Delta(x_i)\|^2\Big]^{1/2} \le H(T)W$$

which, together with (2.1), proves that $C_2(T) \le KH(T)$. □

3. Proof of Theorem 1

For $i \ge 3$, we denote by I_i the interval $[2^{2^{2i}-1}, 2^{2^{2i}}[$. These intervals are disjoint, and $\operatorname{card} I_i = 2^{2^{2i}-1}$. Moreover, $(\log \ell)^{1/2} \le 2^i$ for $\ell \in I_i$. Let $I = \bigcup_{i\ge 3} I_i$. Consider now a sequence $(\alpha_i)_{i\ge 3}, \alpha_i \ge 0$, such that $\sum_{i\ge 3}\alpha_i^2 \le 1/2, \sum_{i\ge 3}\alpha_i = \infty$. For $\ell \in I$, we can find elements z_ℓ of $C([0,1])$, with disjoint supports, such that $\|z_\ell\| = 1$, and $\lambda(\{z_\ell=1\}) \ge \alpha_i^2(\operatorname{card} I_i)^{-1}$ whenever $\ell \in I_i$. Denoting by $(e_\ell)_{\ell\ge 1}$ the canonical basis of c_0, we can define a norm one operator U from c_0 to $C([0,1])$ by $U(e_\ell) = z_\ell$ for $\ell \in I$ and $U(e_\ell) = 0$ otherwise. We set $A = j \circ U \circ \Delta$. The rest of the proof consist in showing that $\pi_2(A) = \infty$, so that Theorem 1 follows from Theorem 2.

We fix $k \ge 2$, and we set $J = \bigcup_{i\le k} I_i$. We consider the set Σ of permutations of J that leave each interval I_i invariant. Consider an element x of c_0, supported by J. For $\sigma \in \Sigma$, we define

$$x_\sigma = (x_{\sigma(\ell)})_{\ell\in J} \in \ell_\infty(J) \subset c_0.$$

Let $y_\sigma = (\operatorname{card}\Sigma)^{-1/2}x_\sigma$. We have

$$\sup_{x^*\in\ell_1,\|x^*\|\le 1}\ \sum_{\sigma\in\Sigma} x^*(y_\sigma)^2 \le \max_{i\le k}\frac{1}{\operatorname{card} I_i}\sum_{\ell\in I_i} x_\ell^2. \tag{3.1}$$

Indeed, to compute the supremum on the left, it suffices to consider coordinate functionals e_ℓ^*, and we have

$$\sum_{\sigma\in\Sigma} e_\ell^*(y_\sigma)^2 = \frac{1}{\operatorname{card} I_i} \sum_{\ell\in I_i} x_\ell^2$$

when $\ell \in I_i$, and $= 0$ otherwise. Thus if we have

(3.2) $$\forall i \le k, \sum_{\ell\in I_i} x_\ell^2 \le \operatorname{card} I_i,$$

the left hand side of (3.1) is ≤ 1, and we have

(3.3) $$\min_{\sigma\in\Sigma} \|A(x_\sigma)\|_{2,1} \le \Big[\sum_{\sigma\in\Sigma} \|A(y_\sigma)\|_{2,1}^2 \Big]^{1/2} \le \pi_2(A).$$

For $i \le k$, and $m \in V(i) = [2^{2i-2}, 3.2^{2i-3}[$, we consider disjoint subintervals $(I_{i,m})$ of I_i. We restrict to the case where $x_\ell = 2^{m+i}$ if $\ell \in I_{i,m}$, and $x_\ell = 0$ if ℓ belongs to no interval $I_{i,m}$. then (3.2) becomes

(3.4) $$\forall i \le k, \sum_{m\in V(i)} 2^{2m+2i} \operatorname{card} I_{i,m} \le \operatorname{card} I_i.$$

Since $(\log \ell)^{1/2} \le 2^i$ for $\ell \in I_i$, we have $A(x) \ge 2^m$ on the set $\{z_\ell = 1\}$ whenever $\ell \in I_{i,m}$. Thus we have

$$\lambda(\{|A(x)| \geq 2^m\}) \geq \alpha_i^2 \text{card } I_{i,m}(\text{card } I_i)^{-1}$$

and thus

$$(3.5) \qquad \sum_{m \in V(i)} 2^m(\lambda(\{|A(x)| \geq 2^m\}))^{1/2} \geq \alpha_i \sum_{m \in V(i)} 2^m(\text{card } I_{i,m})^{1/2}(\text{card } I_i)^{-1/2}.$$

For $i \geq 2$, and $m \leq 3.2^{2i-3}$, we have $2^{2i}-1 \geq 2m + 4i$. Thus we can take $\text{card } I_{i,m} = 2^{-2m-4i}\text{card } I_i$. Then (3.4) holds, since $\text{card } V(i) = 2^{2i-3}$, and (3.5) implies

$$\sum_{m \in V(i)} 2^m(\lambda(\{|A(x)| \geq 2^m\}))^{1/2} \geq \alpha_i/8.$$

Since the intervals $V(i)$ are disjoint, we have

$$\|A(x)\|_{2,1} = \int_0^\infty (\lambda(\{|A(x)| \geq t\}))^{1/2} dt \geq \sum_{m \geq 1}^{\infty} 2^m(\lambda(\{|A(x)| \geq 2^m\}))^{1/2} \geq \sum_{i \leq k} \alpha_i/8.$$

Clearly, for all $\sigma \in \Sigma$ we also have $\|A(x_\sigma)\|_{2,1} \geq \sum_{i \leq k} \alpha_i/8$. From (3.3), we see that $\pi_2(A) \geq \sum_{i \leq k} \alpha_i/8$. Since k is arbitrary, $\pi_2(A) = \infty$. This completes the proof.

References

[1] G. Pisier, Factorization of Operators Through $L_{p\infty}$ or L_{p1} and Noncommutative Generalizations. Math. Ann. 276, 1986, p. 105–136.

[2] M. Talagrand, Regularity of Gaussian Processes, Acta Math. 159, 1987, p. 99–149.

Equipe d' Analyse – Tour 46
Université Paris VI
4 Place Jussieu
75230 Paris Cedex 05

The Ohio State University
Department of Mathematics
231 W. 18th Avenue
Columbus, Ohio 43210